International Union of Geological Sciences
Series A, Number 4

Spilites and Spilitic Rocks

Edited by G. C. Amstutz

Editorial Board
G. C. Amstutz (Chairman), Heidelberg · M. H. Battey, Newcastle-upon-Tyne · F. Fiala, Prague
G. van der Kaaden, Heidelberg · D. S. Korzhinsky, Moscow · E. Lehmann, Garmisch-Partenkirchen
E. Niggli, Bern · A. M. Patwardhan, Chandigarh
A. Rittmann, Catania · G. Rocci, Nancy · F. Saupé, Nancy
T. G. Vallance, Sidney · M. Vuagnat, Geneva

With 138 Figures and 13 Plates

Springer-Verlag New York · Heidelberg · Berlin 1974

Professor Dr. Dr. h. c. G. C. Amstutz
Mineralogisch-Petrographisches Instiut der Universität
Heidelberg, W. Germany

ISBN 0-387-06448-6 Springer-Verlag New York Heidelberg Berlin
ISBN 3-540-06448-6 Springer-Verlag Berlin Heidelberg New York

This work is subject to copyright. All rights are reserved, whether the whole or part of the material is concerned, specifically those of translation, reprinting, re-use of illustrations, broadcasting, reproduction by photocopying machine or similar means, and storage in data banks.

Under § 54 of the German Copyright Law, where copies are made for other than private use, a fee is payable to the publisher, the amount of the fee to be determined by agreement with the publisher.

© by Springer-Verlag Berlin · Heidelberg 1974. Library of Congress Catalog Card Number 73-19100. Printed in Germany. Offsetprinting: Julius Beltz, Hemsbach/Bergstr. Bookbinding: Konrad Triltsch, Graphischer Betrieb, Würzburg

The use of registered names, trademarks, etc. in this publication does not imply, even in the absence of a specific statement, that such names are exempt from the relevant protective laws and regulations and therefore free for general use.

Contents

INTRODUCTION   G.C. AMSTUTZ 1

1. INTRODUCTORY AND GENERAL PAPERS

Some Notes on the Problem of Spilites 9
F. FIALA

Spilitic Magma. Characteristics and Mode of Formation 23
E. LEHMANN

Essai de Caractérisation Chimique des Associations
Spilitiques 39
H. de la ROCHE, G. ROCCI, et Th. JUTEAU

Pyroxenes and the Basalt. Spilite Relation 59
T.G. VALLANCE

2. PAPERS PROPOSING A PRIMARY ORIGIN

A Reappraisal of the Textures and the Composition of the
Spilites in the Permo-Carboniferous Verrucano of Glarus,
Switzerland 71
G.C. AMSTUTZ and A.M. PATWARDHAN

A Series of Magmatism Related to the Formation of Spilite 83
T. BAMBA

Environmental Effects in Magmatic Spilite 113
E. LEHMANN

A Statistical Study of Specific Petrochemical Features of
Some Spilitic Rock Series 127
W. NAREBSKI

Middle Triassic Spilite-Keratophyre Association of the
Dinarides and Its Position in Alpine Magmatic-Tectonic Cycle 161
J. PAMIĆ

Petrogenesis of Spilites Occurring at Mandi, Himachal
Pradesh, India 175
A.M. PATWARDHAN and A. BHANDARI

General Features of the Spilitic Rocks in Finland 191
T. PIIRAINEN and P. ROUHUNKOSKI

The Relationships between Spilites and Other Members of
the Oman Mountains Ophiolite Suite 207
B. REINHARDT

Gradation of Tholeiitic Deccan Basalt into Spilite,
Bombay, India 229
R.N. SUKHESWALA

3. PAPERS PROPOSING AN AUTOHYDROTHERMAL OR AUTOMETAMORPHIC ORIGIN

Vers une Meilleure Connaissance du Problème des Spilites à
Partir de Données Nouvelles sur le Cortège Spilito-
Keratophyrique Hercynotype 253
Th. JUTEAU et G. ROCCI

Spilites of the Lucanian Apennine (Southern Italy) 331
P. SPADEA

Quelques Observations Nouvelles Relatives à la Genèse des
Laves Spilitiques 349
J.-L. TANE

Comments on Spilitization of the Permian Eruptive Rocks of
the Choč Nappe in the West Carpathians, Slovakia 359
J. VOZÁR

4. PAPERS PROPOSING A SECONDARY, DIAGENETIC OR METAMORPHIC ORIGIN

Spilites as Weakly Metamorphosed Tholeiites 365
M.H. BATTEY

On the Mineral Facies of Spilitic Rocks and Their Genesis 373
D.S. COOMBS

The Pillow Lavas of Sakhalin and the Kurile Islands and
Their Significance for the Solution of the Spilite Problems 387
V.N. SHILOV

The Production of Spilitic Lithologies by Burial Metamorphism
of Flood Basalts from the Canadian Keweenawan, Lake Superior ...403
R.E. SMITH

A New Appraisal of Alpine Spilites 417
M. VUAGNAT

BIBLIOGRAPHY

A. Alphabetical List of
   1. Literature on Spilites and Keratophyres....................427
   2. Related Literature (obviously only a selection)............452
B. Geographical Classification of Occurrences...................459
C. Classification According to the (proposed) Geological Age
   (as far as reported...464
D. Classification According to the Years of Publication.........467

SUBJECT INDEX...473

Author Index

AMSTUTZ, G.C., Mineralogisch-Petrographisches Institut, Universität Heidelberg, Berliner Str. 19, D-69 Heidelberg, W. Germany

BAMBA, T., Geological Survey of Japan, Hokkaido Branch, Minami-1, Nishi-18, Sapporo, Japan

BATTEY, M.H., Department of Geology, The University, Newcastle-upon-Tyne NE1 7RU, England

BHANDARI, A., Geological Survey of India, Lucknow, India

COOMBS, D.S., Department of Geology, University of Otago, P.O. Box 56, Dunedin, New Zealand

FIALA, F., Geological Survey of Czechoslovakia, Hradební 9, 11000 Prague 1, CSSR

JUTEAU, Th., Ecole Nationale Supérieure de Géologie, Faculté des Sciences, BP 452, F-54001 Nancy, France

LEHMANN,E., Hindenburgstr. 35, D-81 Garmisch-Partenkirchen, W. Germany

NAREBSKI, W., Collegium Geologicum UJ, ul. Oleandry 2a, 30-063 Kraków, Polen

PAMIĆ, J., Institut za geološka istraživanja, Ilidža kod Sarajeva, cesta br. 3, Sarajevo, Yugoslavia

PATWARDHAN, A.M., Centre of Advanced Study in Geology, Panjab University Chandigarh, India

PIIRAINEN, T., Institute of Geology and Mineralogy, University of Oulu, 90100 Oulu 10, Finland

REINHARDT, B., Koninklijke Shell Exploratie en Produktie Laboratorium, Volmerlaan 6, Rijswijk, Holland

ROCCI, G., Laboratoire de Pétrographie cristalline, Université de Nancy I, CO no. 140, F-54037 Nancy, France

ROCHE de la, H., Centre de Recherches Pétrographiques et Géochimiques, CO no. 1, F-54500 Vandoeuvre-lès-Nancy, France

ROUHUNKOSKI, P., Outokumpu Oy, Exploration, Rovaniemi, Finland

SHILOV, V.N., Mariia Ulianova Street 17, Building 1, Iopi, Moscow V-331, USSR

SMITH, R.E., Division of Mineralogy, C.S.I.R.O., P.O., Wembley, W.A. 6014, Australia

SPADEA, P., Istituto di Mineralogia e Petrografia dell'Università di Catania, I-95128 Catania, Italy

SUKHESWALA, R.N., Geology Department, St. Xavier's College, Bombay-1, India

TANE, J.-L., Institut Dolomieu, Université de Grenoble, Rue Maurice-Gignoux, F-38 Grenoble, France

VALLANCE, T.G., Department of Geology and Geophysics, The University of Sydney, Sydney, N.S.W. 2006, Australia

VOZÁR, J., Geologický ústav Dionýza Štúra, Mlynská dol. 1, Bratislava, CSSR

VUAGNAT, M., Institut de Minéralogie de l'Université, 11 rue des Maraichers, CH-1211 Geneva, Switzerland

Introduction

The idea for the present Spilite Volume was born during the Spilite Symposium at the XXIIIrd session of the International Geological Congress in Prague, 1968. At that time, only a restricted number of petrologists working on spilites was present and, therefore, the group assembled agreed that a Symposium Volume should also include recent papers by many other spilite specialists. At the same time it was agreed that the papers presented at the Symposium should be returned to the authors for changes and additions. This procedure of upgrading and amending the papers has continued until this year (1973) for various technical and editorial reasons. The information presented here is, therefore, up-to-date. To those familiar with the spilite problem it is obvious that the time had come for a review of its state. Also, the existing literature had become so voluminous that a monographic review was necessary. Following a modern trend, the authorship for this review was spread among specialists with variable experience.

For readers not necessarily familiar with the spilite problem, a brief summary is presented here. A short historical note is followed first by the observations, then by the interpretations, finally by some of the major features of scientific logic as they pertain to the problem of the primary or secondary origin of some of the rocks termed spilites and keratophyres.

The history of the spilite problem started when BROGNIARD[1] first introduced the term through his 1827 compendium on rock terms. Then, and until the end of the nineteenth century, the term spilite was largely a field term for aphanitic albite-chlorite or albite-hematite-quartz rocks, often with carbonate amygdules or veinlets. After about 1910 - 1915, especially after BENSON's work, enough microscopic and field work was done to recognize that a group of basaltic rocks contained albite as a main constituent and had to be separated from normal basalts for textural and compositional reasons: texturally, spilites and keratophyres displayed primary differentiation features (along flow lines, breccias, etc.) which could not be considered secondary products. Mineralogically, the albite-chlorite or albite-hematite groupings with or without calcite coincided so well with the primary flow lines, etc. that secondary processes could not be brought in as useful analogues of formation. The coincidence of primary fabrics such as those listed in Table II of AMSTUTZ (1968), especially flow lines, lava diklets, breccia matrices and amygdules with strong mineralogical differences such as those listed in Fig. 1, does not allow the construction of sound hypotheses of secondary formation. On the other hand, some "spilitic rocks" display clear pseudomorphs of the main constituent - plagioclase - and a secondary origin is obvious. Pseudomorphs of early constituents such as olivine, are inconclusive since they can form deuterically or even hydro-

---

[1] All references to be found in the Appendix of this book which contains an extensive and annotated bibliography.

magmatically - the latter term designating those parts of magmas and those processes, in which the presence of accumulated water determined the trend of crystallization.

This trend was clearly stated by P. NIGGLI (compare BURRI and P. NIGGLI, 1945; E. NIGGLI, 1944; P. NIGGLI, 1952, AMSTUTZ, 1954) who summarized the spilite-keratophyre crystallization in 1952 (p.406) as follows, in which he discussed the central parts of the greenstone flow of Michigan, based on a previous paper by CORNWALL (1951): "The mineral paragenesis corresponds to that of spilitic rocks, though there can be no doubt that pyroxenes and olivine actually crystallized at the outset. One has the impression that in the central critical zone a magma rich in volatile constituents, whose early crystallization proceeded on normal lines, assumed the character of a spilitic residual magma. In this stage and under the influence of H_2O, basic plagioclase could no longer crystallize and that already separated out became unstable. Prehnite, epidote, and zoisite, besides albite, take its place and simultaneously chlorite and (Fe-)ores [read: ore minerals] are formed instead of water-free Mg-Fe silicates. This is characteristic of the spilitic keratophyre association (BURRI and NIGGLI, 1945, p.490 et seq.) and often connected with the appearance of iron ores, titanium-iron ores and apatite-iron ores".

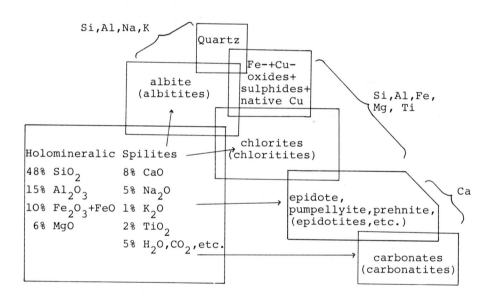

Fig. 1. Differentiation scheme of spilitic rocks

The geometric and mineralogic differentiation pattern of Fig. 1 is ubiquitous in all known spilite and keratophyre provinces. Obviously, a part of the mineral assemblage corresponds to that of the greenschist facies and/or the zeolite facies. But, even though there are transitions between primary spilites and greenschist facies rocks, the similarity is only superficial. To the knowledge of this writer, a separation of the two phenomena is difficult only in metamorphic terranes; but here many other phenomena are obscured by the superposition of the products of the secondary process of metamorphism, which is characterized by recrystallization and non-congruent mobilization. To the degree that the congruent distribution of the spilitic or keratophyric differentiation products is still preserved, an interpretation is not difficult.

The purpose of this book is not to present a one-sided story, but to lead the spilite-keratophyre problem to a better solution - to elevate the discussion to a higher level of scientific logic. The reader will have to decide himself whether this goal was achieved or not. The editorial board has tried hard to preserve the scientific integrity of the contributors and not to demand that they conform to a certain pattern of scientific reasoning. Consequently, the reader himself will have to pass judgement on whether the fact that some contributors ignore certain arguments discussed by previous authors - for example those discussed by BENSON, by P. and E. NIGGLI, by CORNWALL and others - is acceptable or not. The reader is confronted in this book not only by the interesting polarity or "bimodal" nature of the interpretation of spilitic rocks, but also by a typical phenomenon of contemporary literary and scientific writing: previous work and current arguments are often ignored, and an "independent" view is proposed. This "ahistoricism", this break with tradition, is refreshing but contains a pitfall. It remains an open question whether, in the fields of ethics, esthetics and religion, such "revolutions" - as opposed to evolutions - are at times inevitable. In the field of science, however, i.e., the domain of logic, such a hiatus is probably always disasterous. Nevertheless, the editorial board has taken the liberty of presenting all papers of otherwise good quality in this volume and of putting them "on the floor" for discussion.

A review of the vast amount of literatures since 1957 and of the present collection of papers shows that the classification of theories on the origin of spilites is still the same as that presented in a paper on "Spilites and Mineral Deposits" (AMSTUTZ, 1958, p.2-3). It is, therefore, reproduced without change (the references are all contained in the Appendix to this book):

I. Surface weathering theory: TERMIER, P. (1898).

IIa. "Dry" diffusion theory, using analogies to diffusions in metals (analogous to "dry granitization"): PERRIN, R. et ROUBAULT, M. (1941).

IIb. "Wet" diffusion theory, analogous to "wet" granitization. With, or without (IIa and III), considerable additions or subtractions of materials: PERRIN, R. et ROUBAULT, M. (1941).

III. The meta-basalt theory, defining spilites to be altered (regional metamorphism) basalt and thus recommending to drop the term spilite: FAIRBAIRN, H.W. (1934); JOHANNSEN, A. (1939); HENTSCHEL, H. (1952, etc.); and SUNDIUS, N. (1915), (pro parte).

IVa. Sea-water diffusion theory: post-extrusion Na-metasomatism due to upwards diffusion of trapped sea water: PARK, C.F. (1946); DALY, R. (1914); BESKOW, G. (1929); GILLULY, J. (1935), (pro parte).

IVb. Sea-water pressure theory: Hydrous crystallization conditions created by high pressure in ocean bottoms: RITTMANN, A. (1957).

V. Theory of post-consolidation metasomatism through co- or heteromagmatic solutions (partly gradational to deuteric): DEWEY, H. and FLETT, J.S. (1911); GILLULY, J. (1935).

VI. Theory of deuteric metasomatism contemporaneous to the consolidation (autohydrothermal alteration): FLAHERTY, G.F. (1934).

VII. Theory of essentially primary nature of spilites and keratophyres (with gradual, logical transitions to marginal stages corresponding to the theories V and VI): BENSON, W.N. (1915); DALY, R. (1914, pro parte); WELLS, A.K. (1922, 1923); LEHMANN, E. (1941, etc.); NIGGLI, E. (1944); REINHARD, M. and WENK, E. (1951); NIGGLI, P. (1952); BURRI, C. and NIGGLI, P. (1945); VUAGNAT, M. (1946, etc.); OVEREEM, A.J.A. van (1948); HESS, H.H. (1955); PELLIZZER, R. (1954); BATTEY, M.H. (1956); JAFFE, F.C. (1955); VOISEY, A.H. (1939); WENK, E. (1949, etc.); AMSTUTZ, G.C. (1948, 1950, 1951, 1953, 1954, 1957, 1958); and many others.

The subdivision of the book into the following sections needs no explanation because it follows directly from the nature of the problem and is reflected by the preceding list of spilite theories:

Section 1: Papers stating the problem and concept (introductory papers)

Section 2: Papers proposing a primary origin

Section 3: Papers proposing an autohydrothermal or autometamorphic origin

Section 4: Papers proposing a secondary, diagenetic or metamorphic origin.

At this point it may be useful to include some additional observations not found or not readily found in the papers of this volume. The reason for underscoring these observations is the following: unfortunately, there appears to be a tendency in recent literature attempting to wipe out the spilite problem by cutting through its "Gordian Knot". Such a simplified action may have its psychological effect on some unaware workers who are happy to accept any simplifications of complicated problems. However, we consider it to be a blow against the nature of scientific progress, which always calls for more differentiation and more exact analyses. If, for example, the suggestion is made to replace the term diabase by the term spilite, this is properly termed an anachronism because it wipes out decades of progressively specialized research which cannot be ignored.

It is proposed here to recognize the merit of the papers proposing a zeolite facies of diagenetic origin; likewise the existence of replacement features in some "spilitic" rocks is certainly a fact; furthermore there is no doubt that the greenschist facies has many mineralogical features in common with spilitic rocks. However, we reason that the complete analysis of a rock must include all of its properties - including textural congruences and non-congruences, isotopic analyses, regional relationships, etc. On this basis, the conclusion appears to me inevitable that albite basalts and keratophyres with congruent features must be primary and may thus be called spilites and keratophyres. Those with a majority of non-congruent features may be considered albitic metabasalts, meta-andesites or the like, or metamorphosed and albitized equivalents of basalts or of metamorphosed original spilites and keratophyres.

Consequently it is proposed here to reserve the terms spilite and keratophyres for rocks which display enough congruency between their primary igneous textures and their mineral distribution so that a primary origin is the logical conclusion.

Finally, some investigations currently underway yield additional evidence for the justification of a differentiating approach to the spilite-keratophyre problem: 1. To date, basaltic dikes in salt layers and domes have not revealed any albite or albitization, despite the

abundance of Na. 2. Basaltic extrusions and sills in present-day oceans show neither more nor fewer albite phases than fossil ones. 3. Intrusive basalts do not show fewer spilitic phases than extrusive ones.
4. Many spilitic and keratophyric provinces are untouched or hardly touched by metamorphism, and albitite dikes occur in many different rocks without any signs of albitization, beyond normal adinolization in the mm or cm range. 5. A large proportion of very old basalts, gabbros and andesites show no signs of bulk albitization, despite the presence of chlorite, e.g. in most rocks termed diabase in Europe.
6. Spilitic haloes or clouds are normal, well-known transitional hydromagmatic or deuteric phases around entirely enclosed (sealed off) amygdules, nests or streaks of deuteric minerals, etc. 7. Albite, chlorite, epidote, prehnite, calcite, quartz, etc. are entirely normal primary crystallization products of hydrothermal systems observed ubiquitously in veinlets, ore bodies, etc. 8. The common logic of relations does not allow a frequency gap between a "normal" magma and the products of accumulation of its volatile phase. The occurrence of hydromagmatic phases is therefore also required by the logic of the accumulation mechanism. If hydrothermal solutions are produced by the cooling magma, the probability of encountering hydromagmatic phases must not be overlooked. These phases are the primary spilitic and keratophyric portions of basaltic and andesitic provinces.

This introduction would be incomplete indeed without a proper acknowledgement of the enormous editorial help received from the members of the editorial board listed on the title page. Among them, three have shared the largest burden: Professor LEHMANN who, inspite of his 90 and odd years, has helped in clearing many questions; Professor VALLANCE who edited most of the English papers; Dr. PATWARDHAN who sacrificed many hours and days working towards a more complete general bibliography. Finally our thanks go to Mrs. W. ACKERMANN who was never tired in finding out ways and means of pushing the project and who typed the bibliography painstakingly clean and complete.

Last but not least our cordial thanks go also to Dr. KONRAD F. SPRINGER and his team for the help and patience in raising the quality of the book all through the slow growth of it.

Heidelberg, Summer 1973   G.C. AMSTUTZ

1. Introductory and General Papers

Some Notes on the Problem of Spilites

F. Fiala

The author elaborates upon the article by AMSTUTZ (Geologische Rundschau, 1968) devoted to the elucidation of the problem of spilites. First, he presents a brief characterization of two spilite-bearing formations in Bohemia, namely, the Upper Proterozoic (the Barrandian and the Železné hory areas) and the Lower Paleozoic (the Barrandian area). He gives special attention to the relationship between the spilite and basaltoid (diabase) types of these eruptive rocks. Secondly, the author takes a stand on the individual unclear, disputable points of the spilite problem.

In the comprehensive literature the term spilite is differently understood, and therefore Professor AMSTUTZ's suggestion of publishing collected papers in which scientists from various countries would present their views on the subject is welcomed. Such a collection would certainly provide the best basis for further discussion.

In applying the term spilite, two essential views have been followed for many years. According to the earlier view, spilites are regarded as aphanitic diabases, sometimes amygdaloidal (BRONGNIART, 1827; in Bohemia, SLAVÍK, 1908). The second newer view, in which spilites are defined in terms of their mineralogical-chemical genesis (DEWEY-FLETT, 1911), stresses the mineral association, albite + chlorite, and the higher Na_2O content. Contrary to the opinion that the albite of spilites has a secondary character, some authors claim it is of primary crystallization. Others consider spilites to be metamorphic rocks (e.g. KORZHINSKIJ, 1962). At present, the definition and the accurate delimitation of spilites appear to be the most important points. In addition, their genesis and the problem of the possible origin of an albite-chlorite association linking up with the magmatic process, should be discussed. Also, the position and origin of keratophyre rocks remain unclear; they are considered either a primary member of the basalt-spilite-keratophyre suite (under submarine conditions most frequently associated with initial magmas) or secondary products of additional albitization of original quartz-porphyries (rhyolites) and porphyrites (dacites and andesites) often associated with subsequent volcanism.

In the Barrandian area in central and western Bohemia, two groups of submarine volcanics are of importance in the solution of the above-mentioned problems: a) Upper-Proterozoic (Algonkian) and b) Lower-Paleozoic groups.

Upper Proterozoic (Algonkian) Volcanism

Algonkian geosynclinal submarine volcanics have been described by SLAVÍK (1908) in general as spilites in the sense of BRONGNIART's (1827) terminology, the frequent occurrence of fresh augite and basic plagioclase, i.e. their diabase-aphanitic character being, however, pointed out. SLAVÍK recorded the wide range of differentiation among these volcanics (aphanitic, amygdaloidal, granular diabases, variolites, pillow lavas, porphyrites, glassy breccias and tuffs) and suggested (1926) that the differentiation of the pillow lava from the Vodochody area is characterized by the more acid rock of the pillows and a glassy inter-pillow filling. HEJTMAN (1957) noted the incompatibility between the nature of most of these volcanics and DEWEY and FLETT's (1911) concept of spilites. FIALA (1966a, 1967a, 1969) pointed out the rather

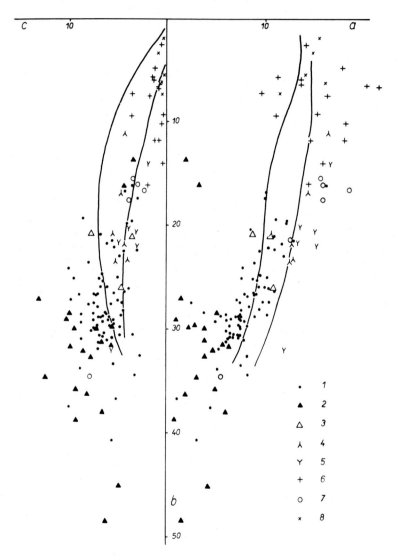

Fig. 1. Differentiation of the Proterozoic volcanics of the Barrandian and the Železné hory areas according to ZAVARITSKIJ, 1950 and ZAVARITSKIJ, 1946. *1* Metabasalts and spilites of the Barrandian area; *2* interpillow matter; *3* varioles; *4* variolitic potash-spilites of the Slatina-Pavlíkov strip; *5* potash-spilites from Dražeň; *6* keratophyres and quartz-keratophyres; *7* spilites of the Železné hory area; *8* quartz-keratophyres of the Železné hory area

composite conditions of this volcanic formation and the association of the typical spilites, which are well definable chemically and mineralogically, with the (meta)basaltoid types; he also discussed (1966a) the main problems relative to spilites, especially their mineral composition and chemistry and their occurrence together with basaltic types. Furthermore, he showed that in spite of regarding the chlorite + albite association as the characteristic feature of spilites, individual authors give considerably different values for the Na_2O content (l.c.1966a, p.13); the data on the critical lower limit of the alk value also differ. FIALA suggested, for the rocks of Bohemia a distinction between the spilitic and basaltic types on the basis of the Na_2O content of about 4%, as did FABRIÈS (1963).

The problem is also complicated by carbonatization, in some cases very strong, of the so-called "greenstones" (Grünsteine), especially those of the amygdaloidal type. Subtraction of $CaCO_3$ or complex carbonate according to the CO_2 content, distorts the picture of the original chemistry of the rock, leading to peraluminate types (PAMIĆ, 1962, p.23). A relatively increased alkali content (Na_2O) often induces the authors to designate the rocks as spilites irrespective of the fact that at least part of the Ca belongs to the original magmatic rock.

If one evaluates the petrographical and chemical conditions of the Upper Proterozoic volcanics of the Barrandian area (ZAWARITSKIJ's, 1946, diagram showing the "spilite" and the "basalt" lines - cf. Fig. 1 - or the alkali/silica wt. % diagram - cf. Fig. 3 - can be used), it becomes evident that most of our volcanics correspond in their chemistry to basalts. The mineral composition is strongly affected by a weak metamorphism (metabasalts). A majority of the rocks of the main volcanic zone in the northern limb of the Barrandian area (*4* and *5*)[1], most rocks of the western part of the southern zone (*6*) and of the Stříbro-Plasy zone (*2*) as well as, farther westward, the majority of the mesozonal amphibolites of the Slavkovský les (Kaiserwald) (*1*) correspond to basalt, their chemistry being fairly basic (SiO_2 between 43.69 - 53.84%, on the average 48-50%). The Na_2O content is usually less than 4%, sometimes even below 3%. The CaO/alk ratio exceeds 2, is sometimes greater than 3. Only some rocks containing well-developed albite and with a Na_2O content which exceeds 4% approach spilites in the strict sense. Even the above-mentioned basaltoid types differ from true basalts in that their feldspars (mostly oligoclase) have a very low K_2O content (0.08 - 0.98%, average 0.37%) and an increased water content; both of which are especially suggestive of spilites. There are gradual transitions between the two types, corresponding approximately to Na_2O contents between 3.5 - 4%.

A relatively small portion of our Proterozoic volcanics attain the alkalinity and the mineral character (albite + chlorite) typical of spilites proper. The southern volcanic zone in the areas of Blovice (*7*), Příbram (*8*), Dobříš (*9*) and Zbraslav (*10*) contains spilites, abundant keratophyre spilites, keratophyres and quartz-keratophyres which, together with less frequently occurring metabasalts, form a continuous differentiation series and which, according to their occurrence, their textures and association with tuffs, are undoubtedly of magmatic origin (contents: 43.09 - 76.40% SiO_2, 0.32 - 6.99% Na_2O, 0.22 - 5.96% K_2O).

Potash spilites of the Slatiņa-Pavlíkov strip in the sedimentary flyschoid suite of the northern limb of the Barrandian area display the characteristics of spilites but have a considerably higher K_2O content

---

[1] The numbers in parentheses correspond to the numbered parts of the differentiation diagram according to NIGGLI - Fig. 2.

(45.28 - 50.96% SiO_2, 3.46 - 6.61% CaO, 2.92 - 3.76% (in tuffs 5.35%) Na_2O and 2.64 - 2.72% (in tuffs 0.45%) K_2O - FIALA, 1966b). Analogous rocks which are strongly metamorphosed and rich in stilpnomelane can be seen in the rocks of the Dražeň area in the Stříbro-Plasy strip (2)

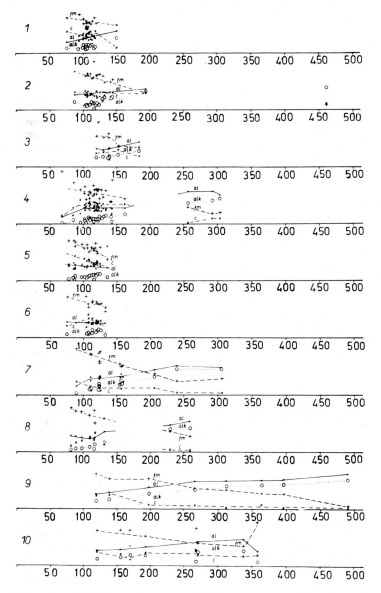

Fig. 2. Differentiation of the Proterozoic volcanics of the Barrandian area shown with NIGGLI values. *1* Amphibolites of the Slavkovský les Mts; *2* volcanic zone of Svojšín, Stříbro and Plasy; *3* potash-spilites of the Slatina-Pavlíkov strip; *4* volcanics of the eastern and middle parts of the main central zone (between Kralupy and Plzeň); *5* volcanics of the western part of the main volcanic zone (west of Plzeň); *6-10* volcanics of the southern zone: *6* western part; *7* the Blovice area; *8* the Příbram area; *9* the Dobříš area; *10* the Zbraslav area

(47.40 - 50.50% SiO_2, 4.16 - 6.57% CaO, 4.58 - 5.67% Na_2O and 1.11 - 2.56% K_2O). In borehole D-1 these rocks show an interesting differentiation into more acidic nests, mostly of a feldspathic nature (56.79% SiO_2, 6.29% CaO, 7.00% Na_2O, 0.62% K_2O, alk = 24.5, k = 0.05) and a more basic groundmass richer in stilpnomelane (38.74% SiO_2, 5.87% CaO, 2.16% Na_2O, 4.24% K_2O, alk = 10.8, k = 0.56).

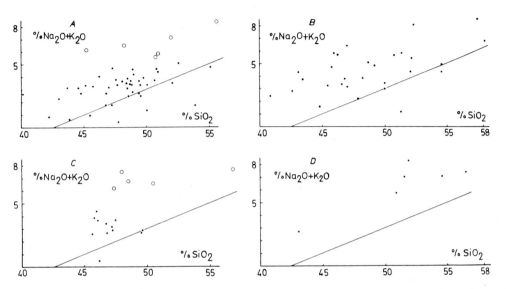

Fig. 3A-D. Alkali silica diagram (wt. %) of the Proterozoic volcanics of the Barrandian and the Železné hory areas. A the main volcanic zone and (circles) potash-spilites of the Slatina-Pavlíkov strip; B the southern volcanic zone; C volcanic zone of Svojšín, Stříbro and Plasy together with (circles) the potash-spilites from Dražeň; D the Železné hory area

A character more basic than that of the main type of the Barrandian metabasalts has been found in the spilites of the predeposit member of the Proterozoic of the Železné hory Mountains (eastern Bohemia), which contain 50.86% SiO_2, 11.40% CaO, 5.27% Na_2O, 0.46% K_2O, their k being 0.05 and CaO/alk = 1.98. Higher up, they are associated with pyrite-bearing quartz-graphitic schists and a deposit of sedimentary carbonate Fe-Mn ore. A still more spilitic chemistry has been established in the Eocambrian spilitic porphyrites, amygdaloidal and strongly epidotized (51.50 - 56.49% SiO_2, 3.44 - 5.51% CaO, 6.26 - 7.60% Na_2O, 0.42 - 0.79% K_2O, alk = 19.2 - 23.3, k = 0.04 - 0.08, CaO/alk = 0.49 - 0.70) which locally are associated with quartz-keratophyres and quartz-albitophyres (70.38 - 77.00% SiO_2, 0.18 - 0.39% CaO, 4.22 - 7.30% Na_2O, 0.33 - 2.04% K_2O, alk = 34.6 - 43.2, k = 0.03 - 0.24, CaO/alk = 0.02 - 0.04 - cf. FIALA, 1967b).

In our Precambrian volcanics, on the whole, an enrichment in Na_2O and a more pronounced spilitic character can be observed in space as one goes from W to E, from the Barrandian towards the Železné hory Mountains, and in time from the Algonkian towards the Eocambrian age; in the Barrandian area, later also during the Lower Paleozoic age.

This Algonkian volcanism belongs to the initial phase of the Cadomian (Assyntian) tectono-magmatic cycle.

Lower Paleozoic Volcanism

In central Bohemia, the intensive Middle and Upper Cambrian subsequent volcanism (WALDHAUSEROVÁ, 1966) produced subaerial porphyrites and quartz-porphyries (paleodacites, paleoandesites and paleorhyolites) in the Křivoklát-Rokycany zone, and basic porphyrites (andesitobasalts) in the Strašice zone. In the first zone, reverberations of this subsequent volcanism extended up to the Lower Ordovician. In the Strašice zone and in the Tremadocian of the Tuklaty area (E of Prague) spilitic members appear; they are evidently connected with subsidence movements of the respective areas and with the Ordovician sea transgression. The rocks of the Tuklaty area contain 43.30 - 46.40% SiO_2, 4.27 - 8.06% CaO, 4.25 - 6.18% Na_2O, 0.28 - 2.28% K_2O, alk = 15.3 - 19.5, k = 0.03 - 0.26, CaO/alk = 0.75 - 1.23.

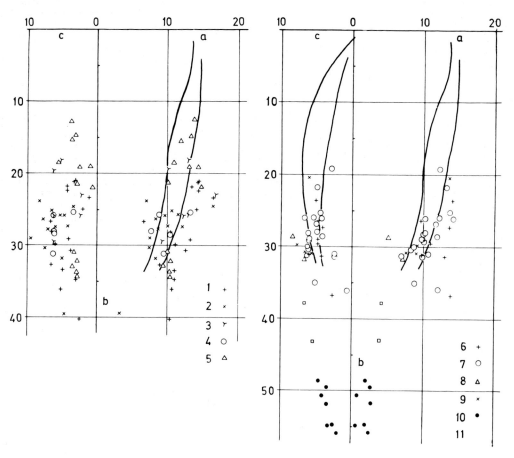

Fig. 4. Differentiation of the Lower Paleozoic diabase volcanics of the Barrandian area according to ZAVARITSKIJ, 1950 and ZAVARITSKIJ, 1946: A Ordovician: *1* spilitic types; *2* earlier aphanitic diabases; *3* later aphanitic diabases (Caradoc); *4* intrusive doleritic diabases; *5* kersantites, minverites and minettes. B Silurian and Devonian: *6* spilitic types; *7* doleritic diabases, partially affected by alkali metasomatism; *8* olivine diabases from Radouš; *9* basaltoid diabases; *10* Silurian ultrabasic rocks; *11* Devonian diabases

The renewed subsidence movements of the mobile zone of the Barrandian area during the Ordovician period accompanied by repeated transgressions and regressions of the sea and alternation of the sedimentary facies of shale and sand were connected with the ascent of magmatic masses. The character of these masses gradually became more basic, from andesite-basaltic to tholeiite-basaltic and olivine basaltic, in the Silurian, locally even ultrabasic. There is no sharp boundary between the subsequent or final Cadomian (Cambrian) and the initial Variscan (Ordovician) volcanisms in the oscillating Barrandian Basin; gradations of spilitized final Cadomian porphyrites (andesitobasalts) into Variscan initial diabase volcanics occur.

That the Ordovician volcanism was predominantly submarine is indicated by the huge development of granulates and granulate tuffs (MĚSKA-FIALA, 1948; FIALA, 1966c), i.e. hyaloclastites in RITTMAN's concept (1958). Compared with the Proterozoic, the Ordovician diabase volcanics show a higher proportion of alkalis and an enhanced total potassium content. Spilitic types are fairly frequent (see ZAWARITSKIJ's diagram in Fig. 4). In some places marginal spilitization of the pillows of the Ordovician pillow lavas can be observed (FIALA, 1968); it is evidently post-magmatic, having taken place after the lava effusion. Elsewhere, however, spilitic facies are developed with pronounced primary textures of intersertal type throughout the volume of the flow, which furnishes evidence that crystallization of a lava primarily rich in Na_2O has taken place here.

The Silurian submarine volcanism of Wenlockian and Lower Ludlovian age is characterized by strong differentiations: normal diabases and slightly spilitized ones, typical spilites, amygdaloidal rocks, pillow lavas, aphanitic and basaltoid diabases, intrusive doleritic types usually automorphically transformed into essexitic and teschenitic diabases, rare teschenite, locally ultrabasic picrites, abundant granulates and granulate tuffs. Compared with the Ordovician volcanics, those of the

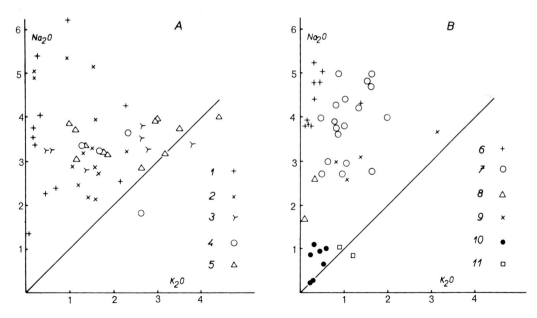

Fig. 5A and B. Alkali (silica diagram)(wt. %) of the Lower Paleozoic diabase volcanics of the Barrandian area. Key the same as for Fig. 4

Silurian age are distinguished (see ZAWARITSKIJ's diagram in Fig. 4 and alkali/silica diagram, Fig. 5) by a higher basicity, a relatively low content of K_2O and a higher proportion of Na_2O. The content of Na_2O being 3.82 - 5.24% and that of K_2O 0.15 - 0.52%, a typical spilite facies often arises. The marginal zones of sills of intrusive dolerite diabases are frequently distinctly spilitized, showing laths of albite (An_{4-8}) in a chloritic groundmass with a normal intersertal texture. At 12 cm from the contact, albite-oligoclase An_{12} with intergranular texture, and fresh augite appear. At a distance of 1-2 m from the contact, a normal doleritic (ophitic) texture with fresh augite and rims of oligoclase to albite-oligoclase around more basic plagioclase cores can be observed. In the spilitized margins of dikes the original, more basic plagioclase phenocrysts have been distinctly replaced by albite. In other cases, especially in minute feldspars of the groundmass, set in a chloritic mesostasis, original crystallization from a magma enriched in water and Na_2O taken up by the magma during its uprise is evident.

Much more intensive to complete spilitization affected the lavas of effusive bodies. Here, typical spilites and spilitic porphyrites developed, in which feldspars (albite to albite-oligoclase An_{2-12}) occur in a chloritic groundmass rich in leucoxene. Their textures are finely or coarsely intersertal, often with long bent laths of albite. The spilitic development was particularly intensive in pillow lavas, especially abundant between Prague and Karlštejn (FIALA, 1966d). Farther westward, transitional oligoclase-bearing types are more frequent.

The youngest Silurian effusions, which are referred to a higher stage of the Lower Ludlow, are formed of basaltoid diabases, partly submarine, partly subaerial and of a dike character; the latter are then often of a fresh basaltic appearance. The marginal portions of the dikes penetrating the Liteň shales (Wenlock) have become green, and basic plagioclase (labradorite of An_{55-69}) is replaced here by oligoclase; chlorite is abundant. This spilitization type of transformation depended on the import of water and the exchange of bases at the area of contact with wet sediments (transvaporization, as defined by SZÁDECZKY-KARDOSS, 1958).

The spilitic character of the margins of doleritic diabases and especially that of the outpoured masses of spilitic diabases suggests that there was a change in the composition of the original olivine-basaltic magma. This change was the result of the taking-up of water and Na_2O by the magma during a submarine effusion and, especially, as early as during its ascent through marine sediments. When the magma was already partly crystallized, basic plagioclases were replaced by albite-oligoclase to albite, and pyroxenes were chloritized. The not yet crystallized magma, enriched in water and Na_2O during its uprise and effusion could solidify directly as spilite. The Silurian sea was substantially deeper than the Ordovician one. A considerable calcite content in amygdales, a high proportion of the calcite component in the Kopanina shales (Lower Ludlow), a great amount of calcite forming the cement of granulate diabase tuffs and the onset of calcite sedimentation after the end of volcanic eruptions during the Ludlow, suggest that CO_2 played a significant part, even during the uprise of the lavas. This fact has certainly also affected the spilitization process.

General Notes

Comparing the knowledge acquired from the study of the volcanics of the Barrandian and the Železné hory areas with the topics presented for discussion, it can be stated:

1. Spilites, characterized by typical magmatic textures and the mineral association, albite (albite-oligoclase) + chlorite, can be formed by primary crystallization from magmatic solutions enriched in H_2O and Na_2O (and often CO_2); in a suitable environment they can also arise by postmagmatic alteration.

2. The use of the term spilite for aphanitic diabases and metabasalts (according to BRONGNIART, 1827) is to be omitted from strict petrographic nomenclature. It is possible to keep the term as a geological one, e.g. for field purposes etc.

3. The association, albite (albite-oligoclase) + chlorite, in spilites is accompanied by increased proportions of Na_2O. As a dividing line, 4% Na_2O can be accepted for occurrences in Bohemia (FIALA, 1966a and 1967a) in accord with FABRIÈS's (1963) standard for Sierra Morena Mountains. This dividing line is not sharp and gradational types with 3.5 to 4% Na_2O and with oligoclase as a main feldspar, sometimes with fresh augite, are abundant. In defining spilites, higher or lower values of Na_2O are given by various authors (PAPEZIK, 1968: 3% Na_2O). But even when a lower content is accepted as a limit, many rocks impoverished in Na_2O, i.e. those of basaltic (metabasaltic) composition, will always exist in the same formation.

4. The value of the ratio $CaO/Na_2O + K_2O$ (wt. %) > 2 is accepted in analogy to that proposed by PAPEZÍK (1968). Some deviations can appear in altered glassy selvedges of the pillows impoverished in alkalis and often carrying abundant calcite amygdales.

5. The association of the spilitic types *sensu stricto* (richer in Na_2O, with albite + chlorite) with the basaltic ones (with higher proportions of Ca, with plagioclase or with pseudomorphs after plagioclase, often with fresh augite) has been recorded in many countries, especially those in which geosynclinal submarine volcanisms (FIALA, 1966a, p.13) are found. These relations can be followed in diagrams, for instance, that according to ZAWARITSKIJ adapted by ZAWARITSKIJ (1946)(Fig. 4), especially along the right side, where there are no vectors, and in the alkali/silica wt. % diagram (Fig. 5).

6. Calculations from analyses of rocks with a higher proportion of CO_2 are often distorted by subtraction of an amount of $CaCO_3$ (or, of a complex carbonate) corresponding to the content of CO_2. The resulting numbers show relatively increased proportions of alkalis, but they often point to a peralumina and not to a magmatic type (PAMIĆ, 1962, p.23). Many bases, however, now bonded in carbonates, belong to the original magma. Besides Ca, Fe and Mg are often bonded in carbonates, e.g. in our Ordovician diabases. The circumstance (determined by Mr. M. HUKA in the chemical laboratory of the Geological Survey (ÚÚG) in Prague) that after treatment with dilute acids (2% HCl or CH_3COOH) a portion of the bases bonded in clay minerals and chlorites also passes into solution, is to be taken into consideration.

7. Spilites do not occur in one type of environment only (AMSTUTZ, 1968, p.942). They frequently occur in mobile zones of a geosynclinal type, but they are not necessarily associated with deep-sea facies; in Bohemia, they seem rather to avoid it. In the Upper Proterozoic of Bohemia the spilites are grouped especially in the strongly mobile southern limb

of the geosyncline, in proximity to the old block of the Moldanubicum. Here, a typical metabasalt-spilite-keratophyre suite occurs. The Silurian spilites of the Barrandian area, very typical chemically and petrographically, are associated with a shallow sea facies. The presence of increased proportions of CO_2, in addition to H_2O and Na_2O, is important in their formation.

8. Spilites are abundant in the pillow lavas. Nevertheless, both terms are not synonymous, as has been assumed by many workers.

9. Data on the symmetrical distribution of the components in spilite pillows and on the increased proportion of Na_2O in their interiors, in contrast to their selvedges (AMSTUTZ, 1968, p.947) are plentiful. In our country, these features can be observed in the Proterozoic pillow lavas, where, under the originally glassy chilled marginal zone, the glowing and mobile interior was thermally insulated; this resulted in the formation of a zone enriched in albite. In our Ordovician diabase pillow lavas (of a basaltic composition, on the whole), the Na_2O content has been enhanced in the marginal zone because of additional sodium coming from outside; the border zones of plagioclase phenocrysts have been replaced by albite. Also, the magma of the marginal zones of sills of our Silurian dolerite diabases was affected by alkalizing and hydrating alterations due to transvaporization from the outside (SZÁDECZKY-KARDOSS, 1958).

10. I would not quite agree with the argument that there are no differences in the chemistry of spilites as compared with that of basalts, which could not be explained by the manner of collecting samples ("On ne peut pas éviter l'impression qu'un échantillonnage complet de toutes les phases pourrait avoir pour résultat une composition basaltique comme le suggèrent AMSTUTZ (1958, p.3) et VALLANCE (1960, p.37") - AMSTUTZ, 1968, p.948). As far as the rocks of a single area are concerned, the differences among the basaltic, spilitic and transitional types are for the most part evident. On the other hand, spilitic chemistry is, with some modifications, that of alkali basalts enriched in water. It is not necessary nor is it possible to speak about independent spilitic magmatic type. The origin of spilites can be explained by differentiation in the framework of the total basaltic composition in the presence of water (the southern volcanic zone of the Barrandian Proterozoic), in many cases from contamination (transvaporization - SZÁDECZKY-KARDOSS, 1958) of the basaltic magma during its upsurge through complexes of wet, Na-rich sediments, sometimes also by alteration (spilitic reaction) of a submarine lava effusion.

11. Transformation of originally dry basaltic magma as a result of the taking up of water and alkalis from the adjacent rocks can occur during the upsurge of the magma through tectonic fissures or even in the upper part of the magmatic chamber. If the magma was affected by such a transformation before starting crystallization or during its initial phases, the resulting magma is the same as if the rock originated from a primary differentiation from the original magma. This also holds true - to some extent - for the alteration of the magma outpouring on the sea bottom or penetrating into sea mud; in this case, however, the crystallization grade and the character of the total effusion are very important. The cover of finely vesicular hyaloclastite fragments formed in the front and on the surface of a submarine lava flow during the process of granulation (FIALA, 1966a, p.16) insulates its still glowing interior, but makes possible, owing to its fissured and granulated nature, import and circulation of hot water and steam, which support alterations and influence crystallization of the lava. Real hydromagmas can form and are capable, as magmatic solutions of lowered temperature, of moving in thin veinlets and of crystallizing as a rock of intersertal texture.

12. The textures and structures of spilites do not differ markedly from those of basalts (AMSTUTZ, 1968, p.945). The intersertal textures with divergent and arborescent modifications (VUAGNAT, 1946) are characteristic. The type of spilitic texture, still maintained in the Soviet literature (POLOVINKINA, 1948, I, p.71; 1966, I, p.69; III, p.148) also appears frequently.

13. The K_2O content is generally very low in the spilites (TURNER-VERHOOGEN, 1960). The k-values (NIGGLI) in our Proterozoic volcanics of the basaltic and spilitic types vary between 0.01 - 0.31, increasing, naturally, in keratophyres and quartz-keratophyres. In our Silurian spilites the k-values are 0.03 - 0.07, exceptionally 0.17. Types enriched in potash do exist, e.g. the potash-spilites of the Slatina-Pavlíkov strip in the northern limb of the Barrandian area and the slightly metamorphosed (stilpnomelane) rocks from Dražeň in the Stříbro-Plasy volcanic zone (western Bohemia); they represent a distinctly defined group with elevated K_2O contents and a relatively high total content of alkalis (see p. 12-13 and graphs in Figs. 2 and 3). These types seem to be primary magmatic ones.

The K_2O content increases in keratophyres and quartz-keratophyres (the southern volcanic zone of the Barrandian Proterozoic, Eocambrian of the Železné hory Mountains). The mutual proportions of alkalis change gradually, being caused, first of all, by differentiation of the magma contaminated during its upsurge. In some cases, the participation of metasomatic processes is to be admitted.

14. The association of spilites with keratophyres and quartz-keratophyres (possibly also andesites and rhyolites) deserves special comment. In some cases (the Proterozoic southern volcanic zone, especially the Blovice, Příbram, Dobříš and Zbraslav areas), the gradation of spilites through keratophyre-spilites into keratophyres, or (Dobříš and Zbraslav areas) even into quartz-keratophyres and quartz-albitophyres can be studied (cf. Fig. 2). Elsewhere, the gradation is less distinct (Železné hory Mountains) or, especially in the zones with prevailing basaltoid types, a distinct gap appears. In some cases, these gaps become successively narrower and filled, owing to an increasing number of analyses. The association of spilitized diabases and spilites with keratophyres is also marked in the initial Devonian volcanism of northern Moravia (BARTH, 1966; FOJT, 1966) as well as in that of albitized metadiabases with keratophyres and quartz-keratophyres in the Upper Silurian volcanics of the Železný Brod crystalline complexes (NE Bohemia, FEDIUK, 1962).

15. In analogy to the analyses cited by VALLANCE (1965, pp.472 and 473, anal. no. 22, 23), Ca-enriched rocks also appear in places among the Bohemian Proterozoic volcanics. Particular instances are represented by basaltoid bytownite porphyrites from Odolena Voda (44.56% SiO_2, 16.42% CaO, 2.69% Na_2O, 0.41% K_2O, 2.13% CO_2 and by some portions of the interpillow glass from the pillow lava of Skomelno (43.84% SiO_2, 15.68% CaO, 0.47% Na_2O and traces of CO_2).

16. The assumption that the younger dikes are richer in hydrothermal and deuteric minerals (AMSTUTZ, 1968) is not always in agreement with facts; it depends on the source of the material. Autometamorphosed rocks can appear during the earliest phases of a volcanic sequence if the uppermost part of the magmatic chamber containing magma enriched volatiles (either diffusing from the lower portions of the magma or taken up from adjacent rocks) was reached by an opening tectonic fissure serving as a magmatic pathway. The magma can be enriched in volatiles during its uprise owing to transvaporization (e.g. in the sills of our Ordovician and Silurian doleritic diabases). The youngest dikes of the Ordovician aphanitic diabases and Silurian-Ludlovian basal-

toid diabases have not been touched by this transformation or have been only slighly affected in their border zones.

17. As far as AMSTUTZ's definition (1968, p.949-950) is concerned: "le mode de formation peut être pris comme résultant d'un transfert et d'une différentiation des constituants dans une phase acqueuse séparée pendant la cristallisation primaire. La nature hydratée de la fusion serait normalement le résultat d'une différentiation primaire ou, par endroits, pourait être due à la contamination par des roches voisines, ou par quelques processus inconnus de fusion d'une partie hydratée du manteau terrestre", I would propose the following modification of the second phrase: the hydrated nature of a magmatic belt can be caused by the primary differentiation or, more often, by contamination (transvaporization) from the rocks surrounding the magmatic chamber or adjoining magmatic pathways or even from the neighborhood of a submarine effusion; it may also result from some unknown events of melting of the hydrated portions of the rocks at depth.

a) All of the above-mentioned processes can produce spilites and keratophyres. Contamination can take place in the upper part of a magmatic chamber where the originally dry basaltic magma comes in contact with the hydrated rocks of the earth's crust or, more often, during its ascent through complexes of sediments soaked with water or salt solutions. The sills of the Silurian doleritic diabases, relatively drier in their interior, with fresh augite and basic plagioclase and with an ophitic texture, are rimmed by a spilitized facies characterized by its mineralogy (albite, chlorite, leucoxene), texture (intersertal porphyritic) and chemistry (Na_2O-enriched).

b) It is difficult to define the types produced by albitization of the small feldspars of the groundmass or by subsequent individualization of albite during recrystallization of the original glass. Even these instances ought to be classified mainly as spilites.

18. Metamorphites of a spilitic composition and sometimes with relics of magmatic texture can originate either by regional metamorphism of the original spilites or by albitization of the original metabasalts and diabases or some other rocks. These rocks should be termed according to the present metamorphic state as "metabasite" (greenstone, Grünstein) or "green schists" (Grünschiefer, roches vertes) or prasinite or amphibolite etc.; these names can also be used for field purposes. After a more detailed study and, if possible, according to chemical analyses, it is often possible to recognize the original character (e.g. metabasalt, metaspilite) and to join it (in brackets) to the first designation together with an indication of the composition, texture and the metamorphic facies. Similar problems can appear, for instance, in the study of various granitoids.

19. The use of a differing nomenclature also causes some lack of clarity. The tendency to use a unified terminology, e.g. suppressing the old terms of paleovolcanics, also has its disadvantages. The effort to apply unified neovolcanic terms to the rocks of early periods sometimes leads to exaggerated generalizations (e.g. terms "paleoandesite", "paleobasalt") while the earlier paleovolcanic terminology, at least in Central Europe, often indicated differences in composition, texture, genesis etc. Where occurrences of "andesites" are given together with spilites *sensu stricto*, it seems to be useful to check the proportions of their alkalis in comparison with the typical Pacific andesites of subsequent orogenic phases.

References

AMSTUTZ, G.C. (1958): Spilitic rocks and mineral deposits. Bull. Missouri School of Mines, Tech. Ser. 96, 11.

AMSTUTZ, G.C. (1968): Les laves spilitiques et leurs gîtes minéraux. Geol. Rundschau 57, 936-954.

BARTH, V. (1966): The initial volcanism in the Devonian of Moravia. Paleovolcanites of the Bohemian Massif, 115-125. Prague: Charles University.

BRONGNIART, A. (1827): Classification et caractères minéralogiques des roches. Paris.

DEWEY, H., FLETT, J.S. (1911): On some British pillow lavas and the rocks associated with them. Geol. Mag. 8, 202-209, 241-248.

FABRIES, J. (1963): Les formations cristallines et métamorphiques du Nord-Est de la province de Séville/Espagne. Essai sur le métamorphisme des roches basique. Sciences de la terre, Mém. (Nancy) 4, 1-267.

FEDIUK, F. (1962): Volcanic rocks of the Železný Brod metamorphic region. Rozpravy Ústředního ústavu geologického (Praha) 29, 116. In Czech with English summary.

FIALA, F. (1966a): Some results of the recent investigation of the Algonkian volcanism in the Barrandian and Železné hory area. Paleovolcanites of the Bohemian Massif, 9-29. Prague: Charles University

FIALA, F. (1966b): Potash-spilites in the Algonkian of the Barrandian area. Ibidem 87-99.

FIALA, F. (1966c): The Silurian diabase volcanism of the Barrandian area. Ibidem 153-165.

FIALA, F. (1966d): Silurské polštářové lávy Barrandienu (Silurian pillow lavas of the Barrandian). Časopis pro min. a geol. (Praha) 11, nr. 3, 267-276. In Czech with English summary.

FIALA, F. (1967a): Algonkian pillow lavas and variolites in the Barrandian area. Sborník geol. věd, geologie, řada G-12 (Praha), 7-65.

FIALA, F. (1967b): The chemism of the Algonkian and Eocambrian volcanites in the Železné hory Mountains. "Geochemistry in Czechoslovakia", transactions of the first conference on geochemistry, Ostrava, 1965. 15-29.

FIALA, F. (1968): Ordovician diabase pillow lavas of the Barrandian area. Věstník Ústředního ústavu geologického (Praha) 43, 169-182.

FIALA, F. (1969): Silurian and Devonian diabases of the Barrandian basin. Sborník geol. věd (Praha), G-17. In print. In Czech with English summary.

FOJT, B. (1966): Keratophyre rocks of the Kies deposits in the Jeseníky area. Paleovolcanites of the Bohemian Massif, 107-114. Prague: Charles University.

HEJTMAN, B. (1967): Systematická petrografie vyvřelých hornin. Praha.

KORZHINSKIJ, D.S. (1962): Problema spilitov i gipoteza transvaporizacii v svete novych okeanologitscheskich i vulkanologitscheskich dannych. Izv. Akad. Nauk SSSR, Ser. Geol. (Moskva) 9, 12-17.

MĚSKA, G., FIALA, F. (1948): Několik poznámek o typech diabasových hornin v Barrandienu. (Quelques remarques concernant les types des roches diabasiques dans le Barrandien). Časopis Národního musea, odd. přír. (Praha) 117, 149-166.

PAMIĆ, J. (1962): Spilitsko-keratofirska associacija stijena u području Jablanice i Prozora (Bosna i Hercegovina). (The spilite-keratophyre association in the area of Prozor and Jablanica). Prirodoslov. istraž. Acta geol. III (Zagreb) knj 31, 5-94.

PAPEZIK, V.S. (1968): Proposed chemical limits of spilite. Written information.

POLOVINKINA, JU.IR. (1966): Struktury i textury izvershennych i metamorfitscheskich gornych porod. I-III. Moskva-Leningrad.

POLOVINKINA et al. (1948): Struktury gornych porod. I. Magmatitscheskie porody. Moskva-Leningrad.

RITTMANN, A. (1958): Il meccanismo di formazione delle lave a pillows e dei cosidetti tufi palagonitici. Bull. Acc. Gioenia Sc. nat. (Catania) 4, 311-318.

SLAVÍK, F. (1908): Spilitische Ergußgesteine im Präkambrium zwischen Kladno und Klattau. Archiv f. naturwiss. Landesdurchforschung Böhmens (Praha) 14, 2.

SLAVÍK, F. (1926): Les "pillow-lavas" algonkiennes de la Bohême. C.R. 14. Congr. geol. intern. (Madrid), 1389-1398.

SZÁDECZKY-KARDOSS, E. (1958): On the petrology of volcanic rocks and the interaction of magma and water. Acta geol. (Budapest) 5, 2.

TURNER, F.J., VERHOGGEN, J. (1960): Igneous and metamorphic petrology. 2nd ed. New York-Toronto-London.

VALLANCE, T.G. (1960): Concerning spilites. Proc. Linnean Soc. N.S. Wales, IXXXV, part 1, 8-52.

VALLANCE, T.G. (1965): On the chemistry of pillow lavas and the origin of spilites. Mineralogical Magazine 34, No. 268, 471-481.

VUAGNAT, M. (1946): Sur quelques diabases suisses. Contribution à l'étude du problème des spilites et des pillow lavas. Bull. Suisse de Min. et Pétr. 26, 115-228.

WALDHAUSEROVÁ, J. (1966): The volcanites of the Křivoklát Rokycany zone. Paeovolanites of the Bohemian Massif, 145-151. Prague: Charles University.

ZAWARITSKIJ, A.N. (1950): Vvedenie v petrochimiyu izverzhennych gornych porod. Moskva-Leningrad. (2nd ed.).

ZAWARITSKIJ, V.A. (1946): Spilito-keratofirovaya formacija okrestnostej Blyavy na Urale. Trudy Inst. Geol. Nauk AN SSSR, Ser. Petrogr. (Moskva) 24.

Spilitic Magma. Characteristics and Mode of Formation

E. Lehmann

Abstract

It is suggested that the lapilli in the Middle Devonian tuff (schalstein) of the Lahn Basin throw light on the question of the existence of a primary spilitic magma. In a largely medium-grained type of lapilli, the mineralogical composition conforms to that of the massive weilburgites, giving evidence of a corresponding melt already in progress at the time of explosive activity and of its origin in depth. No other source or character is conceivable for the co-existing lapilli with deficient or only partially developed crystallization. Moreover, the refractive index of the amorphous material is considerably below that of basaltic glasses, which likewise points to the existence of a weilburgitic hydromagma before the explosion.

Judging from the capacity of forsterite, enstatite and diopside melts at high pressures to dissolve a noticeable amount of H_2O, it is suggested that hydromagmas might have been formed in the high Upper Mantle or in the lower crust zones. Hydrous solid phases, for instance in amphibole peridotites, are referred to as evidence of such an effect. Degassing and ascent are assumed to depend on the release of external pressure under suitable orogenic conditions, especially folding, in synclines.

Attention is called to the localized group of metasomatic spilites in the Lahn district, to the occasional appearance of weilburgites exhibiting semi-spilitic features, and to the characteristic differences between the weilburgites and diabases associated with them. In comparison with the weilburgites the variety is even greater in spilites resulting from burial metamorphism of andesite or basalt.

Weilburgitic Tuffs as a Guide in the Spilite Controversy

In the Lahn syncline (Lahn Basin) of Western Germany, the central rock complex, of upper Middle Devonian and lower Upper Devonian age, consists of so-called "schalstein". Any consistency, however, suggested by that old miner's term, is not reflected in the petrographic character of the rock. In addition to keratophyric and weilburgitic pyroclastics, the author (LEHMANN, 1941) has recognized that coherent bodies of magmatic origin are often intercalated in the tuff deposits. These magmatic bodies, equivalents of the weilburgites, in particular underwent extensive carbonization or chloritization during or following the intrusion. Extensive dissemination of the appropriate magma in the tuffaceous host rock is exemplified by these so-called "injecta" (LEHMANN, 1968, 1970).

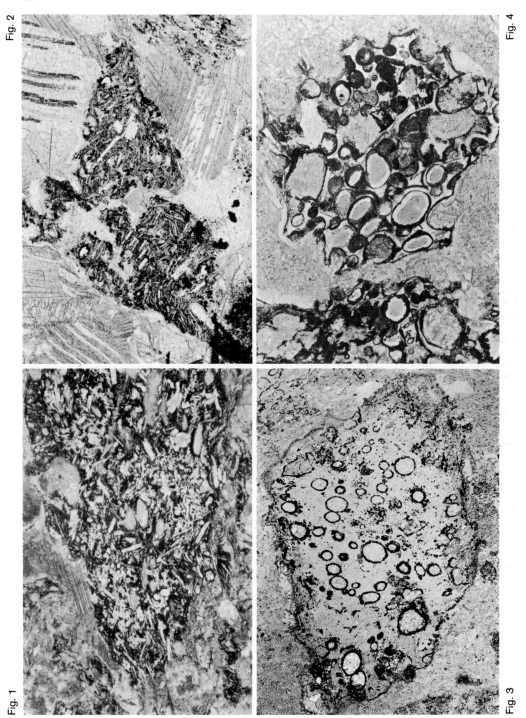

Only the pyroclastic products can be referred to in the following discussion of the spilite problem. But some remarks concerning the tuff on a whole appear to be indispensible. In general, lapilli prevail over sediment fragments, but the quantity of the latter in many places increases to such proportions, that up to 75% of the tuff consists of sedimentary material (limestone, slates and grits) comprising fragments of the underlying rock formations as well as country-rock detritus.

Within the tuffs the pyroclastic material is usually fine-grained; distinct crystals of feldspar are rare, so that it is chiefly the feldspar in the lapilli which should be taken into account in attempting to determine the petrographic character. In some places (e.g. Sangert Berg near Balduinstein, the Fachinger Grundstollen in part the quarry n. of Aumenau), the feldspars are potassic in composition; in other localities, however, (e.g. Alte Berg n. of Laubuseschbach, the Fachinger Grundstollen in part), they are sodic.[1] Thus the alkali variation that characterizes the massive weilburgites, is also reflected in the pyroclastics, providing a clear hint as to their petrographic peculiarity.

The outlines of the lapilli are distinct: they may be rectilinear, rounded, or more or less irregular. Wedged forms occasionally point to incomplete solidification or a degree of plasticity at the time of ejection (Figs. 1-7).

Effects of the Variscan orogenic phases are not wanting, but the degree of deformation varies from place to place and often over short distances. The effects of deformation are most pronounced in the surrounding carbonate aggregates, whereas the lapilli escaped the influence of the stress. The latter only underwent some transformation in so far as the volatiles were able to give rise to reaction (principally chloritization and carbonization).

With regard to the problem that concerns us here, another aspect of the lapilli is more significant. Based on a large number of samples (about 400 thin sections), two principal types and one special type of lapilli can be distinguished. Lapilli I correspond to the weilburgites in their degree of crystallization, i.e. their richness in feldspar, their general mineralogical composition, and their textural character. One can also conclude from the frequency of rodlike feldspar crystals (orthoclase and/or albite) and occasionally also from the presence of indistinctly

---

[1] Because the five locations referred to in the text are far apart, it is impossible to reproduce maps of them. Other localities could also have been sampled. However, the samples were selected on the basis of their freshness, combined with the best possible state of preservation of the feldspar. Consequently exact maps of the locations are not so important. Furthermore two of these localities are no longer accessible.

◄ Fig. 1 and 2. Lapilli I. Crystallization must have been accomplished before the ejection. The spilitic disposition existed originally. Fig. 1. Sangert Berg, ordinary light, magn. x 45. Fig. 2. Alte Berg, ordinary light, magn. x 54

Fig. 3 and 4. Lapilli II. Consolidation in pursuance of the ejection. Abundance of globules filled with calcite, chlorite, occasionally additional quartz, in an optically isotropic matrix. Feldspar is deficient or quite subordinate. Fig. 3. Dürrsteiner Kopf, ordinary light, magn. x 68.5. Fig. 4. Fachinger Grundstollen, ordinary light, magn. x 90

marked phenocrysts of the same compositions, which occur within a matrix mainly composed of chlorite of rather uniform character (diabantite commonly low in iron), that the solidification of these lapilli by and large took place at depth, and thus preceded the setting-in of the explosion.

Lapilli II, however, differ completely in their mineralogical character from type I. Feldspar is lacking or rare. Their principal characteristic is the abundance of an amorphous substance including a variable number of globular bodies, each of which is enveloped by a thin opaque-looking margin. The majority of them are filled with fine aggregates of chlorite, and often associated with other globules containing calcite or occasionally relics of sedimentary materials, e.g. slate or grit (Figs. 3 and 4). The chloritic cores are often zoned, the zones consisting of diverse chlorites, or here and there of chlorite alternating with muscovite (cf. Fig. 28 in LEHMANN, 1970).

Besides these two common types of lapilli, a fairly rare type, lapilli III, should be mentioned, strictly speaking a type in which a comparatively large amount of the amorphous substance, rich in globules, is combined with unusually large feldspar crystals (Fig. 5). This variety was found in the Fachinger Grundstollen (an adit with a NW-SE bearing, about 2 km in length, which cuts mainly tuffs with interlayers of weilburgite). The tuff variety in question was exposed over about 30 meters and probably originated from a separate explosive center.

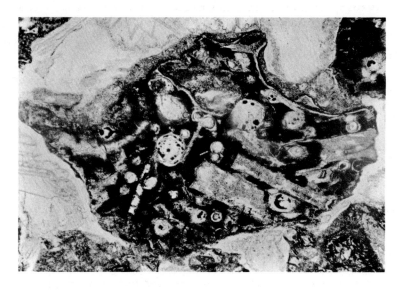

Fig. 5. In lapilli III, pre-ejective crystallization is manifested by large feldspars. The change of conditions due to the ejection resulted in an analogy with lapilli II, i.e. in the formation of an isotropic matrix rich in globules. Fachinger Grundstollen, ordinary light, magn. x 68.5

The weilburgitic nature of lapilli I is substantiated by its mineralogical character. On the other hand, the co-existence of lapilli II with lapilli I in the majority of the samples provides evidence of their common source. Another source can hardly be assumed for the tuff with lapilli III as well. With this in mind, the liquid phase appropriate

to produce rocks of weilburgitic character must have already existed
at the depth at which the explosive conditions originated.

If the liquid originally corresponded with a basaltic magma (see below),
the basaltic character thus must have been changed toward weilburgite
at the time it reached this depth. For lapilli I this follows from the
mineralogical features. In lapilli II and III, it seems to be indicated
by the refractive index of the amorphous (vitreous or colloidal) component (average of 20 determinations n = 1.5594), thus considerably
below n of basaltic glass (McKEE, 1968). Likewise it is below that of
palagonitic glass (LEHMANN, 1970). As to the chlorite in lapilli II
and III, its zoning as well as the variation of n = (1.565 - 1.578) is
opposed to the interpretation as a mere product of devitrification. As
to whether the matrix is basaltic or palagonitic, or how far it differs
from both, the chlorite does not provide a conclusive argument for
identification with palagonite (HENTSCHEL, 1951).

The Derivation of H_2O

Evidence of the abundance of H_2O may be seen in the following features:
1. The extent and the intensity of explosive volcanism in the Lahn
district is inconceivable without a corresponding amount of water. The
quantity of schalstein, including tuff and weilburgite injected therein,
is estimated to be about 3×10^7 m^3, of which perhaps half consists of
tuff. The proportion of sedimentary material brought up in and with this
tuff is often remarkable and much of it seems to originate from considerable depth. This is to be concluded from limestone fragments containing garnet or speckled with minute inclusions inaccessible to microscopic determination. Since no analogous rocks of Middle Devonian age
and no limestones of Lower Devonian age are known at all, it is suggested that the fragments in question were derived from pre-Devonian schists
and limestones. 2. Water plays a major role in most of the widespread
chemical reactions (hydration), and in the extensive mobilization and
transportation of material. Apart from the formation of the globules
in the amorphous matrix of the lapilli, two other examples of solution
activity may be mentioned. Due to partial solution the boundary of a
very fine-grained slate fragment was intensely embayed and distinct
particles were separated from the slate fragment (Fig. 6). The second
example (Fig. 7) shows a fragment of old, probably pre-Devonian limestone in which partial solution is manifested by embayments and the
formation of globules, which are in part chloritic and in part calcitic.

It is clear that not all these processes occurred in the same stage of
evolution of the rocks or at the same depth. The presence of supposedly pre-Devonian limestone fragments among the ejected material suggests
that the magma chamber was deeper than usually assumed. Furthermore,
the advanced crystallization of lapilli I and the large-sized feldspar
crystals in lapilli III could hardly have formed in the course of the
explosive ascent.

There is also no evidence that the chlorite (diabantite) in lapilli I
formed after the explosion. In the post-explosive stage, on the contrary, transformation of the pre-existing diabantite into stilpnomelane
and muscovite is observed. This stilpnomelane seems to be poor in ferric
iron - in the globules its index of refraction is n = 1.565, markedly
below that of the diabantite. Likewise a complete devitrification of
the matrix in lapilli I could not be brought about in the course of the
explosion. Consequently, the only possibility seems to be that the

water needed for the interstitial chlorite in lapilli I had been available before the beginning of the explosion.

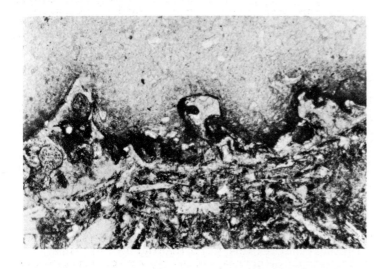

Fig. 6

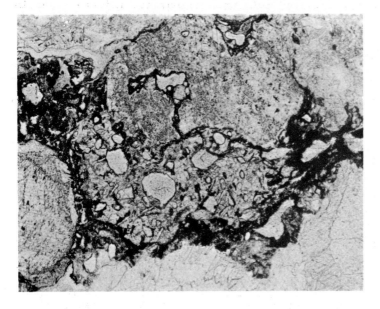

Fig. 7

Fig. 6 and 7. Chemical effects of the injecta on sedimentary components in the tuff. Fig. 6. Partial solution of a fragment of shale (top) is indicated by sinuated outlines, disjoined and occasionally detached and isolated parts of shale (on the left). N. of Aumenau, ordinary light, magn. x 72. Fig. 7. Fragment of old limestone disjoined by immigrating injectum. Formation of globules, mostly filled with chlorite, the small ones with calcite. Fachinger Grundstollen, ordinary light, magn. x 45

Considering the mineralogical analogy of the initial mineral association in lapilli I and, in part, lapilli III with the normal weilburgite, the further conclusion seems to be justified that the weilburgitic magma which ascended mainly after or in part in the intervals of the explosive activity, was in existence at the level from which the explosions proceeded.

The depth of the level of explosion is difficult to assess. The age of the ejected sedimentary fragments being unknown, only the PT-conditions inferred for the crystallization of chlorite may give us a hint as to the depth in question. But these conditions have been investigated for a particular chlorite composition, and thus are appropriate only in a restricted sense for the conditions here. For chlorite of approximately clinochlore character a maximum temperature of stability of 768 ± 7°C at a water pressure (equal to the total pressure) of 3.5 kb was found by FAWCETT and YODER (1966). The chlorite stability field is probably reduced in the presence of iron, but no exact data on the quantitative effect are provided. The temperature of about 770°C would correspond to a depth of about 30 km (Ann. Rep. Geophysical Laboratory; Carnegie Inst. Yearbook No. 40 for 1950-51). Admittedly even the lowering of the temperature by 150-200°C would result in a depth of about 20 km. However, in a crustal level 15-20 km below the surface the derivation of the required water from any superficial or near-surface source seems to be out of the question.

There seems to be only one alternative: the derivation of water from the Upper Mantle. The presence of H_2O below the crust is manifested by the presence of hydrous mineral phases (amphiboles, micas) in certain peridotites. According to WYLLIE (1967, p. 13 ff.) the maximum temperature limit of stability for the amphibole-bearing facies is guessed to lie between about 800 and 950°C at water pressures between about 10 and 30 kb, based on the tremolite stability. GILBERT (1968), after a review of the results arrived at by various authors for many different amphiboles, concluded that no amphibole could be stable below 70 to 100 km depth, if 950°C is a maximum temperature of stability at 30 kb, depending on the geothermal gradient.

At higher temperatures H_2O can be dissolved in the melts of enstatite and olivine (KUSHIRO, 1968, 1969; KUSHIRO et al. 1968); at 10 kb about 10%, and at 20 kb about 15% or more H_2O can enter the liquid phase (KUSHIRO, personal communication). The crystallization of amphibole at lower temperatures may be connected with the hydrous character of the melt.

The crystallization of rocks of mafic or ultramafic composition has been discussed for isobaric conditions (KUSHIRO, 1968, 1969). However, isobaric crystallization is prevented where, due to the onset of orogenic crustal movement, a sudden release of external pressure takes place at localized points of tension or fracturing. Under these conditions degassing of the mantle is intensified, whereas the liquid phase will move towards or ascend along the points of lowered pressure.

The fluids originally contained in a magma will be released by degrees, if the ascent and the emplacement fail to be controlled by tectonic effects (diapiric emplacement). In that case the decrease of temperature and pressure in the magma is adjusted to the external conditions.

Depending upon folding, however, adjustment could be prevented or at least considerably reduced in rate by the degassing previously mentioned. Under such conditions dehydration of the magma after emplacement does not come about if the afflux of fluids continued and exceeded the degree of escape. Then the external pressure in the space of emplacement remains high enough to prevent dehydration. This means that under the persistent influence of mantle degassing due to folding, the magma could continue to retain its H_2O, i.e., preserve its hydromagmatic character, down to a temperature low enough for spilitic crystallization.

It has been suggested that the abundance of fluids and the surplus of pressure required in an explosive activity comparable in extensity with the Middle Devonian tuff (schalstein) in the Lahn district, result from the presence of an additive gas phase. Otherwise the frequency of ejecta exhibiting weilburgitic composition would be incomprehensible.

The Spilite Variation in the Lahn Area

In the previous discussion only one group of spilitic rocks and their PTX conditions are referred to. In the Lahn district another group of spilitic rocks is exposed whose mineralogical features contrast with those of the weilburgites. The metasomatic transformation of these effusive rocks after their crystallization appears to be beyond doubt.

a) This second group is confined to comparatively small and widely separated parts of the area, e.g. south of Weilburg, along the Lahn river from north of Gräveneck downstream to south of Fürfurt, a distance of about 3 km, and southwest of Fachingen, along the railway and the Lahn river. Detailed examination is still to be done. The rocks were at one time called hornblendediabase, but considering the scarcity of hornblende and the local frequency of pyroxene or feldspar phenocrysts, the name porphyrite (equivalent to andesite) seems more correct. Without exception the feldspar is albitized, the pyroxene chloritized and the amphibole, where present, mostly decomposed. The main mineralogical contrast with the weilburgite is the ubiquity of epidote. The massive rocks, especially in the occurrences south of Weilburg, are associated with related tuffs which were also called schalstein, but differ from the common schalstein in their reddish-brown color. This metasomatized rock type is suggested to be contemporaneous with the keratophyres (lower portion of Upper Middle Devonian).

b) A synopsis of the petrographical and petrochemical properties of the weilburgites seems appropriate here, since the relevant data were published by the author in German.

The weilburgites are defined mineralogically by the following essential characteristics: 1. a varied feldspar character resulting in three types of the rock; the first type with orthoclase seems to be only slightly less abundant than the second type with albite only, but the majority of the rocks contain both feldspars in variable proportions; 2. a general deficiency in epidote, pumpellyite, and prehnite; 3. variation of the carbonate content from 0.0% to 30% or more. About 50% of the examples examined contain less than 1% CO_2 (2.3% $CaCO_3$). Based on several observations (LEHMANN, 1941) the exogenous nature of much of the carbonate could be proved; 4. though not essential, the frequency of small idiomorphic biotite crystals may be noted in some occurrences.

Attention may be called to the following chemical data: 1. K_2O grades from 0.5 to 10.3%, Na_2O from 3 to 6%, the total alkali content from about 6 to 11%; 2. MgO varies from 1.5 to 7.5% depending on the abundance of chlorite, i.e. on the relatively more leucocratic or more melanocratic character of the rock; 3. based on compositions with $CO_2 \leqq$ 1% (wt.), the CaO content (incl. $CaCO_3$) does not exceed 4-5%, but 3-4% may be more correct since about 1.3% is required for apatite, if the P_2O_5 average of 0.89% (no. 6 of Table 2) is assumed as a basis.[2] The low CaO content is, with the high alkali content, a distinctive characteristic of the rocks. In the presence of carbonate, however, this characteristic will only be apparent if the rocks referred to for comparison are equally high in carbonate (Table 2).

The outlines of the numerous, usually small bodies of weilburgite scattered through the different levels of tuff layers and slates are similar to those of the "lacolithic series" (JONES and PUGH, 1949) of South Wales. On a geological map many of these failed to be recorded, partly because of their small size, and partly, if contaminated with carbonate, because of their incorporation with schalstein. Dislocation through faulting was often only guessed at by the mapping geologist who took an effusive nature for granted (Fig. 8). The author observed that some small weilburgite bodies, worked for road-material, were exhausted soon after mining started. Nevertheless their lenticular form, with doming and compression of the wall rock, could be recognized in some of the exposures and in several of the small quarries still worked in the first half of the century.

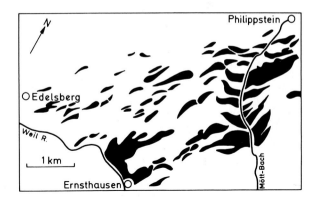

Fig. 8. Bodies of weilburgite in the schalstein complex between the valleys of the Weil R. and the Mött-Bach as registered in the geological map (sheet Weilmünster). The abundance of small-sized bodies suggests analogy with the injecta (see contribution on page 113) rather than derivation from distinct volcanic centers

---

[2]From the carbon and oxygen isotope compositons of $\delta^{13}C$ and $\delta^{18}O$ it has been made evident by M. SCHIDLOWSKI and W. STAHL (N.Jb.Min.Abh. 115 (1971) p.270-275) that the carbonate enclosed in or adjoining to the weilburgites has been derived from sedimentary carbonate rocks. In contrast with the allogenic nature in the rocks from the Lahn Basin, the respective isotope values of the rocks from the Lahn Basin, the respective isotope values of the calcite in the Verrucano spilites from Canton Glarus, Swiss Alps, indicate magmatic affiliation (ibid. p.68/69; see also M.SCHIDLOWSKI, W. STAHL and G.C. AMSTUTZ, Naturwiss. 57 (1970) p.542/43).

c) In only a few rather isolated localities, a rock variety[3] was encountered containing small idiomorphic phenocrysts of clinopyroxene. The amount of pyroxene (as determined with point-counter, 16-19%) is less contradictory to the diabase of the area than the irregularity of distribution, the tendency of clustering (Figs. 9 and 10), and the textural diversity of the rocks from the ophitic texture of the asso-

Fig. 9

Fig. 10

Fig. 9 and 10. Localized accumulation of pyroxene in "semi-spilitic" rocks. Laths of feldspar are often adjusted to the outlines of the accumulations (Fig. 10, boundary on the right). Small quarry (disused) in the district Obernhain, NWN of Arfurt. Crossed Nicols, magn. x 45 (Fig. 9), x 90 (Fig. 10)

---

[3] This variety occurs much less frequently than the normal weilburgite. Only three typical examples were found during more than 10 years of field work.

ciated diabases. The mineralogical composition of the main part of the rocks in question corresponds with that of weilburgite, and originally they were called pyroxene-bearing weilburgites (LEHMANN, 1941). The chemical composition (Table 1, nos. 1-3) of these rocks differs markedly from that of the common weilburgites only in the CaO content, and there is no doubt that the pyroxene accounts for the diversity.

Judging from the striking tendency to agglomerate (developed especially in nos. 2 and 3) and the irregular appearance, there is every reason to believe that the crystallization of pyroxene is separated by a hiatus from the crystallization of the main portion, and that the distribution is incompatible with the behavior in homogeneous melts. It is also improbable that the agglomerates could form in the magma disturbed in the course of ascending. Thus the only alternative seems to be that we are dealing with inclusions of an allothigenic character. In all probability, only a deep-seated femic or ultrafemic rock comes into question as source.

Since the optical properties of the pyroxene ($c:Z = 38-41°$, $2 V\gamma$ about $53°$) do not tell us much, it can only be approximate if analogy with the pyroxene of the gabbro from Coverack Bay, Lizard area (GREEN, 1964, p.152) is assumed. The composition under no. 5 of Table 1 results from substracting 16% of the pyroxene from no. 2 of Table 1. The weilburgite analysis no. 4 is noted for comparison. Though these heterogeneous rocks for which the term "semi-spilites" may be proposed, are without quantitative significance in the Lahn area they are illustrative of a specific variation. It is of more consequence that the chemical analogy eventually could give rise to misinterpretation. However, the mineralogical and textural features are contradictory to the parallelization.

d) In addition to the recorded effects of the fluids another result due to fluid-action may be mentioned. In the top material of the small, long-unused quarry at location no. 2, amygdales have occasionally developed filled with calcite and radiating aggregates of exceptionally large-sized stilpnomelane crystals (length of the cross sections up to 0.7 mm).

In the extremely fine-grained aggregates of the surrounding material, however, the chlorite, without exception, corresponds to diabantite. The interstitial and the amygdaloid chlorites reflect particular stages of the post-effusive history, during which time the continuance of the hydrous character of the magma was effective.

Weilburgite and Associated Diabase

In view of the assumed genetic relation of the weilburgitic magma with a femic (basaltic) or ultrafemic (peridotitic) magma in the preceding paragraph, advocates of metasomatic spilite generation in general may raise the objection that there is no noticeable difference in the result. That is not quite so.

The mineralogical differences between the two trends of spilite (p. 30) and between the weilburgite (non-metasomatic spilite) and diabase have already been referred to (LEHMANN, 1949, 1967). In the Lahn district, one can distinguish two groups of diabase. In the intrusive group, occurring in Middle and Lower Upper Devonian schists, An in the plagioclase varies between 20 and 35%, and epidote, occasionally also prehnite, results from alteration. In the effusive group (low Carboniferous) An grades up to about 50% (in the Dill district about 10% higher). In

Table 1. Chemistry of "semi-spilitic" weilburgites compared to that of weilburgite

No.	SiO_2	Al_2O_3	Fe_2O_3	FeO	MnO	MgO	CaO	Na_2O	K_2O	TiO_2	P_2O_5	H_2O^+	CO_2
1	48.49	13.97	2.66	8.73	0.11	7.58	7.52	4.18	0.66	2.52	0.13	3.86	0.08
2	46.48	13.16	2.53	8.28	0.14	7.83	8.72	3.66	1.10	3.09	0.60	3.73	0.53
3	45.48	12.66	2.10	9.09	0.15	8.25	6.87	3.38	1.40	3.81	0.60	4.09	0.00
4	47.50	15.83	4.45	6.72	0.16	5.94	3.80	5.05	1.56	2.96	0.81	4.64	0.18
5	46.8	15.3	2.8	8.8	0.2	6.5	5.0	4.3	1.4	3.5	0.7	4.6	–

1. Pyroxene-bearing weilburgite. About 200 m SES of Ulm, about 6 km N of the Lahn R. Semi-spilite.
2. Pyroxene-bearing weilburgite. Forest distr. Obernhain, about 2 km NWN of Arfurt on Lahn R.
3. Pyroxene-bearing weilburgite. Valley of the Laubusbach, about 2 km ENE of Münster (S of the Lahn R.)
4. Weilburgite. About 2-5 km E of Braunfels Station, road leading to Burgsolms.
5. Composition resulting from no. 2 minus 16% diopsidic augite of Coverack Bay, Lizard area (GREEN, 1968).

Table 2. Comparison of diabase with weilburgite and spilite

No.	SiO_2	Al_2O_3	Fe_2O_3	FeO	MnO	MgO	CaO	Na_2O	K_2O	TiO_2	P_2O_5	H_2O^+	CO_2
1	48.70	14.73	3.41	8.36	0.12	6.38	8.27	4.26	0.40	1.94	0.10	3.16	0.21
2	48.69	14.97	2.38	8.02	0.13	5.27	3.22	3.93	3.44	2.51	0.90	4.06	0.13
3	42.48	16.52	3.24	8.19	n.d.	11.46	9.16	2.43	1.19	1.74	0.21	2.80	3.30
4	44.53	15.64	3.57	7.60	0.15	3.57	8.11	5.11	1.32	3.03	0.30	3.77	3.60
5	43.34	11.51	4.14	8.20	0.16	9.71	9.63	2.50	0.70	2.41	0.26	3.80	0.26
6	47.98	15.60	3.79	7.89	0.14	5.02	4.28	4.26	2.82	2.83	0.89	4.16	0.35
7	49.65	16.00	3.85	6.08	0.15	5.10	6.62	4.29	1.28	1.57	0.26	3.49	1.63

1. Diabase. Rühlbachskopf, valley of the Helgenbach, about 5.5 km NWN of Braunfels.

2. Weilburgite. Aardeck ruins, valley of the Aar creek, about 3.5 km SW of Diez.

3. Diabase W of the railway tunnel near Eisenroth, Dill Basin (No suitable example available from the Lahn Basin) (BRAUNS, 1909).

4. Weilburgite. Valley of the Weinbach, about 6 km SE of Weilburg.

5. Alkali diabase. Average (10). Prospect quarry, Sydney Basin, New South Wales (WILSHIRE, 1967).

6. Weilburgite. Average (11). All analyses, Lahn Basin ($CO_2 \leqq 1.0$).

7. Spilite. Average (92) (VALLANCE, 1960).

contrast to the pyroxene-bearing weilburgites, the feldspar of the diabases crystallized about contemporaneously with the pyroxene, usually in ophitic intergrowth with it. In the following comparison with weilburgite and spilite, only the intrusive group is referred to.

The chemical differences may be illustrated by three examples (Table 2): nos. 1-2 single analyses, low in carbonate; 3-4 single analyses, medium in carbonate, and 5-7 averages. The differences in CaO has been repeatedly emphasized by the author (1941, 1968) and by VALLANCE (1960). In the diabases, primarily plagioclase, and to a lesser degree, pyroxene, are responsible for the CaO content, while pyroxene is primarily responsible for the MgO and FeO contents. In the weilburgites MgO and FeO are contained in the chlorite. The alkali totals of weilburgite usually exceed those of diabase. For the Lahn rocks, K+Na, Ca, and Mg, and in a markedly lower degree Fe and Ti, characterize the different chemical compositions of weilburgite and semi-spilitic weilburgite as compared with diabase. In the analyzed diabase no. 1 of Table 2 there is about 40-45% plagioclase (An_{35}-An_{20}) and 35% pyroxene, both obviously crystallized from the molten phase. On the other hand, about 5% of the chlorite was formed in the post-magmatic stage, in part with, in part without, visible connection with the pre-existing pyroxene. Thus, as with the mineralogical and chemical properties, the conditions of crystallization of the diabase also contrast with those of the weilburgite. In view of all of these circumstances, the term "dense diabase" (dichter Diabas) thus far used for the latter was rejected by the author (LEHMANN, 1941).

Unlike the former unification of the nomenclature, much the same aim is seen in the proposal to abandon the term "diabase" and to substitute for it "spilite" or "spilitic basalt" (WEDEPOHL, 1969, p.331, p.334). In this way confusion resulting from the different definitions of "diabase" used by American, British, and German petrologists might be avoided. But it seems more likely that acceptance of the proposal would cause an increase in confusion. Even now petrologists engaged in the diabase problem are often in doubt about the precise meaning of "dolerite" and it seems questionable whether in the future every author will be at pains to explain his meaning specifically.

In the diabasic magma of the Lahn area the high temperature conditions evidently continued in the intracrustal reservoirs where consolidation took place. The depth and the nature of the intruded material suggest slow cooling. Small bodies comparable with those of weilburgite are unknown among the diabase intrusions in the Lahn district. In the final stage of diabase crystallization, however, the decrease of temperature could give rise to a rest product containing chlorite instead of pyroxene. Apparently some analogy between the diabase residua and the weilburgitic magma could result if alkalis, especially sodium, were introduced. This final approach to a hydromagmatic character is suggested as a rather common peculiarity of diabase occurring in geosynclines (LEHMANN, 1967). The amounts of the residue, of course, are inadequate in view of the extent of weilburgite or spilite.

Conclusions

Because the diverse weilburgites (sodic, potassic, and transitional), the pyroxene-bearing weilburgites and the diabases (metasomatic spilite not included) co-exist within a comparatively small area (about 25 x 10 km), because all of these rocks formed in the same syncline and in the course of the same orogenic phase, and because the superficial conditions differed to a degree hardly commensurate with the local variation, it is unlikely that this petrographic variety arose during or following the eruption of an originally basaltic magma. Essential conditions might have existed before the ascent, and exceptional conditions likewise are to be assumed after the intracrustal emplacement of the magma. The hydromagmatic character of the original melt is suggested to have been in favor of the variation.

In the Lahn Basin the effects of the gas and vapor phases are not only manifested in weilburgites, metasomatic eruptives, diabase, and keratophyre, but are also apparent in the numerous iron ore bodies intercalated in and superincumbent on the schalstein complex and in the iron impregnation of the Middle Devonian limestone at several places, especially in the vicinity of Gaudernbach (about 6 km WSW of Weilburg) and Schupbach (2.5 km S of Gaudernbach). Considering in addition the loss by exhalation and dispersion in the sediments (recrystallization of limestone), the quantitiy of the gas and vapor phases is suggested to have been out of proportion to the late- and post-magmatic activity of the comparatively small magma bodies. The effects point to a source with a greater capacity for degassing.

Apart from the two trends of spilite generation represented in the Lahn district by homogeneous rocks, another type resulted from secondary alteration of solid rocks by reactions "ranging from local hydrothermal to regional diagenetic burial metamorphism" (VALLANCE, 1969). One example examined in detail (SMITH, 1968) is exposed in an outcrop south of Orange in the Central West of New South Wales. The noticeable variation of the Ordovician volcanic rocks (mainly andesite) straddling a ge-anticline is megascopically indicated by patchy coloration and is confirmed by the mineralogical and chemical composition. CaO for instance grades from 5.00 to 22.40%, generally depending on epidote, pumpellyite, and (rare) prehnite (CO_2 usually low, is not proportional to the CaO variation). MgO is generally low (maximum 5.3%). In the dominant variety Na_2O + K_2O corresponds with spilite (about 5.5 - 6.9%), but is lower in all the other varieties. The example indicates another possibility of spilite generation, but the phenomena contrast with those in the weilburgites to such a degree that parallelization is out of question. In one word: there is no standard solution of the problem involved in spilite generation.

Acknowledgements

Thanks are due to Dr. J. KUSHIRO for kind information, to Prof. Dr. G.C. AMSTUTZ for helpful discussion, and to Prof. Dr. T.G. VALLANCE for critically reading the manuscript.

References

BRAUNS, R. (1909): Beiträge zur Kenntnis der chemischen Zusammensetzung der devonischen Eruptivgesteine im Gebiete der Lahn und Dill. N. Jahrb. Mineral. Beil. Bd. 27, 261-325.

FAWCETT, H.J., YODER, H.S. (1966): Phase relationships of chlorites in the system $MgO-Al_2O_3-SiO_2-H_2O$. Am. Mineralogist 57, 332-360.

GILBERT, M.C. (1968): Reconnaisance of the stability of amphiboles at high pressure. Ann. Rep. 1967-1968 Geophys. Lab. Washington, D.C., 56-67.

GREEN, D.H. (1968): The petrogenesis of the high-temperature peridotite intrusion in the Lizard area, Cornwall. J. Petrol. 5, 132-188.

HENTSCHEL, H. (1951): Die Umbildung basischer Tuffe zu Schalsteinen. N. Jb. Min. Abh. 82, 199-228.

JONES, O.T., PUGH, W.J. (1949): The laccolithic series. Am. J. Sci. 247, 353-371.

KUSHIRO, J. (1968): Composition of magmas formed by partial melting of the earth's upper mantle. J. Geol. Res. 73, 619-630.

KUSHIRO, J., YODER, H.S. (1968): Melting of fosterite and enstatite at high pressures under hydrous conditions. Ann. Rep. 1967-1968 Geophys. Lab. Washington, D.C., 153-158.

KUSHIRO, J., YODER, H.S., NISHIKAMA, M. (1968): Effect of water on the melting of enstatite. Geol. Soc. Am. Bull. 79, 1685-1693.

KUSHIRO, J. (1969): Stability of amphibole and phlogopite in the upper mantle. Ann. Rep. 1968-1969 Geophys. Lab. Washington, D.C., 245-247.

LEHMANN, E. (1941): Eruptivgesteine und Eisenerze im Mittel- und Oberdevon der Lahnmulde. Wetzlar: Scharfe's Druckereien.

LEHMANN, E. (1949): Das Keratophyr-Weilburgit-Problem. Heidelberger Beitr. Mineral. Petro. 2, 9-166.

LEHMANN, E. (1967): Diabasprobleme und problematische Diabase I, Chem. d. Erde 26, 49-86.

LEHMANN, E. (1968): Diabasprobleme und problematische Diabase II, Chem. d. Erde 27, 39-78.

LEHMANN, E. (1970): Zur Schalsteinfrage. Ber. Oberhess. Ges. Natur- u. Heilkunde 37, 5-34.

McKEE, E.H. (1968): Refractive index of glass beads distinguishes Tertiary basalts in the Grays River area, southwestern Washington. Prof. Paper U.S. Geol. Surv. 600, C., 27-30.

SMITH, R.E. (1968): Redistribution of major elements in the alteration of some basic lavas during burial metamorphism. J. Petrol. 9, 191-219.

VALLANCE, T.G. (1960): Concerning spilites. Proc. Linnean Soc. N.S. Wales 85, 8-52.

VALLANCE, T.G. (1969): Spilites again. Some consequences of the degradation of basalts. Proc. Linnean Soc. N.S. Wales 94, 8-51.

WEDEPOHL, H.H. (1969): Handbook of Geochemistry, Vol. 1. Composition and abundance of common igneous rocks. Berlin-Heidelberg-New York: Springer.

WILSHIRE, H.G. (1967): The Prospect alkaline diabase-picrite intrusion, New South Wales. J. Petrol. 8, 97-163.

WYLLIE, P.J. (1967): Ultramafic and related rocks. John Wiley & Sons, London. M.J. O'Hara, Mineral facies in ultrabasic rocks, 7-18.

Essai de Caractérisation Chimique des Associations Spilitiques

H. de la Roche, G. Rocci, et Th. Juteau

Abstract

The authors elucidate the chemical characteristics of the spilitic suite through a graphic representation of 344 published (or to be published) analyses according to the following scheme:

a) data on different spilitic associations (147 analyses)
b) preorogenic hercynian (varistic) volcanism in northern Europe (173 analyses)
c) local variations between the cores and margins of pillows (24 analyses).

Two unconventional graphical representations have been used:

$$(Al/3 - K) \; f \; (Al/3 - Na)$$
$$[Si/3 - (Na + K + 2Ca/3)] \; f \; [K - (Na + Ca)].$$

The application of these to work concerning formation of volcanic rocks is not usual, but allows a clear division between the characteristic transition anorthite - albite in spilitization and that from basic to acidic rocks in magmatic differentiation. In this way, the diversity and uniqueness of the spilite suite have been brought out.

Besides, such a representation makes possible an attempt at an interpretation of spilitic rocks from a chemical point of view. Of the elements taken into consideration, the following values are unchangeable: Al, (Ca + Na) as well as (Si + Ca). Si and Na are thus introduced together, while Ca is released and Al remains in place. In pillow lavas such transformations depend mainly on the exchange of material between the pillows and their chloritic and carbonate matrix.

After a discussion of the problem of too selective a sampling in favor of pillow-cores, the authors reject the hypothesis of an overall composition similar to that of basalts. The spilitization of basalts necessitates a metasomatic introduction of Na and perhaps of Si, with a simultaneous release of Ca.

In the end the relationship of spilitic associations with tholeiitic and alkaline series of rocks is considered. The spilites form, just as the basalts, a continuum: certain spilitic associations have a clearly tholeiitic, others an alkaline character. The link with the paleogeographic conditions of implacement, the presence or absence of a sialic crust, is beyond the scope of hercynian volcanism.

Introduction

Le problème que nous nous proposons de traiter est celui de la particularité chimique des associations spilitiques par rapport aux autres associations volcaniques, et notamment par rapport aux basaltes. Ce problème doit être envisagé au moins sous deux aspects:

a) particularité des tendances de différenciation,
b) particularité des compositions, soit terme par terme le long des tendances de différenciation, soit globalement, sous l'angle de la comparaison entre les "laves spilitiques" et les "laves basaltiques".

A l'origine de notre essai se trouve la publication récente de G.C. AMSTUTZ (1968) selon lequel "l'enrichissement en Na des spilites n'est généralement pas réel mais dû à des erreurs d'échantillonnage liées à la différenciation des spilites et des kératophyres". C'était généraliser une série d'observations et de critiques parfaitement fondées dont on peut trouver trace jusque dans les travaux de SUNDIUS (1930). Mais c'était aussi formuler, à propos d'une "erreur d'échantillonnage" peu contestable, une hypothèse extrême, en rendant cette erreur entièrement responsable des écarts sur le sodium généralement observés entre la composition des spilites et celle des basaltes.

Or l'un de nous (H.R.) publiait au même moment une note de géochimie générale centrée sur un diagramme (Fig. 1) dans lequel les associations spilitiques apparaissaient nettement distinctes des autres associations volcaniques et presqu'entièrement extérieures au domaine des basaltes. C'est ainsi que fut décidé, sur l'incitation de G.C. AMSTUTZ, un examen plus approfondi de ce problème par des méthodes de traitement graphique

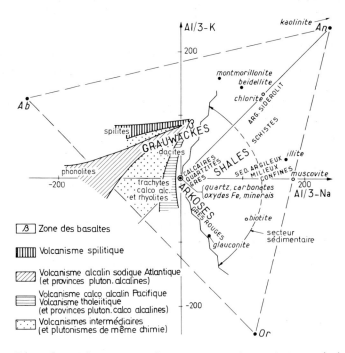

Fig. 1. Diagramme Al, Na, K. Disposition générale des associations volcaniques. Leur opposition par rapport aux associations sédimentaires. D'après H. de la ROCHE (1968)

que nous avons préférées à des calculs de statistique élémentaire pour
la mise en évidence de domaines et de tendances géochimiques.

L'espoir d'éclairer ou de renouveler le problème de la particularité
chimique des spilites - problème fort ancien - ne pouvait reposer que
sur l'emploi de nouveaux paramètres graphiques ou sur l'apport de nouvelles données s'ajoutant à celles de la littérature. Nous nous sommes
engagés simultanément dans ces deux voies, d'une part en utilisant de
nouveaux diagrammes récemment proposés dans un cadre géochimique plus
général (H.R.), d'autre part en faisant appel à un ensemble de données
nouvelles provenant d'une étude systématique du volcanisme préorogénique
hercynien d'Europe du Nord, de la Cornouailles au Massif de Lahn-Dill
(massif schisteux Rhénan)(G.R. et Th.J.).

Nulle critique ne s'élèvera sans doute à propos de l'appel à ces données
nouvelles, présentées par ailleurs en détail dans ce volume (pp. 253-329)
en même temps qu'à celles de la littérature. En revanche l'emploi de
paramètres nouveaux, de préférence à ceux que l'usage a plus ou moins
consacrés, pourrait légitimement susciter la réserve. Il est donc nécessaire de donner une courte explication préliminaire sur le choix des
paramètres et des diagrammes.

Choix des Diagrammes

Les associations minérales caractéristiques des spilites et de leur
cortège sont particulières et bien différentes de celles des roches
basaltiques. L'un de leurs traits principaux est que l'albite se substitue à l'anorthite sans que la "basicité" de la roche évolue sensiblement.

Si l'on veut saisir la particularité chimique des spilites, il faut donc
rechercher des diagrammes dans lesquels la transition "anorthite ⟶
albite", qui s'exprime dans tous les cas par un vecteur, ne soit pas
concordante ou subconcordante avec la transition générale "roche basique ⟶ roche acide". Cette remarque s'impose intuitivement, à défaut
d'une démonstration qu'il serait trop long de mener ici mais dont les
éléments principaux pourront être trouvés dans une publication antérieure
(de la ROCHE, 1964).

De ce fait, les diagrammes les plus classiques (HARKER, NIGGLI, PEACOCK,
JUNG, KUNO, MURATA) apparaissent peu favorables à la mise en évidence
des caractères chimiques particuliers des spilites. Cela explique sans
doute les incertitudes présentes mais aussi, en sens inverse, les observations empiriques assez nettes faites à propos des spilites sur le
diagramme de la Fig. 1, observations qui furent à l'origine de la présente étude.

Le recours à de nouveaux diagrammes se trouvant ainsi justifié, nous
présenterons brièvement ceux dont il sera fait usage dans cette étude,
au nombre de deux.

Unité de Référence Commune aux Deux Diagrammes

Pour l'établissement de ces diagrammes les paramètres sont calculés à
partir des nombres de milliatomes-grammes dans 100 grammes. Le calcul
porte indifféremment sur les roches, d'après les résultats de leur

analyse, ou sur les minéraux, soit qu'ils aient été analysés, soit que l'on se contente de leur formule type et du poids moléculaire correspondant. De cette manière, les minéraux figurent dans les diagrammes chimiques en même temps que les roches et le raisonnement sur les caractères chimiques ne sera jamais dissocié de la connaissance des associations minérales et des paragenèses.

Diagramme (Al/3 - K) f (Al/3 - Na) *(Fig. 1)*

Ce diagramme a été conçu pour faire apparaître le comportement différentiel des alcalins par rapport à l'alumine, et pour opposer la "dégradation sédimentaire" à la "différenciation ignée" (de la ROCHE, 1968). Il n'était pas surprenant que les spilites s'y disposent à l'écart des autres roches volcaniques puisqu'à la suite de WELLS (1923) et de SUNDIUS (1930) on s'accorde à leur reconnaître pour particularités chimiques significatives (nous faisons provisoirement abstraction des objections liées à l'échantillonnage) la richesse en sodium, par rapport auquel le potassium reste à un niveau très bas, ainsi que la pauvreté en aluminium.

Diagramme $[Si/3 - (K + Na + 2Ca/3)]$ f $[K - (Na + Ca)]$ *(Fig. 7)*

Ce diagramme a été conçu primitivement pour la mise en évidence de faibles variations chimiques dans les roches granitiques (de la ROCHE, 1964). Son intérêt plus général pour l'ensemble des roches ignées a été montré ultérieurement (de la ROCHE, 1966, 1968). Le quartz, le feldspath potassique et le plagioclase y apparaissent aux trois pôles d'un triangle équilatéral, reproduisant le triangle Q A P, base de la classification minéralogique des roches ignées. Un paramètre complémentaire (Na + K) sera fréquemment employé en 3e dimension ou "altimétrie", avec référence à une "surface moyenne" au voisinage de laquelle se distribuent les roches ignées communes. Cette surface moyenne est définie dans le plan de la figure par une série de courbes d'isoalcalinité (non représentées à l'échelle de la Fig. 7).

Le vecteur anorthite⟶albite, dont nous avons montré plus haut l'intérêt dans le cas des spilites, se situe ici parallèlement à la 3e dimension. Il se projette dans le plan du diagramme en un très court segment (les points figuratifs de Ab et de An sont très proches et peuvent être considérés en première approximation comme un point unique). Néanmoins, la condition de nette discordance entre transition "anorthite⟶albite" et la transition "roche basique⟶roche acide" se trouve respectée, comme dans le précédent diagramme.

Nature des Données

Les données utilisées portent sur un total de 344 analyses, et se répartissent en trois groupes.

Le premier d'entre eux (groupe A, 147 anal.) englobe de manière indistincte des analyses extraites de diverses publications sur les spilites, dont les 94 analyses que VALLANCE (1960) avait rassemblées dans son essai classique et 53 autres publiées par GILLULY (1935), AMSTUTZ (1954), NAREBSKI (1964) à l'exclusion des séries d'analyses destinées à mettre en évidence les variations chimiques du coeur à la bordure des pillows, DIETRICH (1967), PARROT (1967), PETERLONGO (1968) - sélection de laves spilitiques. Ce groupe est nombreux mais sans nul doute hétérogène, de sorte qu'il ne permet qu'une information statistique générale. De plus, on peut craindre que les effets d'échantillonnage sélectif y exercent une profonde influence.

Le second groupe (groupe B, 173 anal.) est formé par les données nouvelles de ROCCI et JUTEAU (op. cit.) sur le volcanisme préorogénique hercynien d'Europe du Nord.[1] Bien que l'échantillonnage conserve un caractère "typologique", il faut souligner que sa réalisation s'inscrivait dans un projet d'<u>inventaire systématique</u> pétrographique et géochimique. On peut donc en retirer des indications limitées à l'échelle d'un segment orogénique mais plus précises et déjà moins sujettes à la critique que les précédentes, du point de vue des effets de l'échantillonnage sélectif.

Enfin, le troisième groupe (groupe C, 24 anal.) reprend les données publiées par NAREBSKI (1964) et PETERLONGO (1968) pour montrer l'existence de variations entre le coeur des pillows, leur bordure et les matériaux interstitiels. Cette dernière catégorie de données, sur laquelle AMSTUTZ (op. cit.) appuie à juste titre son argumentation, permet une estimation approchée du sens et de l'ampleur de l'effet d'échantillonnage sélectif.

<u>Relations entre Na, K et Al</u>

Diagrammes (Al/3 - K) f (Al/3 - Na)

L'intérêt général d'une étude des relations entre Al, Na et K ressort immédiatement de l'examen de la Fig. 1. Aussi est-ce dans ce diagramme que nous présenterons plus précisément les séries spilitiques.

Avant même de reporter des séries d'analyses sur ce diagramme, la nature des problèmes que nous aurons à résoudre peut être saisie à l'aide de remarques minéralogiques et géochimiques d'ordre général.

Opposition Générale entre les Associations Spilitiques et les Basaltes

On sait la difficulté d'une distinction chimique entre les divers types de basaltes de même que la relative rareté de différenciations chimiques importantes à l'intérieur des domaines basaltiques. A l'inverse, beaucoup d'associations spilitiques présentent une gamme étendue de produits différenciés, jusqu'à l'échelle des coussins des pillows-lavas et de leur enveloppe.

Cette opposition tout à fait classique trouve une expression simple dans notre diagramme. <u>Les basaltes</u> y occupent un domaine restreint, entre

---

[1] Les données concernant particulièrement la région de Schirmeck (Vosges Septentrionales), après avoir été présentées et discutées pour la première fois par JUTEAU et ROCCI (1966), viennent d'être réinterprétées dans une importante publication (FONTEILLES, 1968), parue alors que nous achevions le présent article. Le hasard des circonstance offre ainsi au spécialiste la possibilité de comparer trois méthodes d'interprétation du même groupe de données.

le secteur des plagioclases basiques (Labrador-Bytownite) et le point origine des axes, au voisinage duquel se regroupent les pyroxènes communs, les péridots et les minerais (Fig. 2). Au contraire, les minéraux caractéristiques des associations spilitiques (albite, chlorite, épidote, carbonates, etc.) délimitent un champ beaucoup plus vaste où la différenciation va pouvoir se manifester dans un sens ou dans l'autre.

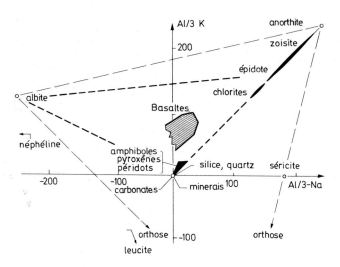

Fig. 2. Diagramme Al, Na, K. Domaine des basaltes et champ des associations spilitiques

Or le domaine restreint des basaltes se trouve bien à l'intérieur de ce champ très large des "minéraux des associations spilitiques" et non loin de sa zone centrale. Il suffirait donc d'une combinaison en proportions adéquates de termes albitiques et de termes chloriteux, épidotiques, carbonatés ou siliceux pour que la composition globale de chaque association spilitique rejoigne les compositions basaltiques banales, contrairement aux opinions souvent avancées à propos de Al, Na et K.

Vu sous cet angle, le problème est bien, comme l'a indiqué AMSTUTZ (op. cit.), celui de la pondération correcte des diverses variétés présentes dans les échantillonnages et dans les calculs de composition globale.

Présentation Graphique des Données sur les Associations Spilitiques. Domaines et Tendances

Les connaissances générales suffisent, comme nous venons de le voir, à dégager l'utilité et la signification du diagramme. Toutefois, ce sont les données numériques et elles seules qui permettront de préciser, dans ce cadre général, les domaines et les tendances correspondant aux associations spilitiques.

Le groupe des données générales (groupe A) et celui des données particulières au volcanisme préorogénique hercynien du Nord de l'Europe (groupe B) sont présentés de manière synthétique, en figures de densité de répartition (Fig. 3 et 4). Le groupe B est repris, point par point (Fig. 5) afin d'illustrer sur son exemple la répartition des principaux types lithologiques. En dernier lieu, on trouvera (Fig. 6) les données spécifiques sur les variations entre le coeur des pillows, leur bordure et la matière interpillow (groupe C).

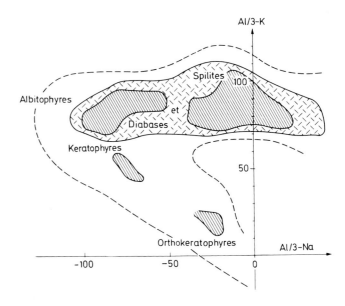

Fig. 3. Diagramme Al, Na, K. Figure en densité de répartition pour le groupe des données générales sur les associations spilitiques (groupe A, 147 analyses)

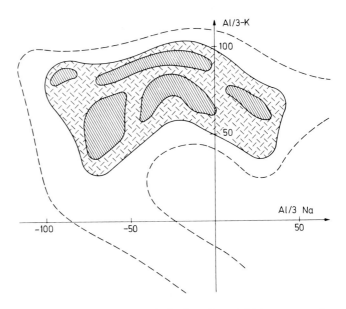

Fig. 4. Diagramme Al, Na, K. Figure en densité de répartition pour les associations spilitiques du volcanisme préorogénique hercynien d'Europe du Nord (groupe B, 173 analyses)

Domaines et Tendances Générales des Associations Spilitiques

Les deux figures en densité de répartition (Fig. 3 et 4) présentent une grande analogie. Les associations spilitiques y forment très distinctement une sorte de chevron, dont les deux branches correspondent à des densités très inégales. La branche supérieure, sub-horizontale, à forte densité, s'enracine à son extrêmité droite dans la zone des basaltes communs et se dirige, en première approximation, vers l'albite. On pourrait la désigner comme tendance de spilitisation, en observant qu'elle présente un angle faible avec la transition "anorthite⟶albite", que nous donnions plus haut comme l'un des traits majeurs des spilites.

La branche inférieure, inclinée vers l'origine des axes, à densité faible ou très faible, est constituée par le groupe des ortho-kératophyres et par les termes de transition vers les kératophyres. On pourrait la désigner comme tendance alcali-potassique en observant qu'elle présente un angle faible avec la "transition albite⟶orthose". On notera de plus que cette tendance alcali-potassique ne paraît se développer qu'au niveau des kératophyres, termes les plus alcalins de la tendance spilitique.

Variations Lithologiques et Variations Régionales

La figure point par point du volcanisme préorogénique hercynien (Fig. 5) précise cette configuration en faisant apparaître la position des différents types lithologiques. Les dolérites "intrusives" des domaines spilitiques tombent dans le champ des compositions basaltiques ou s'en écartent peu. Les spilites (effusions basiques spilitiques), les diabases et diabases quartziques (intrusions diabasiques) et les kératophyres

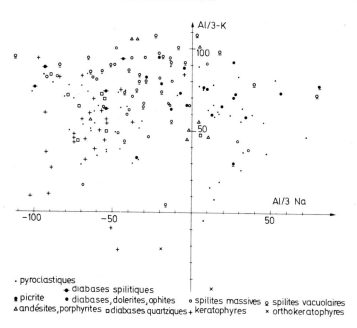

Fig. 5. Diagramme Al, Na, K. Types lithologiques du volcanisme préorogénique hercynien d'Europe du Nord

s'échelonnent le long de la "tendance de spilitisation" en trois domaines successifs non rigoureusement cloisonnés. Les spilites sont déjà nettement à l'écart du champ des compositions basaltiques, le paramètre de différenciation étant ici, pour l'essentiel, Al/3 - Na, dont la valeur est positive pour la plupart des basaltes et négative pour la plupart des spilites. Enfin les ortho-kératophyres dessinent bien avec les kératophyres une seconde tendance très discordante par rapport à la tendance de spilitisation.

Nous n'insisterons pas ici sur un autre aspect important, celui des variations régionales, avec la prédominance d'un type lithologique sur les autres selon la position dans un même segment orogénique. En favorisant la mise en évidence de la différenciation le long de la "tendance spilitique", le diagramme Al/3 - Na, Al/3 - K révèle bien l'existence de ces variations régionales qu'il ne paraît pas possible d'attribuer à un artefact de l'échantillonnage surtout lorsque le géologue collecteur est le même pour les régions ainsi différenciées.

Variations Locales et Effet d'Echantillonnage dans les Pillows-Lavas

Les données sur les variations chimiques systématiques à courte distance entre le coeur et la bordure des pillows (Fig. 6) font apparaître une tendance presqu'uniforme ou du moins très prédominante.

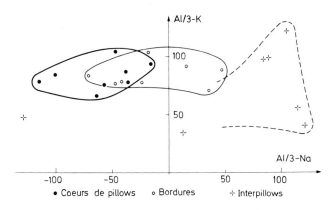

Fig. 6. Diagramme Al, Na, K. Variations locales et effet d'échantillonnage dans les pillows-lavas (groupe C, 24 analyses)

Fait remarquable et contraire à ce que nous avons vu précédemment, le champ des compositions basaltiques se trouve au milieu du parcours de cette nouvelle tendance qui prolonge la "tendance de spilitisation" vers la droite jusque dans le domaine des associations minérales à chlorite, épidote, carbonates, minerais, silice, ... où le paramètre Al/3 - Na reprend une valeur positive égale ou supérieure à celle des basaltes.

Si l'on considère seulement les pillows-lavas, l'échantillonnage et la pondération des parties albitisées (coeurs de pillows) et chloritisées (interpillows) exercent évidemment une influence déterminante. On peut donc légitimement mettre en doute la valeur de Al/3 - Na comme critère de distinction chimique globale entre ces roches et les basaltes, et il en sera probablement de même de la plupart des caractères chimiques, le contenu de fluides ($H_2O + CO_2$) mis à part. Et pourtant, ce problème pourrait être résolu facilement à partir de quelques sondages carottés. Le paramètre Al/3 - Na, bon indicateur de différenciation serait alors à considérer en premier lieu, dans l'optique d'une analyse factorielle.

Plus on élargit l'examen au-delà du seul cas des spilites, jusqu'à l'ensemble des associations spilitiques différenciées comportant des diabases et des kératophyres, plus il devient difficile d'attribuer toutes les différences de composition chimique globale à un simple effet d'échantillonnage. En particulier, l'existence de variations régionales dans la dominante lithologique des associations spilitiques contredit l'hypothèse d'une composition globale uniforme de basalte.

Relations entre Si, Ca, Na et K

Diagrammes $[Si/3 - (Na + K + 2Ca/3)]$ f $[K - (Na + Ca)]$

Les relations d'acidité et de basicité, de silification et de carbonatation, étaient très peu apparentes dans le diagramme (Al/3 - K) f (Al/3 - Na) où l'on voulait dégager, avec un minimum d'interférences, la transition anorthite ⟶ albite et, accessoirement la transition albite ⟶ orthose.

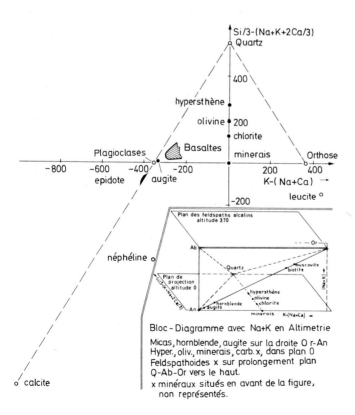

Fig. 7. Diagramme Si, K, Na, Ca. Domaine des basaltes et champ des associations spilitiques

Le second diagramme, un peu plus complexe dans sa structure, va permettre de réintroduire ces relations (Fig. 7). Les plagioclases ont sur ce diagramme un point figuratif unique, du moins en première approximation. La transition anorthite⟶albite correspond ici à un vecteur perpendiculaire au plan de la figure. Elle n'exerce aucune influence, si ce n'est dans cette troisième dimension, lorsque l'on considère en "altimétrie", le troisième paramètre (Na + K). Au contraire, le quartz ou la silice d'une part, et la calcite d'autre part sont aux deux extrêmités opposées d'un très long segment passant par le point figuratif des plagioclases: la silicification et la carbonatation vont en conséquence se manifester nettement et sans interférence a priori avec la "tendance de spilitisation" précédemment définie à partir de la transition anorthite⟶albite.

La chlorite d'une part et l'orthose d'autre part vont jouer un rôle un peu particulier, en raison de leur position très à l'écart de l'axe majeur quartz - plagioclase - calcite. Leur développement pourra donc déterminer des tendances obliques par rapport à cet axe, et se distinguant l'une de l'autre par la valeur de (Na + K), abaissée par la chlorite et élevée par l'orthose.

Spilites et Basaltes

Si l'on considère la position dans ce diagramme du domaine des basaltes communs, on constate qu'il se trouve, comme c'était le cas sur le diagramme Al, Na, K, à l'intérieur du champ délimité par les minéraux des associations spilitiques et qu'il n'en occupe qu'une faible partie (en

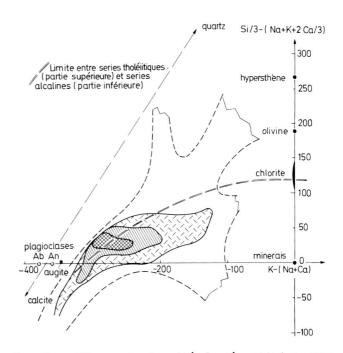

Fig. 8. Diagramme Si, K, Na, Ca. Figure en densité de répartition pour le groupe des données générales sur les associations spilitiques (groupe A, 147 analyses)

hachures sur la Fig. 7). L'hypothèse d'une similitude globale des associations spilitiques avec les basaltes pour les paramètres chimiques considérés ne peut donc pas être exclue a priori.

A un degré de détail supérieur, il faut en outre noter que les différenciations noritiques ou picritiques d'une part et la chloritisation d'autre part risquent de se distinguer mal les unes des autres car les points figuratifs de l'hypersthène et de l'olivine sont assez voisins et peu éloignés de celui de la chlorite. En cas de doute, on séparerait aisément ces tendances au moyen du diagramme Al, Na, K, en raison du caractère alumineux des chlorites.

Présentation Graphique des Données sur les Associations Spilitiques. Domaines et Tendances

Ayant montré quels nouveaux aspects de notre problème pourront être dégagés au moyen du diagramme $[Si/3 - (K + Na + 2Ca/3)]$ f $[K - (Na + Ca)]$ nous y présenterons les données dans l'ordre suivi pour le premier diagramme. Nous traiterons donc successivement les données générales (groupe A) et les données particulières au volcanisme préorogénique hercynien d'Europe du Nord (groupe B), en figures de densité de répartition (Fig. 8 et 9), puis le même groupe B détaillé en figure par points (Fig. 10) et enfin les données du groupe C, spécifiques des variations entre le coeur et la bordure des pillows (Fig. 11).

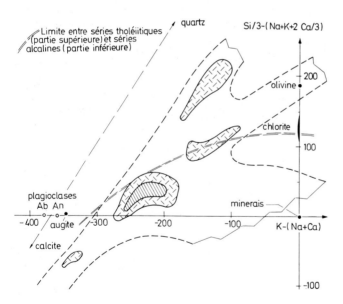

Fig. 9. Diagramme Si, K, Na, Ca. Figure en densité de répartition pour les associations spilitiques du volcanisme préorogénique hercynien d'Europe du Nord (groupe B, 173 analyses)

Domaines et Tendances Générales des Associations Spilitiques

Dans les figures en densité de répartition, les associations spilitiques montrent une distribution centrée sur le domaine des basaltes communs (Fig. 8 à comparer avec la Fig. 7). Il n'y a là rien de surprenant, si l'on se souvient que la transition anorthite ⟶ albite, principale manifestation de la "tendance de spilitisation", n'agit que dans la troisième dimension du diagramme.

Ce qui, en revanche, mérite attention est l'allongement très marqué de cette distribution. L'étirement des associations spilitiques vers le bas du diagramme, dans le domaine des ordonnées négatives, s'explique simplement par le développement de l'épidote et de la calcite. L'étirement en sens inverse apparaît beaucoup plus complexe à la fois dans sa nature et dans son interprétation. Certes, le développement de la chlorite doit être en partie responsable de la configuration axiale, dirigée vers le point figuratif de ce minéral. Mais on ne peut manquer d'être frappé par l'étalement "en éventail" autour de cet axe, avec des courbures inverses qui dénotent l'influence de la silice pour les branches supérieures, et celle des minéraux potassiques pour la branche inférieure.

Il importe de savoir si ce développement en éventail correspond à une série d'oscillations de caractère aléatoire autour d'une "tendance générale" ou s'il exprime au contraire l'existence de plusieurs tendances distinctes correspondant à des types d'associations lithologiques dont il faudrait alors examiner l'extension régionale et la signification géologique.

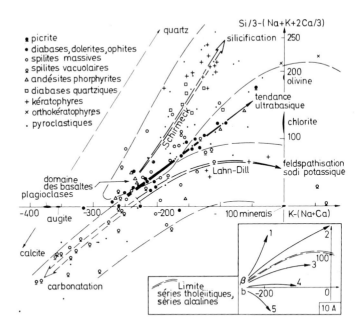

Fig. 10. Diagramme Si, K, Na, Ca. Types lithologiques du volcanisme préorogénique hercynien d'Europe du Nord. Distinction de deux types évolutifs (type Schirmeck et type Lahn-Dill) en liaison avec la paléogéographie. La Fig. 10 A montre schématiquement la disposition dans le même diagramme des séries ignées classiques: β, basaltes; b, basanites; 1, dacites; 2, rhyolites; 3, trachytes quartzifères; 4, trachytes; 5, phonolites

Ajoutons que s'il existait plusieurs tendances divergentes, la "tendance générale" apparente sur la figure pourrait n'être qu'une moyenne des tendances dépourvue de signification géologique précise. Les compilations générales comportent ce risque de confusion, et ne peuvent guère nous apporter de réponse sûre. C'est dans l'étude particulière du volcanisme préorogénique hercynien (groupe B) que nous rechercherons cette réponse.

Variations Lithologiques et Associations Régionales. Mise en Evidence de Deux Types Distincts dans le Magmatisme Initial Hercynotype

L'examen de la Fig. 10 fait apparaître des regroupements, à la fois par faciès pétrographique et par région.

L'ensemble des diabases et des dolérites, sans distinction de provenance, se groupe assez étroitement dans le domaine des basaltes et le long d'un axe partant de ce domaine en direction des points figuratifs de l'olivine ou de la chlorite (cumulats picritiques ou altérations chloriteuses).

Au contraire, les termes plus caractéristiques des associations spilitiques se distribuent selon des rameaux divergents par rapport à cet axe, et ceci en fonction de leur appartenance régionale:

Région de Schirmeck. Les diabases quartziques, les kératophyres, les tufs et les schalsteins prédominent, tandis que les spilites sont relativement subordonnées. Parmi tous les échantillons analysés, un seul spécimen désigné comme "spilite massive" se trouve en dessous de l'axe précédemment défini pour les diabases et les dolérites. A cette exception près, les spilites, les diabases quartziques et les kératophyres se disposent régulièrement le long d'un rameau partant du domaine des basaltes vers le haut, en direction du point figuratif du quartz.

Région de Lahn-Dill. La situation est inverse. A une exception près, les spilites massives sont en dessous de l'axe des diabases et dolérites. Ces spilites et les kératophyres trachytiques qui leur sont associées se disposent le long d'un rameau courbe partant du domaine des basaltes en direction de la zone des minéraux potassiques. Les spilites vacuolaires sont assez dispersées; leur groupement principal prolonge ce rameau en sens inverse vers les ordonnées négatives et les points figuratifs de l'épidote et de la calcite.

Régions de Cornouailles, Devonshire. Les spilites massives et les spilites vacuolaires, principalement représentées dans ces régions, ont la même disposition que celles de Lahn-Dill.

Nous sommes ainsi conduits à opposer deux types d'associations bien distincts, opposition qui nous était déjà apparue lors de l'étude pétrographique de ces régions (JUTEAU et ROCCI, même volume):

a) Le type Schirmeck - C'est un type d'évolution par silicification, sans carbonatation, où nous voyons l'influence sur le magmatisme initial d'un socle sialique épais.

b) Le type Lahn-Dill - Dans ce type, les formations effusives basiques subissent fréquemment une forte carbonatation (spilites à amygdales de calcite). Les kératophyres ne présentent aucune trace de silicification mais une exaltation des alcalins (Na et K) qui leur confère un caractère trachytique. Quant aux roches intrusives basiques, elles montrent des termes picritiques et plus généralement une tendance ultra-basique qui était absente du type Schirmeck. Cette association caractérise vraisemblablement un "magmatisme de fosse" par opposition au "magmatisme de socle" du type Schirmeck.

On n'oubliera pas qu'il existe entre ces deux types des affinités pétrographiques assez grandes pour caractériser un magmatisme initial "hercynotype" (ROCCI et JUTEAU, 1968); l'intérêt du diagramme de la Fig. 10 est la distinction précise, à l'intérieur de ce magmatisme hercynotype, de nuances importantes et de types d'évolution relevant directement de la position paléogéographique ou isopique à l'intérieur d'un même segment orogénique.

Ces observations sur le volcanisme préorogénique hercynien ont probablement une valeur plus générale puisqu'elles expliquent bien la forme "en éventail" de la distribution obtenue dans la Fig. 8 à partir d'une compilation d'analyses provenant de diverses régions.

Variations Locales et Effet d'Echantillonnage dans les Pillows-Lavas

Du coeur à la bordure des pillows, des variations importantes se manifestent dans ce diagramme (Fig. 11), mais de manière beaucoup moins uniforme que dans le diagramme Al, Na, K (Fig. 6).

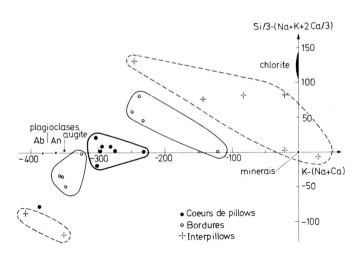

Fig. 11. Diagramme Si, K, Na, Ca. Variations locales et effets d'échantillonnage dans les pillows-lavas (groupe C, 24 analyses)

Les coeurs de pillows sont très groupés et bien localisés à la limite inférieure du domaine des basaltes communs. Leur différence par rapport aux roches basaltiques, ne s'exprime significativement que dans la 3e dimension, la transition anorthite⟶albite apparaissant bien comme la composante essentielle de cette différence.

Les bordures de pillows et les matériaux interpillows s'éloignent du domaine des basaltes communs, le plus souvent en direction de la chlorite, mais parfois aussi en direction de l'épidote et de la calcite.

Données Chimiques et Interprétation des Séries Spilitiques

Les données chimiques tiennent une place importante dans les diverses interprétations qui ont été proposées au sujet des séries spilitiques. Sans reprendre l'ensemble de ce problème, nous nous limiterons à dégager ici les éléments d'interprétation que nos diagrammes mettent plus particulièrement en évidence.

Composition Spilitique et Composition Basaltique

Les basaltes sont, en règle générale, peu différenciés de sorte qu'il est possible de définir assez précisément une composition basaltique. Il n'en va pas de même des spilites, dont on sait qu'elles présentent des différenciations jusqu'à l'échelle du pillow, ni a fortiori des associations spilites-kératophyres où il arrive que les termes acides soient prépondérants.

S'il s'agit de comparer les spilites et les basaltes du point de vue de leur composition chimique, on ne peut guère aller au-delà d'un parallèle entre les basaltes d'une part et les laves spilitiques basiques ou pillows-lavas d'autre part. Il importe déjà d'examiner si une convergence complète apparaît ou non à ce niveau.

Les arguments échangés à ce sujet ont été rappelés au début de cette note. A ceux qui soulignent les différences chimiques nettes (Ca, Na, H_2O, CO_2...) entre basaltes et spilites, d'autres opposent l'invalidité d'une statistique établie sur un échantillonnage incorrect des laves spilitiques. Cet échantillonnage donne bien la composition des spilites en tant que variété pétrographique, mais il néglige les produits de différenciation associés aux spilites (écorces de pillows, interpillows, etc.). Il n'est donc pas représentatif de la formation géologique - lave spilitique - comparée au basalte, autre formation géologique plus homogène. Au total, une compensation des différences est parfaitement concevable, dans le sens d'une convergence chimique quasi complète, à l'exception des fluides, entre laves basaltiques et laves spilitiques.

Cette dernière hypothèse délimite bien le champ de la discussion que nous aborderons, à notre tour, à l'aide de notre présentation graphique des données chimiques. Dans un système graphique quelconque, cette hypothèse suppose que le domaine des basaltes soit situé entre le domaine des spilites - variété pétrographique - et celui des produits différenciés associés à ces spilites. De plus les relations pondérales (barycentriques) doivent se retrouver, de manière cohérente, d'un système graphique à l'autre :

n spilite + p différenciat A + q différenciat B = (n + p + q) basalte

Revenons aux diagrammes présentés dans les pages précédentes. Dans la Fig. 6 (Al, Na, K) à rapprocher de la Fig. 2, le domaine des spilites (coeurs de pillows) et celui des produits les plus différenciés (interpillows) sont bien situés de part et d'autre du domaine des basaltes. Dans la Fig. 11 (Si, Ca, Na, K), à rapprocher de la Fig. 7, le domaine des spilites (coeurs de pillows) est à la limite inférieure de celui des basaltes communs tandis que les produits différenciés se répartissent en deux domaines opposés dont le plus important s'écarte du domaine des basaltes en direction de la chlorite. Dans chacune des deux figures, considérée isolément, il paraît possible que la composition pondérée de la "lave spilitique" tombe dans le domaine des basaltes, mais à la condition qu'il y ait une forte proportion d'interpillows dans la Fig. 6

et, au contraire, une forte proportion de coeurs de pillows dans la Fig. 11. Un calcul approché, prenant pour références les données des Fig. 6 et 11 et le point figuratif d'un basalte moyen, requiert plus de 50% d'interpillows dans un cas et moins de 25% dans l'autre cas. Or il s'agit du même ensemble de données.

Sans avoir la rigueur d'une démonstration mathématique, le raisonnement simultané sur les deux diagrammes conduit ainsi à mettre en doute l'hypothèse extrême d'une similitude globale de composition entre laves spilitiques, marquées par une différenciation plus ou moins intense, et laves basaltiques, peu ou pas différenciées. Sans doute le contraste entre les unes et les autres a-t-il été artificiellement accentué par l'accumulation sélective de données sur les parties albitisées ou "spilitisées" des laves spilitiques. Mais ce que nous venons de voir à propos des pillows-lavas montre que les estimations classiques, bien qu'elles soient quantitativement exagérées, restent qualitativement valables.

Les observations faites sur les deux diagrammes sont complémentaires et s'expliquent bien par une substitution de Na à Ca, avec décalage vers les valeurs négatives de Al/3 - Na (diagrammes Al, Na, K) et convergence pour les paramètres où apparaît la somme Na + Ca (diagramme Si, K, Na, Ca). L'action des fluides et la diversification des associations minérales dans les laves spilitiques (albitisation, carbonatation, chloritisation...) compliquent ce bilan et peuvent même le renverser localement sans en modifier le sens général.

Déplacements de Ca, Na, Si et Invariance de Al dans la Spilitisation

En employant, dans les pages précédentes, le terme de "spilitisation", nous admettions implicitement le rattachement des laves spilitiques à une souche basaltique à un moment de leur histoire que certains auteurs placent au stade magmatique et d'autres au stade post-cristallisation.

Nous ne voyons pas, actuellement, comment les données chimiques pourraient, indépendamment des observations pétrographiques, permettre d'orienter le choix entre évolution magmatique et évolution métasomatique. Ceci devrait être l'objet de futures recherches. Du moins, pouvons-nous, sans aucun choix préalable, essayer de préciser le bilan chimique global de la spilitisation, c'est-à-dire le bilan de l'évolution chimique nécessaire pour passer des compositions basaltiques aux compositions spilitiques.

Nous observons dans le diagramme Si, K, Na, Ca (Fig. 7 et 11) que les spilites massives, les coeurs de pillows et les pillows-lavas dans leur ensemble (notre calcul, fondé sur une proportion de 25% pour les bordures de pillows et la matière interstitielle) demeurent dans le domaine des basaltes communs. Ceci nous autorise à considérer comme "invariants dans la spilitisation" du moins en première approximation, les paramètres Si/3 - (K + Na + 2Ca/3) et K - (Na + Ca). Comme le potassium est très peu abondant et qu'il est dépourvu de rôle apparent dans la spilitisation, ceci revient à écrire que (Ca + Na) d'une part et (Si + Ca) d'autre part sont invariants. Observons en effet que la diminution de Ca au cours de la spilitisation est considérable de sorte que des variations de (Ca + Na) ou de (Si + Ca) ne manqueraient pas d'être sensibles, s'il n'y avait réellement dans les deux cas une excellente compensation par augmentation de Na et de Si.

En revanche, si l'on en juge par l'évolution du paramètre Al/3 - K (Fig. 2, 3 et 6) ou, plus directement, par les données analytiques non reproduites dans cette étude, Al reste dans la même gamme, des basaltes aux spilites et des coeurs de pillows à leur bordure. On en tire l'indication d'une <u>relative invariance de Al</u>.

Nous parvenons ainsi à un bilan chimique global dont l'intérêt principal n'est pas la substitution de Na à Ca, admise de tous, mais le comportement de Si et de Al plus controversé. A partir de ce bilan global, on peut amorcer l'analyse plus complexe des bilans par faciès et des transferts correspondant à la différenciation des masses spilitisées. La formation des spilites sensu stricto, roches albitiques (coeur de pillows, spilites massives, etc.) est le phénomène dominant. Le raisonnement développé pour établir le bilan général de la spilitisation lui est intégralement applicable: substitution atome pour atome de Na à Ca sans variation de Ca + Na, apport de Si lié au départ de Ca sans variation de Si + Ca. L'enrichissement en Si apparaît bien comme l'une des conditions de l'albitisation de sorte que les mouvements de Si contrôlent dans une large mesure le degré de spilitisation.[2]

Les différenciations chloritisées ou carbonatées ont des caractères très contrastés vis-à-vis des "invariants globaux"; (Si + Ca) et (Ca + Na) sont faibles dans les zones chloriteuses et forts dans les zones carbonatées tandis que pour Al, les différences sont de sens inverse. L'exemple de la Fig. 11 relative aux pillows-lavas montre bien comment ces différenciations peuvent exercer les unes par rapport aux autres un rôle compensateur. Dans les deux cas cependant, Si et Na seront libérés, migrant à partir des zones chloritisées ou carbonatées vers les zones d'albitisation.

Il ne paraît donc pas douteux que des échanges à l'intérieur de la masse en voie de spilitisation contribuent largement à l'acquisition des caractères chimiques que l'on attribue parfois exclusivement à des influences extérieures. Mais celles-ci tiennent de leur côté un rôle important et sont mises en évidence par le bilan global (départ de Ca, apport de Na et de Si). On pourrait exprimer cette conclusion d'une autre manière en soulignant que l'action des fluides, évidente dans la spilitisation, ne s'éteint pas, dans un sens (apports) comme dans l'autre (départs) aux limites de la formation spilitisée.

Rattachement des Associations Spilitiques aux Séries Tholeiitiques ou Alcalines

On ne peut aborder le problème du rattachement des spilites soit aux tholéiites, soit aux basaltes alcalins qu'au prix d'hypothèses préalables sur la nature des variations chimiques liées à la spilitisation. En particulier, l'invariance ou la mobilité de Si sont, de toute évidence, importantes à connaître pour cette diagnose.

Comme nous l'avons précédemment indiqué, l'invariance de Si lors de l' expulsion de Ca se serait traduite dans notre Fig. 8 par un décalage général des spilites par rapport au domaine des basaltes communs. Le paramètre Si/3 - (K + Na + 2Ca/3) peut en effet s'écrire Si/3 + Ca/3 - K - (Na + Ca) et il varierait par conséquent comme Ca/3 dans l'hypothèse où Si et (Na + Ca) seraient invariants. N'ayant pas observé de différence significative, nous avons été conduits à considérer que Si variait de

---

[2] Nos observations nous placent sur ce point en désaccord avec M. FONTEILLES (1969), qui présente Si comme invariant dans l'albitisation, alors que nos conclusions rejoignent les siennes en ce qui concerne Na + Ca et Al.

telle sorte que (Si + Ca) reste invariant comme (Ca + Na). Si cette
conclusion est correcte, la spilitisation n'a plus d'influence sur la
position des points figuratifs dans cette Fig. 8 et nous pouvons l'utiliser directement pour notre diagnose.

A partir des données de la littérature (NOCKOLDS et ALLEN, 1953; KUNO,
1968) nous avons pu vérifier que les paramètres utilisés se prêtent bien
à la distinction d'un domaine tholéiitique et d'un domaine alcalin. La
ligne de séparation qui a été tracée sur la figure (tireté double:
domaine tholéiitique au-dessus, domaine alcalin en dessous) ne fera pas
oublier que les basaltes tholéiitiques et les basaltes alcalins forment
un continuum (MANSON, 1967), mais elle sépare nettement les compositions
données pour typiques (KUNO, 1968).

Si l'on considère les spilites dans leur ensemble, on observe qu'elles
se disposent en un domaine mixte, les unes à caractère tholéiitique,
les autres à caractère alcalin, formant comme les basaltes un continuum.
Mais en allant au-delà de cette statistique générale pour revenir à une
étude région par région, il serait certainement possible de déceler des
caractères franchement tholéiitiques pour certaines associations régionales et des caractères alcalins pour d'autres. Les différenciations
siliceuses ou trachytiques accusent mieux encore cette filiation tholéiitique ou alcaline.

D'après l'exemple du domaine hercynien du Nord de l'Europe, Fig. 10, il
semble que le caractère tholéiitique (type Schirmeck) ou alcalin (type
Lahn-Dill) dépende de la présence ou de l'absence d'un écran de socle
sialique à l'emplacement des manifestations du magmatisme initial.

Il existe donc entre diverses associations spilitiques un contraste
chimique de même nature que celui qui oppose les basaltes océaniques
aux basaltes des plateaux. Ce contraste ressortait dès la première
étude chimique du Massif de Schirmeck (JUTEAU et ROCCI, 1966). Les
nouveaux paramètres et les données de comparaison utilisés dans la présente publication en montrent plus clairement la nature et la signification.

Ainsi rejoignons-nous finalement KUNO (1968) lorsqu'il montre par l'étude
détaillée des basaltes japonais que le chimisme des magmas et de leurs
différenciations doit être mis en relation avec l'environnement orogénique et l'éventualité de contaminations sialiques.

References

AMSTUTZ, G.C. (1954): Geologie und Petrographie der Ergußgesteine im
    Verrucano des Glarner Freiberges (Suisse). Thèse, Zurich, 149.
AMSTUTZ, G.C. (1968): Les laves spilitiques et leurs gîtes minéraux.
    Geol. Rundschau, H. 3, 57, 936-954.
DIETRICH, V. von (1967): Geosynklinaler Vulkanismus in den oberen
    penninischen Decken Graubündens (Suisse). Geol. Rundschau, H. 1,
    57, 246-264.
FONTEILLES, M. (1968): Contribution à l'analyse du processus de
    spilitisation. Etude chimique comparée des séries volcaniques
    paléozoïques de la Bruche (Vosges) et de la Brévenne (Massif Central
    français). Bull. B.R.G.M., 2e série, Géologie Appliquée, section II,
    no. 3, 1-54.

GILLULY, J. (1935): Keratophyres of Eastern Oregon and the spilite problem (part 1). Am. J. Sci. 29, 225-252, no. 171.

JUTEAU, TH., ROCCI, G. (1966): Etude chimique du massif volcanique dévonien de Schirmeck (Vosges septentrionales). Evolution d'une série spilite-kératophyre. Sci. de la Terre, t. XI, no. 1, 68-104.

JUTEAU, TH., ROCCI, G. (1974): Vers une meilleure connaissance du problème des spilites à partir de données nouvelles sur le cortège spilito-kératophyrique hercynotype. 253-329 (dans ce volume).

KUNO, H. (1968): Differentiation of basalt magmas. "Basalts", Vol. 2, Interscience Publishers, ed. by H.H. Hess, 623-688.

LA ROCHE, H., de (1964): Sur l'expression graphique des relations entre la composition chimique et la composition minéralogique des roches cristallines. Sci. de la Terre, t. IX, no. 3, 293-337.

LA ROCHE, H., de (1966): Sur l'usage du concept d'association minérale dans l'étude chimique des roches. C. R. Acad. Sci. (Paris) t. 262, série D, 1665-1668.

LA ROCHE, H., de (1968): Comportement géochimique différentiel de Na, K et Al dans les formations volcaniques et sédimentaires. C. R. Acad. Sci. (Paris) t. 267, série D, 39-42.

MANSON, V. (1967): Geochemistry of basaltic rocks: major elements. "Basalts", Vol. 1, Interscience Publishers, ed. by H.H. Hess, 215-270.

NAREBSKI, W. (1964): Petrochemia law publistych Gor Kaczawskich i niektore ogolne problemy petrogenezy spilitow (petrochemistry of pillow lavas of the Kaczawa Mountains and some general petrogenetical problems of spilites). Prace petrograficzne, Prace Muzeum Ziemi, nr. 7, 69-186.

NOCKOLDS, S.R., ALLEN, R. (1953): The geochemistry of some igneous rocks series. Geochim. Cosmochim. Acta 4, 105-142.

PARROT, J.F. (1967): Le cortège ophiolitique du Pinde Septentrional (Grèce). Thèse 3e cycle, Paris.

PETERLONGO, J.M. (1968): Les ophiolites et le métamorphisme à glaucophane dans le Massif de l'Inzecca et la région de Vezzani (Corse). Bull. B.R.G.M., 2e série, section IV, no. 1, 18-94.

ROCCI, G., JUTEAU, TH. (1968): Spilites-kératophyres et ophiolites. Influence de la traversée d'un socle sialique sur le magmatisme initial. Geol. en Mijnbouw. 47 (5), 330-339.

SUNDIUS, N. (1930): On the spilite rocks. Geol. Mag. 67, 1-17.

TURNER, F.J., VERHOOGEN, J. (1960): Igneous and metamorphic petrology. E nd. edition. New York: McGraw Hill.

VALLANCE, T.G. (1960): Concerning spilites. Proc. Linnean Soc. N.S. Wales 85, part I, 1-52.

WELLS, A.K. (1922-1923): The nomenclature of the spilitic suite. Geol. Mag. 59, 346-354 et 60, 62-74.

Pyroxenes and the Basalt. Spilite Relation

T. G. Vallance

Introduction

The term spilite is now 150 years old. Few rock names currently in use have a longer tradition and none continues to be more variously defined. The surprising fact is that, despite a welter of opinions as to what should be diagnostic, the majority of rocks identified as spilitic share a broad community of properties. Most such materials are evidently mafic and are recognized as having modes of occurrence similar to certain basaltic rocks. Distinction from basalts arises from observed contrasts, both mineralogical and chemical.

There is reasonable agreement, for instance, that the characteristic feldspar of spilites is low-intermediate albite (or, in some cases, K-feldspar). Diversity of opinion emerges in matters of detail such as erecting arbitrary limits of soda content as diagnostic criteria for the definition of spilites. However, variations in mineral content and chemical composition, apparently related to such features as local contrasts in rock fabric or proximity to fractures and voids, have been observed in many spilitic bodies. The most obvious examples occur in spilitic pillows (VALLANCE, 1965), but massive spilites may be equally diverse (SMITH, 1968). Albite-rich (and hence soda-rich) variants co-exist with albite-poor materials, and attempts to specify spilites according to limiting compositions ignore this experience of nature.

The present writer prefers to use spilite as a group name for those rocks which are analogous to basalts in their mode of occurrence and broad fabric elements, but differ from basalts in consisting largely of mineral phases of the greenschist facies type such as albite, chlorite, epidote, calcite, etc. Particular variants are then termed albite spilite, albite-chlorite spilite, chlorite spilite, chlorite-epidote spilite, and so forth (VALLANCE, 1969b). Rocks, chemically and mineralogically similar to these spilites but marked by penetrative metamorphic fabrics, are excluded from the group and called schists.

The numerous schemes for the generation of spilites have been recently reviewed elsewhere (AMSTUTZ, 1968; VALLANCE, 1960, 1969b) and need little elaboration here. Suffice it to say the genetic proposals fall into two major groups: a) spilites are derived by crystallization of melts and b) spilites are formed by mineral adjustments in materials already cooled and consolidated. Within each group one finds variety. Thus believers in spilites as primary magmatic rocks invoke either a mafic magma of specific ("spilitic") composition or a basaltic melt. Where appeal is made to the latter it is commonly assumed to have cooled under special conditions or to have had some compositional character, such as particular richness in water, different from those basaltic melts that are known to form basalts on cooling. In all of these cases, characteristic phases like albite and chlorite are taken to have separated directly from melt. Formation of spilites by secondary adjustments in basalts has been variously attributed to late-magmatic (autometamorphic) agencies or to post-magmatic operations.

As CANN (1969) and others have remarked, the idea of specifically
spilitic magma is implausible. No acceptable evidence for the existence
of such a magma is known; certainly no spilitic melts have been recognized
in contemporary volcanic regions. If, then, we leave aside spilite magma,
the genetic problem is simplified to the extent that the parental material
of spilites may be basaltic - either melt or solid.

Recognition of Spilitic Parentage

Any useful proposal for the formation of spilites must offer a means of
identifying parental materials as well as insight into the processes
whereby the parent acquires the characters one associates with spilitic
rocks. Why, for instance, do spilites tend to have mineral associations
contrasted with those of basalts and why should there be such marked
local variations in chemical composition within spilitic bodies when
basalts with closely comparable fabric characters tend to be chemically
homogeneous?

AMSTUTZ (1968) has commented on the significance of sampling in the study
of spilitic rocks; the matter is discussed in some detail by VALLANCE
(1969b). Where heterogeneous spilites, whether pillowy or massive, have
been studied in detail their bulk compositions, determined from extensive sampling, show an approach to basaltic character. Bulk chemistry,
however, appears to have little genetic value beyond reinforcing what
is inferred from fabric analogies - that there is some causal connection
between basalts and spilites.

In the case of basalts, recognition of magmatic character, whether
tholeiitic or alkali basalt or transitional between the two, can be
effected by such means as the nature of modal pyroxenes or normative
compositions based on individual analyses. YODER (1967) has remarked
on the existence of both Qz-normative and Ne-normative types among
spilitic rocks. But the variations in composition found within spilite
bodies would make nonsense of efforts to seek evidence of parentage by
way of the norms of single samples. CANN (1969) reports Ne-normative

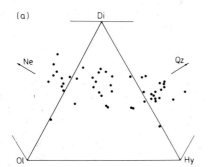 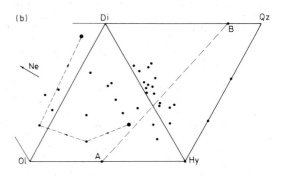

Fig. 1a and b. Plots of normative Ne-Di-Ol-Hy-Qz for basalts and spilites.
a) Average basalt compositions (from COOMBS, 1963).
b) Spilitic rocks of the Nundle district, N.S.W. Individual samples
of massive spilites shown as isolated points. The points A and B, joined
by a dashed line, represent a fragment (corundum-normative) and its
matrix (Wo-normative) from a spilitic breccia. The circled points linked
by dashed lines represent data for a spilitic pillow (VALLANCE, 1960),
the arrows indicating the sense of progression from core to selvedge

to Qz-Hy-C-normative spilites with Hy-Ol-normative basalts in the same dredge haul from the Carlsberg Ridge, Indian Ocean. Normative variety within spilitic pillows is just as extreme (VALLANCE, 1969b). Lest protest be made that pillows are unrepresentative of spilites in general, reference may be made to Fig. 1b.

The points plotted in Fig. 1b relate to data for massive albite-bearing mafic rocks from a Devonian marine association in the Nundle district, New South Wales. The majority of these rocks are fine-grained types - spilites as described by BENSON (1914b); coarser-grained varieties are albite dolerites in BENSON's terminology. Both finer and coarser rocks are closely related in the field and overlap in their compositional diversity. The range of normative Ne-Di-Ol-Hy-Qz in these rocks is of an order with that for all common basalt types (Fig. 1a), except that the spilites tend to have more variety in normative Di. The volcanic pile at Nundle cannot have involved contributions from such widely different parental melts, as might be suggested by the normative characters, because extremes of composition can be found within individual lava bodies.

Pyroxene of Basalts and Spilites. It is well known that basaltic pyroxenes vary in character according to the nature of their hosts. The fact is widely used in basalt diagnosis. Pyroxene also occurs with albite and chlorite in some spilitic rocks and in view of the failure of spilite norms to yield useful answers, it is clearly relevant to consider the nature of spilitic pyroxenes.

Chemical data for bulk pyroxene fractions from six non-pillowy spilitic rocks of the Nundle association are listed in Table 1. General details of field occurrence and petrography have been outlined by BENSON (1914a, b, 1915) and VALLANCE (1969a). These need not be repeated apart from emphasizing that the pyroxenes occur typically as microphenocrysts related subophitically to fine laths of albite. Sample 13242 (no. 4) is a fine-grained albite dolerite (or coarser spilite) retaining some relicts of calcic plagioclase. Despite evident diversity in the compositions of the host rocks, the pyroxene compositions are generally consistent. Only a few representative samples are quoted here; the field of spilitic pyroxenes from Nundle marked in Fig. 2b encompasses data for 16 analyzed pyroxene fractions.

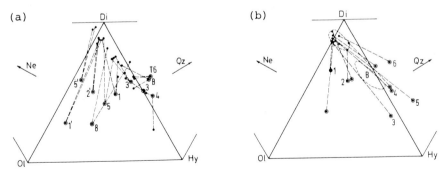

Fig. 2a and b. Plots of normative Ne-Di-Ol-Hy-Qz for selected basalts, spilites and their pyroxenes.
a) Basalts and pyroxenes from Hawaii - 1,3,4,5,8,T6 (MUIR and TILLEY, 1963, Tables 1,2,6,7), from the Mid-Atlantic Ridge - 1',2',3',5' (MUIR and TILLEY, 1964, Tables 1,3), and from Bombay - B (tholeiite R/110, unpubl. data; for pyroxenes see Table 2, this paper).
b) Spilites and pyroxenes from Nundle - 1,2,3,4,5,6 (this paper, Table 1). The field of 8 spilitic pyroxene fractions from Bombay is marked B

Table 1. Spilitic rocks and clinopyroxenes, Nundle, New South Wales

	1 rock	1 cpx	2 rock	2 cpx	3 rock	3 cpx	4 rock	4 cpx	5 rock	5 cpx	6 rock	6 cpx
SiO2	49.28	47.02	52.48	49.73	49.68	49.71	49.38	50.34	52.96	50.21	54.20	48.35
Al2O3	14.34	9.32	13.33	4.04	15.02	4.47	14.96	4.40	14.54	3.19	14.73	6.07
Fe2O3	4.69	2.70	3.08	1.17	2.52	1.08	3.23	0.80	3.02	0.71	3.41	3.57
FeO	7.66	8.61	9.16	10.65	9.71	9.37	8.45	9.36	8.00	11.59	7.65	8.37
MnO	0.20	0.15	0.14	0.29	0.18	0.13	0.17	0.15	0.23	0.24	0.11	0.18
MgO	4.31	11.31	4.72	14.01	5.71	14.64	5.94	14.09	3.74	13.36	3.67	12.55
CaO	7.76	19.16	6.24	17.96	7.23	19.31	9.52	19.50	5.88	19.07	7.87	19.13
Na2O	4.98	0.46	5.51	0.44	3.80	0.62	3.02	0.55	4.48	0.56	4.70	0.69
K2O	0.67	0.05	0.63	0.03	0.54	0.04	0.52	0.06	1.16	0.05	0.41	0.04
H2O+	3.46	0.38	2.41	0.19	3.08	0.23	2.39	0.30	2.89	0.56	1.45	0.42
H2O−	0.44				0.36		0.17		0.07			
TiO2	2.04	1.17	2.19	1.30	1.97	0.92	1.92	0.86	2.16	0.94	2.00	1.20
P2O5	0.29		0.28		0.36		0.36		0.77		0.18	
CO2	0.61		0.00		0.32		0.00		0.43		0.00	
Total	100.73	100.33	100.17	99.81	100.48	100.52	100.03	100.41	100.33	100.48	100.38	100.57

Clinopyroxene formulae (6 O):

	1	2	3	4	5	6
Si	1.758 ⎤	1.870 ⎤	1.853 ⎤	1.875 ⎤	1.894 ⎤	1.810 ⎤
Al	0.242 ⎦ 2.00	0.130 ⎦ 2.00	0.147 ⎦ 2.00	0.125 ⎦ 2.00	0.106 ⎦ 2.00	0.190 ⎦ 2.00
Al	0.169	0.049	0.049	0.069	0.036	0.078
Ti	0.033	0.037	0.026	0.024	0.027	0.034
Fe3+	0.076	0.033	0.030	0.022	0.020	0.101
Fe2+	0.269	0.335	0.292	0.292	0.366	0.262
Mn	0.005	0.009	0.004	0.005	0.008	0.006
Mg	0.630	0.785	0.813	0.782	0.751	0.700
Ca	0.767 1.98	0.723 2.00	0.771 2.00	0.778 2.01	0.770 2.02	0.767 2.00
Na	0.033	0.032	0.045	0.040	0.041	0.050
K	0.002	0.001	0.002	0.003	0.002	0.002

%						
Ca	44.04	38.56	40.36	41.42	40.24	41.93
Mg	36.16	41.83	42.56	41.62	39.21	38.26
∊Fe	19.80	19.61	17.08	16.96	20.55	19.81

1. Sample 13236 (Univ. of Sydney collection), Nundle - Hanging Rock Road, 0.8 km W of Devil's Elbow.
2. Sample 23014, Peel River, 2.4 km S of Bowling Alley Point.
3. Sample 13240, summit of Tom Tiger, Nundle.
4. Sample 13242, 0.8 km NE of Tom Tiger.
5. Sample 41106, below falls, Swamp Creek, Nundle.
6. Sample 23062, 0.4 km SW of falls, Swamp Creek.

Anal: Rocks 1, 3, 4, 5 - HERDSMAN; rocks 2, 6 and clinopyroxenes - N. DE FARIA E CASTRO.

These spilitic pyroxenes are all distinctly calcic augites and can be matched with basaltic augites. Two features, however, call for notice. First is the variation in Al-content, the highest values being found in the pyroxenes of generally fine-grained flow rocks at or near the top of the volcanic pile in the southern part of its outcrop. It remains to be seen whether this apparent spatial restriction is real. The other unusual point about these pyroxenes is their distinctly low TiO_2 content in relation to high tetrahedral Al (cf. LeBAS, 1962) though in this they bear similarity to basaltic pyroxenes from the Rift Zone of the Mid-Atlantic Ridge (MUIR and TILLEY, 1964).

On the basis of features such as observed subophitic relations to albite laths and the assumption that the clinopyroxene was a primary phase derived directly from melt, BENSON (1914b) hypothesized a primary origin for the clear albite in the Nundle spilites. Neither assumption nor conclusion is necessarily valid. Both phases (or either one), as now observed, could be primary materials or have secondary status through formation at the expense of (and pseudomorphing) pre-existing feldspars and pyroxenes. Resolution of this problem is no mere exercise; its relevance to the genesis of spilites should be obvious.

Albite and diopside can crystallize from melts of appropriate composition under both hydrous and anhydrous conditions (YODER, 1967). However, in the cases studied progressive cooling leads to separation of pyroxene before feldspar. This order conflicts with textural relations recorded for spilitic rocks. In addition, as any melt parental to spilite can hardly have been iron-free it is reasonably predictable that a pyroxene equilibrated with albite crystallizing from this melt will be not only a lower-temperature stable variety but compositionally different from a basaltic pyroxene equilibrated with calcic plagioclase. Yet the pyroxenes co-existing with low-temperature albite in the Nundle spilites have generally basaltic pyroxene characters.

YODER (1967) has concluded from experimental study that spilitic mineral associations are produced by sub-solidus adjustment of basaltic material under hydrous conditions. In this regard he suggested the following relation could be relevant:

(Ab_1An_1) + Cpx' + Hyp and/or Ol + H_2O = Cpx" + Ab + Chlor (+ Qtz).
Distribution of normative An on the spilite side of the equation is effected, according to YODER, by solid solution in albite, by enhanced content of Ca-Tschermak molecule ($CaAl_2SiO_6$) in the clinopyroxene, and possibly by formation of an accessory hydrous CaAl silicate-like epidote. In terms of this scheme both albite and clinopyroxene in spilites are secondary phases and spilitic pyroxenes should be distinguishable compositionally from the primary pyroxenes of basalts.

The occurrence of relatively aluminous calcic augites in the Nundle spilites might be taken as encouraging to YODER's prediction, especially as only minor amounts of epidote are present in some of the analyzed host rocks. Further, optical study, at least, of the clear albites suggests low An content (normally An_{0-8}). But one should be careful before deciding that the pyroxenes have Cpx" status. There is no recognizable relation between Ca-Tscherm. content of pyroxene and abundance of epidote. The sample (13242) with relic calcic plagioclase has a pyroxene fraction with Ca-Tscherm. well within the range for the whole suite. In short, the status of the spilitic clinopyroxenes at Nundle is ambiguous; their Ca-Tscherm. contents are within the limits known for igneous pyroxenes. To resolve this problem we must seek an association that offers some internal evidence of spilitic parentage.

Such an association exists in the vicinity of Bombay (SUKHESWALA and POLDERVAART, 1958); the spilitic rocks are described by SUKHESWALA elsewhere in this volume (p.229). Professor SUKHESWALA has kindly supplied material representative of this association. The data presented here constitute only a part of the results of a more detailed study (VALLANCE, 1974).

Both tholeiite (110/R - plotted as B in Fig. 2a and normatively a quartz tholeiite) and spilite (82/R) were taken from the upper part of an essentially tholeiitic flow on Bombay Island, the tholeiite, in fact, from the top. The two rocks display very similar textural characters, and each contains two generations of pyroxene, scattered phenocrysts and a more abundant groundmass phase. The latter in particular has moulded, subophitic relations to feldspars - in the one case labradorite, in the other, albite. Minor interstitial chlorite occurs in the tholeiite; in the spilite, chlorite is more plentiful and forms better-defined interstitial patches. Minor prehnite and some calcite appear in the spilite. Data for two pyroxene fractions, separated in heavy liquids from each rock, indicate a remarkable congruence (Table 2).

These data make it plain the spilitic pyroxenes must have crystallized under conditions similar to those for the basalt pyroxenes. Formation

Table 2. Clinopyroxene fractions from a tholeiite and spilite, Bombay, India

	Tholeiite (110/R)		Spilite (82/R)	
	lighter fract.	denser fract.	lighter fract.	denser fract.
SiO_2	52.31	51.08	52.03	49.77
Al_2O_3	2.61	3.26	2.73	3.19
TiO_2	0.49	0.55	0.50	0.70
Fe_2O_3	0.56	0.19	0.72	0.95
FeO	11.32	16.55	10.97	16.14
MnO	0.18	0.24	0.19	0.24
MgO	16.01	14.47	16.56	14.49
CaO	16.58	12.56	16.00	13.77
Na_2O	0.33	0.42	0.34	0.30
K_2O	0.03	0.10	0.04	0.04
H_2O+	0.16	0.21	0.17	0.34
Total	100.58	99.63	100.25	99.93

Clinopyroxene formulae (6 O):

	Tholeiite (110/R)				Spilite (82/R)			
Si	1.938 }	2.00	1.934 }	2.00	1.928 }	2.00	1.894 }	2.00
Al	0.062		0.066		0.072		0.106	
Al	0.052		0.079		0.047		0.037	
Ti	0.014		0.016		0.014		0.020	
Fe^{3+}	0.016		0.005		0.020		0.027	
Fe^{2+}	0.351		0.524		0.340		0.513	
Mn	0.006 }	2.00	0.008 }	1.99	0.006 }	2.00	0.008 }	2.01
Mg	0.877		0.816		0.914		0.822	
Ca	0.658		0.509		0.635		0.561	
Na	0.024		0.031		0.024		0.022	
K	0.001		0.005		0.002		0.002	

%

Ca	34.61	27.46	33.26	29.18
Mg	46.13	44.00	47.88	42.71
ΣFe	19.26	28.53	18.85	28.11

Anal: N. DE FARIA E CASTRO

of albite in place of labradorite has not been accompanied by shift of pyroxene composition, either in the sense expected in lower-temperature magmatic crystallization or in that of the YODER relation. Ca-Tscherm. in all four pyroxenes is, in fact, almost constant. The spilitic pyroxenes must have crystallized with calcic plagioclase, and the albite now associated with them must be a secondary, pseudomorphous phase.

For the Bombay example the pyroxene data simply confirm what was known from field relations and petrography, that the spilite had a tholeiitic parent. Nature has not been everywhere so accommodating, as was seen in the discussion on the Nundle suite. But if the generation of present albite-clinopyroxenes-chlorite assemblages at Bombay involves no significant adjustment in the pyroxenes and these have evident diagnostic value, it seems reasonable to argue that spilitic pyroxenes elsewhere may also be magmatic phases. That being so, predictions as to the parentage of the Nundle association become possible.

It was seen earlier that despite wide normative spread in the host rocks, the pyroxenes from Nundle occupy a restricted compositional field (Fig. 2b The pyroxenes, in fact, are such as one might expect in a suite of basalts mainly normative olivine tholeiites but perhaps transitional to slightly alkaline types. Preliminary studies of pyroxenes from other spilitic associations (e.g. Porth Dinllän, North Wales, and Cliefden, New South Wales) suggest similar parental materials. In the case of the Nundle rocks it is only fair to observe the tantalizing lack of unequivocal evidence of the former presence of modal olivine (BENSON, 1914b, p.665; VALLANCE, 1969c).

Conclusions

Study of spilitic clinopyroxenes from Bombay and Nundle confirms YODER's argument to the extent that spilitic albite here is secondary. The idea of secondary albite in such rocks is, in fact, at least 50 years old. On the other hand, the clinopyroxenes are likely to be primary relics, metastable in their present situations. Textural characters and the nature of the pyroxenes indicate these spilites formed by adjustment of solid basalts to low-temperature, hydrous conditions. Pyroxenes alone of all the parental phases played a passive role in these adjustments. The compositional gap (Fig. 2b) separating the Bombay and Nundle pyroxenes is believed to be entirely an accident of sampling. The writer feels confident that as more pyroxene-bearing spilites are investigated a complete overlap with the common basaltic pyroxenes will be found. There seems to be no particular reason for restricting the parentage of spilites to tholeiitic material, as suggested by BATTEY (1956).

The onus of proving a primary magmatic or hydromagmatic origin for spilites remains with those who argue that cause. Quite apart from the evidence of pyroxenes the case for primary spilites is supported neither by YODER's (1967) experimental results nor by experience of contemporary volcanism in which there is a singular absence of spilitic produce. The deep ocean basins yield normal basalts, cooled under considerable loads of water. Spilitic rocks dredged from the oceans appear to have had a more complex history than simple magmatic cooling (e.g. CANN, 1969). Relatively rapid chilling of parental melt is indicated by the common textural features of spilites, analogous with those of chilled basalts, and such conditions would be inimical to protracted mineral adjustment with melt during magmatic cooling. For this reason as well as the widespread evidence of volatile-loss by vesiculation (presently exemplified by amygdaloidal character) in spilitic bodies, the present writer has serious doubts as to the efficacy of late-magmatic (autometamorphic or deuteric) operations in spilite genesis. A post-magmatic alteration status appears far more likely for many spilites. Both local hydrothermal and regional burial metamorphic controls are believed to offer competent means of converting solid basalts to spilites.

In a few cases calculated bulk compositions of spilite bodies suggest simply addition of water and, commonly, CO_2 to basalt. These, however, appear to be far less abundant than spilites, the bulk compositions of which are marked by departures from basalt + water. The commonest departures are found in diminished CaO, enhanced Na_2O, and somewhat variable SiO_2 in bulk spilites relative to basalts. Local variations in other components in heterogeneous spilites do not, as a rule, imply any significant overall gain or loss from the concentrations found in basalts (VALLANCE, 1969b).

Conversion of solid basalt to spilite is believed to involve hydrolysis and ionic exchange. Hydrolysis of basalt, and especially rapidly-chilled and somewhat glassy basalt, where the hydrous medium is wholly or in part of marine origin (e.g. trapped pore-water in a buried marine succession) will produce solutions with alkaline pH in which Ca and alkalis are far more soluble than Al, Fe and Mg. These latter tend to remain as residue and in spilitic rocks this residue is chloritic. The pseudomorphous character of spilitic albite, occupying primary plagioclase sites, indicates that albite forms without any intermediate zeolitic stage. Stabilization of albite under conditions of spilitization is likely to require either temperatures above zeolite stability or some control of water chemistry such as enhanced activity of SiO_2. Where albite is the stable sodic phase its growth proceeds by ionic exchange of CaAl from plagioclase with NaSi from aqueous solution adjusted in composition through early hydrolysis and, if of marine provenance, already bearing Na. The distribution of albite in spilites is related largely to the distribution of primary feldspar sites, and high or low rock Na_2O contents reflect accidents of texture. Ca appears to be the most mobile of the constituents of spilites, and its behavior may be controlled in part at least by activity of CO_2. Thus low aCO_2 may determine eventual precipitation of Ca to form hydrous CaAl silicates like epidote, higher aCO_2, formation of calcium carbonate within a spilite body or actual local loss of Ca and perhaps carbonate deposition elsewhere. Clinopyroxenes appear to persist only in relatively carbonate-poor environments. Identification of the specific ancestry of carbonated spilites offers a greater challenge than those materials that retain primary relics.

Acknowledgements

Generous contributions to assist this work have been made by the University of Sydney and the Australian Research Grants Committee. For assistance in the laboratory study the author thanks Mrs. N. DE FARIA E CASTRO, Miss P. BROWN and Mr. J.L. SANDERSON. Special thanks are due to Professor R.N. SUKHESWALA of Bombay for his complete cooperation and generous supply of material.

References

AMSTUTZ, G.C. (1968): Spilites and spilitic rocks. Basalts: The Poldervaart Treatise on rocks of basaltic composition (ed. Hess and Poldervaart). New York. Interscience, Vol. 2, 737-753.

BATTEY, M.H. (1956): The petrogenesis of a spilitic rock series from New Zealand. Geol. Mag. 93, 89-110.

BENSON, W.N. (1914a): The geology and petrology of the great serpentine belt of New South Wales. Part ii. The geology of the Nundle District. Proc. Linn. Soc. N.S.W. 38, 569-596.

BENSON, W.N. (1914b): Idem. Part iii. Petrology. Ibid. 38, 662-724.

BENSON, W.N. (1915): Idem. Part iv. The dolerites, spilites, and keratophyres of the Nundle District. Ibid. 40, 121-173.

CANN, J.R. (1969): Spilites from the Carlsberg Ridge, Indian Ocean. J. Petr. 10, 1-19.

COOMBS, D.S. (1963): Trends and affinities of basaltic magmas and pyroxenes as illustrated on the diopside-olivine-silica diagram. Miner. Soc. Amer., Spec. Pap. 1, 227-250.

LeBAS, M.J. (1962): The role of aluminum in igneous clinopyroxenes with relation to their parentage. Amer. J. Sci. 260, 267-288.

MUIR, I.D., TILLEY, C.E. (1963): Contributions to the petrology of Hawaiian basalts: II the tholeiitic basalts of Mauna Loa and Kilauea. Amer. J. Sci. 261, 111-128.

MUIR, I.D., TILLEY, C.E. (1964): Basalts from the northern part of the Rift Zone of the Mid-Atlantic Ridge. J. Petr. 5, 409-434.

SMITH, R.E. (1968): Redistribution of major elements in the alteration of some basic lavas during burial metamorphism. J. Petr. 9, 191-219.

SUKHESWALA, R.N., POLDERVAART, A. (1958): Deccan basalts of the Bombay area, India. Bull. Geol. Soc. Amer. 69, 1475-1494.

VALLANCE, T.G. (1960): Concerning spilites. Proc. Linn. Soc. N.S.W. 85, 8-52.

VALLANCE, T.G. (1965): On the chemistry of pillow lavas and the origin of spilites. Miner. Mag. 34 (Tilley Vol.), 471-481.

VALLANCE, T.G. (1969a): Albitic mafic rocks of the Nundle District. Geology of New South Wales. J. Geol. Soc. Austr., 235-237.

VALLANCE, T.G. (1969b): Spilites again: some consequences of the degradation of basalts. Proc. Linn. Soc. N.S.W. 94, 8-51.

VALLANCE, T.G. (1969c): Recognition of specific magmatic character in some palaeozoic mafic lavas in New South Wales. Proc. Specialists' Symposia (Canberra, 1968). Geol. Soc. Austr., Spec. Publ. 2, 163-167.

VALLANCE, T.G. (1974): Spilitic degradation of a tholeiitic basalt. J. Petr. 15, in press.

YODER, H.S., Jr. (1967): Spilites and serpentinites. Carnegie-Inst., Ann. Rept. Director Geophys. Lab. 1965-1966, 269-279.

2. Papers Proposing a Primary Origin

A Reappraisal of the Textures and the Composition of the Spilites in the Permo-Carboniferous Verrucano of Glarus, Switzerland

G. C. Amstutz and A. M. Patwardhan

During the past twenty years many new papers on spilite-keratophyre occurrences have appeared. Except for some emphasis on secondary processes of alterations and some interesting papers on new occurrences of primary textures, very little new evidence was added to the previous work by BENSON (1915), P.A. NIGGLI (1952; BURRI and NIGGLI, 1945) and others. A new examination of the Glarus spilites and keratophyres may provide refined criteria on the mode of formation of spilites. Not only the textures which we consider to be primary magmatic, but also the existing clear transitions to rock portions which show secondary features, support the original interpretation of an essentially primary hydromagmatic origin of the spilites in this area.

An additional incentive for the present work was the isotopic results obtained on ^{13}C and ^{18}O from various localities of the Glarus spilites, especially the Matzlenstock (with the "carbonatitic" material)(SCHIDLOWSKI et al. 1970; SCHIDLOWSKI and STAHL, 1971; with additional analyses performed by STAHL for this paper).

First, the types of minerals occurring in these rocks are enumerated, classified according to their textural positions and listed according to decreasing order of abundance (cf. figures):

Albite

a) As bulk constituent, with more or less intergranular matrix, as one or more generations
1. without inclusions, in different generations,
2. with inclusions of opaques, "chlorite", or carbonates;

b) in amygdules, filling or rimming them;

c) as veinlets often grading into a breccia matrix (AMSTUTZ, 1954, Figs. 19, 21, 22; or 1968, Plate I, Type O_b to O_a).

Opaques

a) As distinct grains of medium size;
b) as matrix constituent (mostly as a fine dust);
c) as inclusions in albite;
d) as products of pseudomorphous decompositions of olivine and augite, or of opaques (see figures).

Chlorite

a) Disseminated in the matrix;
b) as possible inclusions in albite;
c) as tails in vesicles and in areas with albite without opaques;
d) as a constituent in the pseudomorphs after mafic minerals;
e) as occasional veinlets with albite or quartz, occasionally starting from amygdules.

Calcite

a) As matrix constituent;
b) as inclusions in albite;
c) as amygdules, breccia matrix or large masses (AMSTUTZ, 1954, Figs. 31-33);
d) as veinlets.

Sphene

Almost ubiquitous in areas without opaques; it has a distinct elongated shape and is often associated with chlorite;

a) intersertal and ophitic with albite;
b) stout subhedral crystals on the inner margins of amygdules.

Pseudomorphs of olivine and/or augite

These may be absent or occur in fair quantities (on the average 2-8 per mm^2; the average diameter being 0.2-0.6 mm).

The following discussion touches upon the essential textural patterns which bear on an interpretation of the origin.

Fluidic Textures; Bent and Undulose Albites; Matrix Inhomogeneities
―――

One of the most apparent features of the rocks described here is their amygdaloidal nature which confirms their eruptive character. Microscopically the fluidic texture indicates flowage; the bending of the albite laths around phenocrysts is especially characteristic (Fig. 1). The bent laths have transverse cracks and show undulose extinction. The bending and breaking of the plagioclases is frequent around vesicles. This flowage has also caused the schlieren-type or patchy distribution pattern of opaques and chlorite (Fig. 2). One can distinguish on a microscopic scale areas with a high concentration of opaques (a solid black matrix) from those with a moderate concentration of opaques (gray matrix); and at times no opaques are seen at all.

This is perhaps the place to clear up a point made by VALLANCE (1969, p.22, lines 2 to 18). First the term "magmatic differentiation" does not only

Fig. 1. Fluidic texture (in a more or less circular pattern) around a pseudomorph of (?) augite. The matrix consists of hematite, albite and chlorite. To the left is a large amygdaloidal patch of albite with some chlorite. Matzlenstock, Glarus; "carbonatitic phase". Magn. 50 x, II Nic.

Fig. 2. Another example of fluidic texture, but with various generations of albite, sometimes in diverging intergrowths, in an almost opaque matrix with some late albite and chlorite. Leglerhüttenkopf, Glarus. Magn. 50 x, II Nic.

include "liquid magmatic differentiation", e.g., gravity separation; the separation accumulation of phases and volatiles during all stages of magmatic evolution - all the way to the hydrothermal stage - can produce heterogeneities, i.e., rocks with contrasting mineral assemblages. What is shown in the scheme of AMSTUTZ (1968, Fig. 1) is nothing more than a very common feature in any hydrothermal or, as a matter of fact, deuteric phase; consequently we observe this separation around amygdules in each and every basaltic rock. Also, this separation is nothing but the consequence of the formation and presence of the hydrothermal stage and its phases. We include this stage in the term "magmatic"; to say that it does not exist or that it is not extensive would be to deny the process of crystallization of a melt at a hydrothermal stage; to assume that it always leaves a magma before affecting it would also be unrealistic. Therefore, VALLANCE's statement (lines 2 - 18 on page 22, 1969) must be due to a misunderstanding. In other words, this sort of separation is a feature present throughout the whole range of igneous petrology, especially where a relatively high concentration of volatiles is present.

The Albite-Chlorite Relationship

In a paper published in 1958 the following statement was made (AMSTUTZ, 1958, p.2): "The assumption of a primary spilitic rock family stands and falls with the assumption that certain silicate melts are able to retain enough volatiles to delay the main crystallization down to the pegmatitic-hydrothermal range, and thus develop hydromagmas and not only hydrothermal fluids".

The presence of the above-mentioned primary distribution patterns of minerals and the absence of crosscutting replacement features support the assumption of a hydromagmatic origin of the rocks described. But still more observations are available.

Numerous microlitic inclusions are commonly found in albite, occupying sometimes the primary cleavage planes in the plagioclases. Microprobe analyses of such inclusions in five albite phenocrysts showed the presence of Mg, Fe, Ti and traces of K in addition to Al and Si. Ca was conspicuously absent in all cases. Consequently, the colorless or sometimes yellowish-greenish inclusions can be chlorite or a mixture of albite, chlorite, etc., but cannot be pumpellyite or any other Ca mineral.

Other spaces which chlorite occupies are lenticular and, besides, contain albite, but no opaques; in addition, chlorite occurs in vesicles, alone or with carbonates and, of course, as a matrix constituent in variable proportions. In these fabric positions chlorite is neither secondary to any other ferromagnesian mineral, nor does it replace albite or any other mineral. The fact that it obviously on occasion does replace plagioclase (AMSTUTZ, 1954, Fig. 28) does not permit us to generalize, especially in view of the small space occupied by most inclusions in albite and of the fact that these inclusions are by no means always chlorite. It is much more probable that these inclusions are primary trapped interstitial "fluids", i.e., trapped portions of "rest melt" or rather rest fluids. LEHMANN (1965) has offered good petrographic evidence for the occurrence of non-metasomatic chlorites in comparable igneous rocks. From an experimental viewpoint, YODER (1967) substantiated the possibility of magmatic chlorites.

The microprobe study was made in order to elucidate the problem of material transfer. As it stands now, it is obvious that the introduction

of Mg, Fe and Ti into the albites would be an additional assumption, if
one accepts a secondary replacement theory. The way chlorite occurs
independently in these rocks indicates its primary nature.

Pseudomorph Structures

Pseudomorph structures are common in many rock phases with high or low
opaques. The often euhedral boundaries of such structures consist either
of opaques with fillings of calcite, quartz and plagioclase, or of a
whole pseudomorph filled with iron oxides, occasionally with a mixture
of iddingsite-bowlingite, and calcite. The fluidic orientation of
plagioclase laths referred to earlier is seen also around these pseudomorph outlines (Fig. 3) and confirms the primary textural relationship
of both, the mafic and the albite phenocrysts. Stout euhedral crystals
of iddingsite with limonite are often seen embedded in the dense opaque
matrix. In spots, they approach interpenetrating and subophitic fabric
with the albite laths.

Fig. 3. A porphyritic spilite with extreme generations of albite
attached in part on the faces of a mafic phenocryst now turned into
iron oxide and bowlingite-iddingsite. Note the large idiomorphic grains
of Fe-oxide. The matrix is albite with late chlorite. Unterkärpf,
Glarus. Magn. 45 x, II Nic.

The Albite-Carbonate Relationship

As reported above, four types of carbonate occur. Of these, types c and
d make up about 95% of the carbonate. Its composition was found to be
calcite. It should be repeated here that types b, c, and d are entirely
transitional. Also there are numerous transitions seen from type a to b.

a

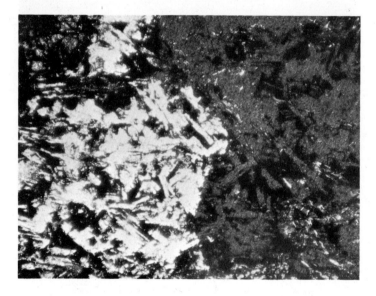

b

Fig. 4a and b. Entirely carbonatized portion of a marginal hematite-spilite variety of the Fukenstock, Glarus; a) without, b) with crossed Nicols. Note the fact that the carbonate grains incongruently replace many albite-crystals - a good example of the total absence of congruent replacement. Magn. 50 x

Consequently, we may conclude that all carbonates belong to the same species.

The textural relationships of the albite and carbonates are perhaps of genetic interest.

The form and abundance of the carbonate inclusions in albite suggest a true entrapping and not a decomposition, as already reported for the other colorless to greenish inclusions. Firstly there is no increase of carbonate in inner (presumably more calcic) zones; secondly, these carbonate inclusions are often connected with the other types of carbonates which are obviously tied to such primary cooling fabrics as amygdules and volcanic brecciation (Fig. Da, Oa, Ob in AMSTUTZ, 1968). There is no textural reason available to our knowledge to conclude that "direct replacement for albites on primary Ca-plagioclase sites occur as pseudomorphs, even to the extent of inheriting twinning characters" (VALLANCE, 1969, p.31). We would prefer explicitly, not to assume any later process such as the exact replacement of such a large amount of material (often 50 - 70, in places up to 90% of the bulk of the rock) without any direct evidence such as patchy relicts, zoning of inclusions, proper proportions of secondary carbonates, transitions to rocks with calcic plagioclase, etc. To illustrate our point we refer to Fig. 4a and b where replacement fabric of carbonates pseudomorphing albites is shown. It is seen by comparing the two figures that the fabric of the albite boundaries with respect to the opaque matrix is entirely different from the fabric of the replacing carbonate. The outlines of the carbonate grains become clear in the crossed Nicols position of this area in the rock. We suppose that the significance of examining congruency of minerals emphasized by us in this paper is to a great extent substantiated with this excellent example.

Along some of the borders or rims of carbonate accumulations of types b, c, and d, the albite is somewhat different from its common equivalent within the major rock mass (Fig. 5). Along these rims it often occurs

Fig. 5. Fluidic hematite-spilite with irregular amygdaloidal patches of calcite rimmed by chlorite. Matzlenstock, Glarus; "carbonatitic phase". Magn. 50 x, II Nic.

without inclusions and perhaps more often without twinning. We are evidently dealing here with an albite formed at a later stage, as pictured in the paragenetic diagram. At this time, the frequently observed albite-filled amygdules, veinlets and breccia matrix - all consisting of massive albite with or without chlorite and calcite - should be mentioned again (AMSTUTZ, 1954, Figs. 21 and 22). The congruency of this albite distribution does not permit one to assume any secondary or even primary process of metasomatism. The only simple explanation is that of a formation by primary crystallization differentiation of these - as also all other - mineral distributions in primary fabric elements. This fact has been summarized in an earlier paper in the "table of congruencies" (AMSTUTZ, 1968).

MERILÄINEN (1961, p.63) and PIISPANEN (1972, p.58-61) also propose magmatic origin of carbonates in the spilites from Finland and further have drawn analogy for this albite-carbonate assemblage in spilites (albitites) from the same assemblage in some carbonatites. The latter author has incorporated a couple of $\delta^{13}C$ results on the calcites in support of the proposal.

The Carbonates in the Light of Isotopic Ratios

On the suggestion of the first author, $\delta^{13}C$ and $\delta^{18}O$ values were obtained of various carbonates in the Glarus spilites in 1969/70. Dr. STAHL was kind enough to make some more and they are here reported in a new graph (Fig. 6). The new values were obtained from samples collected by the first author in 1970 and they confirm accurately the previous results. The plot of the present spilites in Fig. 6 is interesting in so far as the points fall within a narrow field. The fields of variation of $\delta^{13}C$ and $\delta^{18}O$ in carbonatites as given by two different authors fall towards the left of the Glarus spilites in this

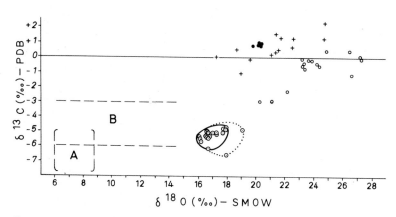

Fig. 6

A field of carbonatites (TAYLOR et al., 1967, p. 423)
B field of carbonatites (DEINES 1970, p. 1220)
⊙ Glarus spilites, Switzerland
+ Schalstein and Weilburgites of Lahn-Dill area, Germany
● Lochseitenkalk (Sedimentary Carbonate), Switzerland
○ Spilites of the Aroser Zone, Switzerland

diagram. At present it is enough to visualize that the values for these spilites differ from those of the sedimentary carbonates from the schalstein of the Lahn-Dill area as well as from the lochseitenkalk of Glarus and fall into the general field of hydrothermal carbonates (SCHOELL and STAHL, 1972; SCHIDLOWSKI and STAHL, 1971). This may be taken as an indication of the primary magmatic nature of these carbonates as already proposed by AMSTUTZ in 1954 on the ground of textural evidence. A careful examination of the textures (e.g., Figs. 19, 20, 29, 30 to 33 in AMSTUTZ, 1954) of the carbonates leads one to the conclusion that the carbonates were formed in and with the final hydromagmatic or hydrothermal crystallization fractionation of the rocks.

Conclusion

In conclusion, one may simply say that the mere co-existence of albite with carbonates and/or chlorite is not per se proof of a secondary nature of any of these minerals or of the assemblage as a whole. In other words, we do not believe that a conclusive genetic interpretation can do without textural criteria. The textural criteria have to match the mineralogic evidence when deriving a genetic interpretation. In our case, the congruence of the primary spaces provided by amygdules, breccia matrix, etc. with the abundance of carbonates and chlorite and their scarcity within the main undisturbed rock mass leads to the simplified paragenetic separation scheme of Fig. 7 which is here repeated for its simplicity and importance.

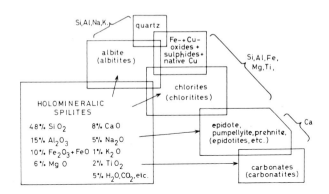

Fig. 7. Differentiation scheme of spilitic rocks

As to the physicochemical problem, we see no reason for assuming that the experimental work has covered the whole range of possibilities in as much as the same assemblage, albite, carbonates and chlorite, is an ubiquitous gangue mineral assemblage in many known hydrothermal vein deposits. In other words, we obviously are dealing with a case of missing experiments and not with a case of impossible phase relations. Chlorites, albite and carbonates are stable over a wide range of conditions - as seen in many natural rocks - and there is no reason why the hydrothermal or hydromagmatic stage of magmatic crystallization should not pass through a low temperature stage and pH-Eh conditions comparable to those existing during low grade metamorphism or even diagenesis (or so-called burial metamorphism). The mineralogical convergence is no reason for assuming a one-track genesis. On the contrary, we propose that this mineralogical convergence - as many other convergences in (our) science - must be overcome by more evidence, clearer criteria and arguments.

We hope to have shown in the very brief review of the Glarus spilites that textural criteria must be used to get out of the spilite dilemma (just as in the case of the granite or the ore genesis dilemma). Such a high degree of congruence as seen between the mineral distribution and the primary fabric elements of the rocks cannot be explained by secondary processes. There is no such thing as a perfect mimetic replacement.

References

AMSTUTZ, G.C. (1954): Geologie und Petrographie der Ergußgesteine im Verrucano des Glarner Freiberges. Publ. No. 5, Vulkaninstitut Immanuel Friedlaender, Zürich, 150.

AMSTUTZ, G.C. (1958): Spilitic rocks and mineral deposits. Bull. Missouri School of Mines, Tech. Ser. no. 96, 11.

AMSTUTZ, G.C. (1968): Spilites and spilitic rocks. In: Basalts: The Poldervaart Treatise on rocks of basaltic composition, H.H. Hess and A. Poldervaart, Ed. Interscience (Wiley), New York, 737-753.

BENSON, W.N. (1915): The dolerites, spilites and keratophyres of the Nundle District. Proc. Linnean Soc. N.S. Wales 40, 121-173.

BURRI, C., NIGGLI, P. (1945): Die jungen Eruptivgesteine des mediterranen Orogens. Publ. No. 3, Vulkaninstitut I. Friedlaender, Zürich, Teil I, 578.

DEINES, P. (1970): The carbon and oxygen isotopic composition of carbonates from the Oka carbonatite complex, Quebec, Canada. Geochim. Cosmochim. Acta 34, 1199-1225.

LEHMANN, E. (1965): Non-metasomatic chlorite in igneous rocks. Geol. Mag. 102, 24-35.

MERILÄINEN, K.(1961): Albite diabases and albitites in Enontekiö and Kittilä, Finland. Bull. Comm. Geol. Finland 195, 1-75.

NIGGLI, P. (1952): The chemistry of the Keweenawan lavas. Jour. Sci., Bowen Volume, 381-412.

PIISPAINEN, R. (1972): On the spilitic rocks of the Karelidic belt in Western Kuusamo, Northeastern Finland. Acta Univ. Ouluensis, Ser. A, Geologica no. 2, 1-73.

SCHIDLOWSKI, M., STAHL, W., AMSTUTZ, G.C. (1970): Oxygen and carbon isotope abundances in carbonates of spilitic rocks from Glarus, Switzerland. Die Naturwissenschaften 57, 542-543.

SCHIDLOWSKI, M., STAHL, W. (1971): Kohlenstoff- und Sauerstoff-Isotopenuntersuchungen an der Karbonatfraktion alpiner Spilite und Serpentinite sowie von Weilburgiten des Lahn-Dill-Gebietes. N. Jb. Miner. Abh. 115, 252-278.

SCHOELL, M., STAHL, W. (1972): Sauerstoff- und Kohlenstoff-Isotopenuntersuchungen an Karbonaten hydrothermaler Lagerstätten. Mitt. Labor f. stabile Isotope Nr. STI 2/72, Bundesanstalt f. Bodenforschung, 22.

TAYLOR, H.P., FRECHEN, J., DEGENS, E.T. (1967): Oxygen and carbon isotope studies of carbonatites from Laacher See District, West

Germany, and Alnö District, Sweden. Geochim. Cosmochim. Acta <u>31</u>, 407-430.

VALLANCE, T.G. (1969): Spilites again: Some consequences of the degradation of basalts. Proc. Linnean Soc. N.S.W. <u>94</u>, 8-51.

YODER, H.S., Jr. (1967): Spilites and serpentinites. Yb. Carnegie Inst. Washington <u>65</u>, 269-279.

A Series of Magmatism Related to the Formation of Spilite

T. Bamba

Introduction

Petrological study on spilite in Japan has been in progress since 1946 by SUZUKI and his co-workers. In 1954, he pointed out that the green rock layers of Jurassic age found in central Hokkaido are occasionally composed of spilitized pillow lavas. The following is a short quotation from SUZUKI (1954b).

"The diabasic pillow lavas found only in the Sorachi Series[1] are made up of numerous separated rounded bodies of compact diabasic rock being filled with fine rude fragments. The bodies take pillow, ellipsoidal, prolate, or spheroidal forms ranging in diameter from 0.3 to 1 m. They show a fine-grained and commonly ophitic texture composed of plagioclase and pyroxene, and the outer margin of each mass is more or less glassy. The rounded masses are frequently cut by radial joints. Partial alteration due to albitization or spilitization may be observable in the rock. The spilitization seems to be caused by auto metasomatism of the lava".

Since that time, discussions on spilite in Japan have continued; the origin of Na in spilite is considered one of the fundamental problems. At all times, the view based on metasomatism proposed by BARTH (1939) seems to be supported in Japan.

Along the Japanese island arc, several geotectonic belts developed from Paleozoic or Mesozoic geosynclines are known. However, the major belts developed in Honshu, Shikoku, and Kyushu are generally affected by regional metamorphism and the green rock layers in the belts are often changed into green schist or amphibolite: this makes the petrological study of the original rocks difficult. On the other hand, in the Hidaka mountains, central Hokkaido, especially in the piedmont of the range, a thick pile of green pyroclastic layers including pillow lava and normal sediments has remained unmetamorphosed except in some restricted metamorphic blocks. Typical pillow lavas, diabasic tuffs with intercalations of chert, limestone, and normal sediments such as slate or sandstone are observed there. Thus the piedmont around the Hidaka mountains offers one of the most excellent fields for the petrological study of a series of volcanic rocks such as diabasic tuff, pillow lava, and intrusive diabase.

Recent investigations in this area suggest that the spilitized facies is confined to several restricted areas which are characterized by intrusions of alkali diabase (BAMBA and MAEDA, 1969).

A model of volcanism leading to the formation of spilite will be presented here based on magmatism in the Hidaka mountains of late Jurassic to Tertiary period.

---

[1] Geosynclinal sediments characterized by green pyroclastic layers of late Jurassic to early Cretaceous period in central Hokkaido.

Three Geotectonic Units of Central Hokkaido

Geosynclinal sediments of the Hidaka mountains, central Hokkaido, are composed of argillaceous rocks and basic pyroclastic layers of Mesozoic era, mostly those of Jura-Cretaceous. These sediments are generally monotonous, but in the upper half of the succession basic pyroclastic layers such as diabasic tuff or pillow lava are rather prominent. The sediments characterized by basic rock layers are called the Sorachi Series, and the monotonous lower member is called the Hidaka System.

They were folded in late Cretaceous to early Tertiary times as the result of the Alpine orogeny and three characteristic geotectonic belts were formed. The central belt, including a metamorphic zone, is called the Hidaka belt, and forms the geologic backbone of Hokkaido (Fig. 1). The eastern part of the belt is composed mainly of diabasic tuff, pillow lava with intercalations of chert accumulated in the geosynclinal phase.

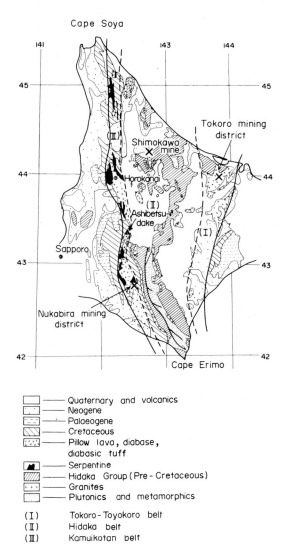

Fig. 1. Three geotectonic belts in central Hokkaido, Japan

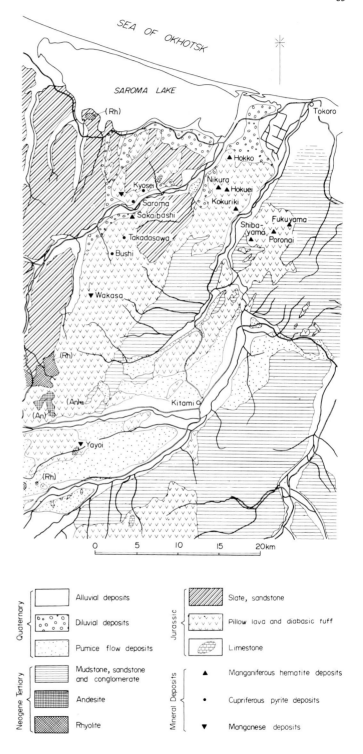

Fig. 2. Geological map of the Tokoro metallogenic province, Japan

Its geologic characters are very different from those of the Hidaka belt proper; thus the eastern part of the Hidaka belt is designated the Tokoro-Toyokoro belt (MINATO et al. 1965, HUNAHASHI, 1957).

On the western side of the Hidaka belt, another geotectonic belt which is characterized by intrusion of serpentine and the presence of various kinds of quartz schist runs parallel to the former. This belt shows features in remarkable contrast to those of the former, and is called the Kamuikotan belt or Kamuikotan metamorphic belt.

Migmatite, gneiss, and hornfels were formed along the axis of the Hidaka mountains. In addition remarkable igneous rocks of syn- or late-kinematic stage were intruded into the above metamorphic rocks. These igneous rocks may be classified into two main rock series; one is gabbroic and the other is granitic. The former is more predominant than the latter, and in general they show distinct schistosities.

Various kinds of mineral deposits related to respective igneous episodes are known. Some of them are very valuable, e.g. the cupriferous pyrite deposit of the Shimokawa mine in the northern part of the Hidaka belt, manganiferous hematite deposits of the Tokoro mining district in the Tokoro-Toyokoro belt, and chromite ore deposits related to serpentine in the Kamuikotan belt. Above all, it has been suggested that the genesis of manganiferous hematite deposits of the Tokoro mining district is closely related to spilitized pillow lava (BAMBA and SAWA, 1967). Therefore the relation between spilitization and mineralization in this district will be touched upon at first in this paper.

Pillow Lava and Spilitization in the Tokoro Mining District

Occurrence of Spilitized Pillow Lava

The Tokoro mining district is situated in northeast Hokkaido. The rocks of the district are thick sediments of Jurassic age and overlying Cenozoic formations. The former consist of slate, sandstone and green pyroclastic rocks intercalated with pillow lavas and chert. The latter are thick marine sediments composed of tuff, mudstone and sand or siltstone of Neogene Tertiary. Large pumice flow deposits of Quaternary age form several hills (Fig. 2).

The strike of the Jurassic formations generally shows a N-E trend; but an E-W trend is found in a mineralized zone (Fig. 3) which is characterized by the presence of thick chert layers. Numerous fractures, and intrusions of ophitic diabase are also noticed in this zone. In addition, manganiferous hematite deposits are found mainly within the zone and the horizon of mineralization is restricted to the boundary between chert and spilitized pillow lava. This zone may be correlated with the deeper member of the geosynclinal sediments, because older amphibolite is frequently found in this zone which is regarded to be up-heaved.

The whole Jurassic system of this district is estimated to be about 6,000 m thick, being composed mostly of green pyroclastic layers such as pillow lava and diabasic tuff with intercalations of chert. The rest consists of slate, limestone and sandstone (ISHIDA et al. 1968).

The typical mode of occurrence of pillow lava in this district is shown in Plate I-1. Pillow lavas are made up of numerous rounded bodies of

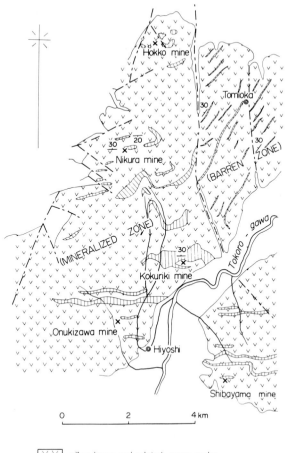

Fig. 3. Tectonic map of the Tokoro mining district, Japan

compact basaltic or diabasic rock. The bodies assume pillow, ellipsoidal, prolate, or spheroidal forms, as noted by SUZUKI (1954b, 1959).

Spilitized pillow lavas occur in limited amount in the mineralized zone, especially in the foot-wall of the manganiferous hematite deposits.

Three Rock Facies Observed in the Pillow Lavas

Petrologically pillow lavas under the microscope can be classified according to the following three facies:
1. glassy facies, 2. variolitic facies, 3. subophitic facies.

The grade of crystallinity varies from glassy to holocrystalline. Among them, the variolitic facies carrying a considerable amount of glass is predominant, and is characterized by the presence of numerous varioles composed of radial aggregates of fibrous plagioclase. The three facies noted above, however, show a continuous gradation between the variolitic facies and the glassy or subophitic one.

1. Glassy Facies (Plate I-2). This is made up of brownish glass with irregular cracks. Occasionally, tortoise-shell structure caused by the development of hexagonal cracks is observed, the areas bounded by the cracks being about 0.3 mm in diameter. Sometimes small numbers of fibrous or embryonic crystallites of plagioclase can be seen under the microscope.

2. Variolitic Facies (Plate I-3 and 4). This is composed of plagioclase and clinopyroxene, with or without a small amount of glass, showing a characteristic texture as presented in Plate I-3 and 4. The dimensions of varioles are commonly about 0.1 mm, rarely up to 0.3 mm. Clinopyroxene is very fine-grained and usually forms cedar-leaf aggregates as shown in Plate I-3. Plagioclase is fibrous or acicular, and commonly occurs in the form of varioles as shown in Plate I-4. Spilitized pillow lavas found in the northern part of the mineralized zone have suffered albitization with subordinate chloritization, as shown in Plate I-5.

3. Subophitic Facies (Plate I-6). Clinopyroxene and plagioclase are the main components associated with a small amount of ilmenite. Clinopyroxene and ilmenite are granular and 0.1 mm in size. Plagioclase is long prismatic and 0.2 mm in length. Accordingly the texture remains subophitic, as shown in Plate I-6.

The three rock facies described above are commonly recognized in an individual pillow. Their distribution therein is, however, irregular and any concentric arrangement is obscure.

Keratophyre

A small amount of keratophyre is associated with pillow lavas in this district, showing a flame-like shape or irregular network. However, the identification of keratophyre in the field is difficult because of its fine-grained appearance and greenish-grey color.

Judging from the mode of occurrence, the keratophyre is considered a kind of segregative vein. Prismatic plagioclase (0.1 - 0.3 mm) is a major component, arranged in subparallel rows as shown in Plate II-7. Subordinate hornblende and chlorite are found as accessory minerals.

PLATE I. Pillow lavas from the Tokoro mining district, Tokoro-Toyokoro belt, Hokkaido, Japan.
1. Occurrence of spilitized pillow lava in the Kokuriki mine. (2-6) Photomicrographs of thin sections of the three rock facies of the pillow lavas. 2. Glassy facies showing tortoise-shell texture (under parallel nicols). gl: glass, pl: plagioclase. 3. Variolitic facies showing cedar-leaf texture composed of clinopyroxene and plagioclase (under crossed nicols). 4. Variolitic facies composed of unaltered prismatic plagioclase and glass (under parallel nicols). 5. Spilite showing variolitic texture composed of albitized plagioclase and glass (under parallel nicols). 6. Subophitic facies composed of unaltered prismatic plagioclase and clinopyroxene (under parallel nicols)

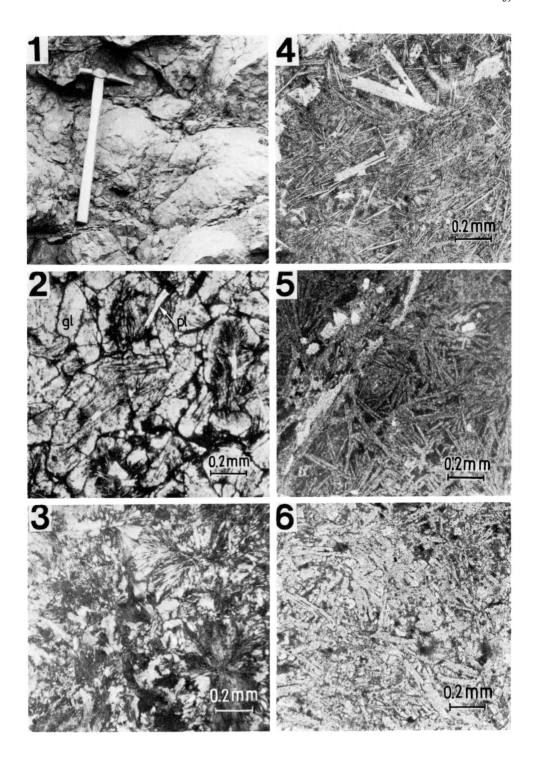

Chemical Composition

Typical examples of the three rock facies of pillow lava, keratophyre and spilitized rocks were selected for the chemical analysis.

Their chemical compositions and norms are given in Tables 1 and 2.

It is noticeable that the chemical compositions of the three rock facies are much alike, in spite of their differences in texture, except that the variolitic facies is comparatively rich in K_2O and the glassy facies is poor in MgO. On the other hand spilitized pillow lavas are notably rich in Na_2O and H_2O in all three rock facies. The increase or decrease of each element as a result of spilitization is given in Fig. 4.

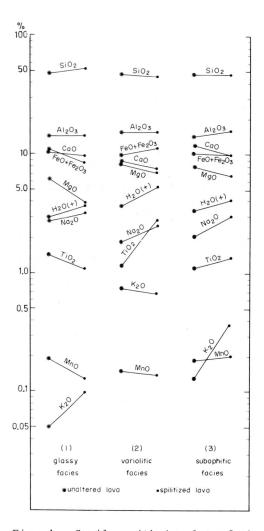

Fig. 4. Semilogarithmic plot of the chemical variations between unaltered pillow lavas and spilitized ones from the Tokoro mining district, Japan

Table 1. Chemical composition of pillow lavas and keratophyre from the Tokoro mining district, Hokkaido, Japan

	(1)	(2)	(3)	(4)	(5)	(6)	(7)
SiO_2	48.88	47.40	47.36	52.94	45.20	47.76	66.10
TiO_2	1.48	1.14	1.10	1.13	2.83	1.39	0.36
Al_2O_3	14.56	15.35	14.22	14.87	15.15	15.90	16.55
Fe_2O_3	3.55	3.11	3.88	7.11	6.84	4.67	4.65
FeO	7.52	6.29	6.78	1.75	4.39	5.13	0.78
MnO	0.19	0.15	0.19	0.13	0.14	0.20	0.06
MgO	6.31	8.53	8.07	4.08	7.27	6.87	0.56
CaO	10.80	11.18	12.05	9.85	7.53	9.81	3.01
Na_2O	2.82	1.82	2.05	3.30	2.61	3.04	6.19
K_2O	0.05	0.75	0.13	0.10	0.69	0.37	0.01
P_2O_5	0.08	0.08	0.08	0.16	0.49	0.12	0.06
$H_2O(+)$	3.06	3.70	3.52	3.80	5.44	4.16	1.40
$H_2O(-)$	0.42	0.44	0.40	0.58	1.36	0.52	0.26
Total	99.72	99.94	99.83	99.80	99.94	99.94	99.99

(1)-(3): fresh pillow lavas from the Kokuriki mine
(1): glassy facies from the outer shell of a unit pillow
(2): variolite facies from the core of a unit pillow
(3): subophitic facies from the core of a unit pillow (BAMBA and SAWA, 1967

(4)-(6): altered pillow lavas from the Hokko mine
(4): glassy facies from the outer shell of a unit pillow
(5): variolite facies from the core of a unit pillow
(6): subophitic facies from the core of a unit pillow (BAMBA and MAEDA, 196

(7): keratophyre from the Tomioka area, Tokoro district (BAMBA and SAWA, 1967)

(Analysts: K. MAEDA, K. MAEDA and M. KAWANO)

Table 2. Norms of pillow lavas and keratophyre from the Tokoro mining district, Hokkaido, Japan

	(1)	(2)	(3)	(4)	(5)	(6)	(7)
Q	1.60	-	0.23	11.43	3.14	0.21	22.92
or	0.30	4.43	0.77	0.59	4.08	2.19	0.06
ab	23.86	15.40	17.35	27.92	22.09	25.72	52.38
an	26.93	31.52	29.22	25.46	27.59	28.65	14.59
c	-	-	-	-	-	-	1.01
wo	10.94	9.82	12.58	9.38	2.88	8.06	-
en	15.71	19.10	20.10	10.16	18.10	17.11	1.39
fs	8.78	6.61	7.79	-	-	3.64	-
fo	-	1.50	-	-	-	-	-
fa	-	0.58	-	-	-	-	-
mt	5.15	4.51	5.62	2.77	6.42	6.77	1.68
il	2.81	2.17	2.09	2.15	5.37	2.64	0.68
hm	-	-	-	5.20	2.41	-	3.49
ap	0.17	0.17	0.17	0.35	1.07	0.26	0.13

The unaltered pillow lavas in this district are clinopyroxene basalt or dolerite in mode; chemically, they are poor in Al_2O_3 and K_2O, and rich in CaO and total iron oxides. Based on the mode and chemical composition, the primary magma from which the pillow lavas were derived is regarded as tholeiitic.

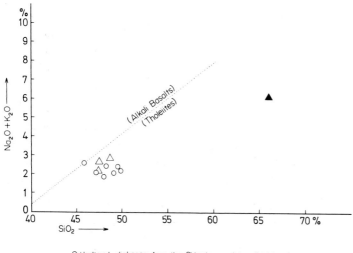

Fig. 5. Alkali-silica variations of unaltered pillow lavas from the Tokoro mining district and those of diabases from the Shimokawa mining district, Japan

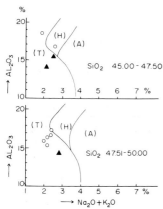

Fig. 6. Al_2O_3-alkali-SiO_2 relationship of the three basalt types from Japan (KUNO, 1960) and the plots of unaltered pillow lavas from the Tokoro mining district and of diabases from the Shimokawa mining district, Japan

Unaltered pillow lavas from the Tokoro mining district are plotted in the tholeiitic fields of either alkali-silica variation (YODER, 1967) or Al_2O_3-alkali-SiO_2 relation (KUNO, 1960), as shown in Figs. 5 and 6. In FMA diagrams illustrating tholeiitic series and calc-alkali series proposed by NOCKOLDS (1954), pillow lavas from the Tokoro mining district show the same trend as the tholeiitic series (Fig. 7).

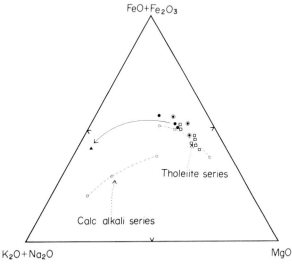

Fig. 7. FMA diagram showing unaltered pillow lavas from the Tokoro mining district and diabases from the Shimokawa mining district, as compared with NOCKOLDS' (1954) average

⊙ Unaltered pillow lavas from the Tokoro district
● Albitized pillow lavas from the Tokoro district
▲ Keratophyre from the Tokoro district
□ Diabases from the Shimokawa mining district
○ Nockolds's average of tholeiite and calc alkali series

Table 3. Chemical composition of ophitic diabase from the Kokuriki mine, Hokkaido, Japan (BAMBA and SAWA, 1967)

SiO_2	45.22
TiO_2	3.00
Al_2O_3	15.42
Fe_2O_3	5.77
FeO	6.03
MnO	0.19
MgO	6.06
CaO	7.03
Na_2O	2.73
K_2O	1.21
P_2O_5	0.41
$H_2O(+)$	4.96
$H_2O(-)$	1.88
Total	99.91

(Analysts: K. MAEDA and M. KAWANO)

With respect to the normative feldspar diagram proposed by YODER (1967), the altered pillow lavas from this district fall in the spilite field, though unaltered ones do not fall in this field (Fig. 8).

Titaniferous Augite-Bearing Ophitic Diabase

Several dikes of coarse-grained ophitic diabase are found along sheared zones or cracks in the mineralized zone. These dikes are composed of plagioclase, clinopyroxene and ilmenite. Their texture is typically ophitic as shown in Plate II-8. The shape of plagioclase is prismatic and it is occasionally replaced by tiny sericites. Clinopyroxene is brownish-purple in color and is regarded as titaniferous augite. Marginal parts of the crystals are often replaced by chlorite.

The chemical composition of the ophitic diabase from the Kokuriki mine is given in Table 3. It is more alkalic, and especially rich in K_2O, compared with the pillow lavas. This rock falls in an alkali rock series.

It is noteworthy that the final magmatism in this district is characterized by intrusion of alkali diabase.

Spilitization and Mineralization

WATANABE (1965) pointed out that the relation between submarine volcanism and mineralization in the geosynclinal basin is one of the most fundamental problems in ore genesis. Manganiferous hematite deposits and overlying cherts should be examined in connection with the genetic study of spilite, because they are closely related to spilitized pillow lava.

Radiolarian chert covering the bedded manganiferous hematite deposits is of considerable thickness, sometimes attaining 100 m. At the bottom of the chert, ferruginous quartz rock occurs occasionally as irregular masses.

Manganiferous hematite ore of the Tokoro mining district is developed between radiolarian chert and spilitized pillow lava. The ore is massive and composed mainly of hematite with subordinate penwithite.

The mode of occurrence supports the view that the ferruginous quartz rock is replaced by hematite ore, viz. the ores occupy the same horizon, and sometimes relicts of ferruginous quartz rock remain within the ore.

A photomicrograph of ferruginous quartz rock is shown in Plate II-9 and that of low-grade ore derived from the rock is shown in Plate II-10. Both contain about 10% iron.

It is considered that this process of mineralization resulted in the formation of workable ore, which is mainly composed of hematite and penwithite containing more than 30% Fe and 10% Mn.

The hypabyssal ophitic diabase found in the mineralized zone is alkaline in composition, different from the older pillow lavas. The hypabyssal ophitic diabase represents the latest phase of the magmatism, and a considerable amount of alkalis may have remained in the hydrothermal solution at the deuteric stage of the intrusion.

The present author came to the conclusion that the pillow lavas in the mineralized zone might have been albitized by hydrothermal solutions derived from the alkaline diabase, and that the ferruginous quartz rock overlying the pillow lava was also mineralized by hydrothermal processes at the same time.

As RÖSLER (1962) suggested in his study of iron ore deposits of the Ostthüringen district in Germany, spilitization of pillow lava and the related iron mineralization both proceeded through the effects of hydrothermal solution caused by diabase intrusions during the latest phase of magmatism in this district.

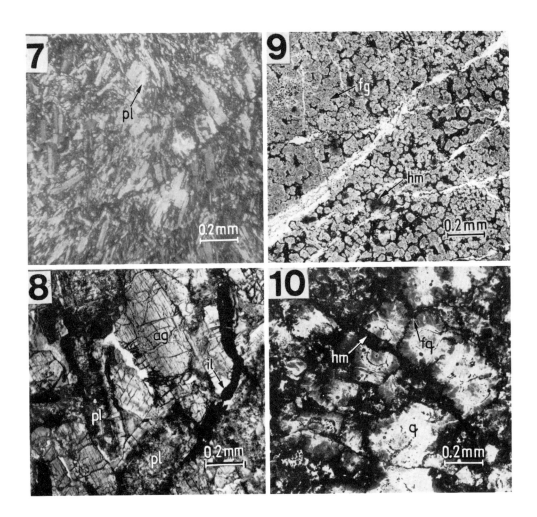

PLATE II. Photomicrographs of thin sections of keratophyre, alkali diabase and ferruginous quartz rock from the Tokoro mining district. 7. Keratophyre. pl: plagioclase (under crossed nicols). 8. Titaniferous augite-bearing ophitic diabase (under parallel nicols). ag: titaniferous augite, pl: sericitized plagioclase, il: ilmenite. 9. Ferruginous quartz rock (under parallel nicols). fg: ferruginous quartz, hm: hematite. 10. Altered ferruginous quartz rock (under parallel nicols). q: quartz, fq: ferruginous quartz, hm: hematite

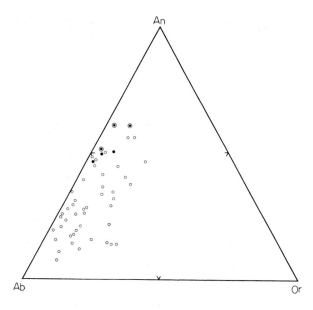

Fig. 8. Normative feldspar of spilites, compiled by YODER (1967) and plots of normative feldspar of pillow lavas from the Tokoro mining district, Japan

Diabase Swarm in the Shimokawa Mining District

Mode of Occurrence

Though the Shimokawa mining district belongs to the axial zone of the Hidaka belt, the plutonic massif represented by gabbros and granites is quite different from that in southern part of the belt, as shown in Fig. 1.

The massif is small and localized there, and cordierite hornfels is found only locally around the massif as metamorphic aureole.

In this district, thick sediments composed of slate and sandstone of Jurassic age remain mostly in an unmetamorphosed state. The sediments are correlated stratigraphically with a deeper horizon than that of the Sorachi Series developed in the Tokoro-Toyokoro belt. The monotonous sediments described above are called Hidaka System. A remarkable longitudinal swarm of diabase bodies associated with lenticular serpentine running with N-S trend is noticed here.

The diabase swarm is restricted to the junction between the two geotectonic units. The eastern unit is characterized by a N-E strike, and the western unit by the contrasting N-W trend. In both units, small massifs of gabbros associated with metamorphic aureoles are found.

The diabase bodies are commonly conformable to the strata dipping steeply to the east. Numerous lenticular bodies of diabase form a swarm within a narrow belt.

The width of the diabase swarm reaches about 1.6 km in the broadest part, and the length about 20 km. Recently it has been demonstrated that the diabase swarm is controlled by echélon structure which is separated into three or four units (Fig. 9).

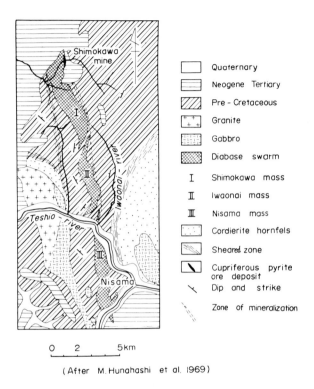

Fig. 9. Geological outline of the Shimokawa mining district, northern Hidaka belt, Hokkaido, Japan

(After M. Hunahashi et al. 1969)

Typical pillow structure is found within the diabase swarm, though it occupies only less than 5% of the whole diabase mass.

Neither alkali diabase dikes nor diabasic tuffs with intercalation of chert such as those known in the Tokoro mining district are found here.

Noteworthy cupriferous pyrite ore deposits are developed along the boundary between the black slate and diabase in the sheared zone in the eastern side of the diabase swarm. The slate plays the role of hanging wall. The ore, containing about 2.5% copper, has been worked since 1942, and total ore production has reached more than one million metric tons.

Discussion on the Origin of the Diabase Swarm and the Genesis of Ore

Origin of the diabase swarm of the Shimokawa mining district has been discussed by MIYAKE (1965) and HUNAHASHI (1969), whose opposing views are given below. First a short note is quoted from MIYAKE (1965):

"The layered cupriferous pyrite deposits of the Shimokawa mine have a characteristically simple geological structure, where the foot-wall consists of thick spilitic and diabasic rocks with numerous thin intercalations of greywackes and slaty rock, and the hanging wall consists of slaty rocks.

These diabasic rocks were previously regarded as dike swarms intruding the Hidaka System at high angles, and the deposits as epigenetic ones having replaced the sheared slaty rocks near the boundary with the basic rocks.

Recent underground and surface geological studies have revealed that some of the basic rocks are subaqueous lava flows with a pillow structure, and the ore deposits rest on the uppermost spilite flows intercalated with thin sedimentary layers. Immediately after the end of the ore deposits, when no sedimentary cover was existent yet, diabase sheets intruded between ore body and the foot-wall, often branching into the ore body forming a distinct chilled margin.

These facts clearly indicate that the deposits are one of the most typical syngenetic sedimentary exhalative deposits originated from a spilitic magma".

On the other hand, HUNAHASHI (1969) objected to MIYAKE's view as follows:

"The diabase swarms in the Shimokawa mining district should be regarded as a kind of hypabyssal dike because they have restricted occurrence controlled by the geotectonics during the initial geosynclinal or initial orogenic phase. Especially echélon structure of the diabase is a peculiar feature related to these intrusive bodies.

Though pillow structure of the diabase appears to be a product of subaqueous lava, it is difficult to identify it as true pillow lava. Since it is commonly found only along the sheared zone, the pillow form of the diabase must be a kind of deformation during syn- or post-intrusion phase. In addition, they cannot find intercalation of chert layer here.

Cupriferous pyrite deposits of the Shimokawa mine developed along the boundary between the diabase swarm and the black slate must be typical hydrothermal deposits of epigenetic origin, because distinguished wall-rock alteration is recognized and a series of mineralization from high temperature to low temperature phase is noticed".

PLATE III. Diabases from the Shimokawa mining district, Hidaka belt, Hokkaido, Japan.
11. Diabase taking pillow form occurred in a gallery of the Shimokawa mine. Outer shell is glassy and interior is variolitic. (12-16) Photomicrographs of thin sections of diabases. 12. Glassy outer shell of the pillow given above. q: fine-grained quartz, pl: plagioclase, gl: glass (under parallel nicols). 13. Variolitic facies occupying the interior part of the pillow shown above. pl: plagioclase, px (gl): glass or clinopyroxene (under parallel nicols). 14. Altered one showing variolitic texture from the Nisama diabase mass. pl: plagioclase subjected albitization and sericitization (under parallel nicols). 15. Variolitic-subophitic textured diabase taking pillow form. pl: plagioclase, px: aggregation of clinopyroxenes (under parallel nicols). 16. Ophitic-textured diabase, which is the most predominant in the Shimokawa diabase swarms. pl: plagioclase, px: augite (under crossed nicols)

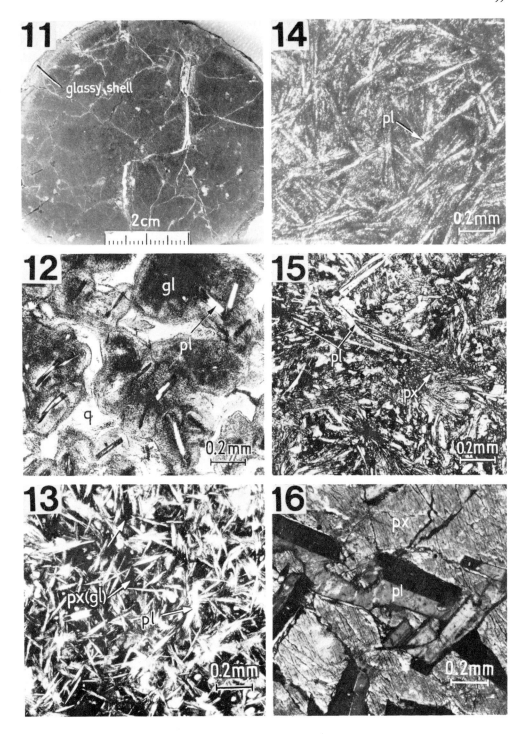

As noted above, confusion concerning the origin of diabase and the genesis of ore in the Shimokawa mine will be inevitable for some time to come between syngenetic and epigenetic theories.

Petrological Properties of the Diabase

The dimensions of individual pillows of diabase from the Shimokawa mine vary generally from 5 cm to 50 cm. The interstices between the rounded bodies are filled by tuffaceous materials containing a considerable amount of quartz. The outer shell of the rounded body is composed of glassy material, about 2 mm thick, as shown in Plate III-11. Microscopically the outer shell is composed of brownish glass including embryos of plagioclase and microcrystalline quartz (Plate III-12), while the interior of the body is composed of plagioclase and clinopyroxene or glass (Plate III-13). Variolitic texture illustrated in Plate III-15 is observed in this part, but no ophitic texture is seen in any pillow or rounded body from the Shimokawa mining district.

Previously the rounded body from the Shimokawa mine was called spilite, but today use of "spilite" for the rock is impertinent because their

Table 4. Chemical compostion of diabases from the Shimokawa mining district, Hokkaido, Japan

	(1)	(2)	(3)	(4)	(5)	(6)	(7)
SiO_2	48.04	49.78	49.02	45.57	47.44	49.75	48.26
TiO_2	1.12	1.37	1.15	1.07	1.15	0.98	1.34
Al_2O_3	16.04	16.54	15.38	16.79	18.68	16.88	17.60
Fe_2O_3	2.52	2.28	2.56	1.88	1.32	1.50	1.69
FeO	6.14	6.43	6.03	7.34	6.72	7.45	7.50
MnO	0.19	0.13	0.16	0.10	0.12	0.08	0.10
MgO	7.91	7.97	7.80	9.13	8.44	6.03	6.63
CaO	12.20	10.06	12.16	11.74	11.26	11.62	11.14
Na_2O	1.82	2.10	1.83	2.24	1.83	2.21	2.24
K_2O	0.17	0.18	0.25	0.30	0.21	0.23	0.21
P_2O_5	0.07	0.14	0.09	0.15	0.11	0.15	0.10
$H_2O(+)$	2.48	2.29	2.32	2.22	1.76	2.13	2.40
$H_2O(-)$	1.06	0.48	0.96	1.30	0.74	0.59	0.61
Total	99.76	99.75	99.71	99.83	99.78	99.60	99.82

(1)-(4): variolite-textured diabase showing pillow form (samples for analysis are prepared as a total pillow)
(1),(2): from the gallery of the Shimokawa mine
(3),(4): from the Nisama area in the southern part of the Shimokawa mining district

(5)-(7): ophitic-textured diabase
(5): from the gallery of the Shimokawa mine
(6): from the Iwaonai diabase swarm
(7): from the gallery of the Shimokawa mine

(Analysts: (1)-(3) K. OTA; (4)-(7) S. ITO)

chemical compositions are poor in Na_2O and the rocks are fresh mineralogically. The present author has a thought that spilite can be merely undertaken as a product of low grade metamorphism, thus he does not agree to use the term "spilite" as a definite rock species.

On the other hand, widespread diabase in the district shows typical ophitic texture under the microscope (Plate III-16). This rock facies is predominant and no variation is observed in the interior part of any individual diabase body. At the edges of a diabase body, a chilled margin is commonly observed; this margin is fine-grained but holocrystalline, and the texture is always ophitic.

Chemical Composition

Hand-specimens for chemical analysis were obtained from four localities in the diabase swarm. One is of a rounded body of pillow structure and the others are common diabase. Seven fresh hand-specimens were selected for chemical analysis.

Their chemical compositions and norms are given in Tables 4 and 5, respectively.

Table 5. Norms of diabases from the Shimokawa mining district, Hokkaido, Japan

	(1)	(2)	(3)	(4)	(5)	(6)	(7)
Q	1.46	3.19	2.40	-	-	2.10	0.03
or	1.00	1.06	1.48	1.77	1.24	1.36	1.24
ab	15.40	17.77	15.49	18.95	15.49	18.70	18.95
an	32.66	35.20	33.02	34.90	42.12	35.45	37.36
c	-	-	-	-	-	-	-
wo	11.48	5.78	11.17	9.38	5.47	8.91	7.23
en	19.70	19.85	19.42	5.80	16.11	15.02	16.41
fs	7.69	7.90	7.36	2.61	7.35	10.98	10.32
fo	-	-	-	11.90	3.44	-	-
fa	-	-	-	5.79	1.73	-	0.03
mt	3.65	3.31	3.71	2.73	1.91	2.17	2.45
il	2.13	2.60	2.18	2.03	2.18	1.86	2.54
hm	-	-	-	-	-	-	-
ap	0.15	0.31	0.20	0.33	0.24	0.33	0.22

No wide variation is observed in their chemical compositions. They are always poor in alumina and alkalis, especially in K_2O, and rich in CaO and MgO, and are very similar to those of the Tokoro-Toyokoro belt.

The chemical compositions are plotted on the Al_2O_3-alkali-SiO_2 diagram (KUNO, 1960) and the MFA diagram (NOCKOLDS, 1954) for comparison with those of recent basalts (shown in Figs. 6 and 7). Most of them fall in the tholeiite fields, though it may be pointed out that some of the diabases from the Shimokawa mining district are slightly nearer the high-alumina basalt area than those from the Tokoro mining district.

The wall-rock alteration represented by silicification, sericitization, chloritization or albitization related to the formation of ore is noteworthy. Sometimes it is difficult to distinguish the original nature of the rocks because of the extensive alteration.

The altered diabase is generally characterized by impregnation with fine-grained pyrrhotite or pyrite. Albitized diabase which is associated with the pyrrhotite ore in the Nisama diabase mass was obtained for chemical analysis. The chemical composition of this rock is characterized by richness in Na_2O and H_2O and poverty in MgO and iron oxides compared with unaltered ones, as given in Table 6.

Table 6. Chemical composition of altered diabase from the Nisama area in the Shimokawa mining district, Hokkaido, Japan

SiO_2	49.94
TiO_2	1.24
Al_2O_3	18.14
Fe_2O_3	2.14
FeO	4.45
MnO	0.12
MgO	5.75
CaO	8.71
Na_2O	4.05
K_2O	0.72
P_2O_5	0.32
$H_2O(+)$	3.24
$H_2O(-)$	0.92
Total	99.74

(Analyst: K. OTA)

Though the chemistry and petrography of this rock (Plate III-14) correspond to those of spilite, the origin of the rock may not be explained as a product of a general spilitization process, but as a product of wall-rock alteration, because it occurs only within a small aureole immediately around the ore bodies.

Spilite-Keratophyre and Serpentine in the Nukabira Mining District

Outline of Geology

The Nukabira mining district is situated in the southern part of the Kamuikotan belt, where a large serpentine body occurs in linear swarms of subparallel or in echélon masses along the fold-axis. Numerous chromite ore deposits are known in the serpentine mass, some of which were mined for many years.

The rocks of the district are pre-Cretaceous sediments such as slate, diabasic tuff, limestone and altered pillow lavas, probably of late

Jurassic age belonging to the Sorachi Series. These formations are intruded by serpentine, including fresh peridotite, as shown in Fig. 10 (BAMBA, 1963). In the eastern part, Cretaceous sediments have developed over the older formations.

Numerous leucocratic dike rocks such as albitite, quartz albitite or rodingite are well developed in the serpentine mass, and are thought to be differentiated dike rocks related to the ultrabasic rocks (SUZUKI, 1953 and 1954a).

Though typical spilitized pillow lavas occur in several parts of the Kamuikotan belt, the pillow lava which is exposed in the middle stream of the Nukabira river as the lowermost member of the diabasic tuff is peculiar, and is characterized by epidote-alteration. On the northeastern side of the Niseu-gawa basin (Fig. 10), keratophyre related to pillow lava is found among the younger formations. Titaniferous augite-bearing diabase dikes are known widely beyond the mining district.

The serpentine mass is commonly approximately conformable to the strata of the Sorachi Series. It occurs mainly within the diabasic tuff and sometimes between diabasic tuff and an underlying formation composed of slate and sandstone, and extends intermittently beyond the mining district, over a distance of about 300 km, running with N-S trend from the northern end to the southern end of the island. Chemical analysis has shown that the serpentine is rich in MgO and poor in FeO + Fe_2O_3, viz. the former is more than 35% and the latter is usually less than 10%.

It is noteworthy that no contact metamorphic phenomena are recognized in the diabasic tuff of the Sorachi Series around the serpentine mass though the intrusion is very large. Thus the serpentine intrusion has

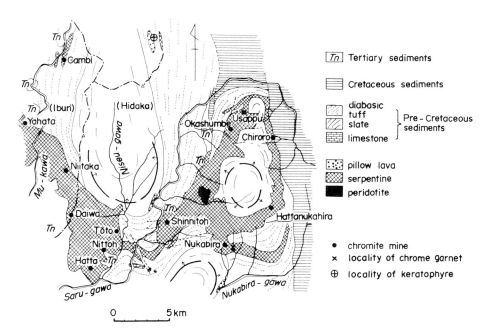

Fig. 10. Geological outline of the Nukabira mining district, southern Kamuikotan belt, Hokkaido, Japan

been regarded as having occurred in the low temperature, crystal-mush state or quasi-solid state.

The Sorachi Series and the serpentine mass of the Kamuikotan belt have been subjected to strong stress from the Hidaka metamorphic zone[2] which was thrust up westward in the early Tertiary period, resulting in the formation of low grade green schist or foliated serpentine. However, some rock facies have remained unmetamorphosed or undeformed, especially surrounding the Iwanaidake peridotite mass, in which unfoliated massive serpentine is developed. A gradual change from peridotite to serpentine is observed here.

Spilite-Keratophyre Suite

Spilitized or epidote-bearing pillow lava, keratophyre and titaniferous augite-bearing diabase are investigated here, because it seems that they are a series of rocks formed in one magmatic cycle.

The occurrence of epidote-bearing pillow lava from the Nukabira mining district and rhythmical banding in a pillow are illustrated in Plate IV-17. The chilled margin of the pillow is glassy and the thickness of the glassy shell varies from 0.5 cm to 1 cm, showing a double or triple rhythmic band (Plate IV-18). Variolitic texture is distinct in the remainder, which is composed of albite, chlorite and a considerable amount of epidote. Sometimes, epidote becomes the most important constituent of this rock.

The chemical composition of this pillow lava is shown in Table 7, and that of the typically spilitized ones from the other localities of the Kamuikotan belt is given in Table 8. The former is poor in alkalis and rich in CaO and Al_2O_3 compared with the latter.

The unique chemistry of this pillow lava is regarded as a local phenomenon caused by strong epidotization which might be considered to be restricted to a lower horizon of the Sorachi Series.

---

[2] The central zone of the southern Hidaka belt is composed of metamorphics such as migmatite and gneiss associated with plutonic complex. This restricted zone is distinguished from the surroundings and is called the Hidaka metamorphic zone.

PLATE IV. Pillow lava, keratophyre and alkali diabase from the Nukabira mining district, Kamuikotan belt, Hokkaido, Japan.
17. Occurrence of pillow lava in the bank of the Nukabira-river. 18. Rhythmic band developing in the margin of the pillow shown above. Dark bands are composed mainly of glass, and the rest consists of varioliticc-textured facies. (19-22) Photomicrographs of thin sections of pillow lava, keratophyre and hypabyssal alkali diabase. 19. Epidote-bearing outer shell of the pillow given above. ep: epidote, gl: glass (under parallel nicols). 20. Spilite from the northern side of the Niseu-gawa basin. pl: albitized plagioclase, px: clinopyroxene (under parallel nicols). 21. Keratophyre associated with the spilite shown above. pl: plagioclase, il: ilmenite (under parallel nicols). 22. Hypabyssal alkali diabase illustrating ophitic texture. ag: augite, pl: plagioclase, il: ilmenite (under parallel nicols)

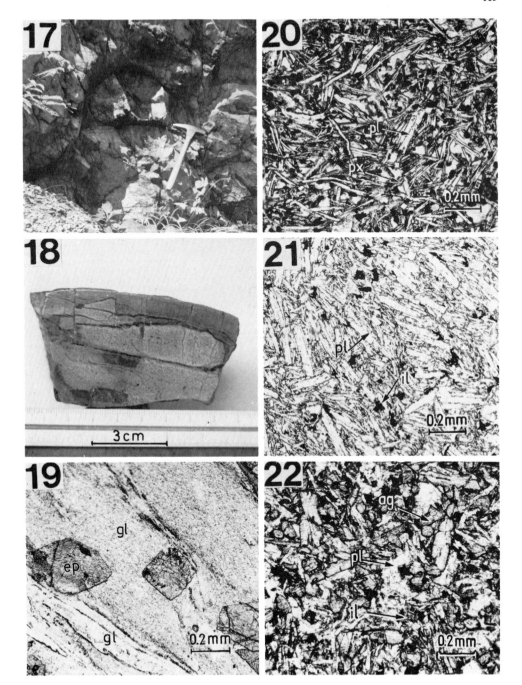

Table 7. Chemical composition of epidote-bearing pillow lavas from the Nukabira mining district, Hokkaido, Japan

	(1)	(2)
SiO_2	46.32	45.31
TiO_2	0.88	0.96
Al_2O_3	15.64	16.68
Fe_2O_3	1.76	4.46
FeO	7.91	6.90
MnO	0.14	0.16
MgO	8.97	7.49
CaO	12.59	13.28
Na_2O	2.23	2.28
K_2O	0.33	0.21
P_2O_5	0.13	0.11
$H_2O(+)$	2.72	2.00
$H_2O(-)$	0.11	0.15
Total	99.73	99.99

(1): variolitic-textured outer shell
(2): variolitic-textured core

Table 8. Chemical composition of typical spilites from the northern part of the Kamuikotan belt, Hokkaido, Japan

	(1)	(2)
SiO_2	47.06	48.95
TiO_2	1.11	1.57
Al_2O_3	14.78	13.93
Fe_2O_3	4.29	5.46
FeO	5.44	6.04
MnO	0.24	0.20
MgO	7.34	6.54
CaO	10.78	11.04
Na_2O	4.17	3.80
K_2O	0.31	0.34
P_2O_5	0.09	0.26
$H_2O(+)$	3.69	1.01
$H_2O(-)$	0.45	0.25
Total	99.75	99.39

(1): spilite from the Horokanai district, Kamuikotan belt, IGI (1956)
(2): spilite from the Ashibetsu-dake district, Kamuikotan belt, SUZUKI (1959)

(Analysts: (1): T. YAMADA and E. OMORI; (2): Y. KATSUI)

On the other hand, the pillow lavas from the Horokanai and the Ashibetsu-dake district (Fig. 1), in the northern part of the Kamuikotan belt, both occur in the uppermost horizon of the green pyroclastic sediments of the Sorachi Series. Petrographically the rocks are much like those of the Tokoro mining district.

Judging from the mode of occurrence and the petrological properties, one may safely say that they belong to a typical spilite as shown in Fig. 11. Such spilite is widespread in the Kamuikotan belt.

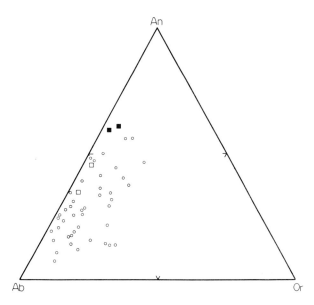

Fig. 11. Normative feldspar of spilites compiled by YODER (1967) and plots of normative feldspar of epidote-bearing pillow lavas and spilites from the Kamuikotan belt, Hokkaido, Japan

■ Epidote-bearing pillow lavas from the Nukabira mining district
□ Spilitized pillow lavas from the northern part of the Kamuikotan belt
○ Spilite compiled by YODER (1967)

Recently, a hand-specimen of keratophyre related to pillow lava was obtained in a bore-core for tunnel building, on the northeastern side of the Niseu-gawa basin (Fig. 10).

The keratophyre is composed of prismatic albite. Neither amphibole nor chlorite, as accessary minerals, is found in the keratophyre, but the petrological properties are much like those of keratophyre from the Tokoro mining district. Photomicrographs of the keratophyre and the related pillow lava are shown in Plate IV-20 and 21.

Titaniferous augite-bearing ophitic diabase occurs widely within the Sorachi Series in the Kamuikotan belt. The ophitic diabase from the Nukabira mining district is generally massive, and occasionally occurs cutting the serpentine mass. Petrographically the rock is homogeneous, composed of clinopyroxene showing a brownish-violet tint, plagioclase and a considerable amount of ilmenite. Its texture is ophitic, but of a finer grain than that from the Tokoro mining district. The chemical composition of this hypabyssal rock is given in Table 9. Because of its chemical composition, the hypabyssal rock is best named "alkali diabase".

Table 9. Chemical composition of alkali diabase from the Nukabira mining district, Hokkaido, Japan

SiO_2	46.49
TiO_2	1.34
Al_2O_3	17.94
Fe_2O_3	4.80
FeO	4.32
MnO	0.20
MgO	6.09
CaO	10.07
Na_2O	3.88
K_2O	0.78
P_2O_5	0.18
$H_2O(+)$	3.04
$H_2O(-)$	0.76
Total	99.89

(Analyst: S. ITO)

Remarkable aggregates of beautiful acicular crystals of pectolite and nephrite occur in serpentine or rodingite along the contact with the alkali diabase mentioned above. This phenomenon has been explained by soda metasomatism due to hydrothermal solutions derived from the differentiated dike rocks related to serpentine (YAGI et al. 1968). However, it is now possible to consider that the alkali diabase plays a role in the formation of the pectolite-nephrite association.

The Relation between Serpentine and Spilite-Keratophyre

Investigation of the relation between serpentine rock series and spilite-keratophyre rock series is one of the most fundamental petrogenetic problems in the Kamuikotan belt.

As mentioned above, the serpentine mass encloses relicts of fresh peridotite in the central part of the mass, and felsic minerals are completely absent from all parts of the ultrabasic rock. No metamorphic effects have been observed around the serpentine mass, which is approximately conformable to the strata of Sorachi Series.

Consequently, as HESS (1955) noted, it is reasonable to consider that the peridotite layer, supplied with juvenile water from the mantle, must have been wholly serpentinized in the sub-Mohorovicic zone. Serpentinized peridotite was probably brought up during the first great deformation of the Kamuikotan belt in a quasi-solid or crystal-mush state. It is difficult to conceive that the ultrabasic rocks are caused by a differentiation process from basaltic magma.

The writer considers that the spilite-keratophyre suite must be derived from the so-called Basaltic Layer (HESS, 1955) lying just above the Mohorovicic Discontinuity.

The serpentine intrusion took place in close relation to the initial great disturbance of the orogenic phase, and the eruption of the spilite-keratophyre suite belonged to the initial magmatism of the geosynclinal phase. Thus the former is apparently younger than the latter in the intrusion epoch. However as DIETZ (1963) suggested, if the serpentine intrusions are ever radiometrically dated, they should prove older than the beds intruded upon.

Therefore, the writer concludes that the two main rock series represented by serpentine and spilite in the Kamuikotan belt must differ from each other in origin.

A Series of Magmatism Related to the Formation of Spilite

The Alpine orogeny in Hokkaido formed three parallel geotectonic belts which are distinguished, on the basis of their tectonic and geologic characteristics, as: 1. Tokoro-Toyokoro belt, 2. Hidaka belt including Hidaka metamorphic zone, and 3. Kamuikotan belt.

Geological outlines and main igneous activities of eugeosynclinal to orogenic phases in the three belts have been briefly described, based on three representative mining districts.

Generalizing from these data there is a feature common to all belts as follows:

Whether the rocks are subaqueous lavas or intrusive dikes, the initial magma in the three belts is regarded as having been tholeiitic. Their chemical compositions are much alike; in fact, essential differences between them can hardly be found. It is likely that pillow lavas have been spilitized by a process closely related to the injection of special hypabyssal dikes characterized by titaniferous augite and high alkalis. As a matter of fact, no spilite has been found in the area where the alkali diabase is absent, as in the Shimokawa mining district. On the other hand, one can easily find spilitized pillow lavas in the area where the alkali diabases are developed, as described in the Tokoro mining district or the Kamuikotan belt.

Consequently, the writer concludes that the initial magmatism belonging to the eugeosynclinal to early orogenic phase of the Hidaka orogenesis was derived from the activities of tholeiitic basalt, associated with segregation veins of keratophyre, and that the ultimate magmatism is represented by the intrusion of alkali diabase, characterized by titaniferous augite. In addition, hydrothermal solutions derived from the alkali diabase dikes in the deuteric stage metasomatized the older pillow lavas to spilites, and the ferruginous quartz rock overlying the pillow lava was also mineralized simultaneously by the hydrothermal effect.

Although the spilitization is most predominant in the Kamuikotan belt, there are hardly any associated iron ore deposits there. The reason for the absence of mineralization in the Kamuikotan belt may be explained as follows:

It is supposed that the development of radiolarian chert and ferruginous quartz rock layers from which ore-forming iron was derived may have necessitated the out-pouring of a large amount of pillow lava. A large amount of pyroclastic materials was accumulated in the Tokoro mining district, whereas in the Nukabira mining district, the activity producing the older pillow lavas was weak and no substantial mineralization occurred, though hypabyssal alkali diabase belonging to the latest phase of the magmatism was rather prominent.

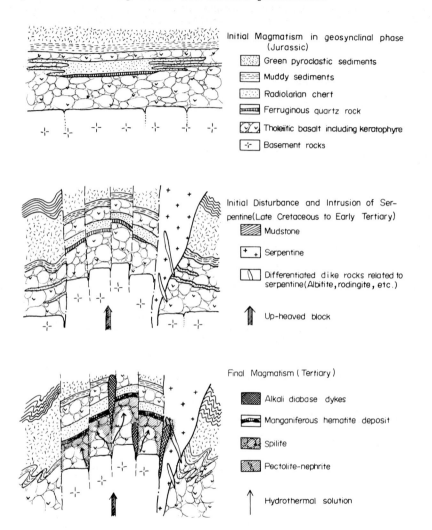

Fig. 12. Schematic diagrams to explain a magmatic series related to the formation of spilite, based on the Tokoro and Nukabira mining districts, Hokkaido, Japan

Acknowledgements

I have received valuable guidance and suggestions on the spilite problem in Japan from Profs. J. SUZUKI, M.J.A., M. HUNAHASHI, K. YAGI, and Y. KATSUI of the Department of Geology and Mineralogy, Faculty of Science, Hokkaido University.

Messrs. K. MAEDA and S. ITO, members of the chemical section of the Geological Survey of Japan, helped me in chemical analysis.

Dr. K. SATO, Director of the Geological Survey of Japan and Dr. M. SAITO, Director of Hokkaido Branch, G.S.J. provided facilities for my investigations and encouragement.

My cordial thanks are expressed to all of them.

References

BAMBA, T. (1963): Genetic study on the chromite deposits of Japan (in Japanese with English abstract). Rep. Geol. Surv. Japan 200, 1-68.

BAMBA, T., SAWA, T. (1967): Spilite and associated manganiferous hematite deposits of the Tokoro district, Hokkaido, Japan. Rep. Geol. Surv. Japan 221, 1-21.

BAMBA, T., MAEDA, K. (1969): Pillow lava and spilitization in the Tokoro district, Hokkaido, Japan (in Japanese with English abstract). Jour. Geol. Soc. Japan 75, 173-181.

BARTH, TOM.F.W. (1939): Die Eruptivgesteine, in: Die Entstehung der Gesteine, written by TOM.F.W. BARTH, CARL W. CORRENS, and PENTTI ESKOLA. Berlin-Göttingen-Heidelberg: Springer 100, 115.

DIETZ, R.S. (1963): Alpine serpentines as oceanic rind fragments. Geol. Soc. America Bull. 74, 947-952.

HESS, H.H. (1955): Serpentine, orogeny, and epeirogeny. 391-407, in Poldervaart, Arie, Editor, Crust of the earth. Geol. Soc. America, Special Paper 62, 762.

HUNAHASHI, M. (1957): Alpine orogenic movement in Hokkaido, Japan. Jour. Fac. Sci. Hokkaido Univ. ser. 4, vol. 9, 415-469.

HUNAHASHI, M. (1969): Report on regional geotectonic investigation of the Shimokawa mining district (in Japanese). 1-10, Metal. Min. Exp. Agency, Japan.

IGI, S. (1956): On the pillow lavas from the Horokanai district, northern Kamuikotan belt, Hokkaido, Japan (in Japanese). Bull. Geol. Comm. Hokkaido 33, 27-28.

ISHIDA, M., HIRAYAMA, K., KURODA, K., BAMBA, T. (1968): "Tanno", 1:50,000, Geological sheet map and explanatory text (in Japanese with English abstract). 1-49, Geol. Surv. Japan.

KUNO, H. (1960): High-alumina basalt. Jour. Petr. 1, 121-145.

MINATO, M., GORAI, M., HUNAHASHI, M. (1965): Hidaka orogenesis in Hokkaido. The geologic development of the Japanese island, 222-239, 442, Tsukiji Shokan, Tokyo.

MIYAKE, T. (1965): On spilitic rocks of the Shimokawa mine and their genetical relations to the ore deposits (in Japanese with English abstract). Mining Geol. Japan 69, 1-11.

NOCKOLDS, S.R. (1954): Average chemical compositions of some igneous rocks. Bull. Geol. Soc. America 65, 1007-1032.

RÖSLER, H.J. (1962): Zur Entstehung der oberdevonischen Eisenerze vom Typ Lahn-Dill in Ostthüringen. Freiberger Forschungshefte, C. 138, 1-50.

SUZUKI, J. (1953): The hypabyssal rocks associated with the ultrabasic rocks in Hokkaido, Japan. Comptes Rendus de la Dix-Neuvieme Session Alger, 131-137.

SUZUKI, J. (1954a): On the rodingitic rocks within the serpentinite mass of Hokkaido. Jour. Fac. Sci. Hokkaido Univ. ser. 4, no. 8, 419-430.

SUZUKI, J. (1954b): On the pillow lavas from Hokkaido, Japan (in Japanese). Bull. Geol. Comm. Hokkaido 26, 11-20.

SUZUKI, J., SUZUKI, Y. (1959): Petrological study of the Kamuikotan metamorphic complex in Hokkaido, Japan. Jour. Fac. Sci. Hokkaido Univ. ser. 4, 10, 349-446.

WATANABE, T. (1965): Submarine volcanism and genesis of ore deposits (in Japanese). Jour. Geol. Soc. Japan 71, 332-336.

YAGI, K., BAMBA, T., OKEYA, M. (1968): Pectolites from Chisaka, Hidaka province and Nozawa mine, Furano, Hokkaido. Jour. Fac. Sci. Hokkaido Univ. ser. 4, 14, 89-95.

YODER, H.S., Jr. (1967): Spilites and serpentinites. Carnegie Institution of Washington, Year Book 65, 269-279.

Environmental Effects in Magmatic Spilite

E. Lehmann

Abstract

Examination of the mineralogical and chemical effects, as a function of separate environmental conditions, was occasioned by a weilburgitic (spilitic) rock deposit in eastern Sauerland. The bifurcation is manifested by shallow intrusive injections in keratophyric tuff as opposed to localized pillow accumulation overlying the same tuff. The mineralogical and chemical characteristics of both varieties (three injecta and the averaged material of a pillow were analysed) are discussed and the existing differences pointed out.

The pillows evidently underwent a notable change of composition after the emplacement. Without a detailed mineralogical examination, pillow analyses are suggested to be a vague basis for discussing the genesis of spilitic rocks.

Introduction

Most spilitic rocks are developed as independent bodies, in part exhibiting structural conformity, in part consisting of accumulations of spheroidal individuals (pillows). Rarely massive rocks are interrupted by the intercalation of single pillows or small pillow aggregations. This paper refers to yet another mode of appearance. Fundamentally the example to be discussed here resembles that already recorded from the environment of Meschede, Sauerland (LEHMANN, 1952). Apart from the fact that the magmatic bodies are usually larger, there is no essential difference from the rock near Meschede. Microscopic study indicates that in both cases a pre-existing keratophyric tuff was infiltrated with magmatic liquid. In other words, the tuff was converted to a mictite. This transformation involves a change in structural, textural and physical properties (destruction of original stratification, coherence of the ejecta, increase in strength of the rocks).

A distinction between large and small (microscopic) injected bodies is of course arbitrary. Transitional sizes exist and they all derive from the same source, but for several reasons it seemed useful to make a distinction. With this in mind the large bodies (about 1 - 10 dm) were called "injecta" (LEHMANN, 1968).

Geology and Megascopic Features

Burg (483.1 m) and Enkenberg (492.5 m), two mountains in the northeastern part of the Rheinische Schiefergebirge, eastern Sauerland, stand on the north side of the rivulet Hoppecke, a tributary of the Diemel River, and are separated from one another by a very narrow valley (Fig. 1). Beginning on the western slope of the Enkenberg

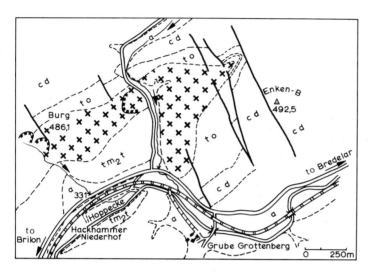

Fig. 1. Simplified sketch map of the geology of the western part of Enkenberg and of Burg Mt., Sauerland (after geologische Karte von Preussen und benachbarten deutschen Ländern, Berlin, 1928)

ridge and following the main axis of an anticline, magmatic rocks, generally mapped as diabases, are exposed. This identification as diabase has not been confirmed for the two localities mentioned. On the Enkenberg side trial-working in a small quarry had just begun when I visited the place for the first time (1966). Two features turned out to be contradictory to diabase: (1) the rock appeared to be unstratified in the main, but in the upper half of the wall a disrupted bank, about 20 cm thick, of a well-stratified and occasionally banded tuff was conspicuous; in addition, red feldspar crystals irregularly distributed in the main rock were unusual and the ensuing examination resulted in the general identification of the rock as a keratophyric tuff; (2) a few blocks, up to about a m^3 in size, of a brick-red rock containing phenocrysts of red feldspar, and occasionally exhibiting grayish veins, were among the debris, but neither in 1966 nor later could their original position be ascertained. For this reason the red rock was not taken into account in the present discussion. A further peculiarity is no less curious; in 1966 a layer of pillows, about 3-4 m thick, was exposed on the top, but had disappeared when I visited the place in 1967. Other pillow accumulations could be found 150 - 200 m farther north, and about 500 m south from the quarry pillow structure was visible on the main road.

In a much larger quarry worked on the Burg side until 1968 (front wall about 150 m in length), samples of the stratified tuff were less often encountered, whereas the red feldspar crystals locally increased in frequency. Another phenomenon, however, is characteristic. Varying in frequency at distinct places of the front-wall, patches and streaks of the second rock contrast with the surrounding tuff by lighter gray color and finer, uniformly grained fabric. Branching, net-like spreading, terminally thinning, accumulation of the red feldspars along boundary planes or sporadic incorporation in the outer parts leave no doubts as to the formation by injection of a liquid phase. Attention finally may be called to the fact that no pillow formation could be discovered in the Burg area. It seems likely that outpouring was prevented by the thickening of the tuff layer as compared to the situation on the Enkenberg side.

Mineralogical and Chemical Properties

The association of the distinctly keratophyric tuff with the effusive spilitic rock at Enkenberg as well as the appearance of the injecta in this keratophyric tuff in the Burg mountain led to an investigation of the common source of the injecta with the pillow accumulations. The geologic analogy, of course, does not apply to emplacement conditions and postemplacement effects, but it is still indicated in the sporadic appearance of the pillow assemblages and in the tendency of the injecta to accumulate in separated zones of the quarry front-wall. On the other hand there are some differences in the megascopic features of the two rocks. Abundance of amygdales, in part carbonatic, in part chloritic, seems to be a characteristic of the pillows, but not of the injecta. (Under the microscope some injecta were shown to contain calcite aggregates or to have undergone intense carbonization. In some examples small chlorite amygdales, often combined with quartz, are numerous.) Large feldspar crystals were more often observed in the pillows than in the injecta; the majority of them are suggested to originate from the tuff, i.e., are xenocrysts.

Injecta

Judging from the feldspar character of the injecta, one might be inclined to believe that these rocks result from keratophyric melt after being re-activated and modified by the introduction of fluids and mafic elements.

The optical properties (n_α = 1.519 - 1.525, n_β = 1.520 - 1.528, n_γ = 1.524 - 1.530; predominance of optically negative character; 2V = 60-74°) indicate that orthoclase, sodic orthoclase, and calcium-bearing alkali feldspars comprise the crystals accessible to optical determination. Analogous feldspars were found in weilburgite from several Sauerland localities (LEHMANN, 1967).

The main alteration of the feldspar consists in carbonization always present to some degree. In extreme cases its association with carbonate fillings of small fractures indicates post-consolidation effect, and it may also be assumed in the case of deficient fracturing. Silification is far less extensive than carbonization; it seems to be confined to the boundary zones with mictite, but even there it is a rare phenomenon. Much the same local restriction applies to sporadic inclusions of allogenic fragments of grits and calcitic aggregates, the latter in part quartz-bearing. Special attention is drawn to the sparse appearance of sericitization which is observed only in small spots. Except for carbonization, the feldspar crystals are surprisingly fresh.

The common chlorite, represented mainly by interstitial aggregates but occasionally also by accumulations of distinct lumps without crystallographic outlines, is characterized by a very light greenish-gray color. $\Delta \leq 0.001$, and $n_\gamma = 1.580 - 1.582$, suggestive of a low iron content. In some injecta, however, besides this chlorite another variety is restricted to distinct zones of irregular amygdaloid forms. Optics: X light greenish-yellow, Z light olive, $Z > X$; $n_\gamma = 1.590 - 1.592$; Δ about $0.020 - 0.025$. In proximity to the mictite, prehnite instead of chlorite appears sporadically in some amygdales. In general the same localization seems to exist for the occasionally increased amounts of calcite and quartz amygdales, the last named sometimes matched with grit fragments and scattered quartz grains resulting from mechanical destruction of the fragments.

The texture of the injecta is generally porphyric but differs in the various examples. In the majority, parallel arrangement of the feldspars is well marked (Fig. 2). Carbonate is sometimes restricted to amygdales, less often to irregular patches differing in size and distribution (Fig. 3). In the marginal zones chlorite amygdales are occasionally frequent.

Three injecta sufficiently separated from one another were analysed to determine the degree of variation, and only core material was used to avoid contamination with tuff materials so far as possible. Judging from the results quoted under 1-3 of Table 1, a moderate variation

Table 1. Chemical composition of 3 injecta from the Burg quarry, eastern Sauerland (Westphalia) and its comparison with weilburgite from Lahn Basin

	Sauerland					Lahn	
	Injecta			Average 1-3	CIPW norm	Average (29)	CIPW norm
	1	2	3	4	4a	5	5a
SiO_2	46.7	45.2	50.1	47.3	or 25.02	45.71	or 21.13
Al_2O_3	15.0	14.8	14.1	15.0	ab 16.77	14.94	ab 29.34
Fe_2O_3	0.2	1.4	1.5	1.0	an 12.51	3.34	an 1.95
FeO	8.1	6.2	5.9	6.7	C 2.35	7.42	C 4.69
MnO	0.07	0.08	0.06	0.07		0.12	Q 1.15
MgO	6.4	5.7	5.5	5.9	en 15.00	5.00	en 12.40
CaO	6.8	7.3	5.4	6.5	hy 8.32	6.36	hy 6.34
Na_2O	1.8	2.2	1.9	2.0		3.48	
K_2O	4.2	4.2	4.3	4.2	mt 1.39	3.53	mt 4.64
TiO_2	2.6	2.5	2.4	2.5	ilm 4.86	2.81	ilm 5.47
P_2O_5	0.7	0.6	0.6	0.6	ap 1.86	0.70	ap 1.68
CO_2	2.3	4.7	3.0	3.3	cc 7.68	2.78	cc 6.35
H_2O	4.8	4.8	4.9	4.8		3.90	
	99.67	99.68	99.66	99.87	Or:Ab:An 51.0:37.0:12.0 (mol. %)	100.07	Or:Ab:An 38.5:58:3.5 (mol. %)

(Anal. 1-3 WEIBEL)

Fig. 2

Fig. 3

Figs. 2 and 3. Textural character of the weilburgite in the injecta, Burg. x 65. The magmatic body of Fig. 2 is thin (about 20 cm) and low in carbonate, while that of Fig. 3 is comparatively thick (about 50 cm) and high in irregularly distributed carbonate

exists for silica, calcium oxide and carbonic acid, probably due to contamination or immigration. The difference in FeO (max. 2.2%) is suggested to be the result of autometasomatic effects.

According to the optical properties and the alkali ratio, there is no doubt that in the injecta the orthoclase component of the feldspar exceeds the albite component. How far the small feldspars influence this result is an open question. Perhaps the surrounding tuff operated as a filter, thus preventing a more general or more intense albitization. In any case, the actual feldspar character is reflected only in the respective norm (4a) and in the molecular proportion of Or:Ab:An when calcite is allotted beforehand in the recalculation of the chemical analyses. If alumina is taken as the standard calculation of the anorthite content, the result would be inadequate compared to the actual content. The normative silica saturation, of course, does not agree with the fact that chlorite is the only mafic component. But quartz is observed in some specimens and may be occult in the matrix of other ones.

Chemical analogy with the composition of weilburgite is evident (5 and 5a). All available analyses were used for the average (5), mainly with regard to CO_2 content, but the difference would be insignificant if one referred only to the samples low in CO_2 (cf. no. 6, Table 2 in the contribution "spilitic magma", this volume, p.35.

Pillows

Conformity of the original feldspar composition in the pillows with that in the injecta is manifested by relicts. In these, measurements of n_γ = 1.530 - 1.533 and extinctions of 12-17° for X : cleavage (010) agree with those found in the injecta.

But in the pillows the initial feldspar was largely replaced by albite as shown by spotty and streaky areas which alternate with homogeneous-looking fields. Occasionally also, a checker-board structure (Fig. 4) appears in the albite. The effects of exsolution and replacement continue in the matrix-feldspars, provided they are large enough to be observed. In comparison with the rod- and threadlike matrix-feldspars (mostly below 50 µ in length), crystals ten times this length already resemble phenocrysts. These micro-phenocrysts and larger-sized crystals are often surrounded by thin rims which, in contrast to the cores, are deficient in the dusty inclusions dispersed throughout the cores. These rims of albite suggest direct crystallization, and probably the same applies to many of the small matrix-feldspars.

In the pillows alteration generally exceeds in degree and extent the effects in the injecta. This applies especially to sericitization, and unusually large crystals (allogenic?) may be carbonized or sericitized throughout. Chlorite, and less often quartz, sometimes joins the replacing calcite, which suggests that the process is a function of temporary and/or local change of conditions of alteration. In these large crystals the secondary albite occasionally contained a slightly higher percentage of anorthite (about 10% An).

The distribution of chlorite in the pillows is irregular, in particular, marginal enrichment or concentric zones of chlorite amygdales could not be observed. The optical properties of the interstitial chlorite do not differ from those recorded for the chlorite in the injecta. In some pillows, however, an additional chlorite is characterized by a more olive green color of the cross section, n = 1.591 - 1.595,

Fig. 4. Albite with checkerboard structure resulting from exsolution of a soda-potash feldspar crystal in a pillow. Enkenberg western slope. x 75

exceptionally 1.599, and Δ (Delta) about 0.003, pointing to an increase in ferrous iron. The aggregates often radiating are sometimes interspersed with small groups of iron hydroxide.

The relatively high carbonate content is distributed irregularly and in variable form: amygdales, compact aggregates, dispersed particles, fillings of thin ruptures. A few aggregates contain sparse grains of quartz. In some samples localized nets and rims surrounding chlorite amygdales consist of quartz, indicating crystallization in a late stage of consolidation.

At high magnification the minute "filty" intergrowth of components in the matrix (Figs. 5 and 6) is barely recognizable. Apart from the rods and threads of feldspar and the interstitial chlorite, the abundance of leucoxene and, in some samples, the frequent appearance of colorless minute particles (diameters up to 5 μ) without visible birefringence suggestive of vitreous character, are worth mentioning.

The fabric of the pillows differs from that of the injecta mainly as follows: 1. The contrast in size between the early crystallized feldspars and the pronounced matrix feldspars is more marked in the injecta, whereas the parallel arrangement of feldspars is wanting in the pillows. This suggests that the final crystallization proceeded in the hydrous silicate melt under the low PT-conditions present during this stage and that albite and chlorite in the main crystallized from this melt. The only matter of dispute may be whether the same period is also involved in the albitization and sericitization of the pre-existing feldspar and in the local crystallization of quartz. 2. Another peculiarity consists in the local abundance of amygdales. Pillows up to 0.3 dm in diameter display amygdales all through, larger ones mainly in outer zones. More of a problem is the occasional restriction of chlorite

Fig. 5

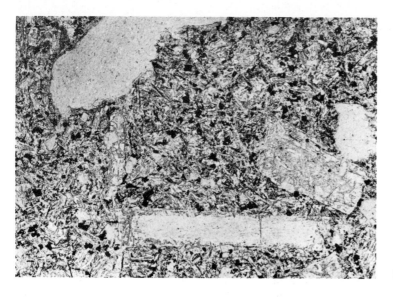

Fig. 6

Figs. 5 and 6. Texture patterns of pillowy weilburgite. Enkenberg western slope. x 70. In Fig. 4 thin rims of albite are visible surrounding the feldspar crystals. Fig. 5, crossed nicols, Fig. 6, ordinary light

amygdales of relatively small areas in the inner parts of the large bodies pointing to a selective appearance of temporary interruption of the release of volatiles in the late-magmatic or post-magmatic stage.

A fresh-looking pillow with a maximum diameter of about 1 m was used for chemical analysis (Table 2, no. 1). The material was taken from several parts of the core about 20 cm inside the margin. The localized amygdaloidal chlorite accumulations just mentioned were excluded in hopes of obtaining a sample more representative of the average than one including pieces rich in chlorite.

Compared with the composition of the injecta (analysis no. 2, Table 1), the ferrous iron content and the total alkali content diverge most markedly. The increase in FeO, indicated mineralogically by the appearance of the additional chlorite variety in the amygdales of the pillow, is probably due to introduction of FeO with volatiles. No notable difference exists regarding the ferric iron. The proportion of alkali in the pillow analysis is about the inverse of that in analysis no. 2 (injecta), again in accordance with microscopic observations. This contrast is intensified in the extreme members of the weilburgitic series in the Lahn district (nos. 3-5, Table 2). There, however, the rocks in question are isolated from one another, and the alkali differentiation preceded the magma emplacement, thus no information is to be expected regarding the present question. In fact, in the Lahn district highly potassic and highly sodic compositions are encountered among both, the massive and the pillowy bodies.

In the CIPW norm the pillow (Table 2, no. 1) is undersaturated in silica, whereas saturation exists for the average of the injecta (no. 2). This statement cannot be generalized, however, because oversaturation of injectum no. 3 (Table 1) is due to the quartz content. Injecta nos. 1 and 3 are actually undersaturated: the molecular norm of no. 1 gives 8% normative orthosilicate, that of no. 2, 12% (against 16% for the pillow analyses). Nevertheless, in the pillows, too, the SiO_2 content may vary. On the whole the silica content seems to be less significant, since occasionally grains of quartz appearing as inclusions originating from the tuffaceous country rock or from quartz-bearing limestone may be present in these rocks. Less ambiguous is the increase in the amounts of MgO and FeO in the pillows. In the analysed example, the totals of ortho- and metasilicate in the molecular norms are 30.5 against 24.5, 25.3, and 25.1 for injecta nos. 1, 2 and 3 of Table 1, respectively, and the difference seems above rather than below the amount in the analysed pillow.

As opposed to these variations, the constant contents in Al_2O_3 and TiO_2 deserve attention. They support the conclusion already drawn from the geological appearance, i.e., that the distinct pillow accumulation and injecta originated from the same hydromagma.

Conclusions

1. The common source of the injecta appearing mainly in the Burg quarry and of the pillow accumulations in the western part of Enkenberg is supported by the geological position and by the feldspar character. The objection that in the pillows albite is increased is unfounded since numerous relicts of the original feldspar are preserved, giving evidence of the original conformity with the composition of the injecta.

Table 2. Chemical composition of a pillow (core) from Enkenberg (1), as compared with that of the injectum (2), Anal. WEIBEL, and of the weilburgites from the Lahn basin (3-5). CIPW norms of the rocks (1a-5a). 3. Soda weilburgite from Hauser Berg, Balduinstein, Anal. GÖTZ. 4. Weilburgite average (29). 5. Potash weilburgite from Rehbach Valley, E of Braunfels Station, Anal. GÖTZ

	1	2	3	4	5		1a	2a	3a	4a	5a
SiO_2	45.5	47.3	44.75	45.71	45.26	or	10.01	25.02	3.34	21.13	36.14
Al_2O_3	15.4	15.0	14.75	14.94	14.71	ab	32.00	16.77	45.06	29.34	11.53
Fe_2O_3	1.3	1.0	2.00	3.34	3.80	an	11.08	12.51	14.73	1.95	1.67
FeO	9.0	6.7	7.31	7.42	8.80	C	2.45	2.35	–	4.69	5.41
MnO	0.08	0.07	0.13	0.12	0.07	Q	–	–	–	1.15	1.38
MgO	6.9	5.9	3.28	5.09	5.67	di	–	–	1.08	–	–
CaO	5.3	6.5	9.36	6.36	3.81	en	12.50	15.00	1.80	12.40	14.10
Na_2O	3.8	2.0	5.34	3.48	1.36	hy	8.32	8.32	2.11	6.34	7.39
K_2O	1.7	4.2	0.56	3.53	6.08	ol	5.67	–	7.73	–	–
TiO_2	2.3	2.5	3.47	2.84	2.57	mt	1.86	1.39	3.02	4.64	5.57
P_2O_5	0.05	0.6	0.47	0.70	0.98	ilm	4.91	4.86	6.54	5.47	4.86
CO_2	2.3	3.3	4.33	2.73	1.75	ap	1.34	1.86	0.48	1.68	2.35
H_2O	5.7	4.8	3.93	3.90	5.03	cc	5.85	7.68	9.80	6.35	4.00
	99.78	99.87	99.68	100.07	100.13						
Or	19.5	51.0	5.0	38.5	72.5						
Ab	65.5	37.0	73.0	58.0	24.5						
An	15.0	12.0	22.0	3.5	3.0						
	(mol. %)		(mol. %)								

The objection that the soda-potash feldspars in the injecta may be incorporated crystals originating from the keratophyric tuff is also thought to be unfounded. The equally high contents in soda and potash found in three separate injecta would be hard to understand were the feldspars allogenic. Feldspars of analogous character (Figs. 7 and

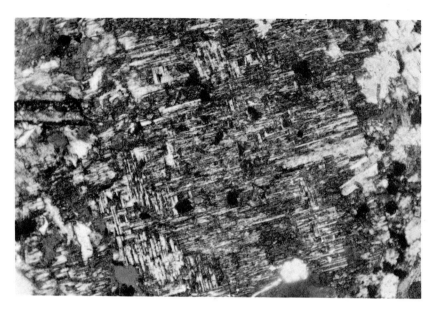

Fig. 7. Microperthite in weilburgite. E. Bontkirchen, Sauerland. x 100

Fig. 8. Unmixing of soda-potash feldspar in weilburgite. Stein-Berg, southern border of Ostwig (R. Ruhr), Sauerland. x 220

8) were found in weilburgite from other Sauerland localities (LEHMANN, 1967).

2. The pillows were submitted to compositional changes of an incomparably higher degree than the injecta. The effect is especially manifested by the albitization of the feldspar. Whereas the alkali total in the pillow analysis is only 0.5% below that of the injecta analyses, the alkali ratio in the respective analyses is almost inverse. Since an overwhelming portion of the feldspars was involved in the replacement reaction, it is to be assumed that the exchange of the alkali ions took place at a comparatively late stage of consolidation or after the period of consolidation. Consequently we cannot evade the question of the source of the Na needed in the exchange reaction.

Judging from the prevalence of potassium in the injecta and in the feldspars of the tuff-mictite complex (LEHMANN, 1968, p.72), an origin of sodic volatiles from the respective melts is most unlikely. The only possible answer seems to be the assumption of a magma existing in a more deeply seated or neighboring reservoir which might differ in character but be high enough in sodium. However, in every attempt to give an explanation, we are thrown on hypothesis. In addition to the uncertainties involved in the question of alkali migration (MEHNERT, 1960), there are many others in the problem with which we are concerned, for instance, the possible dependence on effects resulting from gravitative differentiation, different composition of the fluid phase, and different reaction of the respective magma and the fluids with the adjacent permeated rocks.

3. It is to be concluded from the frequent appearance of exsolution phenomena in the tuff (mictite) and in the injecta, that the majority of the feldspar crystals initially consisted of uniform, solid solutions. Disregarding the An-content and assuming correspondence with the solvus as found for $Or_{45}Ab_{55}$ in the system $NaAlSi_3O_8-H_2O$ (BOWEN-TUTTLE, 1950), the crystallization temperature ought to be <u>above</u> 660-500°, depending on the composition. For $Or_{30}Ab_{70}$ ORVILLE (1963) found the crest of the solvus to be near 680° at 2 000 atm. In the vapor phase co-existing with feldspars and rather concentrated in K and Na, the proportion K: (K + Na) was 0.260 at 670°, but 0.16 at 400°. Thus in the higher temperature range K, in the lower range Na is increased. This result evidently is confirmed by the feldspar character and the chemistry of the injecta and the pillow, respectively. There is no reason, in my opinion, for denying the definition of the liquid phase as a hydrous silicate melt.

The same definition is thought to hold true for the stage of chlorite crystallization from this liquid phase in the temperature range <u>below</u> 660-500°. Judging from the optical properties the main chlorite in the pillows does not essentially differ from that in the injecta, whereas in the case of the chlorite occasionally filling vesicles and showing other properties (iron enrichment), an influence of the fluids or volatiles is suggested to be more justified than a change in composition of the liquid phase.

4. Knowledge of the average pillow composition was sought in order to compare the mineralogical and chemical properties with those of the associated injecta. Since in the Enkenberg example no regularity in the change of composition seemed to exist for concentric zones, one could expect no new information from a detailed examination of a

single pillow. But zoned variation, especially divergence of composition between the core and the margin, is recorded from numerous localities, by many authors (HOPGOOD, 1962; NICHOLLS, 1939; VALLANCE, 1960, 1965; VUAGNAT, 1946, 1959). In a detailed chemical examination of pillows from a number of localities, VALLANCE (1965) showed that the variation of the single zones as well as that between the cores and margins is often considerable and includes not only the alkali and MgO + FeO contents but also those of Ca, Fe^{3+}, Ti, and Si. Pyroxene or traces of it often seem to exist, whereas the place of original glass is commonly said to be taken by microcrystalline aggregates containing chlorite, epidote or prehnite. In spite of these mineralogical divergences in the British examples studied by VALLANCE, chemical analogy occasionally exists with the examined Enkenberg pillow. On the other hand, only two (nos. 24 and 32) of the spilite analyses listed by VALLANCE (1960) resemble those of injecta from the Burg quarry in their potassic character, whereas in the Lahn district the potassic type is by no means rare.

Thus, one must agree with SHAND (1949, p.210) that "chemical analysis fails to indicate the mineralogical composition" and "does not provide a suitable basis", if the discussion is aimed at the genetic question. In the present example, not only the eruptive itself but even the associated rocks should be considered. So long as the injecta were overlooked and their country rock was considered to be diabase, the chemical analysis of a pillow, if available, would only have been suggestive of analogy with common spilite. Without knowing the character of the feldspar preceding albitization, the specific character of this spilite would be hidden, and without paying attention to the connection of the pillow accumulations with the injecta, considerations of the genesis of this spilite probably would be misleading.

Due to the accumulation of distinct bodies, circulation of vapors and hydrous solutions along the pillow boundaries is facilitated, irrespective of the source of the vapors or solutions. Their chemical reaction with the pillows and with the interspaced sedimentary materials often combines with physical effects (AMSTUTZ, 1954, p.96; LEHMANN, 1941, p.377). Consequently, the objection raised by SHAND to the exclusive reliance on chemical analyses of igneous rocks in general, also holds true for spilitic rocks; but the risk of misleading conclusions is still enhanced when using analyses of pillows only, or even only of parts of pillows.

Acknowledgements

The author thankfully acknowledges suggestive criticism by Prof. T.G. VALLANCE. Thanks are due to Prof. AMSTUTZ for helpful discussion and to E.M. SCHOT for style improvement. The analyses of the injecta and the pillow were kindly carried out by Prof. M. WEIBEL.

References

AMSTUTZ, G.C. (1954): Geologie und Petrographie der Ergußgesteine im Vorrucano des Glarner Freiberges. Publ. No. 5, Vulkaninst. Immanuel Friedlaender, 150.

BOWEN, N.L., TUTTLE, O.E. (1950): The system $NaAlSi_3O_8$-$KAlSi_3O_8$-H_2O. Jour. Geol. 58, 459-511.

HOPGOOD, A.M. (1962): Radial distribution of soda in a pillow of spilitic lava from the Franciscan, California. Am. J. Sc. 260, 383-396.

LEHMANN, E. (1941): Eruptivgesteine und Eisenerze im Mittel- und Oberdevon der Lahnmulde. Scharfe Wetzlar, 391.

LEHMANN, E. (1952): Beitrag zur Beurteilung der paläozoischen Eruptivgesteine Westdeutschlands. Z. D. Geol. Ges. 104, 219-237.

LEHMANN, E. (1967): Diabasprobleme und problematische Diabase. Chem. d. Erde 26.

LEHMANN, E. (1968): Diabasprobleme und problematische Diabase II. Chem. d. Erde 27, 39-73.

MEHNERT, K.R. (1960): Das Problem des Alkalihaushalts im Orogen. Geol. Rundschau 50, 124-131.

NICHOLLS, G.D. (1939): Autometasomatism in the lower spilites of the Builth volcanic series. Q. Jour. Geol. Soc. London 114, 137-162.

ORVILLE, F.M. (1963): Alkali ion exchange between vapor and feldspar phases. Am. Jour. Sc. 261, 201-237.

SHAND, S.J. (1949): Eruptiv rocks. 2nd ed. London-New York.

VALLANCE, T.G. (1960): Concerning spilites. Presidental Address, Proc. Linnean Soc. New South Wales 85, 1-51.

VALLANCE, T.G. (1965): On the chemistry of pillow lavas and the origin of spilites. Min. Mag. 33, (Tilley volume), 171-181.

VUAGNAT, M. (1946): Sur quelques diabases suisses. Contribution à l'étude du problème des spilites et des pillow lavas. Schweiz. Min. Petr. Mitt. 26, 117-228.

VUAGNAT, M. (1959): Les basaltes en coussins d'Aci Castello et au Val di Noto. Rendic. Sec. Min. Italiana 15, 311-321.

A Statistical Study of Specific Petrochemical Features of Some Spilitic Rock Series

W. Narebski

Abstract

The paper presents a statistical examination of sets of chemical analyses of six spilitic rock series of different geographic position and age. It is shown that the frequency distribution patterns of the major elements generally correspond with normality, while most trace elements display the lognormal distribution pattern. All spilitic associations under examination differ in petrochemical features from other related rock series, representing statistically specific populations. Nevertheless, the obtained statistical data strongly suggest various origins of individual spilitic rock series. Thus the writer fully supports the opinion of the polygenesis of spilites, i.e., that they may be the products of magmatic, metasomatic and metamorphic processes. Basing on a quantitative analysis of the spilite reaction, a new reaction scheme is proposed, applicable to interpreting the origin of secondary varieties of these rocks. It is also suggested to restrict the term "spilite" to initial volcanites exhibiting specific chemical, mineral and textural features.

Introduction

The general opinion among the most experienced students of the spilite problem is that four essential criteria should be used in the recognition of these rock series: mineralogy, chemical composition, fabric and the mode of geological occurrence. Having the opportunity to participate in the Spilite Symposium during the XXIII Session of the International Geological Congress in Prague, the writer could fully appreciate VALLANCE's statement (1960) that "much of the confusion associated with the name spilite derives from the fact that various workers have emphasized different features".

There is no reason to repeat the 140-year history of the term spilite and of the evolution of opinions on its content, since this has been done in detail by VALLANCE (1960) and AMSTUTZ (1968). Several authors (VALLANCE, 1960; LEBIEDINSKI, 1964; AMSTUTZ, 1968; VIELINSKII, 1968) as well as the present writer (NAREBSKI, 1968a) have also expressed opinions on the petrogenetical theories of the origin of spilites, i.e., that they can be products of various geochemical processes, developing under similar physicochemical conditions in the geosynclinal environment.

The goal of this paper is to contribute to the problem of diversity of origin (polygenetism, heterogenetism) of spilites using simple statistical methods: an interpretation of chemical variability of elements within some typical spilitic rock series, a comparison between them and other related basic rocks, and a test of the correlation between chemical elements in individual associations of spilites.

It should be remembered that, following a long period during which the exact treatment of analytical data was ignored, we have recently observed a renaissance of the application of mathematical methods to the geological sciences, initiated in mineralogy and petrology by the pioneer works of NIGGLI (1923), LOEVINSON-LESSING (1925), CLARKE (1924) and RICHARDSON and SNEESBY (1922). The general development of statistical analysis of geological problems is described by MÜLLER and KAHN (1962) and its application to sedimentary petrology by GRIFFITHS (1962). The statistical approach to geochemical problems was initiated with several papers by AHRENS (1954-1966) as well as by SHAW and BANKIER (1953), VISTELIUS (1958, 1960), RODIONOV (1964), etc. BONDARENKO (1967) recently published a valuable book on the statistical methods of investigation of volcanogenic complexes. Quantitative data on the frequency distribution of elements and the correlation between them in various magmatic rock series are contained in the papers of PODOLSKII (1962, 1963), IVANOV and KOSKO (1965) and KUTOLIN (1967, 1968). It is hoped that large-scale application of statistical analytic methods to ever-increasing numbers of analytical data on spilites will focus attention on the not generally appreciated role of geochemical criteria in solving essential petrological problems of these rock series.

General Characteristics of Statistically Examined Spilitic Rock Series

The selection of proper analytical material for statistical examination is not a simple task. As follows from theoretical premises, the most valuable materials are those collected at random but according to a special sampling pattern. It is, however, very difficult to accomplish this. Moreover, the examined sets should include a large number of chemical analyses, carried out possibly by one laboratory since considerable and non-systematic errors may often mask the natural variations of concentration (AHRENS, 1964).

Bearing this in mind, the writer tried to select sets of data which would fulfill such requirements. Because of an insufficient number of analyses it was impossible to examine statistically such well-known and thoroughly investigated suites as those of New Zealand (BATTEY, 1956), the Alps (VUAGNAT, 1946, 1949; JAFFEE, 1955; AMSTUTZ, 1954), Ural Mts (ZAVARITSKII, 1946), the Crimea (LEBIEDINSKII and MAKAROV, 1962), etc. It should be emphasized that the author intended to avoid any accumulation of analytical data from various spilitic associations differing in geographic and tectonic position, geological age and petrogenetic evolution. The latter procedure, as well as being somewhat inappropriate from a theoretical point of view, results in artificial averaging of specific geochemical trends in individual spilitic suites. Consequently, after examining a large number of papers, six sets of chemical analyses of spilitic rock series were selected.

Two of these series, investigated petrochemically by the writer (NARĘBSKI, 1964, 1968 a,b), are products of two main periods of initial submarine, volcanic activity in the Caledonian geosyncline of the West Sudetes (TEISSEYRE, 1968a). The older one is represented by the thick Upper Cambrian greenstone complex of the Góry Kaczawskie Mts (complete Polish name of the Kaczawa Mts), the pillow varieties of which have been investigated by the author. Interpretation of analytical and geological data indicates that this submarine activity was two-stage in character. The earlier volcanisms developed as thick paleobasaltic flows exhibiting only a slight spilitic tendency, while the later-stage products are distinctly spilitic in character, being also associated with keratophyres. From available data the writer supposes that the spilitic evolution of this series is due to alkali-volatile diffusion differentiation of homogeneous liquid magma, which took place in very elongated fissure chambers, corked by solidified lava after the first-stage submarine effusions.

The spilitic series of the adjacent Rudawy Janowickie Mts is probably Upper Silurian in age (TEISSEYRE, 1968b). It consists of massive metadiabases, albite amphibolites, greenschists and prasinites, accompanied by sodic keratophyres, as well as by tuffs and calcareous tuffites, altered into striped amphibolites. The latter represent a transition series to underlying carbonate rocks. It is not yet clear whether distinct Na-enrichment of these rocks is due to magmatic or other processes (NARĘBSKI, 1968; TEISSEYRE, 1968b).

Another Lower Paleozoic series examined statistically in this paper is represented by basic members of the Cambrian spilite-keratophyre formation of the Western Sayan Mts (Central Asia, USSR), described in detail by VIELINSKII (1968). In his interesting, exhaustive monograph, he presents numerous geological, petrographical and mineralogical data and a set of about 150 chemical analyses of the rocks in question, 71 of which refer to the basic members. Thus it is the largest and most valuable source of analytical material used for statistical examination in this paper. It would be impossible to discuss even briefly all of the interesting considerations of VIELINSKII except for some of the most important data and opinions. Of utmost theoretical importance is the discovery of high-temperature albite in spilites of one of the subformations of this series, determined by means of X-ray diffractometric and optical methods. Consequently the rocks of this spilite-keratophyre subformation are, in VIELINSKII's opinion, of magmatic origin. It should be emphasized that the mechanism of differentiation proposed independently by VIELINSKII is very similar to that described by the present writer for spilitic series of the Góry Kaczawskie Mts (NARĘBSKI, 1964). According to VIELINSKII, the fractionation of primary homogeneous, volatile-rich melt would proceed under increasing oxygen pressure, being due to the interaction of gravity and diffusion processes, and would result in the formation of a spilitic fraction in the upper part of the magma chamber, abounding in volatiles, alkalies and ferric iron. Among the initial volcanites of the Western Sayans there also occur rocks of the "spilite-diabase subformation" of late magmatic origin, containing low-temperature autometasomatic albites. The last-stage magmatism of this region is represented by rocks of the "porphyritic formation". In VIELINSKII's opinion they are products of submarine eruptions of an alkali-low and volatile-deficient portion of the parent magma from the lower part of the differentiated chamber.

Pre-Cambrian spilitic series are represented in this paper by metabasites of the Hecla Hoek Succession occurring in the Hornsund region, Vestspitsbergen (BIRKENMAJER and NARĘBSKI, 1960; NARĘBSKI, 1960, 1966; SMULIKOWSKI, 1965) and by greenschists and albite diabases of the NW part of the Karelides (Fennoscandian Lapland) in Finland (MERILÄINEN, 1961) and Norway (GJELSVIK, 1958). The Spitsbergen meta-volcanites form two stratigraphically equivalent Skålfjellet and Vimsodden series, the latter being less metamorphosed and distinctly spilitic in character. Such two-stage development of initial volcanism is observed often (Góry Kaczawskie, W Sayans). Since in general only one of these phases produces distinctly spilitic rocks, this fact must be taken into account in petrogenetic considerations.

The origin of greenstones and associated albite diabases and albitites occurring in the NW part of the Karelian chain is unclear. GJELSVIK (1958) suggested that albitization in them cannot be connected with metamorphism but rather with late magmatic processes, whilst MERI-LÄINEN (1961, p.67) stated that "in all probability the differentiation of the parent olivine-basaltic magma, assisted by volatiles as well as tectonic movements, have also given rise to spilitic and keratophyric extrusives". Some differentiation took place also during the emplacement of magma within albite-diabase sills.

The last rock series statistically examined here consists of basic members of the Devonian spilite-keratophyre association of Schirmeck massif in the Vosges Mts, investigated in detail by JUTEAU and ROCCI (1966). This "hercynotype" initial volcanic geosynclinal series (ROCCI and JUTEAU, 1968), composed of lavas and tuffs, exhibits distinct late-magmatic silification phenomena which results in an acidic character of even its most basic members. Consequently, from abundant chemical data only 13 analyses of effusive rocks, containing less than 55 wt. % silica, could be selected for statistical examination: 6 microlitic diabases, 3 dolerites, 2 spilites, 1 porphyric diabase and 1 quartz diabase. All other analyses are of keratospilites, keratophyres and pyroclastic rocks. Considerable diversity among initial volcanic rocks occurring within the Hercynian belt of Central Europe is thought to be connected with hybridization of the parent basaltic magma by anatectic melts from acid basement rocks and siliceous detrital geosynclinal deposits (ROCCI and JUTEAU, 1968).

Methods of Statistical Examination

The essential advantage of examining the sets of chemical analyses representing rock associations using statistical methods lies in the possibility of computing not only their average elemental contents, but also of determining important and only recently well-appreciated petrochemical features such as the range of variability within these associations and of quantitative comparison of these variabilities in different suites. Moreover it is well known that the use of arithmetic mean and standard deviation of an element in a comparative or

Table 1

Al_2O_3 content (ordered) wt. percent	Cumulative frequency, percent (cell midpoint)	Al_2O_3 content (ordered) wt. percent	Cumulative frequency, percent (cell midpoint)
13.1	2.5	15.5	52.5
14.1	7.5	15.7	57.5
14.4	12.5	15.7	62.5
14.6	17.5	15.8	67.5
14.7	22.5	15.9	72.5
14.9	27.5	15.9	77.5
15.1	32.5	16.1	82.5
15.1	37.5	16.1	87.5
15.1	42.5	16.6	92.5
15.2	47.5	17.1	97.5

Number of samples - 20, cell interval $\frac{100}{20} = 5$

correlative study is fully correct only if the frequency distribution of the latter is normal or closely approaches normality.

Thus the first step of any statistical examination of analytical data is to estimate the frequency distribution of individual elements. This important statistical feature can be easily approximated using the normal (arithmetic-) or lognormal-scale probability paper. This procedure is almost the only method for examining small sets of data, i.e., 10-20 chemical analyses, which represent the most common sample size for the majority of petrological papers. The data are as usual ordered from the smallest to the largest (without grouping into classes), and individual values are plotted on the probability paper according to percent cumulative frequencies, corresponding to the cell midpoints between adjacent cumulative frequencies. This method is illustrated (Table 1) by the percent cumulative frequencies for Al_2O_3 distribution in the spilitic rock series of the Rudawy Janowickie Mts.

All cumulative frequency distribution diagrams for the examined rock series except for that of the West Sayan formation have been constructed using this graphical method.

If the number of analyses is large enough, as in the case of the West Sayan spilitic series (n = 70), the data are first divided into classes, the number and interval of which are so adjusted as to obtain the clearest picture of the distribution. For convenience the number of classes should be odd and should correspond approximately to $\sqrt{n}$ (n - number of samples), usually amounting to 5, 7 or 9. Using data arranged in frequency tables which are divided into convenient classes, it is easy to construct frequency distribution histograms for individual elements, illustrating qualitatively their distribution patterns (Figs. 1 and 2). Since practically no one empirical frequency distribution curve corresponds exactly to a theoretical one, (e.g. normal, lognormal, Poisson's), it is necessary to apply a statistical test of goodness of fit, whereby it may be concluded if the former matches some of the latter.

Among several proposed statistical testing methods, the most universal, particularly for small sets of analyses, is the computation of the first four moments of the distribution under examination (GRIFFITHS, 1962; SHARAPOV, 1965; RODIONOV, 1964; BONDARENKO, 1967). It should be remembered that by means of moments, we can compute four essential para-

Table 2

Class limits wt. percent	Class midpoint X_i	Frequency f	t	ft	ft^2	ft^3	ft^4
2 - 3	2.5	2	-3	- 6	18	-54	162
3 - 4	3.5	7	-2	-14	28	-56	118
4 - 5	4.5	11	-1	-11	11	-11	11
5 - 6	5.5	19	0	0	0	0	0
6 - 7	6.5	13	+1	13	13	13	13
7 - 8	7.5	8	+2	16	32	64	128
8 - 9	8.5	7	+3	21	63	189	567
totals		67		19	165	145	1049
Arithmetic means of the columns i.e. moments around the assumed mean $\bar{X}_0$				0.29 n_1	2.46 n_2	2.17 n_3	15.66 n_4
Class interval (C) = 1, assumed mean $(\bar{X}_0)$ = 5.5							

meters of the distribution curve: arithmetic mean, variance (standard deviation), skewness (asymmetry) and kurtosis (peakedness).

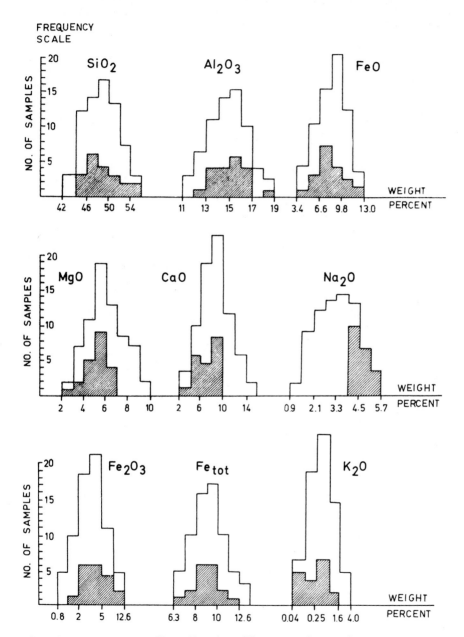

Fig. 1. Frequency distribution diagrams for major elements in basic rocks of the West Sayan spilite-keratophyre formation. Chemical data after VIELINSKII (1968). Cross hatched segments of histograms include samples containing more than 4 wt. % Na_2O. Percent contents of SiO_2, Al_2O_3, FeO, MgO, CaO and Na_2O in linear scale, while those of Fe_2O_3, Fe_{total} and K_2O in logarithmic one

The procedure under consideration may be illustrated by constructing the frequency distribution histogram and computing the moments for MgO distribution in the West Sayan spilites. These calculations are simplified considerably by working with an assumed mean ($\bar{X}_O$) and in units of the cell interval ($t = \frac{X_i - \bar{X}_O}{C}$) as well as by using the grouped frequency table method (Table 2).

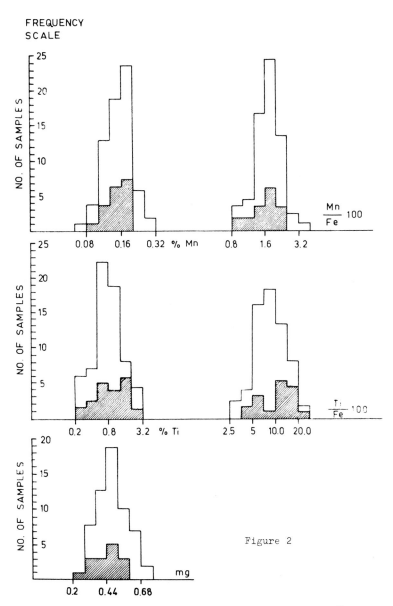

Figure 2

Fig. 2. Frequency distribution diagrams for Ti, Mn, NIGGLI mg, $\frac{Mn}{Fe}100$, and $\frac{Ti}{Fe}100$ in basic rocks of the West Sayan spilite-keratophyre formation. Cross hatched segments of histograms include samples containing more than 4 wt. % Na_2O. Contents of elements and ratio values in logarithmic scale

From the above data, we can easily compute the first four central moments (about the estimated mean $\bar{X}$):

First central moment: $m_1 = C(n_1 - n_1) = 0$

$\bar{X} = \bar{X}_0 + C \cdot n_1 = 5.5 + 0.29.1 = 5.79$

Second central moment (= variance):
$m_2 = C^2 (n_2 - n_1^2) = 1 (2.46 - 0.08) = 2.38$
$S = 1.54$ (standard deviation)

Third central moment:
$m_3 = C^3 (n_3 - 3n_2 n_1 + 2n_1^3) = 1 (2.17 - 2.12 + 0.05) = 0.10$

Fourth central moment:
$m_4 = C^4 (n_4 - 4n_3 n_1 + 6n_1^2 n_2 - 3n_1^4) = 1 (15.66 - 2.42 + 1.33 - 0.03) = 14.54$

From the above moments we compute the conventional measures characterizing the asymmetry (skewness) and peakedness of the frequency distribution curve:

Asymmetry (skewness):
$Sk = \dfrac{m_3}{S^3} = \dfrac{0.10}{3.55} = 0.03$

Kurtosis (peakedness):
$K = \dfrac{m_4}{S^4} - 3 = 2.56 - 3 = -0.44$

Using these statistics and theoretical tables, we may finally determine whether the asymmetry and peakedness of the curve under consideration do not exceed the critical values of these features for the given number of samples. We find in statistical tables that, for this sample size (n = 67), the critical values of Sk and K at the 5% significance level are 0.56 and 1.00, respectively. Since the computed values for our empirical curve are distinctly lower than the tabulated values, we may conclude that the frequency distribution of MgO in the West Sayan spilitic rock series closely approaches normality.

Similar distribution histograms and computations have been carried out for all the other elements in the West Sayan suite (Tables 3 and 4, Figs. 1 and 2). Moreover, the frequency distribution patterns of major and minor elements in other spilitic rock series displaying smaller sample sizes has also been determined using the method of moments. The computed statistics and corresponding critical values of skewness and kurtosis for the given sample sizes at the 5% and 1% significance levels are detailed in Tables 3 and 4.

Preliminary Statistical Comparison of the Examined Spilitic Rock Series with a Related Magmatic Rock Formation

Before discussing the results of statistical examination of the spilitic rock series in question, let us compare their most specific petrochemical features with those of another possibly chemically related basic rock association. A detailed statistical study of the latter formations has been performed recently by KUTOLIN (1967, 1968), the results of which are detailed in Table 5. Preliminary comparison of frequency distribution statistics for some basaltic rock formations with the data obtained by the writer indicates that spilitic rock series in general differ from the former in that the CaO content is lower than average and the Na_2O content, higher than average, and their concentrations are more dispersed. It is evident that among

Table 3. Average content and statistics of frequency distributions for major elements of the examined spilitic rock series

$\overline{X}$ - arithmetic mean of oxide content (in wt. %)
S - standard deviation
$\overline{lgX}$ - arithmetic mean of logarithms of oxide content
SlgX - standard deviation of logarithms of oxide content
$\tilde{X}$ - geometric mean of oxide content (in wt. %)
$V_\%$ - coefficient of variation
Sk - skewness (asymmetry) of the distribution curve
K - kurtosis (peakedness) of the distribution curve

	Western Sayans (n = 70)					
	$\overline{X}$	$\tilde{X}(\overline{lgX})$	S(SlgX)	$V_\%$	Sk	K
SiO_2	48.48		3.66	7.6	0.11	-0.63
Al_2O_3	14.97		2.19	14.5	0.03	-0.90
Fe_2O_3		3.12(0.494)	(0.244)	49	0.13	-0.46
FeO	8.35		2.20	26	0.11	-0.27
Fe_{tot}	9.01		1.55	17	0.51	0.04
MgO	5.79		1.54	27	0.03	-0.54
CaO	8.24		2.62	31	0.04	-0.93
Na_2O	3.37		1.04	32	-0.29	-0.51
K_2O		0.34(1.535)	(0.419)	78	0.11	-0.60
mg	0.477		0.108	23	0.03	-0.50
5% level					0.56	1.00
1% level					0.73	1.34

	Spitsbergen (n = 18)					
	$\overline{X}$	$\tilde{X}(\overline{lgX})$	S(SlgX)	$V_\%$	Sk	K
SiO_2	49.25		2.36	4.8	0.29	-1.28
Al_2O_3	15.83		2.33	15	0.03	-0.27
Fe_2O_3		3.87(0.588)	(0.178)	30	0.10	0.15
FeO	5.96		1.81	30	0.11	-1.00
Fe_{tot}	7.73		1.30	17	-0.15	-0.02
MgO	5.94		1.01	18	0.33	-0.42
CaO	8.90	bimodal distribution				
Na_2O	3.56		0.90	25	-0.30	-1.00
K_2O	1.07		0.47	44	-0.05	-0.88
mg	0.518		0.077	15	0.08	-0.24
5% level					0.95	1.41
1% level					1.27	1.95

Distribution is considered to be normal if Sk ≤ $Sk_{5\%}$ and K ≤ $K_{5\%}$ and to be not normal if Sk > $Sk_{1\%}$ or K > $K_{1\%}$

Table 3 (cont.)

X̄   - arithmetic mean of oxide content (in wt. %)
S   - standard deviation
lgX̄ - arithmetic mean of logarithms of oxide content
SlgX - standard deviation of logarithms of oxide content
X̃   - geometric mean of oxide content (in wt. %)
V%   - coefficient of variation
Sk   - skewness (asymmetry) of the distribution curve
K   - kurtosis (peakedness) of the distribution curve

	Rudawy Janowickie Mts (n = 22)					
	X̄	X̃(lgX̄)	S(SlgX)	V%	Sk	K
SiO_2	51.61		3.16	6.2	0.11	-0.65
Al_2O_3	15.32		0.96	6.3	0.14	-0.54
Fe_2O_3		3.52(0.547)	(0.150)	27	0.42	0.36
FeO	7.48		1.54	20	0.43	-1.42
Fe_{tot}	8.67		1.53	18	-0.18	-0.20
MgO	5.75		1.47	26	0.09	0.05
CaO	6.72		1.81	27	0.02	-0.50
Na_2O	4.35		0.98	23	0.13	-0.85
K_2O		0.35(1.545)	(0.296)	54	0.32	-0.66
mg	0.458		0.075	17	0.09	0.97
5% level					0.90	1.40
1% level					1.20	1.92

	NW Karelides (n = 21)					
	X̄	X̃(lgX̄)	S(SlgX)	V%	Sk	K
SiO_2	50.14		2.27	4.5	0.06	-0.81
Al_2O_3	13.77		1.67	12	0.03	-0.17
Fe_2O_3		4.47(0.65)	(0.224)	37	0.10	-0.66
FeO	8.98		2.24	25	0.07	-0.80
Fe_{tot}		9.82(0.99)	(0.088)	8.9	0.03	-0.82
MgO	6.47		1.77	27	0.23	-0.80
CaO	7.24	bimodal distribution				
Na_2O	4.11		1.01	24	0.20	-1.05
K_2O		bimodal distribution				
mg	0.474		0.101	21	0.23	-0.57
5% level					0.91	1.40
1% level					1.22	1.93

Distribution is considered to be normal if $Sk \leq Sk_{5\%}$ and $K \leq K_{5\%}$ and to be not normal if $Sk > Sk_{1\%}$ or $K > K_{1\%}$

Table 3 (cont.)

$\bar{X}$ - arithmetic mean of oxide content (in wt. %)
S - standard deviation
$\overline{lgX}$ - arithmetic mean of logarithms of oxide content
SlgX - standard deviation of logarithms of oxide content
$\tilde{X}$ - geometric mean of oxide content (in wt. %)
$V_\%$ - coefficient of variation
Sk - skewness (asymmetry) of the distribution curve
K - kurtosis (peakedness) of the distribution curve

	Góry Kaczawskie Mts (n = 18)					
	$\bar{X}$	$\tilde{X}(\overline{lgX})$	S(SlgX)	$V_\%$	Sk	K
SiO_2	47.43		3.14	6.6	-0.13	-0.54
Al_2O_3	14.94		1.79	12.0	-0.14	-0.98
Fe_2O_3	4.99		2.32	46	0.09	1.96
FeO	5.71		2.32	40	-0.10	1.97
Fe_{tot}	8.20		1.92	23	-0.13	-1.04
MgO	6.01		1.91	31	-0.18	-0.80
CaO	8.90		2.32	26	-0.14	-0.96
Na_2O	4.23		1.19	28	0.35	-0.84
K_2O		0.40(1.60)	(0.300)	50	0.02	-1.26
mg	0.492		0.101	21	-0.03	-1.93
5% level					0.95	1.41
1% level					1.27	1.95

	Vosges (n = 13)					
	$\bar{X}$	$\tilde{X}(\overline{lgX})$	S(SlgX)	$V_\%$	Sk	K
SiO_2	52.67		2.91	5.5	-0.44	-1.05
Al_2O_3	15.90		1.05	6.9	-0.10	-0.84
Fe_2O_3	no data					
FeO	no data					
Fe_{tot}	7.93		0.67	8.5	-0.05	-0.95
MgO	4.99		1.10	22	0.05	-1.39
CaO	3.92		2.38	61	0.11	-1.12
Na_2O	4.71		1.17	25	0.09	-1.66
K_2O		1.17(0.068)	(0.233)		-0.17	-0.78
mg	0.490		0.084	17	-0.11	-0.85
5% level					1.05	1.42
1% level					1.40	-

Distribution is considered to be normal if $Sk \leq Sk_{5\%}$ and $K \leq K_{5\%}$ and to be not normal if $Sk > Sk_{1\%}$ or $K > K_{1\%}$

Table 4. Average content and statistical parameters of frequency distributions of some trace elements in the investigated spilitic rock series (explanations of symbols as in Table 1)

Element	Rudawy Janowickie Mts (n = 21)				Góry Kaczawskie Mts (n = 17)				Spitsbergen (n = 18)			
	$\tilde{X}$ ppm	$\frac{S \lg X}{\lg X}$	Sk	K	$\tilde{X}$ ppm	S	Sk	K	$\tilde{X}$ ppm	$\frac{S \lg X}{\lg X}$	Sk	K
Ti	5010	$\frac{0.164}{3.700}$	0.25	-0.50	14040	3740	-0.31	-1.09	8320	$\frac{0.224}{3.920}$	0.09	-0.54
Mn	1010	$\frac{0.096}{3.044}$	-0.02	-0.97	1040	320	-0.34	-0.06	980	$\frac{0.186}{2.990}$	0.10	-0.24
					$\tilde{X}$	$\frac{S \lg X}{\lg X}$	Sk	K				
Cr	65	$\frac{0.447}{1.815}$	-0.30	-0.65	249	$\frac{0.167}{2.375}$	0.02	-0.71	232	$\frac{0.179}{2.366}$	0.12	-0.86
Ni	24	$\frac{0.351}{1.376}$	0.16	-0.78	82	$\frac{0.242}{1.913}$	-0.27	-0.88	53	$\frac{0.126}{1.723}$	-0.28	-0.96

Element	Western Sayans (n = 69)				NW Karelides (n = 19)				Vosges (n = 12)			
	$\tilde{X}$ ppm	$\frac{S \lg X}{\lg X}$	Sk	K	$\tilde{X}$ ppm	$\frac{S \lg X}{\lg X}$	Sk	K	$\tilde{X}$ ppm	S	Sk	K
Ti	7700	$\frac{0.253}{3.884}$	0.05	0.17	9230	$\frac{0.250}{3.965}$	-0.37	-0.74	7700	1400	–	–
Mn	1500	$\frac{0.118}{3.176}$	-0.22	-0.03	1180	$\frac{0.220}{3.072}$	-0.02	-0.24	1500	200	–	–

Table 5. Arithmetic mean, standard deviations and coefficients of variation (X) for major elements in basalts of various magmatic rock formations (after KUTOLIN, 1968)

Oxide	Basalts of the andesitic formation (n = 355)			Basalts of the continental olivine-basalt formation (n = 276)			Tholeiitic basalts of the oceanic olivine-basalt formation (n = 110)			Alkaline olivine basalts of the oceanic olivine-basalt formation (n = 118)		
	$\overline{X}$	S	$V_\%$	$\overline{X}$	S	$V_\%$	$\overline{X}$	S	$V_\%$	$\overline{X}$	S	$V_\%$
SiO_2	50.86	1.96	3.9	47.78	2.21	4.6	49.15	1.33	2.7	45.83	2.40	5.2
TiO_2	1.04	0.43	42	2.22	0.74	33	2.09	0.52	25	3.06	0.86	28
Al_2O_3	17.73	1.73	9.8	15.33	1.87	12	15.09	1.04	7.0	14.99	2.13	14
Fe_2O_3	3.92	1.71	44	4.09	2.05	50	3.35	1.36	41	3.90	1.58	41
FeO	6.33	2.22	35	7.51	1.88	25	7.56	1.28	17	8.43	1.77	21
MnO	0.17	0.11	65	0.15	0.15	100	0.17	0.02	12	0.14	0.09	64
MgO	5.37	1.70	32	6.99	1.98	28	7.75	1.23	16	7.65	2.58	34
CaO	9.81	1.25	13	9.00	1.32	15	10.61	0.70	6.6	10.02	1.44	14
Na_2O	2.69	0.68	25	2.85	0.78	27	2.23	0.40	18	2.91	0.45	16
K_2O	1.00	0.54	54	1.31	0.55	42	0.30	0.15	50	1.18	0.24	20

(X) computed by W. NARĘBSKI
-- oxides, the content of which is characteristic for the given formation

$\tau \dfrac{Al_2O_3 - Na_2O}{TiO_2}$	14.45	5.62	4.79	3.95

Table 6. Comparison of parameters of CaO and Na$_2$O frequency distributions in the investigated spilitic rock series with those of the continental olivine-basalt formation using Student's t and Fisher's F tests

Rock series	n	CaO $\bar{X}$	CaO s^2	Na$_2$O $\bar{X}$	Na$_2$O s^2	F CaO	F Na$_2$O	F$_{0.05}$	F$_{0.01}$	t CaO	t Na$_2$O
Western Sayans	71	8.24	6.86	3.37	1.08	3.95	1.76	1.34	1.51	–	–
Rudawy Janowickie	22	6.72	3.28	4.35	0.96	1.87	1.58	1.57	1.87	5.85	5.86
Góry Kaczawskie	18	8.90	5.38	4.23	1.41	3.08	2.30	1.64	2.00	–	–
Spitsbergen	18	bimodal		3.56	0.81	–	1.33	1.64	2.00	–	2.54
NW Karelides	21	bimodal		4.11	1.02	–	1.64	1.61	1.97	–	4.75
Vosges	13	3.92	5.66	4.71	1.37	3.25	2.24	1.76	2.20	–	–
Continental olivine-basalt formation (after KUTOLIN, 1968)	276	9.00	1.74	2.85	0.61						

The two compared sets of data belong to the same normal population if $F \leq F_{0.05}$ and $t < 1.96$, and to different populations if $F > F_{0.01}$ and $t > 2.58$

——— significant at the 5% level
- - - significant at the 1% level

KUTOLIN's data (Table 5) the statistics of CaO and Na_2O distribution in basalts of the continental olivine-basalt formation approach most closely our data for the spilitic rock series.

To determine whether some of the latter series belong to the olivine-basalt formation, a preliminary comparison of CaO and Na_2O frequency distribution statistics has been carried out using Fisher's "F" and Student's "t" tests (Table 6). As follows from the results obtained from the six examined spilitic series, only the Spitsbergen metabasites, and there only as regards the frequency distribution of sodium, can be conditionally and at the lowest probability level assigned to "normal" basalts. All other spilitic series examined, no matter what their local peculiarities, which will be discussed later, display distinctly specific petrochemical features and ought to be considered as statistically separate populations.

Average Content and Frequency Distribution of Elements in the Examined Spilitic Rock Series

Inspection of the obtained diagrams and of the results of statistical testing clearly indicates that there is a distinct difference in the frequency distribution patterns of major and minor elements in the rock series studied. While the distributions of "rock-forming" elements (except trivalent iron and potassium) closely approach normality (Table 3, Figs. 1 and 3), most of the trace elements display a lognormal distribution pattern (Table 4, Figs. 2 and 5). This phenomenon, which is concordant with the general data on the distribution of chemical elements in volcanic rocks (BONDARENKO, 1967), is of great theoretical importance. In spite of the diversified opinions on the causes of different distribution laws (JIZBA, 1959; VISTELIUS, 1960; SHAW, 1961; ROGERS and ADAMS, 1963; RODIONOV, 1964; BUTLER, 1964), based on this conclusion, when comparing the chemical compositions of various volcanic and spilitic rock series, we can safely apply simple, well-known statistical tests, the power of which is maximal for populations displaying normal frequency distributions. Concerning the theoretical explanations of distribution laws of elements in rocks, the difference between the major and minor elements is supposed to depend on their relative concentrations in solidifying magma (JIZBA, 1959; SHAW, 1961; ROGERS and ADAMS, 1963) and on their distribution among rock-forming minerals (RODIONOV, 1964). Generally abundant elements found in several rock-forming minerals (e.g. silica, alumina, calcium) show nearly normal frequency distributions, while trace elements, often concentrated in one or very few minerals (e.g. zircon, titanium, chromium) approach lognormal ones. Consequently the latter elements, especially those sensitive to crystal fractionation (Cr, Ni) separate early from the melt, displaying distinctly concave curves of their variation trends against an evolution index of rocks (TURNER and VERHOOGEN, 1960, Figs. 30 and 43). It should be remembered that according to BUTLER (1964) such a variation pattern corresponds to lognormal frequency distribution, while the more complicated concentration trends of major elements result in normal frequency distributions. We may conclude, therefore, that there is in fact no essential discordance among the various theoretical explanations of the distribution laws of elements.

Examining the statistics of the distribution of major elements in the examined spilitic rock series, we observed distinct similarities in their specific petrochemical features, i.e., lowered calcium content, increased sodium concentration and high coefficients of variations of both elements. The content of iron and magnesium was relatively constant, the only exception being the Vosges series, which exhibited increased acidity.

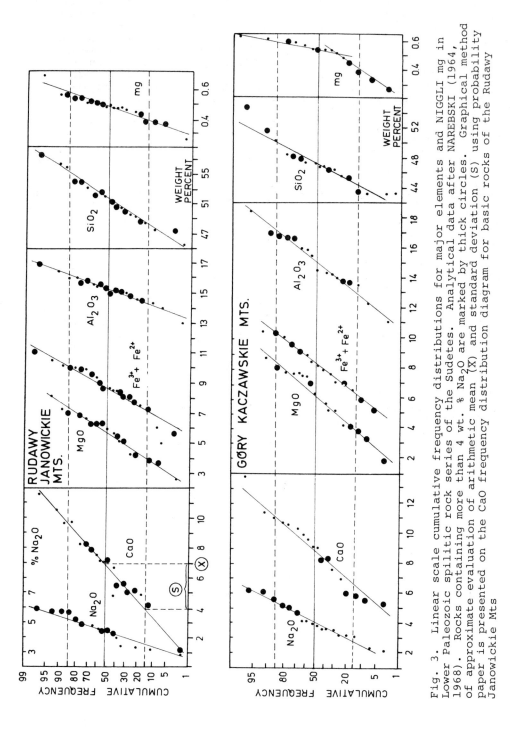

Fig. 3. Linear scale cumulative frequency distributions for major elements and NIGGLI mg in Lower Paleozoic spilitic rock series of the Sudetes. Analytical data after NAREBSKI (1964, 1968). Rocks containing more than 4 wt. % Na$_2$O are marked by thick circles. Graphical method of approximate evaluation of arithmetic mean (X) and standard deviation (S) using probability paper is presented on the CaO frequency distribution diagram for basic rocks of the Rudawy Janowickie Mts

A comparison of Lower Paleozoic rock series from the Rudawy Janowickie Mts and the Góry Kaczawskie Mts, which are territorially adjacent and only slightly different in age, was very instructive. The latter mountain range is evidently distinguished from the former by much more variable ferric and ferrous iron contents as well as by lower silica and higher calcium contents and NIGGLI mg ratio. Much more pronounced differences are observed in their trace element contents (Table 4). The spilitic suite of the Góry Kaczawskie Mts, in accordance with its more basic character, is distinguished by large amounts of Ti, Cr and Ni and by an increased Ti/Fe ratio, characteristics approaching those typical for trachybasaltic rocks (Tables 8 and 9). On the other hand, the spilitic rocks of the Rudawy Janowickie Mts are exceptionally low in Ti, Cr and Ni.

It should be emphasized that, as one would suppose from the obtained data (Table 4), large quantities of titanium and small quantities of alumina are not characteristic geochemical features of spilites, as was hypothesized from the SUNDIUS averages, computed from 19 analyses of spilites from Britain, Australia and Fennoscandia (TURNER and VERHOOGEN, 1960, p.262). This conclusion is supported not only by our statistical computations but also by the data of DEWEY and FLETT (1911), WELLS (1923) and VALLANCE (1960) and by the average composition of the Ural spilites (SHARFMAN, 1968) which are very low in titanium (Table 7)

Table 7. Average chemical composition and τ ratio of spilites

	Average spilite after				Average Ural spilite after SHARFMAN (1968)
	VALLANCE (1960)	DEWEY and FLETT (1911)	WELLS (1923)	SUNDIUS (1930)	
SiO_2	49.65	48.58	46.01	51.22	52.45
TiO_2	1.57	1.77	2.21	3.32	0.66
Al_2O_3	16.00	14.58	15.21	13.66	15.05
Fe_2O_3	3.89	1.89	1.35	2.84	2.66
FeO	6.08	7.65	8.69	9.20	7.54
MnO	0.15	0.46	0.33	0.25	0.14
MgO	5.10	6.36	4.18	4.55	5.91
CaO	6.62	9.80	8.64	6.89	7.23
Na_2O	4.29	4.02	4.97	4.93	3.64
K_2O	1.28	0.43	0.34	0.75	0.40
P_2O_5	0.26	0.19	0.61	0.29	0.42
CO_2	1.63	-	4.98	0.94	-
τ	7.46	5.97	4.63	2.63	17.29

Average τ ratios for the examined spilites

Western Sayans	8.99
Rudawy Janowickie Mts	13.22
Góry Kaczawskie Mts	7.65
Spitsbergen	8.83
Karelides	6.27
Vosges	8.67

and which usually display the same Al_2O_3 content as other basaltic rocks.

The content of Al, Na and Ti in volcanic rocks was recently discussed by RITTMANN (1967), who regards the ratio $\tau = (Al_2O_3 - Na_2O) : TiO_2$ as an important petrogenetical indicator. Basalts which originate from upper mantle material at great depths are characterized by low τ values, while basic rocks of orogenic suites resulting from anatectic remelting of crustal substance display high τ ratios. This may be exemplified by KUTOLIN's values for the four basaltic formations

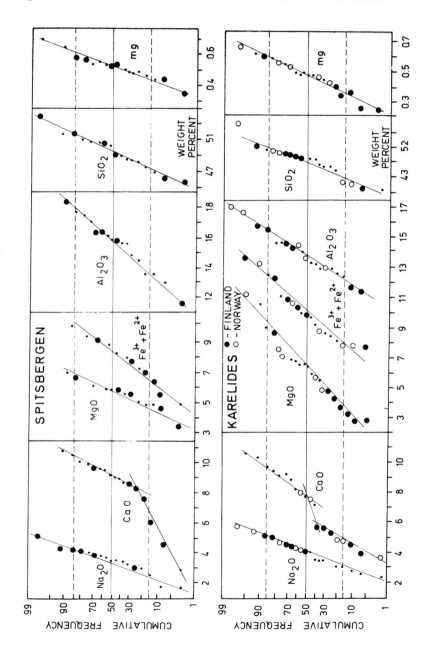

(Table 5, bottom). Corresponding τ values for the examined spilite suites are given in Table 7 together with the data for average spilite composition.

In the writer's opinion, the observed diversity and intermediate values of the τ ratio, as well as all of the above-presented geochemical differences between individual associations, are convincing arguments of their possible origin from various parental magmas in the course of different but convergent petrogenetic processes. It is possible, following RITTMANN's suggestion (personal communication),

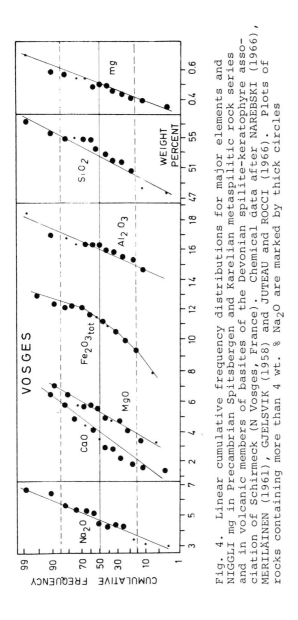

Fig. 4. Linear cumulative frequency distributions for major elements and NIGGLI mg in Precambrian Spitsbergen and Karelian metaspilitic rock series and in volcanic members of basites of the Devonian spilite-keratophyre association of Schirmeck (N Vosges, France). Chemical data after NAREBSKI (1966), MERILÄINEN (1961), GJELSVIK (1958) and JUTEAU and ROCCI (1966). Plots of rocks containing more than 4 wt. % Na_2O are marked by thick circles

146

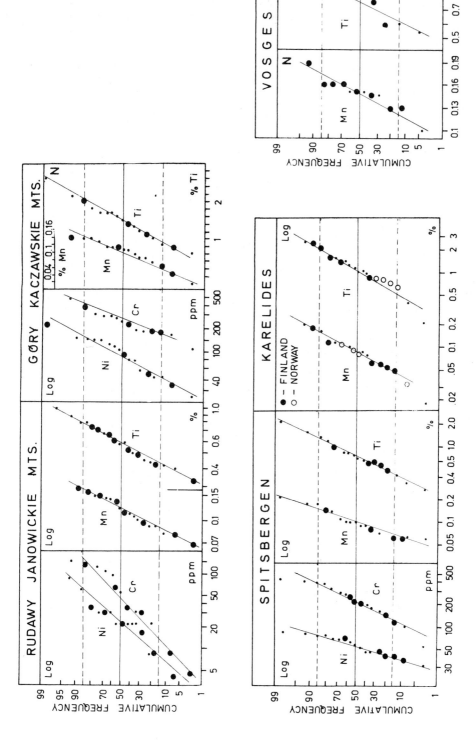

that spilitic associations displaying a high τ ratio are of secondary origin. They could be altered "high-alumina basalts" of final volcanism connected with a preceding orogenic phase, albitized during the next geosynclinal state (e.g. Hercynian, after the Caledonian period). Thus the Rudawy Janowickie Mts and Ural Mts suites really represent geosynclinal regions which were subjected to more than one orogenic phase.

Very interesting and instructive is the examination of the position of distinctly spilitic varieties within individual rock series. In order to illustrate these petrogenetically important observations, the plots of Na-enriched and Ca-deficient rocks on frequency distribution diagrams are marked by different symbols (Figs. 1-6).

As can be seen from the distribution of areas of Na-enriched members within the histograms for all rocks of the W Sayan spilitic series, the former are lower in CaO, MgO and K_2O, slightly enriched in Fe_2O_3 and Ti and exhibiting a higher Ti/Fe ratio. Moreover on the average they display lower NIGGLI mg numbers (Figs. 1 and 2).

Inspection of the position of plots of rocks containing more than 4 wt. % Na_2O within the cumulative frequency distribution diagrams for the Sudetic spilitic associations, clearly confirms previously reported differences between the Rudawy Janowickie Mts and Góry Kaczawskie Mts series (Figs. 3, 5 and 6). In the former the plots of the sodium-enriched members are evenly dispersed along the linear diagrams (except CaO), while the definitely spilitic varieties of the Góry Kaczawskie suite are evidently lower in CaO and MgO and a little higher in SiO_2 and Al_2O_3. Thus, as regards Ca and Mg deficiency, there is a clear analogy between sodium-rich members of both Cambrian associations - of West Sayans and of Góry Kaczawskie Mts. It should be emphasized that this analogy is not accidental. As was already mentioned, petrogenetic conclusions of VIELINSKII (1968) and of this writer (NAREBSKI, 1964) are very close, explaining the spilitic evolution of the parent magmas by diffusion differentiation of homogeneous magma, causing upward migration of volatiles, alkalies and easily fusible elements (e.g., ferric iron).

A somewhat different conclusion may be drawn from examining the cumulative frequency distribution diagrams for major elements of Precambrian Spitsbergen metabasites and Karelian greenschist - albite diabasic rock series. The most characteristic petrochemical features of these territorially not-very-distant suites, is bimodality of CaO distribution, the Na-enriched members forming a separate, calcium-deficient group. Moreover the Norwegian and Finnish spilitic members differ in that the MgO content of the latter is lower. Consequently the soda-rich basites of Finnmark display higher NIGGLI mg numbers. These petrochemical data would indicate a complicated and probably multistage evolution of parent magma of these series, resulting in the development of several individualized differentiation products. Some are due to a differentiation within sills, actually often composed of several rock types (MERILÄINEN, 1961). Unlike previously described associations the differentiation of these series could be connected, at least partly, with fractional crystallization, since Na-enriched members of the Spitsbergen metabasites are distinctly

---

◄ Fig. 5. Logarithmic (Log) and linear scale (N) cumulative frequency distributions for some trace elements in various spilitic rock series. Chromium and nickel contents in ppm, Mn and Ti, in wt. %. Analytical data after NAREBSKI (1964, 1966, 1968), MERILÄINEN (1961), GJELSVIK (1958) and JUTEAU and ROCCI (1966). Plots of rocks containing more than 4 wt. % Na_2O are marked by thick circles

Table 8. Average titanium (iron and manganese) iron ratios and their standard deviations in the investigated spilitic rock series

Ratio	Western Sayans		Rudawy Janowickie		Góry Kaczawskie		Spitsbergen		Karelides		Vosges	
	$\tilde{x}$	$\frac{SlgX}{lgX}$	$\tilde{x}$	$\frac{SlgX}{lgX}$	$\tilde{x}$	$\frac{SlgX}{lgX}$	$\tilde{x}$	$\frac{SlgX}{lgX}$	$\tilde{x}$	s	$\tilde{x}$	s
$\frac{Ti \cdot 100}{Fe}$	8.50	$\frac{0.206}{0.929}$	5.54	$\frac{0.158}{0.743}$	17.6	$\frac{0.131}{1.245}$	10.28	$\frac{0.224}{1.010}$	10.12	3.84	9.78	1.26
			$\bar{x}$	s	$\bar{x}$	s			$\tilde{x}$	$\frac{SlgX}{lgX}$		
$\frac{Mn \cdot 100}{Fe}$	1.69	$\frac{0.122}{0.227}$	1.34	0.33	1.31	0.38	1.35	$\frac{0.213}{0.130}$	1.02	$\frac{0.248}{0.068}$	1.95	0.21

Explanations of symbols as in Table 1.

Table 9. Average titanium (iron and manganese) iron ratios in basic rocks (45 - 55% SiO_2) of various magmatic formations (after ABRAMOVICH, 1964)

Ratio	Andesite-dacite formation	Andesite-basalt and spilite-keratophyre formation	Trachybasaltic formation
Ti/Fe·100	10.4	7.4	15-16
Mn/Fe·100	2.16	1.67	no data

lower in Ni and Cr, and to lesser extent in Ti. The same is true of
the titanium content in Norwegian soda-rich rocks of the Karelian zone
(Fig. 5).

The distribution of major elements in the basic effusive members of
the Devonian spilite-keratophyre association of the Vosges suggests
a simple origin of this silica-rich suite, representing initial volcanites of the Central European Hercynian chain. Its only distinctive
petrochemical feature is positive skewness of the iron frequency
distribution curve. This phenomenon may be connected with the paragenetic relation of well-known iron deposits to this submarine volcanism (JUTEAU and ROCCI, 1966, p.95). It should also be emphasized
that the Vosges series in question is considered to represent a
characteristic "hercynotype" spilite-keratophyre association, in
contrast to the more basic "alpinotype" ophiolitic one (ROCCI and
JUTEAU, 1968). These authors suggest that the more acid character
of the former is due to contamination of basic magma with siliceous
material from the basement and detrital geosynclinal deposits.

On the Titanium/Iron and Mangenese/Iron Ratios in Spilitic Rocks

ABRAMOVICH and his co-workers recently published a series of papers
on the petrogenetic significance of the Ti/Fe and Mn/Fe ratios in
igneous rocks (ABRAMOVICH et al. 1963; ABRAMOVICH and VYSOKOOSTROVSKAYA,
1964). They feel that spilitic associations are usually lower in Ti
than corresponding rocks of the calc-alkaline series and, like other
volcanic suites, display no variation in the Mn/Fe ratio during their
magmatic evolution. It should be emphasized that an increase in the
Mn/Fe ratio in acid differentiates of gabbroic magma is supposed to
be connected with fractional crystallization differentiation in deep-
seated chambers (ABRAMOVICH et al. 1963).

Generally the Ti/Fe ratio seems much more variable and thus more
valuable for petrogenetic considerations. This is also the case for
the rock series under consideration (Table 8).

Statistical interpretation of the variability of both ratios within
individual rock series suggests specific geochemical features of
various suites. Examination of the distribution of sodium-enriched
members within the cumulative frequency diagrams is also instructive -
it illustrates the variability of the Ti/Fe ratio. In both Sudetic
associations, Na-rich rocks generally display a high titanium-iron
ratio (Fig. 6). Because of rather dispersed plots on the cumulative
frequency distribution diagram for Ti in these rocks (Fig. 5), this
phenomenon should be related to the lower iron content in the sodium-
enriched members of these spilitic rock suites, as should the third
Lower Paleozoic association from West Sayan (Fig. 2). On the other
hand no such regularity in the variation of the Ti/Fe ratio is observed in either Precambrian spilitic series. The plots of the Na-
enriched rocks are evenly dispersed along the linear cumulative frequency distribution diagrams of this ratio (Fig. 6). It should be
noted that spilites of the Karelian zone in Norway (Finnmark) generally display lower Ti/Fe and Mn/Fe ratios than those of northern Finland. This phenomenon may be due to the fact that the latter are
represented exclusively by intrusives (albite diabases), which on
the average are higher in Ti and Mn than comagmatic effusive rocks
(ABRAMOVICH et al. 1963, 1964).

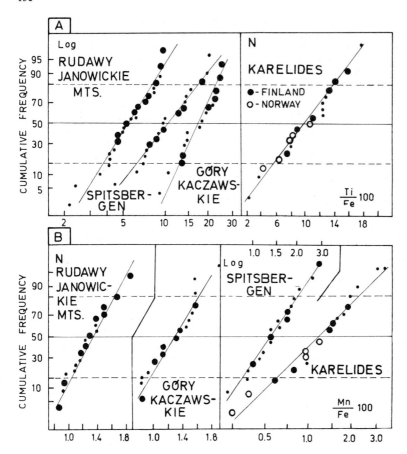

Fig. 6. Logarithmic (Log) and linear scale (N) cumulative frequency distribution diagrams for: A, titanium/iron and B, manganese/iron ratios in the examined spilitic rock series

As mentioned previously, the variations of the Mn/Fe ratio in the examined spilitic rock series exhibit no evident regularities and are therefore of much less distinct petrogenetical importance. On the other hand, even these preliminary studies indicate that a statistical examination of the variation of Ti/Fe ratio can be a valuable geochemical criterion for tracing the evolution of individual spilitic rock series.

Correlation Between Selected Elements and Indices in the Examined Spilitic Rock Series

Several years ago PODOLSKII (1963) applied linear regression analysis to determine the paragenesis of major elements in rocks of the spilite-keratophyre association in general. For computations, he collected 100 analyses of spilites and keratophyres of different ages (from Precambrian to Mesozoic) and from occurrences in the USSR, Fennoscandia, Britain, USA and Australia. This statistical examination resulted in determining positive correlations between elements constituting mafic minerals (Fe^{2+}, Mn, Mg, Ca and Al) except Si, which in this collective population exhibits only one significant correlation coefficient - with Na. PODOLSKII felt that this fact could be considered a confirmation of the correctness of ESKOLA's "spilite reaction" and thus of the opinion on secondary or late-magmatic origin of spilites.

In view of the previously stated opinion, however, it is dangerous to unite into one population the chemical analyses of rocks which, on the average, display similar chemical features and trends, but which belong to quite different associations. It should be also emphasized that PODOLSKII (1963) did not include any data concerning the frequency distribution patterns of the elements, which is a necessary preliminary operation in correlation analysis.

Taking into consideration all such doubts, the present writer has computed correlation coefficients between elements and indices, being of essential petrogenetical importance for the discussion on the origin of spilites: SiO_2, Na_2O, CaO, Al_2O_3, iron oxidation ratio (Fe_2O_3/Fe_2O_3 + FeO), NIGGLI mg, Mn/Fe ratio and the most sensitive indicators of fractional crystallization, Cr and Ni. The results of these computations, presented in Table 10, fully confirm the previously obtained data on the different behavior of elements during evolution of individual spilitic rock series.

Significant positive correlation between Na_2O and SiO_2 was observed in only two of the six investigated series (Spitsbergen and the Vosges) and nearly significant (at the 90% probability level) in two others (West Sayans and Góry Kaczawskie Mts). For the second pair of elements, theoretically covariant in "spilite reaction", there were only two significant correlation coefficients between CaO and Al_2O_3 - one positive (Vosges) and one negative (West Sayans). All remaining coefficients were negative but nonsignificant (Rudawy Janowickie, Góry Kaczawskie) or could not be computed because of bimodality of calcium frequency distributions. Si and Na display covariance in the spilite-keratophyre and andesite-basalt formations (PODOLSKII, 1963; IVANOV and KOSKO, 1965), being significantly antagonistic in the intrusive alkaline rocks of the Kola penninsula (PODOLSKII, 1962, Table 10).

A series of computations was carried out to check the relation of some elements and ratios to the NIGGLI mg number, which is a convenient index of fractional crystallization of basic magma. The following most important results were obtained:

1. significant negative correlation between Na_2O and mg in the Karelian and Vosges spilites;

2. highly significant covariance of Cr, Ni and mg in the Rudawy Janowickie Mts rocks and a less significant one in the Spitsbergen metabasites;

3. positive correlation between Mn/Fe ratio and mg in the West Sayan and Rudawy Janowickie associations.

Statistical testing of the relation between Na_2O and the iron oxidation ratio, most probably a proper index of alkali-volatile diffusion differentiation (NARĘBSKI, 1964; VIELINSKII, 1968), has shown that the correlation coefficient between them may vary from significantly positive (Góry Kaczawskie, West Sayans) to significantly negative (Rudawy Janowickie).

The above data indicate that even this preliminary correlation analysis of several spilitic rock suites revealed considerable differences in

Table 10. Coefficients of correlation between some elements and parameters in the investigated spilitic rock series

No of samples	Western Sayans 70		Rudawy Janowickie 22		Góry Kaczawskie 18		Spitsbergen 17		Karelides 21		Vosges 13	
$r_{5\%}$ $r_{1\%}$	0.234	0.308	0.42	0.54	0.47	0.59	0.48	0.61	0.43	0.55	0.55	0.68
SiO_2-Na_2O	0.20		0.18		0.41		0.50		0.33		0.89	
Al_2O_3-CaO	-0.31		-0.21		-0.27		Ca bimodal		Ca bimodal		0.68	
Na_2O-mg (NIGGLI)	-0.03		-0.20		0.33		0.30		-0.69		-0.61	
$\dfrac{Fe_2O_3}{Fe_2O_3 + FeO} - Na_2O$	0.21		-0.37		0.60		non determined		0.18		no data	
Cr-mg (NIGGLI)	no data		0.73		0.16		0.49		no data		no data	
Ni-mg	no data		0.80		0.02		0.55		no data		no data	
$\dfrac{Mn.100}{Fe} - mg$	0.47		0.40		no correlation		no correlation		no correlation		no correlation	

The probability of underlined significant correlation coefficients is greater than 95% ($r \geq r_{5\%}$), while those marked by dashed lines correspond to the level greater than 90%.

Coefficients of correlation between Si, Na, Al and Ca in some magmatic rock formations

	Spilite-keratophyre formations (after PODOLSKIJ, 1963)	Andesite-basalt formation of Kamchatka (after IVANOV and KOSKO, 1965)	Alkaline intrusives of the Kola (after PODOLSKIJ, 1962)
Si/O - Na/O	0.46	0.57	-0.63
Al/O - Ca/O	0.21	0.03	-0.08

covariance of the most characteristic chemical elements and their petrochemical indices. It is supposed that these differences are of definite petrogenetical importance, indicating indirectly the role of some differentiation mechanism in the spilitic evolution of individual rock series. Thus the results of correlation analyses seem to confirm the writer's and VIELINSKII's opinions on the significant role of alkalivolatile diffusion-differentiation of homogeneous magma in the origin of the sodium-rich fraction of the Góry Kaczawskie Mts and the West Sayan spilite associations. It may be also supposed that fractional crystallization was an important differentiation factor during the evolution of parent magmas of the Rudawy Janowickie Mts and the Spitsbergen metabasites. The same conclusion may be suggested for the origin of sodium-enriched rocks of the Karelides and Vosges, based on a strongly negative Na_2O-mg correlation coefficient. Nevertheless this assumption should be confirmed by additional data, e.g., by statistical analysis of Cr-mg and Ni-mg relations in these rock series. Moreover highly significant positive correlation coefficients between the pairs Na_2O-SiO_2 and Al_2O_3-CaO for the effusive basites of the Vosges series would suggest some role of the exchange albitization reaction, accompanied by a removal of Ca and Al. This is strictly connected with the controversial problem of the so-called "spilite reaction" illustrating the process of origin of secondary varieties of these rocks.

On the Proper Presentation of the "Spilite Reaction"

During the last two decades the classical presentation of the "spilite reaction" (ESKOLA et al. 1937) - $CaAl_2Si_2O_8 + Na_2CO_3 + 4 SiO_2 \longrightarrow 2 NaAlSi_3O_8 + CaCO_3$ - has been criticized first by TURNER (1948) and then by HENTSCHEL (1960) as not complying with the equal volume law, while albitization in spilites usually does not alter the microstructure of these rocks. Consequently TURNER (1948) proposed the following reaction for the process under consideration:

$2 CaAl_2Si_2O_8 + Na_2O + 2 SiO_2 \longrightarrow 2 NaAlSi_3O_8 + 2 CaO + Al_2O_3$

or using the labradorite formula and writing both introduced and removed elements in ionic form:

$NaCaAl_3Si_5O_{16} + Na^+ + Si^{4+} \longrightarrow 2 NaAlSi_3O_8 + Ca^{2+} + Al^{3+}$.

On the basis of this reaction SHTEINBERG (1964, p.83) calculated that if a basic rock of specific gravity 2.9 contained initially 60 wt. % labradorite (An_{50}), its complete albitization would result in introducing 6.7 wt. % SiO_2 and 3.4 wt. % Na_2O with simultaneous removal of 5.6 wt. % Al_2O_3 and 5.9 wt. % CaO. Consequently a completely albitized basalt of such composition would contain approximately 7 wt. % Na_2O and ca. 55% silica. However, a detailed comparison of chemical analyses of basalts, diabases, albite diabases and spilites reveals only negligible differences in the CaO and Na_2O content, the amounts of SiO_2 and Al_2O_3 being nearly constant. Consequently the silica-enrichment and alumina-impoverishment of the feldspathic components of the rocks in question should be compensated by corresponding gain and loss of these elements from their mafic minerals, i.e., by alteration of pyroxene into Si-lower and Al-higher actinolite, chlorite and epidote. Thus, a reaction of deanorthitization and carbonization of basic plagioclase coupled with uralitization, chloritization and epidotization of diopside may be presented as follows (NARĘBSKI, 1964):

$$10\ CaMgSi_2O_6 + 4\ CaAl_2Si_2O_8 + 8\ CO_2 + 6\ H_2O \longrightarrow Ca_2Mg_5Si_8O_{22}/OH/_2 +$$
660 cc. 400 cc. 266 cc.

$$2\ Ca_2Al_3Si_3O_{12}/OH/ + Mg_5Al_2Si_3O_{10}/OH/_8 + 8\ CaCO_3 + 11\ SiO_2.$$
268 cc. 209 cc. 296 cc. 249 cc.

From the above data on the molal volumes of initial and resultant secondary minerals, the decomposition of diopside and calcic plagioclase by carbon dioxide-bearing solutions results in the formation of typical greenschist paragenesis without any appreciable volume change, provided that a part of the newly formed calcite and silica is removed. Thus the final products of such alteration show a slight relative enrichment in soda, which is, however, incommensurate with the increased amount of this element in typical spilites. Another imperfection of the above reaction scheme lies in the quantitative relation of diopside and anorthite, which in basaltic rocks amounts to approximately 1:1 (60% labradorite An_{50} + 30% pyroxene). Thus the most adequate reaction scheme, illustrating both qualitatively and quantitatively the coupled spilitization of basalt, may be expressed:

$$10\ CaAl_2Si_2O_8 + 10\ CaMgSi_2O_6 + 3\ Na_2CO_3 + 7\ H_2O + 7\ CO_2 \longrightarrow$$
1 000 cc. 660 cc.

$$4\ Ca_2Al_3Si_3O_{12} + Mg_5Al_2Si_3O_{10}/OH/_8 + Ca_2Mg_5Si_8O_{22}/OH/_2 +$$
536 cc. 209 cc. 266 cc.

$$6\ NaAlSi_3O_8 + 10\ CaCO_3.$$
600 cc. 370 cc.

Since the sum of molal volumes of initial basaltic minerals is ca. 1 660 cc., while that of secondary spilitic ones ca. 1 610 cc. except calcite, 15% of the latter must be left in the rock if the reaction is to proceed without volume change. It should be emphasized that the coupled spilitization reaction proposed in this paper needs considerably lower amounts of introduced sodium to accomplish a complete basalt-spilite transformation as compared with the reactions of ESKOLA and TURNER. While in these authors' reactions the molal ratio of albite/anorthite is 2 and 1, respectively, in the coupled one it is 0.6. Moreover the proposed reaction clearly explains the frequent occurrence of chlorite "inclusions" within albite grains in the rocks studied ("gefüllte Feldspäte" of AMSTUTZ, 1954; "Rahmen-Feldspäte" of HENTSCHEL, 1960). HENTSCHEL (1960, p.39) pointed out that they cannot be products of metasomatic alteration of feldspars but, being sometimes accompanied by epidote and calcite or other calcic minerals, simply occupy free space left after the removal of the anorthite component of primary calcic plagioclase.

All of these features of spilitic rocks, except distinct Na-enrichment, are also explained by a simpler reaction, which illustrates coupled saussuritization and chloritization only, and which requires no introduction of sodium but a removal of some part of resultant calcite and silica ("prasinite reaction"):

$$5\ CaMgSi_2O_6 + 4\ CaAl_2Si_2O_8 \cdot NaAlSi_3O_8 + 5\ H_2O + 5\ CO_2 \longrightarrow$$
330 cc. 800 cc.

$$2\ Ca_2Al_3Si_3O_{12}/OH/ + 4\ NaAlSi_3O_8 + Mg_5Al_2Si_3O_{10}/OH/_8 + 5\ CaCO_3 + 9\ SiO_2.$$
268 cc. 400 cc. 209 cc. 148 cc. 203 cc.

In the writer's opinion, the internal redistribution of Ca and Al during the alteration of basalts, accompanied by partial removal of mobile calcite, may be responsible for a negative correlation coefficient between these elements in nearly all of the spilitic rock series examined.

Some Petrogenetical Remarks and the Problem of Nomenclature of Spilitic Rocks

Thorough petrological analysis of rapidly increasing detailed field and laboratory data strongly suggests that the problem of genesis of various similar rock types should not be raised in an uncompromising manner, but rather should be viewed as, possibly, the convergence of different processes which lead to very similar final products.

The writer fully accepts and supports some authors' opinions (VALLANCE, 1960; TURNER and VERHOOGEN, 1960; LEBIEDINSKIJ, 1964; VIELINSKII, 1968; AMSTUTZ, 1968; NARĘBSKI, 1968) that the principle of petrogenetic convergence applies unquestionably to spilites, since there is no one hypothesis which could account for all the occurrences of these rocks.

Preliminary statistical study of some selected spilitic rock series presented here clearly indicates that they are polygenetic (heterogenetic according to Soviet authors) associations. Differential distribution of Na-enriched members within various element frequency distribution diagrams for individual spilitic rock series, varying contents and ratios of trace elements and considerable diversity of covariant pairs of elements strongly support this opinion. Despite these differences, all of the obtained and statistically tested data clearly indicate general sodium-enrichment of each of the examined spilitic associations, irrespective of the sample size. Consequently the writer cannot fully agree with Prof. AMSTUTZ's opinion that the theory of Na-enrichment of spilites is due mainly to biased sampling, i.e., to collecting possibly the most typical and freshest rocks, containing no other mineral concentrations except albite. Though the sampling procedure obviously influences the results of petrochemical and statistical examinations, this factor can hardly change the essential features of the rocks under investigation, provided the number of samples is sufficiently large.

It is hoped that the further application of statistical methods to the examination of ever-increasing sets of analytical data on the rock series under consideration will be very helpful in solving some of the most fascinating problems of their origin and on the relation of spilites, keratospilites and keratophyres. Nevertheless, even with the available preliminary data, at least three genetic types can be distinguished among the initial volcanites: 1. magmatic spilites (and associated, more acidic rocks) containing primary, possibly high-temperature albite and, as observed by AMSTUTZ, displaying distinct coincidence between distribution of mineral phases and the primary fabric of lavas; 2. autometasomatic spilites (and associated intermediate and acidic rocks), also of magmatic origin but subjected to deuteric albitization; 3. metamorphic spilites (and keratospilites + keratophyres), products of non-spilitic submarine geosynclinal effusions but deanorthitized according to the previously proposed reaction scheme under conditions of greenschist facies. Therefore they should be lower in sodium.

Coexistence of these varieties even within the same geosynclinal sequence has been established beyond any doubt (West Sayans, Spitsbergen, etc.). Consequently there appears an important new petrological problem to find simple, definite criteria for distinguishing various genetic types of spilites and accompanying more acidic rocks. At present, specific structural features (AMSTUTZ, 1968), occurrence of high-temperature albite (VIELINSKII, 1968) as well as results from statistical work, presented in this paper, can be diagnostic for magmatic spilites. The same refers to C and O isotope composition calcite components as reported recently by SCHIDLOWSKI, STAHL and AMSTUTZ (1970) and SCHIDLOWSKI and STAHL (1971). Unquestionably, the magmatic origin of some spilite-keratophyre formations motivated VIELINSKII (1968, p.139) to defend firmly the use of the terms spilite, keratospilite and keratophyre as corresponding to the rocks of contemporaneous Pacific geosynclinal zones: hawaiites, mugearites and sodic trachytes.

According to the available data, spilitic series may originate not only from different petrogenic processes but also from various parent basaltic magmas: olivine-basaltic (West Sayans, Góry Kaczawskie Mts) or tholeiitic (Spitsbergen). Nevertheless all true spilite occurrences are restricted exclusively to geosynclinal environments, because both tectonic and bathymetric conditions therein are extremely favorable for the origin of larger masses of sodium-enriched rocks. Consequently, the writer feels that there is an urgent necessity to limit the use of the terms "spilite", "keratospilite" and "keratophyre" exclusively to rocks which are genetically connected with geosynclinal initial volcanisms. All other rocks of similar chemical and mineral composition but formed under completely different geological conditions should be named simply albitized or albite basalts, diabases, andesites, rhyolites, etc. It is well known that such albite rocks occur in the upper parts of some differentiated basic rock sills (e.g. DZIEDZICOWA, 1958; HENTSCHEL, 1960). Such terminological delimitation is absolutely necessary to avoid any petrological and geological misunderstandings and if the terms "spilite", "keratospilite" and "keratophyre" are to have unequivocal meaning.

Acknowledgements

The author is deeply indebted to Prof. A. RITTMANN for reading the manuscript and for valuable suggestions.

References

ABRAMOVICH, I.I., VYSOKOOSTROVSKAYA, E.B., DOROFEYEVA, E.F. (1963): On manganese-iron ratio in magmatic rocks. Geokhimya 1963, 11, 996-1001 (in Russian).

ABRAMOVICH, I.I., VYSOKOOSTROVSKAYA, E.B. (1964): Titanium-iron ratio in rocks of heterogeneous magmas. Geokhimya 1964, 7, 641-645 (in Russian).

AHRENS, L.H. (1954): The lognormal distribution of the elements. Geochim. et Cosmochim. Acta vol. 5, 49-73.

AHRENS, L.H. (1963): Lognormal-type distributions in igneous rocks-V. Geochim. et Cosmochim. Acta vol. 27, 877-890.

AHRENS, L.H. (1964): Element distributions in igneous rocks-VII. A reconnaissance survey of the distribution of SiO_2 in granitic and basaltic rocks. Geochim. et Cosmochim. Acta vol. 28, 271-290.

AHRENS, L.H. (1966): Element distribution in specific igneous rocks-VIII. Geochim. et Cosmochim. Acta vol. 30, 109-122.

AMSTUTZ, G.C. (1954): Geologie und Petrographie der Ergußgesteine im Verrucano des Glarner Freiberges. Vulkaninstitut Immanuel Friedländer, Publ. 5, Zürich, 150.

AMSTUTZ, G.C. (1968): Spilites and spilitic rocks: in basalts, the Poldervaart Treatise on rocks of basaltic composition. Ed. H.H. Hess, Intersci. Publ., vol. 2, 737-753.

BATTEY, M.H. (1956): The petrogenesis of a spilitic rock series from New Zealand. Geol. Mag., vol. 93, 89-110.

BIRKENMAJER, K., NARĘBSKI, W. (1960): Precambrian amphibolite complex and granitization phenomena in Wedel Jarlsberg Land, Vestspitsbergen. Studia Geol. Polon., vol. 4, 37-82.

BONDARENKO, V.N. (1967): Statistical methods of examinations of volcanogenic complexes. "Nedra" Moscow, 134 (in Russian).

BUTLER, J.R. (1964): Concentration trends and frequency distribution patterns in igneous rock types. Geochim. et Cosmochim. Acta vol.28, 2013-2024.

CLARKE, F.W., WASHINGTON, H.S. (1924): The composition of the earth crust. U.S. Geol. Surv. Profess. Papers 127.

DEWEY, H., FLETT, J.S. (1911): British pillow lavas and the rocks associated with them. Geol. Mag. 563, Doc. 5, vol. 8, 202-209, 241-243.

DZIEDZICOWA, H. (1958): Metasomatism of the Permian "Melaphyres" from Swierki (Lower Silesia). Ann. Soc. Geol. Pologne, vol. 28, 79-106.

ESKOLA, P., VUORISTO, U., RANKAMA, K. (1937): An experimental illustration of the spilite reaction. Comm. Geol. Finl. Bull, No. 119, 61-68.

GJELSVIK, T. (1958): Extremely soda rich rocks in the Karelian zone, Finnmarksvidda, Northern Norway. A contribution to the discussion of the spilite problem. Geol. Fören. Stockholm Förh, vol. 80, 381-406.

GRIFFITHS, J.C. (1962): Statistical methods in sedimentary petrography. In: Milner H.B., sedimentary petrography, vol. I, 565-617.

HENTSCHEL, H. (1960): Basischer Magmatismus in der Geosynklinale. Geol. Rundsch., vol. 50, 33-45.

IVANOV, D.N., KOSKO, M.K. (1965): Linear parageneses of the main rock-forming elements of the andesite-basalts of Kamchatka. Doklady Akad. Nauk USSR, vol. 164, 6, 1363-1365 (in Russian).

JAFFEE, F.C. (1955): Les ophiolites et les roches connexes dela region du Col des Gets (Chablais, Haut Savois). Schweiz. Min. Petr. Mitt., vol. 35, 1-150.

JIZBA, J.V. (1959): Frequency distribution of elements in rocks. Geochim. et Cosmochim. Acta vol. 16, 79-82.

JUTEAU, Th., ROCCI, G. (1966): Etude chimique du massif volcanique dévonien de Schirmeck (Vosges Septentrionales). Evolution d'une série spilite-kératophyre. Sci. de la Terre, vol. XI, 1, 68-104, Nancy.

KUTOLIN, V.A. (1967): Certain problems involved in petrochemistry and petrology of basalts. Doklady Akad. Nauk USSR, vol. 176, 3, 683.

KUTOLIN, V.A. (1968): Statistical petrochemical criteria of formational appartenance for basalts and dolerites. Doklady Akad. Nazk USSR, vol. 178, 2, 434-437 (in Russian).

LEBIEDINSKII, V.I. (1964): Genesis and classification of the spilite-keratophyre formation. Petrographic processes and the problems of petrogenesis. Reports Soviet Geol. XXII Sess. Intern. Geol. Congr., Probl. 16, Moscow, 31-43 (in Russian, English abstract).

LEBIEDINSKII, V.I., MAKAROV, N.N. (1962): Volcanism of the Upper Crimea. Izd. A.N. USSR, Kiev, 207.

LOEVINSON-LESSING, F.J. (1925): On the separation of basalts and andesites. Izvestia Geol. Kom., vol. 44, 4.

MERILÄINEN, K. (1961): Albite diabases and albitites in Enontekiö and Kittilä, Finland. Bull. Comm. Geol. Finlande No. 195, 75.

MÜLLER, R.L., KAHN, J.S. (1962): Statistical analysis in the geological sciences. 483, J. Wiley and Sons.

NARĘBSKI, W. (1964): Petrochemistry of pillow lavas of the Kaczawa Mountains and some general petrogenetical problems of spilites. Prace Muzeum Ziemi (Travaux de Musée de la Terre), vol. 7, 69-206, Warszawa.

NARĘBSKI, W. (1966): Geochemistry of elements of the iron group in amphibolites of the Hecla Hoek Succession in Wedel Jarlsberg Land, Vestspitsbergen. Arch. Mineral., vol. XXVI, 1-2, 167-214, Warszawa.

NARĘBSKI, W. (1968a): Über die petrogenetische Bedeutung von Spurenelement-Paragenesen der Eisengruppe in den initialen Vulkaniten einiger kaledonischer Geosynklinalen. Freiberg. Forschungsh. C 231, 259-265, Leipzig.

NARĘBSKI, W. (1968): Geochemistry and the problem of origin of metabasic rocks of the Rudawy Janowickie Mts (E Karkonosze). Bull. Acad. Polon. Sci. ser. sci. geol. geogr. vol. XVI, 1, 1-7, Warszawa.

NIGGLI, P. (1923): Anwendungen der mathematischen Statistik auf Probleme der Mineralogie und Petrologie. N.J. Miner. Beil. Bd. XLVIII, 167-222.

PODOLSKII, J.V. (1962): Linear parageneses of the major elements in alkaline rocks of the central part of the Kola penninsula. Doklady A.N. USSR, vol. 146, 2, 443-446 (in Russian).

PODOLSKII, J.V. (1963): Linear parageneses of the major elements in rocks of the spilite keratophyre formation. Doklady A.N. USSR, vol. 152, 4, 975-978 (in Russian).

RICHARDSON, W.A., SNEESBY, G. (1922): The frequency distribution of igneous rocks. Miner. Mag., vol. 19, No. 97, 303-313.

RITTMANN, A. (1967): Die Bimodalität des Vulkanismus und die Herkunft der Magmen. Geol. Rundsch. Bd. 57, 1, 277-295.

ROCCI, G., JUTEAU, Th. (1968): Spilite-keratophyres et ophiolites; influence de la traversée d'un socle sialique sur le magmatisme initial. Geol. en Mijnbouw. vol. 47, 5, 330-339.

RODIONOV, D.A. (1964): Frequency distribution functions of elements and minerals in igneous rocks. Nauka, Moscow, 101 (in Russian).

ROGERS, J.J., ADAMS, J.A.S. (1963): Lognormality of thorium concentrations in the Convay granite. Geochim. et Cosmochim. Acta vol. 27, 7, 775-783.

SCHIDLOWSKI, M., STAHL, W., AMSTUTZ, G.C. (1970): Oxygen and carbon isotope abundances in carbonates of spilitic rocks from Glarus, Switzerland. Die Naturwissenschaften 57, 542-543.

SCHIDLOWSKI, M., STAHL, W. (1971): Kohlenstoff- und Sauerstoff-Isotopenuntersuchungen an der Karbonatfraktion alpiner Spilite und Serpentinite sowie von Weilburgiten des Lahn-Dill-Gebietes. N. Jb. Miner. Abh. 115, 252-278.

SHARAPOV, I.P. (1965): Application of mathematical statistics in geology. Nedra, Moscow, 259 (in Russian).

SHARFMAN, V.S. (1968): On the average chemical composition of spilites. Doklady A.N. USSR, vol. 180, 1, 202-203 (in Russian).

SHAW, D.M. (1961): Element distribution laws in geochemistry. Geochim. et Cosmochim. Acta vol. 23, 2, 116-134.

SHAW, D.M., BANKIER, J.D. (1953): Statistical methods applied to geochemistry. Geochim. et Cosmochim. Acta vol. 5, 111-113.

SHTEINBERG, D.S. (1964): On the chemical classification of effusive rocks. Trudy Inst. Geol. Uralsk. Fil. A.N. USSR wyp $\underline{72}$, 106, Sverdlovsk (in Russian).

SUNDIUS, N. (1930): On the spilite rock. Geol. Mag. $\underline{67}$, 1-17.

TEISSEYRE, J. (1968): On the Old-Paleozoic volcanism in the West Sudetes. Acta Geol. Polon, vol. XVIII, 1, 239-256, Warszawa.

TEISSEYRE, J. (1968b): On the age and petrogenesis of metavolcanic rocks of the Rudawy Janowickie and Lasocki Ridge. Bull. Acad. Polon. Sci. ser. sci. geol. geogr., vol. XVI, 1, 9-15, Warszawa.

TURNER, F.J. (1948): Mineral and structural evolution of metamorphic rocks. Geol. Soc. Amer. Mem. $\underline{30}$.

TURNER, F.J., VERHOOGEN, J. (1960): Igneous and metamorphic petrology. 2nd ed., New York: McGraw-Hill.

VALLANCE, T.G. (1960): Concerning spilites. Linnean Soc. N.S. Wales Proc., vol. 85, 1, 8-52.

VIELINSKII, V.V. (1968): Cambrian volcanism of the West Sayans. Nauka, Novosibirsk, 153 (in Russian).

VOLK, W. (1958): Applied statistics for engineers. New York: McGraw-Hill.

VISTELIUS, A.B. (1958): Paragenesis of sodium, potassium and uranium in volcanic rocks of Lassen Volcanic National Park, California. Geochim. et Cosmochim. Acta vol. 14, 29-34.

VISTELIUS, A.B. (1960): Skew frequency distributions and fundamental law of geochemical processes. Jour. Geol., vol. 68, 1-22.

VUAGNAT, M. (1946): Sur quelques diabases suisses. Contribution à l'étude du problème des spilites et des pillow lavas. Min. Petr. Mitt., vol. 26, 116-228.

VUAGNAT, M. (1949): Variolites et spilites. Comparaison entre quelques pillow lavas brittaniques et alpines. Arch. Sci. Fasc. 2, Genève, 223-236.

WELLS, A.K. (1923): The nomenclature of the spilitic suite. Pt. II: The problem of spilites. Geol. Mag., vol. 60, 62-74.

ZAWARITSKII, V.A. (1946): The spilite-keratophyre formation in the region of Blyava deposit in the Ural Mountains. Trudy Inst. Geol. Nauk 71, Moscow, 1-83 (in Russian).

Middle Triassic Spilite-Keratophyre Association of the Dinarides and Its Position in Alpine Magmatic-Tectonic Cycle

J. Pamić

Abstract

Middle Triassic igneous rocks of the Dinarides have all of the characteristics of spilite-keratophyre associations. Volcanic rocks represented predominantly by spilite, keratophyre and quartz keratophyre with subordinate basalt and andesite are followed by pyroclastic rocks. Both volcanic and pyroclastic rocks are conformably interbedded with Ladinian sediments. Contemporaneous hypabyssal and abyssal rocks (albite granite-porphyry, albitite, albite diabase, gabbro, diorite, granodiorite and albite granite) cut sediments older than Ladinian. Volcanics are preponderant over hypabyssal and abyssal rocks. Diagrams based on 187 available chemical analyses illustrate the normal course of crystallization from initial basalt to final quartz keratophyre. On the basis of the facts presented a primary origin of albite in these rocks is inferred.

KATZER (1906) recognized two magmatic-sedimentary complexes of the Mesozoic epoch in the central region of the Dinarides (the area of Bosnia and Hercegovina), but he used no particular terms to distinguish between them. The first of the two complexes is Middle Triassic (Ladinian) and is characterized by the presence of "melaphyre" and "porphyrite". The second complex is Jurassic and is characterized by the presence of large ultramafic massifs followed by amphibolite, gabbro and diabase. The two Mesozoic magmatic-sedimentary complexes are treated as a single geologic unit called the "Diabase-Hornstein Formation" by subsequent Austrian and German investigators (AMPFERER-HAMMER, 1921; PILGER, 1939, 1940, etc.). Various authors working in different or even in the same areas speak of this formation as belonging either to Triassic or to Jurassic.

Ćirić (1954) has cleared the confusion concerning the two Mesozoic magmatic-sedimentary complexes of the Dinarides. He distinguishes between two distinctly individualized complexes: a porphyrite-chert series belonging to the Middle Triassic epoch and a diabase-chert formation of Jurassic age. Neither of the two terms is, strictly speaking, quite appropriate; recently published papers show clearly very complex compositions for both their igneous and sedimentary members. In view of recent research it may be advisable to drop these two terms. It is more convenient to call these complexes simply magmatic-sedimentary formations (KUZNECOV, 1960); the porphyrite-chert series is a magmatic-sedimentary formation of Middle Triassic age and the diabase-chert formation is one of Jurassic age.

The Middle Triassic magmatic-sedimentary formation of the Dinarides has recently been investigated in detail in many localities, particularly as regards its igneous members. The results obtained make it possible to offer some general conclusions on the basis of published research (BARIĆ, 1957; BEHLILOVIC and PAMIĆ, 1963; BERCE, 1954; DOLAR-MANTUANI, 1941, 1942; DUHOVNIK, 1953; FANINGER, 1965; GERMOVŠEK, 1953, 1959; GOLUB and ŠIFTAR, 1965; JURKOVIĆ, 1954; KARAMATA, 1952, 1960, 1960a,

1961; LUKOVIĆ, 1952; MAJER and JURKOVIĆ, 1958; MARIĆ, 1928, 1936; MARIĆ and GOLUB, 1965; MILADINOVIĆ, 1964; MILADINOVIĆ and ZIVKOVIĆ, 1962; NIKITIN, 1930; PAMIĆ, 1957, 1960, 1960a, 1961, 1962, 1962a, 1963, 1963a; PAMIĆ and BUZALJKO, 1966; PAMIĆ and JURIĆ, 1962; PAMIĆ and MAKSIMOVIĆ, 1967; PAMIĆ and PAPEŠ, 1967; PELLERI, 1942; PETKOVIĆ, 1960; TRUBELJA, 1962, 1963, 1963a; TRUBELJA and ŠIBENIK-STUDEN, 1965; TRUBELJA and SLIŠKOVIĆ, 1967; TUĆAN, 1928).

Triassic rocks are widely spread in the Dinarides. They are connected in the outer Dinarides with the crests of the larger, deeply eroded, anticlines or they occur along stronger faults. Triassic rocks are in particular widespread in "the zone of Paleozoic schists and Mesozoic limestones" (PETKOVIĆ, 1961) which divides the Dinarides almost symmetrically in the outer and in the inner belt. This zone is wider in the central and eastern parts where Triassic rocks cover an area of several thousand square kilometers; their northern margins encroach upon "the central ophiolite zone" (PETKOVIĆ, 1961), building up larger structural forms (the areas of Vareš, Zlatibor, etc.). Triassic rocks also frequently occur in the extreme northwestern parts of the Dinarides adjacent to the southern Alps.

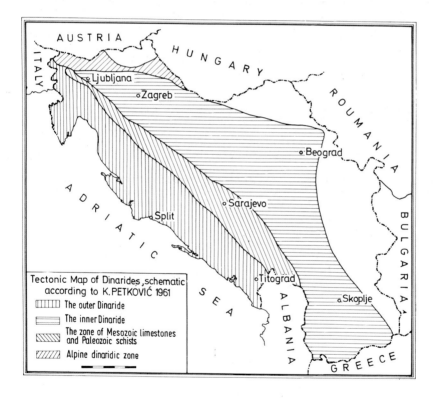

In general, the Triassic sequence has uniform facies characteristics throughout its area of occurrence (HERAK, 1962). The Lower Triassic is represented by Werfenian rocks: sandstone, shale and limestone. The Anizian consists predominantly of limestone and dolomite. The Ladinian is characterized by volcanic-sedimentary facies which vary in detached structures; for example, extrusive rocks are completely absent in the imbricated Triassic structure of the Trebević Mountain (near Sarajevo). On the other hand, facies characteristics also vary within a single structure. Thus in the large, imbricated anticline of Vareš (JOVANOVIĆ, 1960), the facies development of Ladinian varies along the strike; it is represented in some places by alternating series of cherts, limestones, shales and tuffs, in other places we find only volcanic rocks or only massive limestones with chert nodules and lenses etc. The Upper Triassic is almost everywhere represented by dolomites and limestones.

The Age of Magmatic Activity

At the beginning of this century several authors established a Middle Triassic (Ladinian) age of magmatic activity. KATZER (1906) came to this conclusion for the area of Bosnia and Hercegovina, BUKOWSKY (1904) for the Cukali zone, and KERNER (1916) for the Dalmatian coastal area of the Adriatic sea. Some recent authors are of the opinion that the magmatic activity was of longer duration; BEŠIĆ (1950, 1953) considers that the volcanic rocks of Montenegro were formed during the Skythian and Anizian stages, ČELEBIĆ (1967) claims that the magmatic activity took place from the Skythian to Ladinian stages. Such conclusions are based mainly on field observation, and they are influenced by the fact that the Triassic areas are very complicated in structure. Imbricated structures, sometimes with quite reduced separate members, can frequently be seen and this is why volcanic bodies are sometimes apparently connected with sediments older than Ladinian.

A Ladinian age of magmatic activity has recently been proved in many localities (KARAMATA, 1960; PETKOVIĆ, 1960; PAMIĆ, 1960). This conclusion is supported by several important facts:

1. Volcanic rocks occur as single flows or successions of distinctly separated sheets or flow units conformably among Ladinian sediments which are paleontologically defined.

2. Composite flows are frequent and in such cases volcanic sheets alternate either with pyroclastic or with sedimentary rocks of Ladinian age. Tuffs have not been found anywhere in the Skythian and Anizian series.

3. Xenoliths of Anizian limestones containing diagnostic fossils are found in volcanic rocks in many localities even where they are in direct contact with Skythian sediments (influenced by tectonic movements).

4. Phenomena of contact metamorphism can be frequently seen in Skythian and Anizian sediments when they are cut by intrusive or hypabyssal rocks.

Geosynclinal Environment

The exact beginning of the alpine magmatic-tectonic cycle in the Dinarides cannot be established. A precise subdivision of the Paleozoic era has not been carried out, particularly as regards to boundary between the Permian and the older Paleozoic periods. It has been definitely observed in numerous localities that sediments of Permian age gradually pass to Skythian with clear conformity, and it is probable that the beginning of the alpine cycle falls within the Permian, at least in its higher parts. KARAMATA (personal communication) is of the opinion that the Mesozoic geosyncline of the Dinarides represents in fact a regenerated Hercynian geosyncline.

The geosynclinal environment was common to the entire Dinaridic belt during Lower Triassic time; it was manifested in clastic sedimentation followed by frequent but minute movements of the sea bottom (the sedimentation of Werfenian beds). This was followed by a tendency toward stabilization of the sedimentation environment during the Anizian stage because limestones and dolomites were then deposited.

Large masses of lava poured forth on the sea bottom during the Ladinian age. Some of the volcanic phases were explosive and caused the formation of pyroclastic rocks; they were solidified and interbedded in the volcanic-sedimentary formation. Finally, both volcanic and pyroclastic rocks were conformably inserted in the marine sediments of Ladinian age. All this points to the submarine character of volcanic activity, which explains the presence of pillow lavas, sometimes brecciated, and volcanic rocks with a great quantity of amygdaloids.

Both extrusive and pyroclastic rocks were formed simultaneously with Ladinian sediments. The submarine flows were followed with contemporaneous subvolcanic and intrusive rocks placed in sediments older than Ladinian.

The large masses of limestone and dolomite, about 1 000 m thick, were deposited during the Upper Triassic epoch. The tendency of the geosynclinal environment to stabilize, which was already expressing itself in the Anizian stage was only interrupted during the Ladinian stage.

Such a geosynclinal environment must have existed in the entire Dinaridic area because of the almost identical facies characteristics. But we must bear in mind that the Dinaridic geosyncline was still not differentiated during the Triassic age. The differentiation occurred at the end of the Upper Triassic or in the beginning of the Jurassic age; two distinctly separate belts, miogeosynclinal (the outer Dinarides) and eugeosynclinal (the inner Dinarides), begin to appear in the Lower Jurassic age.

Volcanic Rocks

The results of recent research show that Triassic volcanic rocks are very different in composition. Two main groups can be distinguished: the normal, subalkaline group and the alkaline-rich group containing predominantly albite but sometimes a greater quantitiy of potash feldspar.[1]

---

[1]There is much confusion in the terminology used in the classification of Triassic volcanic rocks. Some authors use the term paleotypal, and some, cenotypal volcanic rocks. Quartz porphyrite was used for rocks containing about 0.5% CaO, dacitoandesite for the rocks in which albite is the only representative of the feldspar group, the term melaphyre for the basic volcanic rocks containing albite etc. In this way, two or even more terms referred to a rock of the same composition. Thus, we find it necessary to define some of the terms used in this paper. Normal, subalkalic volcanic rocks are termed cenotypal. The terms spilite and keratophyre are used to describe the basic and intermediate extrusive rocks containing albite, regardless of its origin. Quartz keratophyre is applied to quartz-bearing volcanic rocks containing either albite $\pm$ potash feldspar (LEWINSON-LESSING, 1928). Potash-rich basic volcanic rocks are called poeneites (according to DE ROEVER, 1942).

Both groups of volcanic rocks have the same general structural and textural characteristics. Amygdaloidal structure is common and the quantity of amygdaloids varies from several to 50% or more. Amygdaloids can be present in all kinds of volcanic rocks; however, they are more frequent in intermediate and particularly in basic members.

All Triassic volcanic rocks have both hypocrystalline and holocrystalline porphyritic textures: pilotaxitic, hyalopilitic, etc. Albite-bearing extrusive rocks including spilites show also almost all of the varieties of porphyritic texture. Curved growth forms of albite are common, and albite microlites can be found in almost all varieties. Intersertal texture is particularly frequent in spilites; owing to a gradual decrease of glass, they pass to ophitic spilites which can barely be distinguished from hypabyssal albite diabases.

Pyroxene is a characteristic femic constituent of both rock groups, in particular of basic and intermediate rock varieties. It is represented usually by augite which is well developed in subalkalic rocks but frequently feather-shaped in spilites. The alteration of pyroxene to chlorite is common. Nevertheless, chlorite without pyroxene relicts is the most frequent constituent of albite-bearing extrusive rocks. It can be concluded that the abundance of chlorite increases gradually, going from basic to acid volcanic rocks. Other kinds of pyroxene alteration are subordinate; it is sometimes changed into a fine-grained aggregate of chlorite, epidote and calcite, or into greenish amphibole (usually fibrous uralite).

The group of normal, subalkaline rocks is represented by basalt, andesite, dacite and rhyolite depending on mineral composition. Transitional varieties between basalt and andesite and between andesite and dacite are also fairly frequent. Plagioclase composition varies from bytownite (in basalt) to oligoclase (in rhyolite). It is frequently fresh and zoned, but sometimes metamorphosed with variable intensity into fine-grained aggregates of calcite, prehnite, clinozoizite, sericite and albite. Such changes are particularly evident in the basic members which can hardly be distinguished from some spilites. But plagioclases of the groundmass are usually quite fresh.

The second group is composed of albite-containing extrusive rocks: spilite, keratophyre and quartz keratophyre with mutual transitions (spilite-keratophyre and keratophyre-quartz keratophyre). It is of particular genetic interest that the transitional rocks to normal subalkalic ones have been recognized: andesine-keratophyre containing fresh andesine phenocrysts embedded in pilotaxitic groundmass of oligoclase-albite and chlorite, and andesine spilite with coarser andesine inserted in albite-ophitic mass.

Albite is certainly the most important member in the mineral association of the second group of rocks. It occurs as phenocrysts as well as microlites in the groundmass. In the latter case it is always quite fresh. But albite phenocrysts show great diversity in regard to the content of secondary minerals. On the one hand, there are fresh albite phenocrysts; at the other extreme, however, sometimes only their outlines are preserved and the space of primary plagioclase is filled by a very fine-grained aggregate of secondary minerals. Clinozoizite is frequent, but calcite, prehnite and sericite are common. Nevertheless, chlorite is the most frequent inclusion of albite phenocrysts. Its quantity varies and it is usually inserted in the form of thin flakes and wedges along the cleavage fissures. But most frequently all of these secondary minerals are included in albite phenocrysts with distinct preponderance of chlorite. However, we must bear in mind that albite in the groundmass is quite fresh. Although the described phenomena can be found in each of the groups of albite-containing extrusive rocks, they are more clearly expressed where the percentage of silica is lower.

Apart from the sodium-rich extrusive rocks we also find here rock varieties containing potash feldspar. It is fairly common in keratophyres and in quartz keratophyres, where it occurs with albite or even without it. Poeneites ("potash spilites") have been also identified, and they occasionally contain very high quantities of K_2O (about 10%). Potash feldspar is also present in intersertal and ophitic spilites as narrow rims enclosing albite, and it appears sometimes in needle-like microlites in the interstices. Basic volcanic rocks of the Dinarides with variable quantities of albite and potash feldspar are very similar to weilburgites of the Lahn-Dill described by LEHMANN (1949, 1957).

Pyroclastic rocks following extrusive ones are represented predominantly by crystal tuffs. There are also two groups of tuffs; plagioclase fragments are characteristic of the first and albite fragments of the second.

Hypabyssal and Abyssal Rocks

The markedly volcanic character of Ladinian magmatic activity has also produced corresponding hypabyssal and abyssal rocks. Hypabyssal rocks occur as small bodies placed in sediments older than Ladinian, but also in the form of veins cutting volcanic bodies. Quartz albitite, ankerite albitite, albite granite-porphyry and different varieties of albite diabase have been recognized.

There are several intrusive bodies; the largest massif is the Jablanica gabbro covering the surface of about 15 km^2. Intrusive rocks are also represented by different rock types: gabbro, diorite, quartz diorite, granodiorite and albite granite; there are also transitional rock varieties between gabbro and diorite etc. The mineral composition of abyssal rocks does not show an essential difference from common gabbro, diorite etc.

Distribution and Chemistry of Rocks

All of the obtained results indicate a constant association of widely divergent rock types in the Triassic areas of the Dinarides. The distribution of the main groups of rocks can be estimated on the basis of available chemical analyses:

Normal, subalkalic volcanic rocks 18%
Alkali-rich volcanic rocks 60%
Abyssal and hypabyssal rocks 22%

The relation is still more favorable for the second group mentioned because some areas consisting predominantly of normal, subalkalic volcanic rocks as well as of intrusive ones are documented with many more analyses than are the areas consisting predominantly of alkali-rich volcanic rocks. Quartz keratophyre and spilite-keratophyre occur in approximately equal proportion.

On the basis of these details it can be concluded that Ladinian igneous rocks of the Dinarides have all the characteristics of a spilite-keratophyre association which is usually connected with geosynclinal regions.

The petrochemical interpretation is based on 187 available chemical analyses. However, two factors complicate the classification problem: a) the number of available analyses is not balanced in regard to separate parts of the Dinarides (there are no data for the eastern parts of the Dinarides in western Serbia), b) the calcite content caused predominantly by the assimilation of limestone varies to a great extent (0-50%).

The results are given on NIGGLI's triangles illustrating the course of variation of the major elements. The analyses of abyssal rocks are presented in Fig. 1-3; the normal course from gabbro to grano-diorite can be seen on the QLM triangle and the moderately expressed tendency of Na-differentiation on the k-π diagram.

Because volcanic rocks are predominant in the association, the trends of their variation (Figs. 4-6) are of great interest. It is characteristic that almost all the points are gathered in a comparatively narrow field and that they show distinct continuity from normal basalt to extremely acid quartz keratophyre (QLM triangle). This is in general agreement with the corresponding triangle of abyssal rocks. The scattering of points is characteristic for the k-π and γ-mg triangles. However, on the k-π triangle the greatest number of points is grouped along the Ne-Cal-line and in the field characteristic for the normal course of differentiation of basalt magma. The points on the γ-mg diagram are mainly gathered on the line Mg-Fe.

The last 3 diagrams (Figs. 7-9)[2] contain data for the same volcanic rocks but in relation to their geographical distribution. It is evident (QLM triangle) that, going from the northeastern parts (the inner Dinarides) to the southwestern (the outer Dinarides), and to the alpine-dinaridic zone, we pass gradually from basalt to final quartz keratophyre. This suggests that the course of differentiation varies from the eugeosynclinal towards the miogeosynclinal parts of the Dinarides.

---

[2]Data of chemical analyses for the presented diagrams have been used from the papers of the following authors: DUHOVNIK, 1953; FANINGER, 1965; GERMOVŠEK, 1953, 1959; GOLUB and ŠIFTAR, 1965; KARAMATA, 1952, 1960; LUKOVIĆ, 1952; MAJER and JURKOVIĆ, 1958; MARIĆ, 1928, 1936; MARIĆ and GOLUB, 1965; MILADINOVIĆ and ŽIVKOVIĆ, 1962; NIKITIN, 1930; PAMIĆ, 1957, 1960a, 1961, 1962, 1962a, 1963; PAMIĆ and BUZALJKO, 1966; PAMIĆ and PAPEŠ, 1967; PELLERI, 1942; PETKOVIĆ, 1960; TRUBELJA, 1962, 1963, 1963a; TRUBELJA and ŠIBENIK-STUDEN, 1965; TRUBELJA and SLIŠKOVIĆ, 1967; and TUĆAN, 1928. The author is grateful to V. KNEŽEVIĆ-DJORDJE-VIĆ who placed at his disposal 27 unpublished chemical analyses of volcanic rocks from Montenegro.

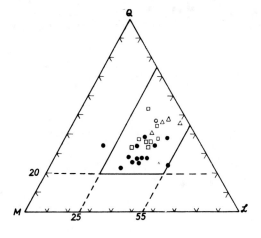

Fig. 1

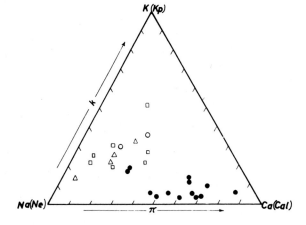

△ Albite granite
○ Granodiorite
□ Quartz diorite and diorite
● Gabbro

Fig. 2

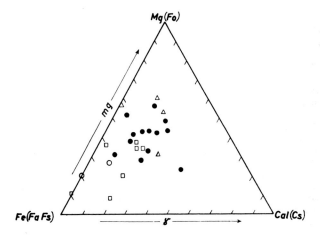

Fig. 3

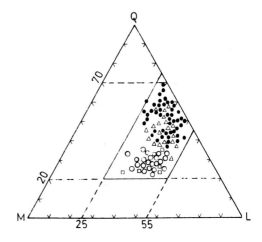

Fig. 4

- Quartz keratophyre and keratophyre
- Spilite and metadiabase
- Basalt and diabase
- Andesite, dacite and rhyolite
- Poenite

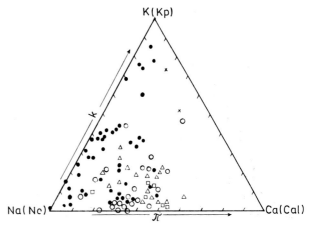

Fig. 5

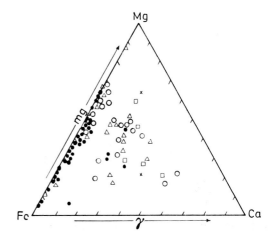

Fig. 6

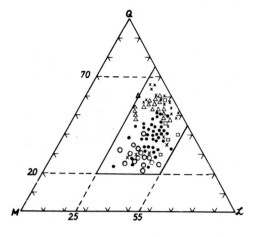

Fig. 7

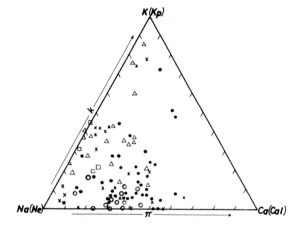

○ *The inner Dinarides*
● *The northern parts of the transitional zone*
△ *The southern parts of the transitional zone*
□ *Sea-shore area of Adriatic*
× *Alpine-dinaridic zone*

Fig. 8

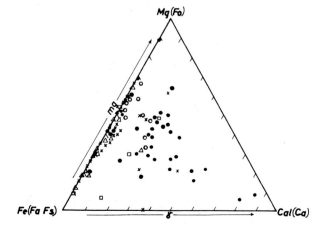

Fig. 9

Petrology

Petrological considerations of the Ladinian spilite-keratophyre association of the Dinarides involve very complex problems. In all cases the most subtle question is how to explain the origin of albite-containing extrusive rocks. As is well known, there are two basically opposite opinions about the primary and secondary origin of albite. The two different hypotheses have also been applied to the explanation of the genesis of the albite-containing extrusive rocks of the Dinarides; BARIĆ (1957) and TRUBELJA (1962) have referred to the classical spilite reaction in their genetic interpretation. MARIĆ and GOLUB (1965) and KARAMATA (1960) are the adherents of the hypothesis of albitization, but they connect it carefully with either late-magmatic or post-magmatic processes. TAJDER (1956), studying albite-containing extrusive rocks in the Panonian area, is of the opinion that both albite and chlorite were formed by magmatic processes. The present author, studying the problem in several localities, has come to the conclusion that albite-containing extrusive rocks represent only the members of the series of igneous rocks which may have originated from a primary basaltic magma by the process of fractional crystallization. The following are the most important facts supporting this interpretation:

1. Albite-containing extrusive rocks including potash-rich varieties are associated in many localities with normal, subalkaline rocks, and their link is proved by the presence of transitional varieties (andesine keratophyre and andesine spilite). Albite-containing rocks have also been determined among hypabyssal and abyssal rocks produced by the same magmatic activity.

2. Both groups of extrusive rocks have the same textural and structural characteristics. The preservation of the texture with few albite phenocrysts, sometimes curved and embedded in a groundmass containing albite microlites, suggests the normal course of crystallization of separate sodium-rich lava.

3. Pyroclastic rocks are also of very different composition and it is significant that albite-bearing tuffs are very common. Spilite and keratophyre fragments are common in lithic tuffs; the albite in the fragments has the same physiographic characteristics as in volcanic rocks. All this suggests that albite and albite-containing extrusive rocks have already been formed before the ejection of pyroclastic material.

4. No traces of albitization could be observed in sediments adjacent to volcanic rocks. The marks of metasomatic reactions have been observed only in some areas of intensive mineralization caused by post-magmatic processes.

The evolution of primary basalt magma, with a moderately increased quantity of alkalis, occurred in a submarine environment, thus causing a particular "hydromagmatic" environment (BURRI and NIGGLI, 1945). These conditions, followed by the intensive assimilation of limestone, gave rise to the lowering of melting points of crystallizing lava. This interpretation is in agreement with results of experimental petrology on phase relations under pressure of water vapor (TUTTLE and BOWEN, 1950; YODER, 1952).

In view of the interpretation presented, it may seem difficult to explain the presence of the "secondary" minerals (clinozoisite, etc.). But if we suppose that clinozoisite, calcite and prehnite in albite phenocrysts are proofs of eventual metasomatic reactions, i.e. albitization, then we must bear in mind that albite in groundmass is always quite fresh. This suggests that the formation of "secondary" minerals (together with chloritization of pyroxene) must have occurred in a later stage (not even late-magmatic) which preceded the solidification of the groundmass. The results of recent experimental petrology show that some typical "secondary" minerals (calcite) can be produced at high temperature in the presence of water (WYLLIE and TUTTLE, 1959).

REFERENCES

AMPFERER, O., HAMMER, W. (1921): Ergebnisse der geologischen Forschungsreise in Westserbien. Denkschr. Akad. Wiss. math.-nat. Kl. 98, 11-43, 44-56.

BARIĆ, Lj. (1957): Eruptivi iz okolice Sinja u Dalmaciji uz kraći osvrt na eruptivne pojave kod Knina, Vrlike i Drniša. II. Kongr. geol. Jugosl., 255-263.

BEHLILOVIĆ, S., PAMIĆ, J. (1963): Ladinska vulkanogena tvorevina u dolini Drežanke (Hercegovina). Geol. glas. 7, 39-44.

BERCE, B. (1954): Kremenov porfirit v ožji okolici rudnika Sv. Ana nad Tržičem. Geol. 2, 179-190.

BEŠIĆ, Z. (1950): Prilog poznavanju starosti porfirita sjeverne Crne Gore. Geol. an. Balk. poluos. 13.

BEŠIĆ, Z. (1953): Geologija sjeverozapadne Crne Gore. Cetinje.

BUKOWSKY, G. (1904): Erläuterungen zur geologischen Detailkarte Blatt Budua. Geol. Feichsanst. Wien.

BURRI, C., NIGGLI, P. (1954): Die jungen Eruptivgesteine des mediterranen Orogens. Zürich.

ČELEBIĆ, Dj. (1967): Geološki sastav i tektonski sklop terena paleozoika i mezozoika izmedju Konjica i Prozora s naročitim osvrtom na ležišta Fe, Mn rude. Pos. izd. Geol. glas. 10, 1-139.

ĆIRIĆ, B. (1954): Neka zapažanja o dijabazno-rožnačkoj formaciji Dinarida. Ves. Zav. geol. geofiz. istr. 11, 31-88.

DOLAR-MANTUANI, L. (1941): Keratofirske kamenine v Kamniški in Korski dolini. Zbor. prir. dr. 2, 52-56.

DOLAR-MANTUANI, L. (1942): Triadne magmatske kamenine v Sloveniji. Raspr. mat.-prir. r. AZU. 2, 427-480.

DUHOVNIK, J. (1953): Prispevek h karakteristiki magmatskih kamenin Črne Gore, njihova starost in razmerje do triadnih kamenin v Sloveniji. Geol. 1, 182-218.

FANINGER, E. (1965): Petrokemične tabele wengenskih magmatskih kamenin. Geol. 8, 249-262.

GERMOVŠEK, C. (1953): Kremenov keratofir pri Veliki Pirešici. Geol. 1, 135-168.

GERMOVŠEK, C. (1959): Triadne predornine severovzhodne Slovenije. Slov. akad. znan. umje. 1, 1-53.

GOLUB, Lj., ŠIFTAR, D. (1965): Eruptivne stijene južnih padina planine Ivanščice (Hrvatsko Zagorje). Acta geol. 4, 341-350.

HERAK, M. (1962): Trias de la Yougoslavie. Geol. vjes. 15/1, 301-310.

JOVANOVIĆ, R. (1960): Dijabaz-rožnjačka formacija u Bosni i Hercegovini. Simp. probl. alps. inic. magm., ref. 3.

JURKOVIĆ, I. (1954): Augitsko-labradorski andezit kod Orašina jugoistočno od Bakovića. Geol. vjes. 5-7, 111-126.

KARAMATA, S. (1952): Opšta karakteristika melafira okoline Vareša (Bosna). Spom. M. Kišpatića., 237-243.

KARAMATA, S. (1960): Melafiri Vareša. Simp. probl. alp. inic. magm., ref. 5.

KARAMATA, S. (1960a): Prilog poznavanju magmatizme dijabaz-rožnačke formacije. Zbor. Rud.-geol. fak. 6, 85-102.

KARAMATA, S. (1961): Produkti i tipovi trijaske magmatske aktivnosti u jugoistočnoj Crnoj Gori. III. Kongr. geol. Jugosl. 1, 357-361.

KATZER, F. (1906): Über die historische Entwicklung und den heutigen Stand der geologischen Kenntnis Bosniens und der Hercegovina. Glas. Zem. muz. BH. 18, 37-68.

KERNER, F. (1916): Erläuterungen zur geologischen Karte der im Reicharate vertretenen Königreiche und Länder der österr.-ung. Monarchie. No. 124, Sinj und Spalato.

KUZNECOV, G.A. (1960): Magmatic formations and their classification. Rep. XXI-st Sess. Norden. 13, 94-95.

LEHMANN, E. (1949): Das Keratophyr-Weilburgit Problem. Heidelb. Beitr. Min. Petr. 2, 1-166.

LEHMANN, E. (1957): Exkursion zum Studium des Weilburgits im Lahngebiet. Fortschr. Min. 35, 89-108.

LEWINSON-LESSING, F. (1928): Some quaries on rock classification and nomenclature. Ref. Neues Jahrb. 2, 336.

LUKOVIĆ, S. (1952): O pojavi kvarckeratofira u kanjonu Tare. Zbor. rad. Geol. Rud. Fak. 2, 119-125.

MAJER, V., JURKOVIĆ, I. (1958): Petrološke karakteristike diorita Bijele Gromile južno od Travnika u srednjobosanskom rudogorju. Geol. vjesn. 11, 129-142.

MARIĆ, L. (1928): Masiv gabra kod Jablanice. Vijesti Geol. zav. 2, 1-65.

MARIĆ, L. (1936): Amfibolski porfirit sa Vratnika nad Senjom. Glas. Hrvat. prir. dr. 41-48, 149-155.

MARIĆ, L., GOLUB, Lj. (1965): Magmatizam Šuplje Stijene, Velike Ljubišnje i donjeg slivnog područja Tare i Pive u Crnoj Gori. Acta geol. 4, 111-166.

MILADINOVIĆ, M. (1964): Geološki sastav i tektonski sklop šire okoline planine Rumije u Crnogorskom primorju. Pos. izd. Geol. glas. Sarajevo.

MILADINOVIĆ, M., ŽIVKOVIĆ, M. (1962): Magmatske stene Crnogoskog primorja. Geol. vjes. 15/1, 75-92.

NIKITIN, V. (1930): Prilog karakteristici eruptivnih stena iz okoline Bara. Geol. an. Balk. poluos. 10, 35-70.

PAMIĆ, J. (1957): Petrološka studija efuzivnih stijena u oblasti Ilidža-Kalinovik, I. Područje Igmana i sjeveroistočnih padina Bjelašnice. Geol. glas. 3, 171-180.

PAMIĆ, J. (1960): Kontaktnometamorfne pojave u trijaskim sedimentima južno od Prozora. Geol. vjes. 13, 197-212.

PAMIĆ, J. (1960a): Osnovne karakteristike trijaskih vulkanita u širem području Kalinovika. Simp. probl. alp. inic. magm., reg. 12.

PAMIĆ, J. (1961): Rezultati mikroskopskih i kemijskih ispitivanja granitskih stijena s južnih padina Prenja u Hercegovini. Geol. glas. 5, 263-269.

PAMIĆ, J. (1962): Spilitsko-keratofirska asocijacija stijena u području Jablanice i Prozora. Acta geol. 3, 5-80.

PAMIĆ, J. (1962a): Petrološka studija efuzivnih stijena u oblasti Ilidža-Kalinovik, II Područje izvorista rijeke Željeznice. Geol. glas. 6, 45-59.

PAMIĆ, J. (1963): Trijaski vulkaniti okolice Čevljanovića i kratak osvrt na trijaski vulkanizam u zoni Borovica-Vareš-Čevljanovići. Geol. glas. 7, 9-20.

PAMIĆ, J. (1963a): Osvrt na problem vulkanogene-sedimentnih formacija u Dinaridima na području Bosne i Hercegovine. Geol. glas. 8, 5-27.

PAMIĆ, J., BUZALJKO, R. (1966): Srednjotrijaski spiliti i keratofiri u okolici Čajniča (jugoistočna Bosna). Geol. glas. 11, 55-78.

PAMIĆ, J., JURIĆ, M. (1962): Razvoj trijasa južno od Jajca. Geol. glas. 6, 107-110.

PAMIĆ, J., MAKSIMOVIĆ, V. (1967): Srednjotrijaski kvarc albitski dijabazi u skitskim sedimentima Bijele kod Konjica. Geol. glas. 12, in press.

PAMIĆ, J., PAPEŠ, J. (1967): Produkti ladiničke magmatske aktivnosti u području Kupreškog Polja. Geol. glas. 12, in press.

PELLERI, L.C. (1942): Sulle rocce dioritiche degli scoli Pomo e Mellisello nel Mare Adriatico. Period. Min. Roma. 13/2, 191-199.

PETKOVIĆ, K. (1961): Tektonska karta FNRJ. Glas. SAN. 22, 129-139.

PETKOVIĆ, M. (1960): Trijasko metalogenetsko područje Vareš. Disertacija. Beograd.

PILGER, A. (1939): Die Stellung der dinarischen Schiefer-Hornstein Formation. Zbl. Min. etc. B. 5, 182-190.

PILGER, A. (1940): Magmatismus und Tektonik in den Dinariden Jugoslawiens. Zbl. Min. etc. B. 9, 257.

ROEVER, W.P.DE (1942): Olivine-basalts and their alkaline differentiates in the Permian of Timor. Geol. Exped. Lesser Sunda Islands IV, 209-289.

TAJDER, M. (1956): Albitski efuzivi okolice Voćina i njihova geneza. Acta geol. I, 35-48.

TRUBELJA, F. (1962): Albitizirane stijene iz okolice Bosanskog Novog. Geol. glas. 6, 23-29.

TRUBELJA, F. (1963): Granitske stijene okolice Čajniča. Geol. glas. 7, 21-25.

TRUBELJA, F. (1963a): Efuzivne stijene okoline Čajniča s kratkim osvrtom na srodne stijene iz područja rijeke Lim. Geol. vjes. 15/2, 475-500.

TRUBELJA, F., SIBENIK-STUDEN, M. (1965): Efuzivne stijene iz doline rijeke Vrbasa i graniti Komara. Glas. Zem. muz. BH. 3/4, 99-103.

TRUBELJA, F., SLIŠKOVIĆ, T. (1967): Stratigrafski položaj i mineraloški sastav magmatskih stijena nacionalnog parka Sutjeska. Bull. Scie., in press.

TUĆAN, F. (1928): Andezitska erupcija u hercegocačkom kršu. Vijesti Geol. zav. 2, 178-188.

TUTTLE, O.F., BOWEN, N.L. (1950): High-temperature albite and contiguous feldspars. Journ. Geol. 58/5, 572-583.

WYLLIE, P.J., TUTTLE, O.F. (1959): Melting of calcite in the presence of water. Amer. Min. 44, 453-459.

YODER, H.S., Jr. (1952): The $MgO-Al_2O_3-SiO_2-H_2O$ system and related metamorphic facies. Amer. Journ. Scie., Bowen Volume, 569-627.

Petrogenesis of Spilites Occurring at Mandi, Himachal Pradesh, India

A. M. Patwardhan and A. Bhandari

The subject of the present paper is the petrogenesis of the Sukheti Khad exposures of the Mandi Traps described earlier by McMAHON (1882) and FUCHS (1967). An effort has been made to define the suite to which they belong on the basis of detailed petrographic study and chemical data. McMAHON (1883), GUPTA (1966) and KANWAR (1968) have briefly referred to the petrographic and structural disposition of similar rocks occurring in Dalhousie and Dharamsala areas of the same province.

Geological Setting

FUCHS (op.cit., p.122) has made mention of the geology around Darang (31° 48' N 77° 1' E) as well as of the section on the road from Mandi towards Kulu (32° 6' N 77° 14' E). MATHUR and EVANS (1964) have given a detailed account of the Tertiary geology of the Sundernagar (31° 31' N 76° 53' E) - Mandi-Jogindernagar (32° 0' N 76° 40' E) - Dharamsala (32° 13' N 76° 20' E) section, confining their description exclusively to the southern side of the Main Boundary Fault (KRISHNAN, 1960, p.65) which separates the Tertiary sequence from the pre-Tertiary rocks. On the northern side, along the Main Boundary Fault occur discontinuous exposures of basic igneous erruptives which have been given names like Dalhousi Traps, Dharamsala Traps, Darang Traps, and Mandi Traps, depending on the localities in which they are found. Associated with these exposures of basic erruptives there is always a sequence of quartzite, limestone-quartzite, and limestone-quartzite-phyllite in the above-mentioned areas; they have been given different stratigraphic ages. In the Mandi-Darang section this sequence of limestone-quartzite-phyllite has been said to be belonging to the Shali Formation (FUCHS, op.cit., p.127). A horizon of salt beds called 'Lokhan Formation' occurs between the Traps and the Tertiary rocks. These salt beds occur in discontinuous exposures from Guma towards Mandi and have also been grouped within the Shali Formation by SRIKANTIA and SHARMA (1969) after detailed regional mapping. Fig. 1 contains a map of the Mandi area including its geological formations and the locations of the trap rocks dealt with below.

It is apparent that, because these trap rocks occupy a position near the thrust line between the Shali Formation and the Tertiary at Mandi, the original flow sequence has been involved in deformative movements, thereby making sampling based on any vertical section impossible. However, on the basis of the amygdaloidal surfaces, which occur occasionally, the presence of more than one flow seems plausible. Not all of the trap exposures in their extension towards Jogindernagar provide this opportunity of studying some hard, compact and less deformed rocks as the Sukheti Khad exposures. Yet, they are distinguishable in the field from other phyllites of the Shali rock units and are recognized as traps. The phyllitic structure being the re-

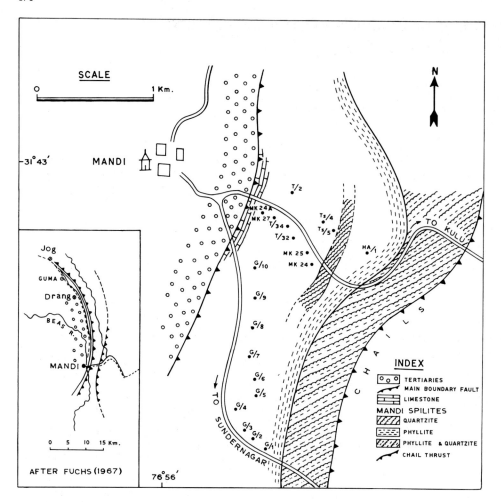

Fig. 1. Geological map around Mandi showing the location of samples whose chemical analyses are included in Table 1

sult of deformative movements is more pronounced towards outcrop margins where the rocks are also rich in chlorite. Some chlorite is, however, present in almost all the samples collected, and a gradation exists from incipiently altered rocks to those which can megascopically be called phyllites.

Mineralogy and Micro-Fabric

McMAHON (op.cit.) for the first time on the basis of textures and specific gravity established these rocks to be of igneous origin, and later WEST (1938, p.96) determined their petrographic similarity with the Panjal volcanics of Kashmir (WADIA, 1961, p. 220). Mineralogically, the rocks are simple and contain albite, augite, chlorite, opaques, sphene, epidote, calcite, palagonite, quartz,

and muscovite in variable proportions. The occasional presence of
altered glass and microlites together with the zeolites in the amygdules gives them a typical erruptive appearance. The mineralogical
assemblages most commonly encountered are albite-augite, albite-augite-chlorite, albite-augite-chlorite-opaques-calcite. Quartz,
calcite, epidote, chlorite occur mostly in the veins and do not contribute to the host fabric. Most of the rocks show the presence of
subhedral patches of kaolinite material which seems to have been
formed after an earlier potash feldspar. Pseudomorph structures consisting of opaque aggregates, chlorite and small crystals of quartz
sometimes suggest earlier presence of olivine. Frequently such
patches of pseudomorphs show a shattered appearance in the chlorite
matrix, and the pressure shadows thus formed are filled with quartz.

The mineral assemblage is typical of the spilites and spilitic rocks.
The essential genetic criteria discussed below are derived from the
fabric relationships of the minerals just enumerated and pictured
on Plate I. "The degree of geometric coincidence is taken as the
degree of contemporaneity" (AMSTUTZ, 1968, p.750). The albite-augite relationship is, in general, intersertal and in many cases
ophitic and poikilo-ophitic. In these cases the albite crystals
tend to be needle-shaped and are seen implicated with the pennate
and dendritic augite. In many cases augite is granular, has a
denser brown color towards the margins and is also faintly pleochroic. Broken crystals of augite are seen healed with chlorite.
In most of the samples in which the primary fabric is preserved,
the albite laths are free of carbonate inclusions but do contain
inclusions of chlorite. A few of these laths are bent and show
undulose extinction. One often comes across albite microlites and
medium grained laths set in an aphanitic matrix of altered glass,
chlorite and opaques. Besides being implicated with augite and
occurring poikilitically within the above mentioned matrix, albite
also occupies other primary spaces within the rock. These albites
are comparatively broader, untwinned and invariably show wavy extinction. Chlorite inclusions are present in these albites too.
Within these spaces as well as on their margins an interesting
sequence of crystallization of albite, pyroxene, chlorite is usually
seen to form a typical structure which may be described as "colloform". It is here proposed that chlorites which form such structures in these rocks are primary and that more than one generation
of primary albite exists in these rocks. The marginal colloform
structures should then represent a transitional stage.

Such primary albite-augite relationships have been described as intersertal, subophitic, ophitic, and poikilo-ophitic or only as intergrowths by BENSON (1915, p.143), COX (1915, p.328), ESKOLA (1925),
REED (1950, p.115), FIALA (1966, p.26), MIDDLETON (1960, p.202),

---

PLATE I (see p. 178 and 179)
1. Albite-augite intergrowth, poikilo-ophitic texture. Needle-shaped ▶
albite crystals. Albite laths with chlorite inclusions. 50
2. Poikilo-ophitic and intersertal texture of albite and granular
augite. Fluidic curving of albite crystals. 50
3. Feather shaped and dendritic augite implicated with acicular albite.
Amygdules with albite and chlorite. 50
4. Bent albite crystals set in a matrix of opaques and albite microlites. Also seen is intersertal granular augite with albite phenocrysts. 50
5. Colloform chlorite around augite-albite. The white areas are
occupied with albite. 50
6. Amygdaloidal spilite. Amygdules with albite, chlorite and sometimes
calcite. 50

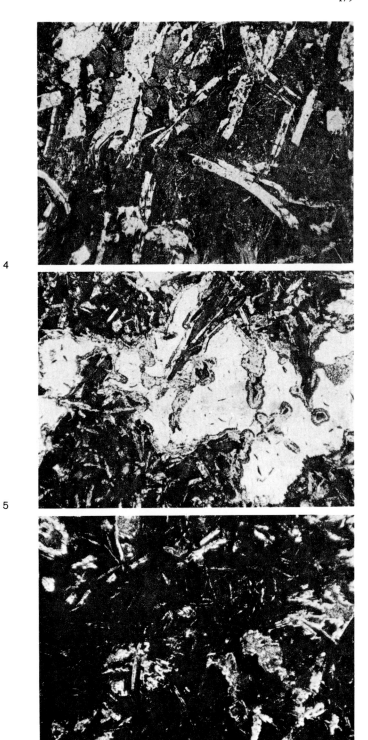

4

5

6

SCOTT (1951, p.429), BARTH (1966, p.119), BAMBA and SAWA (1967, p.6). Rocks with variable amounts of augite and albite showing primary implications have been termed spilite-augitite and pyroxene-albitite by VON ECKERMANN (1938) and DE ROEVER (1940, p.4; 1940a, p.108, 181; 1941, p.3). AMSTUTZ (1954) describes two generations of albite in the Swiss spilites and in a later paper (AMSTUTZ, 1965, p.149) points towards the occurrence of perfect congruence in fabric between hydromagmatic minerals like chlorite, epidote, zeolites, quartz and calcite and the ore minerals. This congruence he interprets as proof of a cogenetic, in this case, deuteric origin.

VALLANCE (1965, 1968, 1969) prefers not to interpret the subophitic to ophitic albite-augite relationships in spilites as cogenetic and holds the occurrence of albite even in such primary spaces to be post-magmatic. An exact replacement of a calcic feldspar by albite through a post-magmatic process does not seem to be an acceptable proposal. Experimental results have indicated (YODER, 1967) the possibility of a magmatic co-crystallization of albite-pyroxene-chlorite assemblage, and until we know more about the stability fields of pyroxenes in hydrothermal conditions it is only logical to recognize the albite-augite congruency based on petrographic evidence.

The albite-chlorite assemblage in spilites has also been considered primary (VAN OVEREEM, 1948; VUAGNAT, 1949; LEHMANN, 1952, 1965; REED, 1957), and such spilites from Germany have been called weilburgites (LEHMANN, 1949).

Not all of the spilites at Mandi have preserved the primary fabric described above. We attribute the transitions to noncongruent portions to alterations of autohydrothermal origin and to late tectonic effects. Some authors (e.g. FIALA, 1967, p.9; ZAVARITSKY, 1946, 1960, p.679) consider the albite in spilites to be of "late magmatic", "pneumatolitic" or of "deuteric" origin, but use the term "secondary". It is here suggested that the term "secondary" be used only for processes not related to any effects connected with magmatic crystallization, even if it is late-magmatic.

Bulk Chemistry

The chemical nature of these rocks and the suite to which they belong is illustrated from the data in Table 1 and the subsequent diagrams. While all of the rocks analyzed fall in the field of alkali-basalt (Fig. 2), it is of interest to note that they have either a high Na content or Na + K content. Fig. 3 illustrates the spread of normative Or - Ab - An for the Mandi rocks. Plotted also in this diagram is the variation line by FAIRBAIRN (1934) together with the points of average spilites (A, E) of SUNDIUS (1930) and VALLANCE (1960), respectively. It is evident from the diagram that the present rocks as well as points A to E differ considerably in normative feldspar composition in being far less calcic than the average metabasalts (M) and do not show any particular trend which could suggest their origin as a result of degradation of basalts. The spread of the Mandi rocks in the diagram indicates that they are even richer in alkalis (especially potash) than the average spilites (A, E) and 'somewhat potassic spilites' (B, C). Fig. 4a shows the compositional field of these rocks with respect to total iron, magnesium and total alkalis. The Mandi spilites are comparatively poorer in Fe and somewhat richer in MgO. Their spread, however, does not

Fig. 2. Total alkali-silica diagram showing Mandi spilites whose analyses are in Table 1 with respect to KUNO's (1959) line dividing the tholeiitic (below the diagonal line) and alkali basalts (above line)

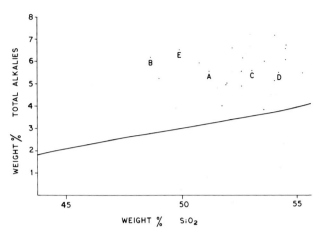

Fig. 3. Or-Ab-An spread in the Mandi spilites
M-T variation line FAIRBAIRN (1934)
M=Metabasalts
P=Plateau basalt
B'=Basalt
A'=Andesite
T=Trachyte
D'=Deccan basalt

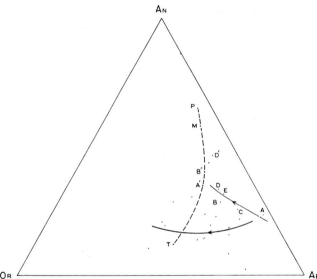

show much difference with respect to the total alkali content (Na_2O + K_2O) when compared with the spread of the reference points of spilites. Fig. 4b is significant as it gives the spread of these rocks taking into consideration the total lime with respect to soda and potash separately. Fig. 5a and 5b illustrate the relation of the ratios $CaO/Na_2O + K_2O + CaO$ and $Na_2O/Na_2O + K_2O$ given in Table 1 with respect to silica. It appears that there has been an overall decrease in the first ratio and an increase in the latter with respect to increase in silica. However, it does not hold true rigidly for each and every numerical value of these ratios. The rocks under consideration neither represent any fractionation series nor do they belong to specific flows and, therefore, it is not intended here to show any variation to that effect, the only aim being to illustrate their chemical nature with respect to constituents which define the spilite suite. When a wide range of variation in the percent of silica is observed (36.9 - 63.5%) in rocks accepted to be spilites (VALLANCE, 1960), it is evident that low lime and high alkali content are the inherent characteristics of the rocks belonging to this suite and their abundance need not be rigidly proportional to the increase in silica content.

Table 1

Sample Nos. Oxides	1 TS/4	2 G/4	3 MA/1	4 G/1	5 MK/24	6 T/2	7 G/8	8 G/10	9 GS/8	10 G/7	11 T/32	12 GS/1
SiO_2	49.01	50.81	51.88	52.03	52.13	52.37	52.66	52.66	52.75	53.60	53.64	54.07
TiO_2	1.41	1.18	1.29	1.18	1.53	1.06	1.76	0.47	1.41	2.11	0.89	0.71
Al_2O_3	13.75	14.61	13.32	13.32	12.89	14.08	15.04	15.47	15.92	14.61	15.47	14.18
Fe_2O_3	2.31	4.78	0.32	4.47	3.11	1.11	0.69	0.89	2.09	2.93	3.83	3.17
FeO	7.79	6.56	9.56	8.64	7.60	8.60	10.41	8.02	9.14	10.09	5.76	5.68
MnO	0.25	0.33	0.25	0.28	0.35	0.33	0.22	0.33	0.25	0.33	0.28	0.55
MgO	5.32	8.64	8.73	7.65	6.65	6.15	7.39	7.31	6.23	4.57	7.56	4.82
CaO	4.85	0.24	1.88	1.25	3.89	4.22	1.08	3.50	3.67	4.30	2.72	8.38
Na_2O	2.33	4.43	1.91	2.89	1.73	4.86	2.67	3.14	2.94	2.61	4.07	4.93
K_2O	2.90	1.54	2.15	2.10	2.40	1.82	1.18	2.35	3.27	0.89	2.00	0.36
P_2O_5	–	–	0.10	–	0.05	–	–	–	–	–	–	–
H_2O^+	6.40	5.96	8.05	3.31	3.99	6.09	7.72	6.59	2.40	3.07	3.11	1.57
CO_2	2.93	1.57	0.73	–	1.90	–	0.40	–	–	–	–	–
	99.25	100.65	100.27	99.83	98.22	100.79	100.82	100.73	100.07	99.11	99.33	100.62

Ratios

	1	2	3	4	5	6	7	8	9	10	11	12
CaO/Na_2O	2.08	0.05	0.98	0.43	2.19	0.86	0.40	1.11	1.25	1.64	0.66	1.70
Na_2O/Na_2O+K_2O	0.45	0.74	0.47	0.56	0.42	0.73	0.70	0.57	0.47	0.74	0.67	0.90
CaO/Na_2O+K_2O+CaO	0.48	0.04	0.32	0.20	0.48	0.40	0.22	0.40	0.37	0.55	0.31	0.61

Table 1 (cont.)

Sample Nos. Oxides	13 GS/5	14 T/34	15 M/	16 MK/25	17 GS/3	18 MK/24A	19 TS/5	A	B	C	D	E
SiO_2	54.07	54.50	54.51	54.51	54.54	55.39	55.90	51.22	48.60	53.15	54.20	49.65
TiO_2	0.82	0.90	1.06	1.18	0.82	0.71	1.29	3.32	1.94	1.50	1.31	1.57
Al_2O_3	15.04	13.32	13.32	14.61	15.47	12.89	14.18	13.66	16.10	14.39	17.17	16.00
Fe_2O_3	1.77	1.18	1.17	0.45	2.02	3.43	1.57	2.84	7.60	1.28	3.48	3.85
FeO	7.22	8.56	8.83	9.45	7.26	5.26	8.79	9.20	4.00	9.33	5.49	6.08
MnO	0.30	0.28	0.37	0.33	0.33	0.29	0.29	0.25	0.34	0.14	0.15	0.15
MgO	6.32	8.31	8.14	6.07	6.48	8.14	7.98	4.55	3.66	4.74	4.36	5.10
CaO	3.80	2.11	2.27	2.21	3.32	2.60	1.15	6.88	6.20	7.04	7.92	6.62
Na_2O	4.57	4.71	4.43	2.83	5.37	3.43	2.72	4.93	4.50	4.58	3.67	4.29
K_2O	2.65	1.86	1.73	3.80	1.32	2.10	2.20	0.75	1.76	1.01	1.11	1.28
P_2O_5	–	–	0.10	0.10	–	–	–	0.29	0.34	0.19	0.28	0.26
H_2O^+	3.51	2.63	2.76	4.77	3.80	4.26	3.32	2.90	2.54	2.02	0.86	3.49
CO_2	–	–	–	–	–	–	–	0.94	1.45	0.10	–	1.63
	100.07	98.40	98.60	100.31	100.73	98.50	99.39	100.72	99.60	99.60	100.00	99.97
CaO/Na_2O	0.83	0.45	0.51	0.78	0.62	0.75	0.42	–	–	–	–	–
Na_2O/Na_2O+K_2O	0.63	0.71	0.72	0.43	0.80	0.62	0.55	–	–	–	–	–
CaO/Na_2O+K_2O	0.34	0.24	0.27	0.25	0.33	0.32	0.20	–	–	–	–	–
CaO												

A = average spilite (SUNDIUS, 1930, p.9).
B = somewhat potassic spilite, Wellington, New Zealand (REED, 1957, p.37).
C = spilite somewhat high in potash, eastern orogen (GILLULY, J. (1935) Am. Jour. Sci., vol. 23, p.235)
D = andesite (NOCKOLDS, S.R. (1954). Average chemical composition of some igneous rocks. Bull. Geol. Soc. Am. 65, p. 1007)
E = average spilite (VALLANCE, 1960)

Discussion

The mineralogy and chemical nature of these rocks are well within the norms of spilites. The textural evidence is the only but sufficient proof that we are dealing with primary spilites. We stated earlier that these rocks have been grouped together within the Shali Formation around Mandi. It is also important to record here that the associated rock units of the Shali Formation do not show any signs of meta-

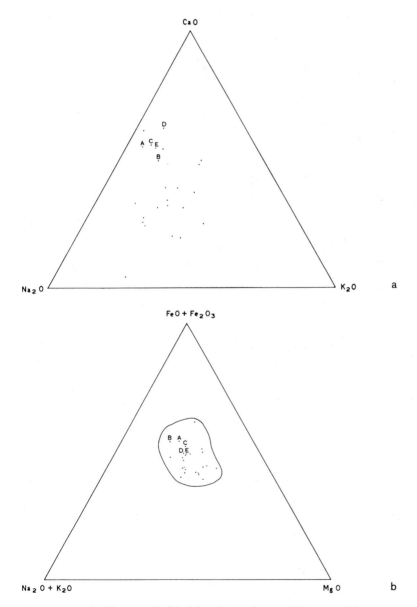

Fig. 4. a) Chemical field of Mandi spilites with respect to iron, magnesia and alkalis.
b) Chemical field of Mandi spilites with respect to lime and alkalis

morphism other than those of mild mechanical deformation and rupture. Thicker outcrops of limestones and quartzites exposed near Guma (towards Jogindernagar), where basic erruptives are also interbedded, contain preserved sedimentary structures. The only rocks in the vicinity which show pronounced metamorphic phase changes belong to the Chail Formation, towards the east, with which granites are associated. The Chail Formation has a thrusted contact with the Shali Formation. This field disposition of the Mandi spilites may point towards a possible submarine environment, but the attainment of primary fabric relationship during diagenesis among albite, augite and chlorite seems a remote possibility. WELLS (1923, p.73), by citing examples of basalt and mugearite errupted from the same vent and exposed to the modifications arising from submarine erruption, states that the action of sea water on rock magmas is practically without effect. It is interesting to note that those who regard spilites as post-magmatic secondary products propose interaction of water in the process and therefore call it "hydrous degradation of solid basalt". We fail to understand why the same cannot be achieved

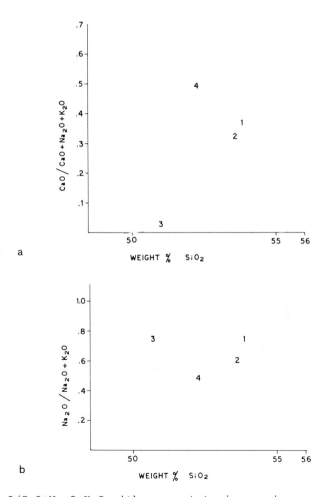

Fig. 5. a) Variation of $CaO/CaO+Na_2O+K_2O$ with respect to increasing silica.
b) Variation of Na_2O/Na_2O+K_2O with respect to increasing silica

before or during consolidation (RITTMANN, 1958; ZAVARITSKY, 1960) in case of an underwater erruption. If these post-magmatic alterations were to take place, they would certainly affect the primary fabric of minerals, especially the feldspars. We do not subscribe to such changes, therefore, at least so far as the albite-pyroxene contemporaneity is concerned. SUNDIUS (1930, p.16) distinguishes spilites as special rock types with low Al_2O_3 and high FeO TiO_2, thereby promoting soda-rich composition of feldspar and also affecting the eutectic line for pyroxene and plagioclase causing early crystallization. However, he also states that "as to the time of separation and generation of the secondary minerals we do not know anything with certainity". High contents of water and CO_2, according to him, are not inherent in the spilite magma. WELLS (1923, pp.72,74) postulated a magma different from the normal basaltic and did not recognize spilites as having been differentiated from the basaltic magma because occurrence of basalts and spilites associated with each other was not known at that time. FLAHERTY (1934) envisages a spilite magma inherently different from normal basaltic, but attributes the albite to residual soda-rich solutions in the consolidating flow. DUSCHATKO and POLDERVAART (1955) also envisage a spilite magma giving rise to albitite as a product of differentiation. BATTEY (1956) made an interesting contribution by recognizing a close affinity between spilites and certain tholeiitic basalts and attributed the congruency of albite, augite, and chlorite to the influence of water in proposing the existence of a series from dry tholeiitic basalt to spilite. VALLANCE (1969, p.12-15) showed that mean composition of pillow spilites has a close similarity with that of basalts. His data have confirmed once again that characteristic heterogeneity can be found in spilite exposures. Textural evidence in many spilites, as in the case of the Mandi spilites, does not, however, permit us to agree that this heterogeneity is only a result of low grade metamorphism of basaltic rocks and that it could not be an inherent primary character attained through differentiation (BENSON, 1915; E. NIGGLI, 1944; P. NIGGLI, 1952; VAN OVEREEM, 1948; CORNWALL, 1951; BURRI and P. NIGGLI, 1945; AMSTUTZ, 1958). Many examples of differentiated spilites are known (e.g. FIALA, 1966, p.14) and spilite exposures often present heterogeneous compositions. The soda-rich content of certain members (one finds even albitites) need not be viewed as increased soda content of the magma, as proposed by NICHOLLS (1959, p.159), because such rocks may only be representing differentiated portions. Consequently, it is not necessary in our opinion to fix the lower limit of the soda content (4%, FIALA, 1967, p.28) to define spilites. One may even find rocks which contain a calcic plagioclase, as transitions must exist even in magmatic rocks. The mean soda content of some spilite pillows (2.8%, VALLANCE, 1969, p.12) is far lower than that of their cores (4.3%) and than that of the "holomineralic spilite" (AMSTUTZ, 1968, p.742; 5%). This difference in soda content is hardly produced by anything other than differentiation. It appears logical to refer to spilites as primary differentiated portions of a basic magma - perhaps a normal basaltic which had the potential to give rise to rocks richer in sodic feldspar. Extended stability of pyroxenes, primary crystallization of albite, or even the presence of primary chlorite and carbonates may be referred to as characteristics of spilites.

Conclusions

The mineral albite and consequently the soda content of the Mandi spilites are primary. Albite, augite, and a part of the chlorite show a primary congruent fabric. The occurrence of epidote, calcite and chlorite in veins noncongruent with the primary fabric can be attributed to autohydrothermal solutions. Recognition of the congruency of the minerals is essential for a correct genetic interpretation of spilites.

Basic rocks attaining spilitic compositions through any post-magmatic process may perhaps be referred to as spilitic rocks; but terms such as metabasalts, propylites, etc. more adequately describe the actual origin and composition of such rocks.

Acknowledgements

The authors thank Prof. I.C. PANDE, Director, C.A.S. in Geology, Panjab University, Chandigarh, for all the facilities. They also extend their thanks to Prof. Dr. G.C. AMSTUTZ, Director, Mineralogisch-Petrographisches Institut, Heidelberg, for going through the manuscript critically and for his valuable suggestions.

References

AMSTUTZ, G.C. (1954): Geologie und Petrographie der Ergußgesteine im Verrucano des Glarner Freiberges. Vulkaninstitut Immanuel Friedländer Publ. 5, 150, Zürich.

AMSTUTZ, G.C. (1958): Spilitic rocks and mineral deposits. Bull. Missouri School of Mines, Tech. Series No. 96, 11.

AMSTUTZ, G.C. (1965): Some comments on the genesis of ores; Symposium. Problems of magmatic ore deposition II, 147-150, Prague.

AMSTUTZ, G.C. (1968): Spilites and spilitic rocks. In: Basalts, 2, 737-753. Ed. Hess and Poldervaart.

BAMBA, T., SAWA, T. (1967): Spilite and associated manganiferous hematite deposits of the Tokoro district, Hokkaido, Japan. Geol. Survey Japan Rep. 221, 1-21.

BARTH, V. (1966): The initial volcanism in the Devonian of Moravia. Paleovolcan. Bohemian massif (Praha), 115-125.

BATTEY, M.H. (1956): The petrogenesis of a spilitic rock series from New Zealand. Geol. Mag. 93, 89-110.

BENSON, W.N. (1915): The dolerites, spilites and keratophyres of the Nundle district. Linnean Society N.S. Wales Proc. 40, 121-173.

BURRI, C., NIGGLI, P. (1945): Die jungen Eruptivgesteine des Mediterranen Orogens. V.III, Vulkaninstitut Immanuel Friedländer, 654, Zürich.

COX, A.H. (1915): The geology of the district between Abereidy and Abercastle (Pembrokenshire). Q.J.G.S. (Lond.) 71, 328.

CORNWALL, H.R. (1951): Differentiation in magmas of Keweenawan series. Journ. Geol. 59, 151-172.

DUSCHATKO, R., POLDERVAART, A. (1955): Spilite intrusion near Ladron peak, Socorro County, New Mexico. Bull. Geol. Soc. Am. 66, 1097-1108.

ESKOLA, P. (1925): On the petrology of eastern Fennoscandia I. The minderal development of basic rocks in the Karelian Formation. Fennia 45, 1-93.

FAIRBAIRN, H.W. (1934): Spilites and average metabasalts. Am. Journ. Sci. 27, 92-97.

FIALA, F. (1966): Some results of the recent investigations of the Algonkian Volcanism in the Barrandian and the Zelezne hory areas. In Paleovolcanites of the Bohemian massif. Praha, 9-29.

FIALA, F. (1967): Algonkian pillow lavas and variolites in the Barrandian area. Geologie 12, 7-64.

FLAHERTY, G.F. (1934): Spilitic rocks of southeastern New-Brunswick (Canada). Jour. Geol. 42, 875.

FUCHS, G. (1967): Zum Bau des Himalayas. Springer: Wien-New York, 211.

GUPTA, L.N. (1966): Darrang Traps and enclosed quartzofelspathic inclusions of the Dharamsala area. Indian Mineralogist. 7, 55-60.

KANWAR, R.C. (1968): Some observations on the traps occurring near Ghanyara, H.P., India. Vasundhara 4, 55-59.

KRISHNAN, M.S. (1960): Geology of India and Burma. Higgingbothams. 604. Madras.

KUNO, H. (1959): Origin of Cenozoic petrographic provinces of Japan and surrounding areas. Bull. Volcanogique, Ser. 2,20, 37-76.

LEHMANN, E. (1949): The keratophyr-weilburgit problem. Heidelberg. Beiträge Min. und Petr. 2, 1-66.

LEHMANN, E. (1952): The significance of hydrothermal stage in the formation of igneous rocks. Geol. Mag. 89, 61-69.

LEHMANN, E. (1965): Non metasomatic chlorite in igneous rocks. Geol. Mag. 102, 24-35.

MATHUR, L.R., EVANS, P. (1964): Oil in India. Brochure. Int. Geol. Congr. 22nd session, 85, New Delhi.

McMAHON, C.A. (1883): On microscopic structures of some Dalhousie rocks. Rec. Geol. Survey, India 16, pt. 3, 35-49.

McMAHON, C.A. (1882): Traps of Mandi. Rec. Geol. Survey, India 15, 155-164.

MIDDLETON, G.V. (1960): Spilitic rocks in southwest Devonshire. Geol. Mag. 97, 192-207.

NICHOLLS, G.D. (1959): Autometasomatism in the lower spilites of the Builth Volcanic series. Q.J.G.S. (Lond.) 114, 137-161.

NIGGLI, E. (1944): Das westliche Tavetscher Zwischenmassiv und der angrenzende Nordrand des Gotthardmassivs. Schweiz. Min. Petr. Mitt. 24, 58-301.

NIGGLI, P. (1952): The chemistry of the Keweenawan lavas. Am. Journ. Sci. (Bowen Volume), 381-441.

REED, J.J. (1950): Spilites, serpentines and associated rocks of the Mossburn dist., Southland (New Zealand). Trans. Roy. Soc. N.Z. Wellington, 78, 106-126.

REED, J.J. (1957): Petrology of the lower Mesozoic rocks of the Wellington, New Zealand. Bull. Geol. Survey N.Z. 57, 34-39.

RITTMANN, A. (1958): Geosynclinal volcanism, ophiolites and Barramiya rocks. Egyptian Journ. Geol. 2, 61-65.

DE ROEVER, W.P. (1940): Über Spilite und verwandte Gesteine von Timor. Koninklijke Nederlandische Akad. Wetenchappen. Proc. XLIII, 5, 1-7.

DE ROEVER, W.P. (1940a): Geological investigations in the southwestern Moetics Region (Netherlands Timor). Diss. Amsterdam, 244.

DE ROEVER, W.P. (1941): Die permischen Alkaligesteine und die Ophiolithe des Timorschen Faltengebirges. Proc. Nederland. Akad. Wetenchappen. XLIV, 8, 1-4.

SCOTT, B. (1951): A note on the occurrence of intergrowth between diopsidic augite and albite, and of hydrogrossular from King Island, Tasmania. Geol. Mag. 88, 429-431.

SRIKANTIA, S.V., SHARMA, R.P. (1969): Shali formation - a note on the stratigraphic sequence. Bull. Geol. Soc. India. 6, 94.

SUNDIUS, N. (1930): On spilitic rocks. Geol. Mag. 67, 1-17.

VALLANCE, T.G. (1960): Concerning spilites. Proc. Linn. Soc. N.S. Wales. IXXXV, 8-52.

VALLANCE, T.G. (1965): On the chemistry of pillow lavas and the origin of spilites. Min. Mag. 34, 471-481.

VALLANCE, T.G. (1968): Recognition of specific magmatic character in some Palaeozoic mafic lavas in N.S. Wales. Special Publ. Geol. Soc.

VALLANCE, T.G. (1969): Spilites again: some consequences of degradation of basalts. Proc. Linn. Soc. N.S. Wales. 94, 8-51.

VAN OVEREEM, A.J.A. (1948): A section through the Dalformation (S.W. Sweden). Leiden, Holland, 131.

VON ECKERMANN, H. (1938): A contribution to the knowledge of late sodic differentiates of basic eruptives. Jour. Geol. 46, 412-437.

VUAGNAT, M. (1949): Variolites et spilites. Archiv. Sci. 2, Fasc. 2, 223-236, Genève.

WADIA, D.N. (1961): Geology of India. 536. McMillan: New York - London.

WELLS, A.K. (1923): The nomenclature of the spilitic suite. II. The problem of the spilites. Geol. Mag. 60, 62-74.

WEST, W.D. (1938): General report for 1937. Rec. Geol. Survey India 73, 96.

YODER, H.S., Jr. (1967): Spilites and serpentinites. Yearb. Carnegie Instn. Washington, 65, 269-279.

ZAVARITSKY, V.A. (1946,1960): The spilite-keratophyre formation in the region of the Blyava deposit in the Ural mountains. Trudy Inst. Geol. Nauk. 71, Petr. Ber. 24, 1-83.

General Features of the Spilitic Rocks in Finland

T. Piirainen and P. Rouhunkoski

Introduction

An examination of the geological map of Finland (Fig. 1) shows that in eastern and northern Finland the basic effusives and intrusives are extensively distributed over the Karelian schist zone. Both the intrusives and effusives vary in composition from ultra-basic to basic. Among the basic rocks, spilites, i.e. rocks with albitic plagioclase, abound. Since both a volcanic and a hypabyssic milieu is exposed here, these rocks provide abundant material for studying

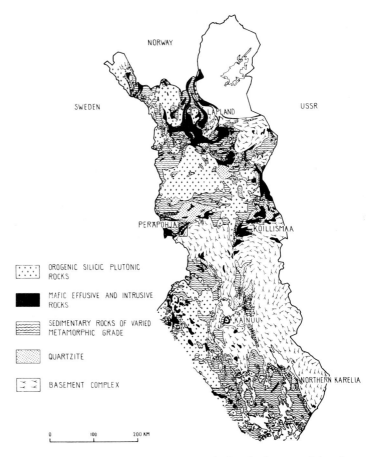

Fig. 1. Geological map of eastern and northern Finland (generalized after geological map of Finland drawn by AHTI SIMONEN, 1960)

the spilite dilemma. Consequently, spilites have twice been strongly to the fore in the course of geologic investigations in Finland; the first time was at the beginning of this century and the second in the 1960's.

As the general geologic mapping in the first half of this century proceeded from south to north, the distribution, mineral and chemical composition and structure of the rock types in question were roughly established (FROSTERUS, 1902; FROSTERUS and WILKMAN, 1920; HACKMAN, 1910; HACKMAN and WILKMAN, 1925, 1929; VÄYRYNEN, 1928). More light was shed on these rocks by the investigations of ESKOLA (1925), which were later to be of great significance for our understanding of the basic rocks of the Karelian schist zone. At that time, it was generally held that metamorphosis and metasomatosis played an important part in the formation of these rocks, as is indicated by the term metabasite which was generally applied to them. More definitely, metabasites were uralite diabases, greenstones and greenschists. It was considered that the occasional Na excess compared with normal basalts was caused by Na metasomatosis. ESKOLA (1925), among others, emphasized that in all of the spilitic rocks he had seen, for which a crystallization sequence could be established, plagioclase was the first mineral, and, consequently, it must have crystallized primarily rich in Ca. On the other hand, observations were made which showed that magmatic processes were significant in the formation of rocks from Kainuu which contain abundant albite and some amphibole and calcite. VÄYRYNEN (1928) called these rocks karjalites. The dikes of similar features appear in Koillismaa and central Lapland, where they are termed albite diabase (HACKMAN, 1927; MIKKOLA, 1941). Since observations were made which supported the metasomatic origin of the spilites on the one hand and the magmatic origin on the other, laboratory experiments were carried out to decide the matter. Albite and calcite were obtained from anorthite by heating it in a soda solution (ESKOLA et al. 1935). The problem was considered solved and the conclusion at this stage of the spilite studies was that "spilites are basalt stewed in a soda solution".

MERILÄINEN (1961) initiated a new phase in the studies on the spilites in the Karelian schist zone by describing a large number of albite-diabases and albitites with variable composition from central and northern Lapland. According to him, the structural features of the rocks as well as the relatively large variation in their composition indicated that magmatic processes played a decisive role in their formation.

But spilitic rocks were also encountered in prospecting, e.g. in the course of a search for uranium in northern Karelia (PIIRAINEN, 1968, 1969) and in the studies on iron and manganese ores in central Lapland (NUUTILAINEN, 1968; PAAKKOLA, 1971). On the other hand, systematic mapping of the Karelian schist zone provided the opportunity for a study on the Karelian basic intrusives and effusives (PIISPANEN, 1971). These studies are still underway; nevertheless, the results obtained so far indicate that the problem, which was thought to have been solved, is still topical and is of both petrologic and ore geologic interest.

Distribution of Spilites

The studies carried out so far indicate that spilitic rocks are encountered in the schist areas of northern Karelia, Kainuu, northern Bothnia, Peräpohja, Koillismaa and central Lapland in a zone which begins in SE Finland as a narrow belt and extends north-westwards to northern Finland where it widens to cover an extensive area (Fig. 1). The above-mentioned schist areas form the Karelian schist zone.

The bedrock in all these sub-areas is composed of a pre-Karelian basement complex and of Karelian schists. The Karelian schists and the pre-Karelian basement are clearly distinguishable from each other and each has its own geologic problems.

Gneisses of different types predominate in the pre-Karelian basement complex. The main types are granite, granodiorite, trondhjemite and mica gneisses. Among the gneisses, distinctive schists can be discerned locally. These are mainly strongly metamorphosed basic, intermediate and acid lavas and tuffs. This obviously represents a very deep section through an ancient mountain chain which reveals the syn- and late-orogenic plutonism as well as probably also the subsequent volcanism. According to age determinations by KOUVO and TILTON (1966), the main phase of the folding of the mountain chain took place about 2.8 billion years ago.

Two stratigraphic units, Jatuli and Kaleva, have long been distinguished in the Karelian schist zone. Their age relation as well as their importance has changed with time. VÄYRYNEN (1928), a distinguished student of the Karelian schist zone, considered Jatuli to be older than Kaleva. According to him, Jatuli is composed of well-sorted and chemically far-weathered sediments, ortho-quartzites, Fe-Al-rich mica schists and limestones. Kaleva, on the other hand, was formed by rapid sedimentation and is composed of poorly sorted and chemically only slightly weathered sediments which are mainly graywacke schists. These two different sedimentary associations are separated from each other by basic magmatism in such a way that the effusives are mainly located between the evolution series, i.e. Jatuli and the revolution series, i.e. Kaleva, whereas the intrusives, which obviously represent the feeding channels of the effusives, cross-cut Jatuli and the basement (Fig. 2). Since the revolution series is followed in many places by acid plutonism, it can be concluded that basic effusives and intrusives represent initial magmatism (the initial volcanism of STILLE, 1938). The acid plutonism represents syn- and late-orogenic magmatism of the same orogeny, the age of which is about 1.8 billion years according to KOUVO and TILTON (1966).

At the time of VÄYRYNEN's studies, the Karelian schist zone was considered to be a mountain chain of its own. Opinions have since changed and the Karelian schist zone and the Svecofennian schist zone further west have been joined together as formations of the Svecofenno-Karelian orogeny in such a way that the Svecofennian schist zone represents the geosyncline proper and the Karelian one the marginal region of the geosyncline (SIMONEN, 1960). The same bipartition with basic magmatic phases which is discernible in the eastern parts of the Karelian schist zone seems to continue over the Svecofennian side (METZGER, 1959). In the Svecofennian area, the syn- and late-orogenic plutonism have developed much further. Also the metamorphic facies conditions have reached higher levels than in the Karelian zone in which recrystallization took place over extensive areas under the conditions of albite-epidote-amphibolite and greenschist facies. Spilites are known only within the Karelian schist zone, but it is likely that they have a much larger distribution than that suggested by the studies

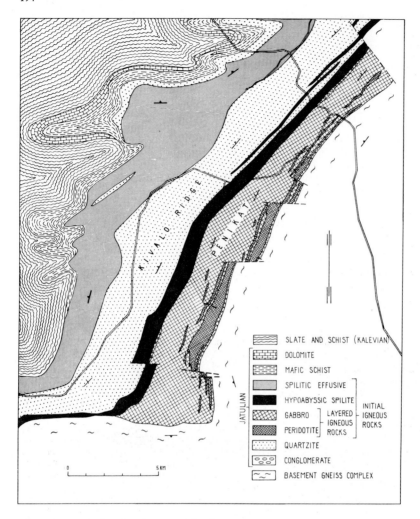

Fig. 2. Geological map of the Penikat-area mainly after HÄRME (1949) and KUJANPÄÄ (1964). The location of the area is shown in Fig. 1

carried out so far. It can be expected that their metamorphosed equivalents will be found in the Svecofennian schist area under the graywacke schists.

Initial Igneous Rocks in the Karelian Zone

On the basis of their mineral and chemical composition and structure, the basic intrusives and effusives in the Karelian zone can be divided into two groups, i.e. tholeiitic and spilitic. Both groups have their own rock associations which, however, are closely connected with each other in every schist area to such an extent that sometimes it is difficult in the field and even in the laboratory to tell which rock

is of which type. In general, the differences are so distinct that
a division such as that mentioned above is already feasible on the
basis of studies carried out in a limited area.

Tholeiitic Association

The tholeiitic association is composed of lavas, dikes and stratiform
intrusions. Tuffs, breccias and amygdaloidal rocks are associated
with the lavas. Otherwise, the lava and dike rocks bear a considerable resemblance to each other. They are fine-to medium-grained rocks
and differ in this respect from the plutonic rocks.

The mineral composition of the lava and dike rocks varies from that
of hornblende-labradorite rocks to those of hornblende-epidote-albite
and chlorite-epidote-albite rocks (Tables 1 and 2). In places, they
contain ophitic structures as relicts. Since they are also very
similar in chemical composition to each other and to tholeiitic basalt,
they are beyond doubt tholeiitic metamorphic rocks, i.e. metabasites,
which is the general term for the basic igneous rocks in the Karelian
schist zone. The hornblende-labradorite rocks are basalts which have
undergone uralitization (Fig. 3). The hornblende-epidote-albite rocks

Fig. 3. Hornblende-labradorite rock with the tholeiitic composition
(uralite diabase). Thin section, with analyser. 44 x

represent greenstones which attained their mineral composition under
the conditions of an albite-epidote-amphibolite facies (Fig. 4), whereas the chlorite-epidote-albite rocks are greenschist recrystallized
under the conditions of a greenschist facies.

In addition to tholeiitic lava and dike rocks, there are plutonic intrusions with a tholeiitic composition (Tables 1 and 2) which pene-

Table 1. Mineral composition of the rocks of the tholeiitic association in percentages by volume

	1	2	3	4	5	6	7	8	9	10	11
Olivine	47.6	4.9	–	–	–	–	–	–	–	–	–
Hyperstene	–	35.4	64.0	31.7	–	–	–	–	–	–	–
Diopside	4.1	28.5	4.4	4.6	7.7 }	33.3	29.7	–	–	–	–
Amphibole	–	–	–	3.6	18.1	2.4	32.3	51.9	59.5	47.1	41.6
Biotite	–	–	–	–	–	–	–	–	1.4	3.6	–
Serpentine	38.5	19.6	–	–	–	–	–	–	–	–	–
Chloride	–	–	–	–	–	–	–	–	–	–	25.2
Talc	9.7	11.6	16.2	–	–	–	–	–	–	–	–
Plagioclase	–	–	10.0	55.9	67.1	61.0	34.7	41.6	9.3	12.1	19.0
Epidote	–	–	5.4	4.2	7.1	–	–	4.8	27.6	32.4	7.6
Sphene	–	–	–	–	–	–	–	–	1.0	–	2.8
Quartz	–	–	–	–	–	1.6	–	–	–	–	2.6
Other minerals	–	–	–	–	–	1.7	3.3	1.7	1.2	4.8	1.2

1. Dunite, Peräpohja (KUJANPÄÄ, 1964)
2. Peridotite, Peräpohja (KUJANPÄÄ, 1964)
3. Pyroxenite, Peräpohja (KUJANPÄÄ, 1964)
4. Pyroxene gabbro, Peräpohja (KUJANPÄÄ, 1964)
5. Anorthosite, Peräpohja (KUJANPÄÄ, 1964)
6. Diabase, Koillismaa (Anal. 5, Table 2)
7. Diabase, northern Karelia (Anal. 6, Table 2)
8. Amphibole-labradorite rock (uralite diabase), northern Karelia (Anal. 7, Table 2)
9. Amphibole-epidote-albite rock (greenstone diabase), northern Karelia (Anal. 8, Table 2)
10. Amphibole-epidote-albite rock (greenstone diabase), Peräpohja (Anal. 9, Table 2)
11. Lava-born amphibole-chlorite-albite-epidote rock (greenstone), Peräpohja (Anal. 10, Table 2)

Table 2. Chemical composition of the rocks of tholeiitic association

	1	2	3	4	5	6	7	8	9	10
SiO_2	40.96	53.16	52.52	49.25	50.06	49.61	49.58	50.25	50.25	49.72
TiO_2	0.35	0.18	0.20	0.12	0.94	0.88	0.84	1.56	0.97	1.13
Al_2O_3	2.05	7.36	17.10	20.14	15.21	13.80	17.23	13.19	13.53	13.06
Fe_2O_3	1.16	1.97	0.31	1.45	2.27	2.12	2.00	4.94	1.90	1.84
FeO	10.26	7.30	6.63	3.02	9.60	8.81	7.34	9.05	9.43	12.07
MnO	0.16	0.19	0.17	0.08	0.18	0.17	0.14	0.20	0.21	0.20
MgO	34.12	22.08	10.15	8.70	7.65	8.70	6.35	6.32	7.68	8.19
CaO	0.42	5.02	8.18	13.43	11.77	12.38	12.24	10.00	10.55	6.49
Na_2O	0.11	0.93	2.16	1.18	1.60	1.71	1.92	1.76	2.61	2.64
K_2O	0.00	0.29	0.38	0.22	0.29	0.20	0.24	0.30	0.31	0.05
P_2O_5	0.02	0.72	0.94	0.04	0.10	0.04	0.03	0.20	0.09	0.11
CO_2	—	—	—	—	0.00	0.16	0.18	0.12	0.00	0.00
H_2O	10.25	1.11	1.40	2.37	0.56	0.77	1.33	1.96	2.13	4.06
H_2O^-	0.20	0.04	0.06	0.03	0.04	0.09	0.09	0.03	0.05	0.06
	100.06	100.35	100.20	100.03	100.27	99.44	99.51	99.88	99.71	99.62

1. Serpentinite, Peräpohja. Anal. E. Lindqvist (KUJANPÄÄ, 1964)
2. Pyroxenite, Peräpohja. Anal. E. Lindqvist (KUJANPÄÄ, 1964)
3. Pyroxene gabbro, Peräpohja. Anal. E. Lindqvist (KUJANPÄÄ, 1964)
4. Anorthosite, Peräpohja. Anal. H.B. Wiik (HÄRME, 1949, p.28)
5. Diabase, Koillismaa. Anal. R. Saikkonen (PIISPANEN, 1971)
6. Diabase, northern Karelia. Anal. R. Saikkonen (PIIRAINEN, 1969, p.25)
7. Amphibole-labradorite rock (uralite diabase), northern Karelia. Anal. R. Saikkonen (PIIRAINEN, 1969, p.25)
8. Amphibole-epidote-albite rock (greenstone diabase), northern Karelia. Anal. R. Saikkonen (PIIRAINEN, 1969, p.25)
9. Amphibole-epidote-albite rock (greenstone diabase), Peräpohja. Anal. Ojanperä and Saikkonen (KORKALO, 1971)
10. Lava-borne amphibole-chlorite-albite-epidote rock (amygdaloidal greenstone), Peräpohja. Anal. Ojanperä and Saikkonen (KORKALO, 1971)

Fig. 4. Hornblende-epidote-albite rock with the tholeiitic composition (greenstone diabase). Thin section, with analyser. 32 x

trated the space between the sediments and the basement. In the schist area of Peräpohja, Jatulian quartzites form a roof for the intrusives, whereas in Koillismaa the roof seems to be composed of volcanites. Locally, the intrusives are several kilometers wide. They are differentiated in such a way that the bottom parts are composed of ultrabasic rocks, the central parts of gabbros and the upper edge of anorthosites. Albite-biotite rocks occur in the margins of the intrusives and sometimes attain a width of several hundred meters. Their mode of formation is not yet fully understood. Ultramafic rocks in the schist area of Peräpohja include some chromite ores, and Cu-Ni ores have been encountered in association with the ultrabasic rocks in Koillismaa. In addition to their occurrence in ultrabasic rocks at the bottom of the intrusives, Cu-Ni ores also occur as intercalates in gabbros. Between the gabbros and anorthosites, vanadium-bearing titanium iron ores have been encountered.

Spilitic Association

The rocks of the spilitic association occur in a manner similar to that of the tholeiitic association, i.e. as lavas, dikes and intrusives. Compared with the tholeiitic association, the lavas and dikes are more variegated, whereas the intrusives do not reach dimensions such as those of the tholeiitic stratiform bodies.

The lavas are often pillow lavas and the dikes obviously represent their feeding channels. They vary in composition within approximately the same ranges, i.e. from pure hornblende and hornblende-chlorite rocks through hornblende-albite-epidote and chlorite-albite-epidote rocks to albite-hornblende, albite-chlorite, albite-biotite and almost pure albite rocks (Tables 3 and 4). In addition to hornblende and chlorite, diopside and titanomagnetite may also occur as dark

Table 3. Mineral composition of the rocks of the spilitic association, in percentages by volume

	1	2	3	4	5	6	7	8	9	10
Olivine	21.4	–	–	–	–	–	–	–	–	–
Pyroxene	15.3	–	53.2	–	–	–	11.4	–	–	–
Amphibole	5.1	73.0	34.0	97.7	59.8	17.8	14.6	–	–	0.3
Biotite	–	–	–	–	2.4	6.7	–	9.5	–	4.1
Chlorite	–	20.0	5.1	–	–	–	–	29.5	–	6.8
Serpentine	45.6	–	–	–	–	–	–	–	–	–
Sphene	–	–	1.4	0.9	2.7	2.6	6.1	0.2	–	–
Leucoxene	–	–		–			–	–	–	–
Ore minerals	–	3.0	–	–	1.1	7.6	7.9	9.1	2.2	7.3
Albite	–	4.0	6.1	0.7	33.8	62.5	58.7	57.3	88.2	66.6
Epidote	–	–	–	–	–	2.7	–	–	–	0.1
Carbonate	–	–	–	–	–	–	–	0.4	7.0	3.6
Apatite	–	–	–	0.4	–	–	1.0	–	–	–
Quartz	–	–	–	–	–	–	–	–	–	10.6
Other minerals	–	–	0.2	0.3	0.2	0.1	0.3	–	2.6	0.6

1. Metaperidotite, northern Karelia
2. Lava-borne amphibole-chlorite rock, central Lapland (Anal. 2, Table 4)
3. Pyroxenite, northern Karelia (Anal. 3, Table 4)
4. Hornblendite, Koillismaa (Anal. 4, Table 4)
5. Amphibole-albite rock (albite diabase), northern Karelia (Anal. 5, Table 4)
6. Albite-amphibole rock (albite diabase), northern Karelia (Anal. 6, Table 4)
7. Albite-amphibole-pyroxene rock (albite gabbro), Peräpohja (Anal. 7, Table 4)
8. Lava-borne albite-chlorite-biotite rock (weilburgite), central Lapland (Anal. 8, Table 4)
9. Lava-borne albite rock (variolite), Koillismaa (Anal. 9, Table 4)
10. Albite-chlorite-biotite rock (albitite), northern Karelia

Table 4. Chemical composition of the rocks of the spilitic association

	1	2	3	4	5	6	7	8	9	10
SiO_2	39.53	43.83	50.12	50.12	53.28	52.33	55.42	53.31	61.36	48.76
TiO_2	0.04	0.90	0.90	1.08	1.73	2.06	1.66	1.20	0.65	0.55
Al_2O_3	0.74	7.13	4.64	5.10	9.28	12.46	13.11	16.39	17.73	13.46
Fe_2O_3	4.79	4.24	2.45	3.40	2.78	9.84	5.43	2.74	0.59	—
FeO	2.16	10.76	6.03	7.21	6.27	6.39	4.28	11.19	0.36	1.02
MnO	0.13	0.17	0.15	0.14	0.23	0.12	0.07	0.21	0.10	0.05
MgO	37.40	18.33	14.50	15.34	11.82	3.87	4.43	3.11	0.17	6.20
CaO	1.09	9.26	16.79	13.55	7.90	4.41	7.81	1.04	4.86	9.25
Na_2O	0.20	1.10	1.58	1.33	3.30	6.59	6.25	5.87	9.40	6.80
K_2O	0.05	0.36	0.19	0.25	0.61	0.68	0.31	0.34	0.68	0.44
P_2O_5	0.20	—	0.08	0.09	0.16	0.18	0.35	0.38	0.07	0.07
CO_2	0.55	—	0.00	0.17	0.10	0.03	0.00	0.27	3.61	13.48
H_2O^+	12.61	3.51	1.12	1.74	1.99	0.81	0.59	3.72	0.60	—
H_2O^-	0.50	0.13	0.12	0.04	0.11	0.03	0.05	0.04	0.04	0.35
	99.99	99.72	99.52	99.56	99.56	99.80	99.76	99.81	100.22	100.43

1. Serpentinite, Peräpohja. Anal. A. Heikkinen (NUUTILAINEN, 1968, p.23)
2. Lava-borne amphibole-chlorite rock, central Lapland. Anal. H. Lönnroth (MIKKOLA, 1941, p.239)
3. Pyroxene-amphibole rock (pyroxenite), northern Karelia. Anal. R. Saikkonen (PIIRAINEN, 1969, p.30)
4. Amphibole rock (hornblendite), Koillismaa. Anal. R. Saikkonen (PIISPANEN, 1971)
5. Amphibole-albite rock (albite diabase), northern Karelia. Anal. R. Saikkonen (PIIRAINEN, 1969, p.30)
6. Albite-amphibole rock (albite diabase), northern Karelia. Anal. R. Saikkonen (PIIRAINEN, 1969, p.30)
7. Albite-amphibole-pyroxene rock (albite gabbro), Peräpohja. Anal. P. Ojanperä (NUUTILAINEN, 1968, p.23)
8. Lava-borne albite-chlorite-biotite rock, central Lapland. Anal. P. Ojanperä (PAAKKOLA, 1971, p.53)
9. Lava-borne albite rock (variolite), Kuusamo, northeast-Bothnia. Anal. R. Saikkonen (PIISPANEN, 1971)
10. Albitite, northern Lapland. Anal. A. Heikkinen (MERILÄINEN, 1961, p.38)

minerals. With the exception of albite and epidote, calcite and occasionally paragonite are met with as light minerals. As accessories, the rocks may contain quartz, apatite, turmaline, pyrrhotite, pyrite, chalcopyrite and chalcocite. The structure of the volcanic rocks is frequently pilotaxitic and sometimes also porphyric. Spherulitic structures are encountered in albite-rich rocks. The structure of the dike rocks is either ophitic or hypidiomorphic.

Besides dikes and lavas, the spilitic association also forms stratiform intrusives. These occur in the same way as those of the tholeiitic intrusives, often between the pre-Karelian basement and Jatuli or as sills in Jatulian quartzites. In places they may reach a width of almost a kilometer. The intrusive exhibit a differentiation series from ultrabasic, mainly hornblende-bearing rocks, through hornblende-albite and albite-hornblende rocks to rocks in which albite predominates but which also contain variable amounts of biotite, hornblende, chlorite and quartz (Tables 3 and 4). Epidote may be abundant, especially in the hornblende-albite rocks, as individual grains and small inclusions in albite.

The intrusives are cut by carbonate-rich dikes of which two types can be discerned, i.e. albite-carbonate dikes and sulphide-bearing quartz-carbonate dikes. In the marginal parts of the former, there is often magnetite.

Hornblende occurs in the ultrabasic rocks as porphyroblasts and includes relicts of olivine and pyroxene. Besides the formation of hornblende, other alterations have also taken place in the ultrabasic rocks giving rise to biotite, chlorite, talc, serpentine and carbonates. The ultrabasic rocks grade upwards into hornblende-albite rocks in which hornblende occurs as idiomorphic crystals surrounded by poikilitic albite (Fig. 5). In the albite-hornblende rocks underlain by the rocks rich in hornblende the order of crystallization is the reverse

Fig. 5. Hornblende-albite rock with the spilitic composition (albite diabase). Thin section, with analyser. 6 x

(Fig. 6), and albite occurs as idiomorphic crystals enveloped by poikilitic hornblende. The crystallization of hornblende has continued, however, after the conclusion of the main phase of the magmatic crystallization, with the result that hornblende exhibits porphyroblastic features.

Fig. 6. Albite-hornblende rock with the spilitic composition (albite diabase). Thin section, with analyser. 5 x

An examination of the chemical composition in different parts of the intrusives brings to light an outstanding feature in the distribution of Na and Ca. Calcium is abundant in the lower parts of the intrusives, in ultrabasic rocks (Table 4, anal. 3 and 4), in which it is mainly incorporated in diopside and hornblende, and in cutting dikes in which it is bound to carbonates. Sodium predominates over calcium in the upper parts of the intrusives where it is mainly incorporated in albite (Talbe 4, anal. 6).

The rocks of the spilitic association also contain ores. In central Lapland, exhalative-sedimentogeneous iron and manganese ores are encountered (PAAKKOLA, 1971) in association with volcanic rocks, and liquimagmatic magnetite ores are met with in connection with plutonic rocks (NUUTILAINEN, 1968). Also the sulphides show an extensive distribution in spilites. In a hypabyssic environment, abundant quartz-carbonate dikes occur with a mineral paragenesis of pyrite-pyrrhotite-chalcopyrite-cobalt pentlandite, i.e. the same mineral paragenesis as that encountered in the ores of the Outokumpu type. Some dikes are rich in magnetite and others in chalcocite.

The Genesis of Spilites

As was mentioned at the beginning of this paper, two theories have been presented as a possible explanation for the formation of the spilites in the Karelian schist zone. According to the first, they are basalts stewed in a soda solution (ESKOLA, 1925). The second theory presumes them to be primary magmatic rocks (MERILAINEN, 1961). Both theories can be supported by ample arguments. The former theory is favored on the one hand by laboratory experiments (ESKOLA et al. 1935) in which albite and calcite were produced by heating anorthite in a soda solution, and on the other hand by field observations which indicate a close association between the spilites and normal basalts (ESKOLA, 1925). The sodium metasomatic alterations in the environment of spilites also support this view (VÄYRYNEN, 1928). The latter theory can also be supported by field observations which show that the spilites form a differentiation series from ultrabasic rocks to albitites and albite-carbonate dikes, as well as by microscopic observations indicating that the albite in various differentiates is a primary mineral (VÄYRYNEN, 1928; MERILÄINEN, 1961).

Both theories are partly right. Rocks with approximately similar chemical and mineral composition were formed in different ways. The saying, "there are spilites and spilites" applies to these rocks as much as the saying, "there are granites and granites" applies to granites. There are two main groups in the spilites of the Karelian schist zone, i.e. a metamorphic-metasomatic and a magmatic group, both of which are closely entangled with each other.

It is not absolutely necessary to assume a spilitic parental magma proper with sodium predominating over calcium in order to produce magmatic spilites. Field observations support the view that the presence of volatiles, notably CO_2 and H_2O in tholeiitic (PIIRAINEN, 1969; PIISPANEN, 1971) or olivine basaltic (MERILÄINEN, 1961) parental magma, has controlled the course of crystallization in such a way that spilites were formed. At the beginning, olivine and pyroxene crystallized in a normal manner as is indicated by the ultrabasic rocks so often associated with spilitic intrusives. On the other hand, ultrabasic rocks occur as dikes and individual intrusives which indicate the liquation of the ultrabasic phase from the parental magma (PAAKKOLA, 1971; cf. NUUTILAINEN, 1968). Subsequently, the crystallization continued with diopside-albite, hornblende-albite and almost pure albite rocks. Calcium was incorporated to some extent in the lattice of pyroxene and amphibole but not in that of plagioclase. The surplus of calcium formed as a result of this was concentrated into a carbonate phase of its own, which subsequently crystallized as a dissemination and a network of veins.

Volatiles were heterogeneously distributed in the magma. The parts devoid of volatiles crystallized as diabases and normal basalts. After the magmatic spilites had crystallized, liquids remained which activated the recrystallization of diabases and normal basalts thus producing greenstones under the conditions of an albite-epidote-amphibolite facies and greenschist under the conditions of a greenschist facies. At the same time, the chemical composition might also have changes, giving rise to rocks resembling magmatic spilites in chemical composition. In any case, adinolization is common in the environment of the magmatic spilites due to which, not only diabases and basalts, but also any rock might have turned into albite-rich rocks.

References

ESKOLA, P. (1925): On the petrology of eastern Fennoscandia. I. The mineral development of basic rocks in the Karelian formations. Fennia 45, no. 19.

ESKOLA, P., VUORISTO, U., RANKAMA, K. (1935): An experimental illustration of the spilite reaction. Compt. Rend. Soc. Géol. Finlande 10, 61-68. Bull. Comm. géol. Finlande 119.

FROSTERUS, B. (1902): Bergbyggnaden i sydöstra Finland. Deutsches Referat: Der Gesteinsaufbau des südöstlichen Finnland. Bull. Comm. géol. Finlande 13.

FROSTERUS, B., WILKMAN, W. (1920): Vuorilajikartan selitys D3, Joensuu. General geological map of Finland, 1:400 000.

HACKMAN, V. (1910): Suomen geologinen yleiskartta. The general geological map of Finland. Lehti-Sheet C6, Rovaniemi.

HACKMAN, V. (1927): Studien über den Gesteinsaufbau der Kittilä-Lapmark. Bull. Comm. géol. Finlande 79.

HACKMAN, V., WILKMAN, W. (1925): Suomen geologinen yleiskartta. The general geological map of Finland. Lehti D6, Kuolajärvi.

HACKMAN, V., WILKMAN, W. (1929): Suomen geologinen yleiskartta. Lehti D6, Kuolajärvi.

HÄRME, M. (1949): On the stratigraphical and structural geology of the Kemi area, Northern Finland. Bull. Comm. géol. Finlande 147.

KORKALO, T. (1971): Peräpohjan liuskealue Tervolan Värejärven alueella. Manuscript at the Department of Geology, University of Oulu.

KOUVO, O., TILTON, G.R. (1966): Mineral ages from the Finnish Precambrian. Jour. Geology, v. 74, no. 4.

KUJANPÄÄ, J. (1964): Kemin Penikkain jakson rakenteesta ja kromiiteista. Manuscript at the Department of Geology, University of Oulu.

MERILÄINEN, K. (1961): Albite diabases and albitites in Enontekiö and Kittilä, Finland. Bull. Comm. géol. Finlande 119.

METZGER, A. Th. (1959): Svekofenniden and Kareliden; eine kritische Studie. Medel. Åbo Akad. Geol. Min. Inst. 41.

MIKKOLA, E. (1941): Kivilajikartan selitys D7-C7-D7. Summary: Explanation to the map of rocks. General geological map of Finland, 1:400 000.

NUUTILAINEN, J. (1968): On the geology of the Misi iron ore province, Northern Finland. Ann. Acad. Sci. Fennicae, Ser. A III, 96.

PAAKKOLA, J. (1971): The volcanic complex and associated manganiferous iron formation of the Porkonen - Pahtavaara area in Finnish Lapland. Bull. Comm. géol. Finlande 247.

PIIRAINEN, T. (1968): Die Petrologie und die Uranlagerstätten des Koli - Kaltimogebietes im finnischen Nordkarelien. Bull. Comm. géol. Finlande 229.

PIIRAINEN, T. (1969): Initialer Magmatismus und seine Erzbildung in der Beleuchtung des Koli - Kaltimogebietes. Bull. Geol. Soc. Finlande 41, 21-45.

PIISPANEN, R. (1971): Karjalaisen jakson spiliittisistä kivistä
Kuusamon liuskealueen länsiosassa. Manuscript at the Department
of Geology, University of Oulu.

SIMONEN, A. (1960): Pre-Cambrian stratigraphy of Finland. XXI. Int.
Geol. Congr. Norden, Port 9, 141-153.

STILLE, H. (1938): Die Großfelder der Erdkruste und ihr Magmatismus
(kurzes Vortragsprotokoll), Sitzungsber. d. Preuss. Akad. d.
Wiss., Phys.-math. Kl. 51-52.

VÄYRYNEN, H. (1928): Geologische und petrographische Untersuchungen
im Kainuugebiet. Bull. Comm. géol. Finlande 78.

The Relationships between Spilites and Other Members of the Oman Mountains Ophiolite Suite

B. Reinhardt

Abstract

In the ophiolites of the Oman Mountains, a group of primary spilitic rocks occurs in the volcanic and subvolcanic parts of the suite. The relationship of these rocks with other fresh, subvolcanic rocks suggests that they have crystallized from a hydrous residual melt that approached basaltic composition.

The mineral assemblage characteristic of spilites is also found to occur in the interstitial spaces of associated subvolcanic rocks of normal hornblende-gabbroid mineralogic composition. It is suggested that the development of the spilitic residual liquids in the interstitial "pores" of the hypabyssal rocks was essentially caused by high water pressure in the magma. Consequently, the oxygen pressure was also increased, and titanium and iron oxides crystallized instead of titanium- and iron-rich silicates. The final differentiation product was a hydrous basaltic liquid enriched in silica and having a relatively low crystallization temperature.

The spilitic liquids may have been separated from the hypabyssal rocks by a filter flow mechanism. The separated liquids then formed a large part of the subvolcanic dikes that fed the overlying extrusive pillow and breccia lavas. During the extrusion some of the silica may have diffused into sea water.

It is inferred from the associated sediments that spilitic lavas of the Oman Mountains "geosyncline" were extruded over a long period, ranging from the late Permian to late Cretaceous eras.

Introduction

In the Oman Mountains the rocks belonging to the ophiolite suite can be subdivided into two categories:

a) Volcanic rocks associated with the "geosynclinal" sediments that collectively form the Hawasina Nappes. Some of these extrusives are found at the base of individual thrust sheets, where they are conformably overlain by Permo-Triassic sedimentary sequences. Such lavas are, therefore, believed to represent part of an original simatic crust, on top of which the Hawasina sediments were deposited.

Others are lava flows interbedded with Hawasina sediments, or with dikes and sills penetrating these sediments. On the basis of biostratigraphic dating, it is inferred that these lavas were extruded from latest Permian to earliest Jurassic time. Lithoclastic Hawasina sediments of the same age occasionally contain volcanic debris suggesting erosion of nearby basic igneous rocks.

The greater part of the volcanic rocks of this category consists of spilites.

b) A great variety of ultrabasic rocks, and coarsely granular, medium-grained ophitic and fine-grained basic rocks collectively form the Semail Nappe[1]. This nappe is the highest exposed tectonic unit in the Oman Mountains overlying the Hawasina Nappes and consists mostly of ophiolites with very few sediments. The top part of the Semail Nappe, where locally preserved, consists essentially of subvolcanic to volcanic spilites. Rare siliceous lime mudstones and cherts interbedded with the pillow lavas contain pelagic foraminifera which point to a Cenomanian to Coniacian age of the sediments and volcanic rocks.

The History of the Ophiolite Emplacement into the "Geosyncline"

From a palinspastic reconstruction of the Hawasina and Semail Nappes it appears that the Oman Mountains "geosyncline" was situated to the north and northeast of the present mountain range. In a recent publication (REINHARDT, 1969) it has been suggested that the emplacement of the ophiolite suite into the "geosyncline" took place over a period of more than 100 million years and covered at least the entire time span of the deposition of the Hawasina sediments. The emplacement started in the Permian (or even earlier?) with the development of an "oceanic ridge". From the median volcanic feeder of this ridge, basaltic extrusives were produced at the surface. A part of these lavas is now found associated with the Hawasina sediments, either as lava flows or as sedimentary debris (see a) in foregoing section). It is thought that large quantities of the basaltic melt produced during that period did not reach the surface, but were cooled within the feeder at a greater depth. The result of this was the crystallization of hypabyssal and gabbroic rocks within the volcanic feeder systems. The lateral walls of the feeder are thought to have been formed by peridotites of the Upper Mantle. During the early part of the "geosynclinal" period (i.e. from late Permian to early Jurassic) discontinuous patches of shallow water carbonates were deposited on the volcanic crest of the oceanic ridge.

It has been suggested that after this long period of relative stability, the emplacement of the plutonic ophiolite members, which now form the bulk of the Semail Nappe, started. The mechanism of this emplacement has been explained as a slow "eversion" of the volcanic feeder system, which amounted to a spreading of the ocean floor away from the ridge and the generation of new oceanic crust. This tectonic process was accompanied by further volcanic activity. The resulting new oceanic crust consisted of the upwelling Mantle peridotites and

---

[1] In his paper on the geology of Oman, LEES (1928) suggested that the "Hawasina Series" and the "Semail Igneous Series" both formed large low-angle thrust sheets or nappes. Opposing views have been expressed by MORTON (1959) and WILSON (1969). Recent and more intensive field work by the writer and his colleagues, however, supports the hypothesis already put forward by LEES. Additional support for this view is given by GREENWOOD (1968).

the gabbros from the feeder. These rocks were thus emplaced in a solid state and became covered by subvolcanic and volcanic rocks as they reached the surface. This was accomplished at least by Cenomanian time as indicated by the age of the sediments associated with the lavas of the newly formed simatic crust.

During Santonian and Campanian time, large fragments (thousands of cubic kilometers) of this newly formed simatic crust were detached

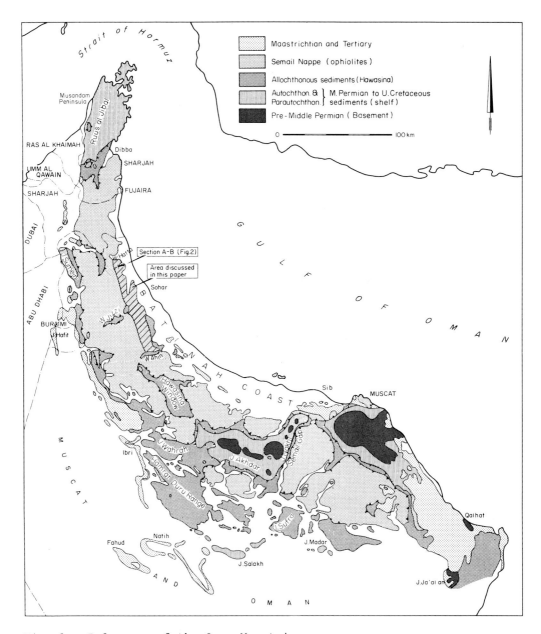

Fig. 1. Index map of the Oman Mountains

from the ocean floor and emplaced as the present Semail Ophiolite Nappe on the edge of the Arabian shield.

The ophiolites exposed in this enormous tectonic unit (Fig. 1) are thus far unique in that the nappe transport did not entirely disrupt the original relationships that existed between the volcanic and the subvolcanic rocks on the original oceanic ridge. For this reason, the Semail Nappe provides an excellent field for the study of spilites and their relationships with other members of the ophiolite suite.

Spilites and Altered Rocks with Similar Mineralogic Composition

In the Semail Nappe a subdivision has been made into two groups of rocks, both of which have a "spilitic" mineralogy. The main petrographic differences between these groups are summarized in Table 1.

Table 1. Some differences observed between "primary spilites" and altered rocks in the ophiolites of the Oman Mountains

Group 1 ("true" spilites)	Group 2 (altered volcanic, hypabyssal and gabbroic rocks)
fine-grained rocks with volcanic textures such as vesicles, variolithic texture, pillow structure, volcanic breccia, flow lamination, etc. often associated with hyaloclastic rocks.	not confined to volcanic textures. Many other textures, such as coarsely granular, coarsely ophitic, fine-grained ophitic, intersertal, porphyritic, pegmatitic, etc.
plagioclase forms thin laths that may branch at both ends.	tabular or hypidiomorphic grains of plagioclase are also common.
plagioclase shows no obvious signs of albitization.	plagioclase is cloudy and speckled with microliths of epidote, zoisite, mica, etc.
mostly low epidote content.	usually high content of pistacite-epidote.
may have very high content of hematite (plagioclase laths may be embedded in a hematite matrix).	usually low hematite content.
hornblende (if present) forms branches (see Figs. 14 and 15).	no branches of hornblende observed.

The rocks belonging to group 1 are suggested here to be spilites in the sense of AMSTUTZ (1968, p.949); they are suggested to have crystallized from a melt. In the Semail Nappe, their occurrence is restricted to the volcanic and subvolcanic "cover" of this nappe. The aim of this paper is to describe their relationships with <u>unaltered</u> diabases, dolerites and gabbros that underlie them. The association of spilites with <u>altered</u> rocks is not discussed.

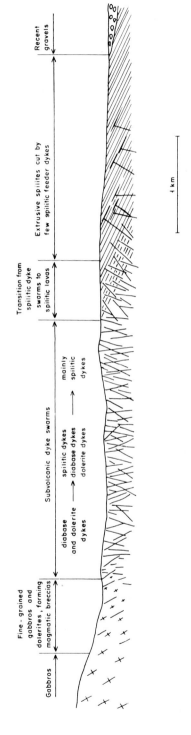

Fig. 2. Cross section through the volcanic and subvolcanic top part of the Semail ophiolite nappe in Northern Oman (see section line on Fig. 1)

The rocks belonging to the second group are only locally found in the Oman Mountains. They are commonly associated with faults and shear planes. It is thought that they represent altered gabbros, dolerites and diabases, because they grade over short distances into such rocks that still contain calcic to intermediate plagioclase and primary magmatic pyroxene and hornblende. These unaltered rocks are much more common and have been briefly described by REINHARDT (1969) as "gabbros, hornblende-gabbroid types with parallel texture, dolerites" and "diabases". The rocks of group 2 are thus regarded as products of slight, mostly localized hydrothermal metamorphism. They are excluded from the following discussion.

Field Relationships

The following description is based on observations made on the east side of the Oman Mountains in the area between Wadi Hatta in the north and Wadi Ahin in the south (Figs. 1 and 2). In this region, the dorsal part of the Semail Nappe is well exposed and shows from west to east the most complete sequence of ophiolites found in Oman: peridotites - gabbros - hypabyssal rocks - extrusive rocks. All of these units are exposed in belts that trend from north to south. The uppermost unit represents a cover of well-bedded spilitic pillow and breccia lavas about 3 000 meters thick.

Most of the lavas are composed of albite, chlorite, hornblende, hematite, occasionally clinopyroxene (salite), quartz and minor accessories such as zeolites, epidote and other as yet unidentified minerals.

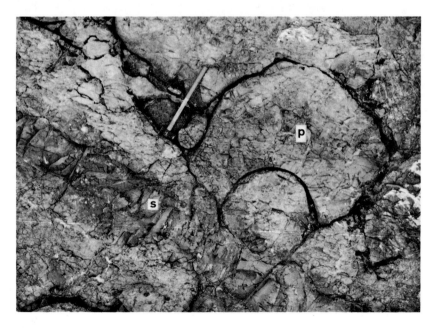

Fig. 3. Spilitic pillow (p) fed by a spilitic sill (s). The color difference between pillow and sill is real and is due to variation in the proportion albite + quartz/chlorite + hornblende + pyroxene. The mafic minerals are better represented in the sill than in the pillow

Vesicles filled with quartz, calcite and zeolites are very common especially among the breccia lavas. Pseudomorphs of calcite and hematite after early-formed pyroxene phenocrysts are sometimes also found.

Towards the base of the lavas an increasing number of dikes cut through the volcanic layers. Occasionally, these dikes branch into sills that penetrate the lava piles parallel to the layering. In pillow lavas it can be observed that the sills and dikes represent the feeders of the pillows (Fig. 3). The feeding sills and dikes that are connected with the spilitic pillow lavas are themselves also spilites. However, the relative proportions of the main constituents are slightly different in the pillows as compared with the sills and dikes: the former contain more albite, whereas the sills are considerably richer in chlorite, hornblende, sometimes clinopyroxene and another mafic 9 Å-mineral that has not been more precisely determined. These proportional variations cause the striking color differences observed between the darker feeders and the lighter lavas. In addition, textural variations can also be observed: the pillows may often be vesicular, whereas the sills and dikes are not.

At the base of the purely extrusive suite there is a further increase in the number of subvertical dikes, until isolated parts only of pillowed sequences are occasionally preserved. This is the transition zone from the volcanic to the hypabyssal sequence. The dikes of this zone are still spilites for the major part, but further below the contact zone, unaltered diabase dikes that are associated with the spilitic dikes become increasingly more important (Fig. 13). Here,

Fig. 4. Isolated pillow magmatically reworked in a younger diabase dike of the subvolcanic feeder zone. The pillow shows secondary reaction rims acquired during reworking. The primary volcanic microtextures are largely destroyed and a metamorphic texture has developed instead. The main constituents are still albite, chlorite, quartz, hornblende and ilmeno-magnetite, but the epidote content is somewhat increased as compared to the unaltered pillow lavas

Table 2. The compositions of plagioclase in the various rock types

		sample No. Rht	rock name	geological situation
		317	quartz-rich porphyrite	vein cutting hornblende gabbros (Rht 318 + 319)
		334	quartz-rich diabase	matrix in hypabyssal magmatic breccia (see Rht 332)
		415	quartz-rich porphyrite	" " " " "
Rocks w. ophitic and intersertal texture	fine-grained	365	diabase	dark component in hypabyssal magmatic breccia (see Rht 367)
		371a	"	dyke in hypabyssal dyke swarm (subvolcanic)
		377	"	" " " " " (assoc. w. Rht 376)
		456	"	base of > 30 m thick lava flow
		86	hornblende diabase	isolated outcrop
		288	hornblende diabase	rock sequence overlying gabbros
		289	" "	fine-grained layer in subvolcanic rock sequence (see Rht 290)
		336	" "	dyke cutting serpentinites
		371	" "	dyke cutting Rht 370
	coarse	330	dolerite	rock unit underlying extrusive sequence
		336a	"	dark component of hypabyssal magmatic breccia (see Rht 366b+367)
		370	"	dark "Schlieren" in sequence of ophitic rocks (see Rht 371)
		376	"	dyke in hypabyssal dyke swarm (subvolcanic, see Rht 377)
		416	"	" " " " " "
Gabbro-pegmatites and -porphyrites		224	gabbro pegmatite	coarse variety of Rht 223
		252	gabbro pegmatite	dyke cutting peridotites
		263	hornblende gabbro porphyrite	dyke cutting troctolites
		341	hornblende gabbro pegmatite	dyke cutting serpentinites
		366b	hornblende gabbro porphyrite	vein cutting dolerite (Rht 366a)
		367	hornblende gabbro porphyrite	component of hypabyssal magmatic breccia (see Rht 366)
		395	hornblende gabbro porphyrite	spot sample from layered sequence of hypabyssal rocks
GABBROS	parallel texture	87	eucrite gabbro	isolated gabbro outcrop
		290	eucrite gabbro	gabbro layer in sequence of diabases (see Rht 289)
		298	eucrite gabbro	intrusion into Rht 297
		332	olivine gabbro	component in hypabyssal magmatic breccia (see Rht 334)
		373	leucocratic gabbro	dyke cutting peridotites
		431	eucrite gabbro	dyke in hypabyssal dyke swarm (subvolcanic)
		666	olivine gabbro	dark layer from layered gabbro sequence
	coarsely granular texture	43	olivine gabbro	dyke cutting peridotites
		219	" "	" " troctolites
		223	eucrite gabbro	" " "
		226	" "	spot sample from layered gabbro sequence
		230	troctolite gabbro	country rock cut by dykes of Rht 231
		231	eucrite gabbro	10-50 m thick dykes cutting Rht 230
		242	" "	country rock overlying serpentinites
		264	" "	isolated outcrop
		297	plagioclase-lherzolite	country rock near contact w. gabbros; intruded by Rht 298
		318	hornblende gabbro	matrix of hypabyssal magmatic breccia (see Rht 319)
		319	hornblende gabbro	component in hypabyssal magmatic breccia (Rht 318)
		337	eucrite gabbro	dyke in hypabyssal dyke swarm (subvolcanic)
		379	hornblende gabbro	country rock (large gabbro sequence)
		391	olivine gabbro	dark layer from layered gabbro sequence
		714	" "	country rock overlaying Rht 718
		718	troctolite gabbro	" " " peridotites

of the ophiolite suite

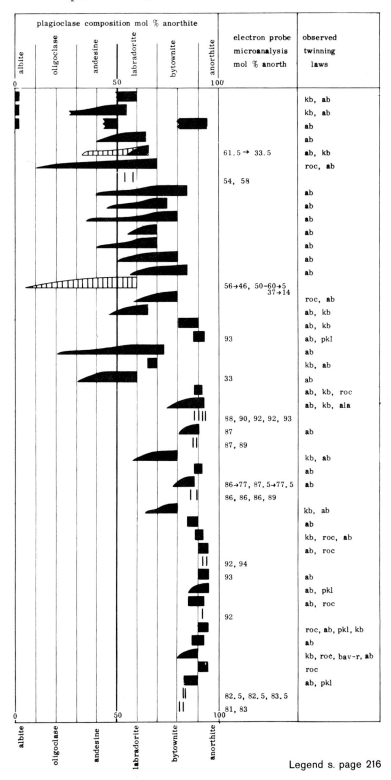

Legend s. page 216

the lavas have disappeared and only very occasionally can isolated pillows be found that have been magmatically reworked and now occur as components included in the dikes (Fig. 4). In contrast to the spilitic pillows described above, these isolated pillows are rich in pistacite and show signs of secondary alteration involving the partial destruction of the primary textures. The age relationship between the two varieties of associated dikes - diabases and spilites - is variable: spilites are sometimes obliquely cut by diabases and vice versa. This suggests that they overlap each other in time. In most cases the diabase dikes in contact either with older or with younger spilite dikes are themselves also unaltered. They contain the primary magmatic minerals (zoned calcic to intermediate plagioclase laths, diopsidic clinopyroxene and magmatic brown to green hornblende.

Still further below the contact with the lavas, the fresh diabases, together with dolerites and hornblende-gabbroid types form the great majority of the dikes, and spilitic dikes are absent. These relationships strongly suggest that:

a) the contact zone between dike swarms and lavas is of primary magmatic origin and represents the transition from the subvolcanic feeder zone to the superficial extrusive sequence.

b) the spilitic rocks found amongst the lavas and in the upper part of the feeder zone are not products of alteration. Their plagioclase laths have not been albitized, but originally crystallized as albite. This is inferred from the preserved calcic composition of the plagioclase found in the closely associated diabase dikes.

c) the spilitic dikes that feed the extrusive layers at the surface can collectively be followed into the subvolcanic dike swarms, where they are of gradual transition from spilite mineralogy to diabase mineralogy within the individual dikes. Unfortunately, no individual dike could be followed over distances long enough to confirm this. On the other hand, transitional rock members between spilite and diabase are commonly observed in the subvolcanic dike zone. They are discussed in the following section.

The spilitic dikes thus have their "roots" in the subvolcanic feeder zone.

Legend to Table 2

■ measured optically; width of block shows compositional variation

||| measured with microanalyzer; each vertical line corresponds to one grain measured or to one measurement within an inhomogenous grain

$60 \rightarrow 45$ measurements of zoned grain with microanalyzer indicating composition of center and rim

normally zoned grains; highest part of the graph indicates the compositional variation of the centers and main parts of measured grains; the thin tongue to the left indicates composition of the rims

■ corroded grains, usually surrounded by plagioclase of more acid (anorthite-poor) composition

twinning laws: ab = albite-law, bav-r = Baveno-right, pkl = pericline-law, ala = Esteral-law, kb = Karlsbad-law, roc = Roc Tourné

Petrographical Observations

Plagioclase

There is considerable variation in the composition of the plagioclase from the coarsely granular to the fine-grained rocks of the Oman Ophiolite Suite. This is illustrated in Table 2.

The coarsely granular gabbros contain rather homogenous plagioclase of very calcic composition (Fig. 5). Only three samples of coarsely granular gabbro (Rht 43, 242 and 337) have been studied that show a weak normal zonation of the plagioclase.

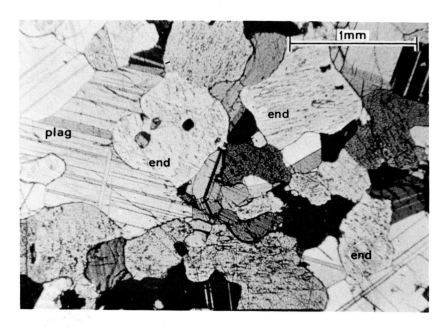

Fig. 5. Coarse-grained eucritic gabbro composed of endiopside (end) and calcic plagioclase (plag). Plagioclase contains more than 90% An. Rht 231; crossed polarizers

The plagioclase of the gabbros with parallel texture is tabular parallel to (010) and is commonly zoned. The lath centers are still very calcic but the rims are labradorite and bytownite.

The same tendency is even more pronounced in the porphyrites and pegmatites, where the rim of zoned plagioclase may even be as sodic as oligoclase.

In the typical rocks of the subvolcanic level - the dolerites and diabases (Figs. 6 and 7) - the strongest variation in plagioclase composition is observed. The centers of the plagioclase laths are labradorite and bytownite, but the grain rims are commonly oligoclase and andesine.

Fig. 6. Zoned plagioclase phenocrysts (pl) in a light-colored porphyritic vein cutting a dolerite. The plagioclase composition varies from 47% An. (center) to 30% An. (rim). The mesostasis between the corroded plagioclase phenocrysts is light-green to colorless fibrous hornblende (ho). Rht 366; crossed polarizers

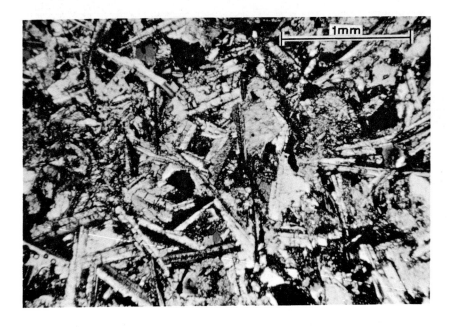

Fig. 7. Intersertal diabase with plagioclase laths ranging in composition from 65% An. (center) to 33.5% An. (rim). Rht 371 A; crossed polarizers

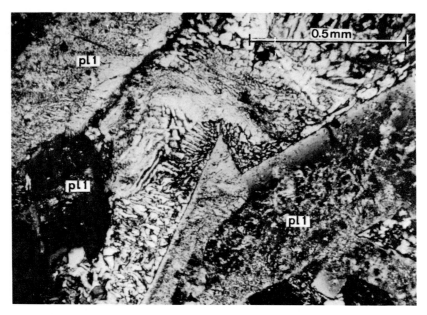

Fig. 8. Quartz-rich diabase with more than one generation of plagioclase. The large phenocrysts (pl 1) are zoned and have compositions ranging from 55% An. (center) to 25% An. (rim). The fine grained graphic intergrowth between these crystals is composed of quartz and albite. Rht 334; crossed polarizers

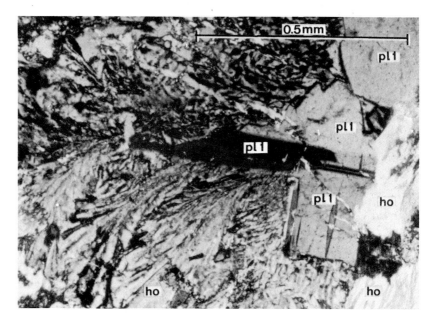

Fig. 9. Quartz-rich porphyrite consisting of zoned phenocrysts of calcic plagioclase (pl 1) with compositions ranging between 90% An. (center) and 75% An. (rim). The fine divergent branches surrounding pl 1 are composed of albite and quartz associated with hornblende (ho) and chlorite. Rht 395; crossed polarizers

Some of the diabases, dolerites and porphyrites found in the subvolcanic dike swarms may be regarded as transitional members that grade into spilites. The early-formed plagioclase of these rocks is also calcic to intermediate, but the latest plagioclase found in the interstitial spaces is albite which is sometimes associated with quartz (Figs. 8 and 9). When quartz is present, it commonly shows a graphic intergrowth with albite, suggesting that both minerals have crystallized from an eutectoid liquid. The early-formed, more calcic plagioclase laths may be corroded along the contacts with the latest-formed crystallization products. Some of these intermediate rock members are listed at the top of Table 2.

The gradual transition from calcic plagioclase to albite in a sequence of unaltered igneous rocks suggests that the presence of albite is the result of magmatic differentiation. The appearance of albite in the interstitial spaces of our ophitic rocks adds weight to the suggestion that the subvolcanic level is where the spilitic residual liquids have been generated, the process involved being one of fractional crystallization.

The Hydrous Mafic Minerals

In the unaltered feldspathic rocks of the ophiolite suite, hornblende and chlorite are the most important hydrous mafic minerals. The first to appear is brownish to greenish hornblende in some of the granular gabbros. There it commonly surrounds the diopsidic clinopyroxene or forms small blebs within or close to the edge of the pyroxene grains. The transition zone between both minerals consists of a fine intergrowth (Fig. 10) which suggests that hornblende has crystallized

Fig. 10. Endiopsidic clinopyroxene (end) grading into hornblende (ho) towards the periphery. The gradual transition consists of an intimate intergrowth of the two minerals. Gabbro, Rht 254; crossed polarizers

during a late magmatic stage and is not a product of secondary alteration (The term "uralite" is, therefore, not appropriate).

In gabbros that show a tendency towards the development of planar parallel textures, hornblende becomes an even more important product of primary crystallization. Fig. 11 shows laths of fresh calcic plagioclase surrounded by both diopsidic clinopyroxene and brown basaltic hornblende. Here also, the pyroxene has crystallized before hornblende.

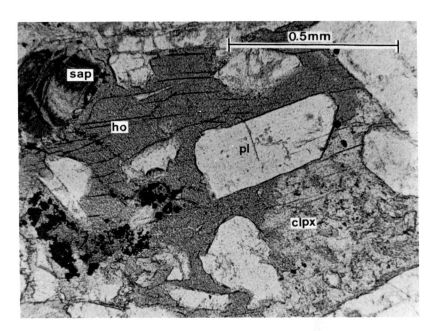

Fig. 11. Brown basaltic hornblende (ho) which surrounds and intergrows with diopsidic clinopyroxene (clpx). Plagioclase (pl) forms euhedral laths that precede the mafic minerals. Saponite (sap) forms pseudomorphs after early-developed orthopyroxene. Gabbro, Rht 332; parallel polarizers

In the dolerites, diabases and associated hypabyssal ophitic and porphyritic rocks, hornblende, apart from forming millimeter-sized xenomorphic grains between the laths of fresh plagioclase, is also found as fine, fibrous material in the interstitial "pores". There, it is commonly associated with chlorite, albite and ilmeno-magnetite (Fig. 12 and also Fig. 6).

Fig. 13 shows a spilitic dike (a) containing albite, hornblende, chlorite, ilmeno-magnetite and some quartz (see also Figs. 14 and 15), cutting other subvolcanic dike rocks. The neighboring dike rocks are coarse ophitic diabases and dolerites (b and c) which contain calcic to intermediate plagioclase, diopsidic clinopyroxene and brown to green magmatic hornblende. The accessories are quartz, ilmeno-magnetite and sphene. Moreover, the interstitial spaces of the darker wall rock (b) are filled with fibrous hornblende, chlorite and some albite that forms the rim of the zoned plagioclase laths (see Rht 370, Table 2). Thus, the minerals that are typical for the spilitic dike are also found as latest crystallization products in the residual pores of the associated ophitic rocks.

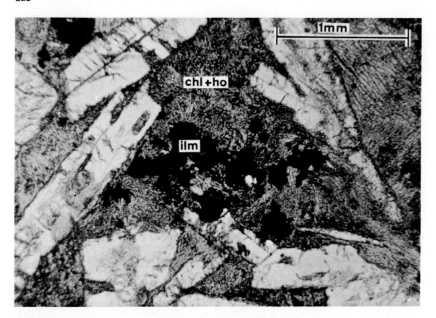

Fig. 12. Diabase with fine-grained mesostasis of fibrous hornblende (ho), chlorite (chl) and ilmeno-magnetite (ilm). Rht 376; parallel polarizers

It is inferred from these observations that the increase of the hydrous minerals hornblende and chlorite in the sequence gabbro to diabase is gradual and linked with the crystallization of increasingly sodic plagioclase. Both albite and the hydrous mafic minerals appear first in the interstitial spaces of the subvolcanic ophitic rocks.

The Ti and Fe Ores

In the plutonic members of the suite, chromium picotite is the only ore commonly found in the ultramafic members. Gabbros may also contain minute portions of it, as well as rare, small grains of magnetite. In the hypabyssal rocks, however, up to 10 vol % of skeletal ilmenomagnetite may be present (Fig. 16) and the extrusive rocks contain equally high amounts of hematite and minor portions of ilmenite and magnetite.

Conclusions

The Source of the Spilitic Residual Liquids

The field relationships suggest that the subvolcanic diabase dike swarms are the source area for the spilitic feeder dikes. In addition, petrographical observations suggest that spilitic residual liquids

Fig. 13. Spilitic dike (A = Rht 368) containing hornblende, chlorite and albite and cutting dolerite (B = Rht 370) and diabase (C = Rht 369). The interstitial spaces of the dolerite (B) are filled with fibrous hornblende, chlorite and albite

are formed at this level by the fractional crystallization of the minerals found in the gabbro - diabase sequence. The crystallization products of the residual liquids present in the "pores" of fresh dolerites and diabases contain precisely the same minerals and some of the microtextures that characterize the "primary spilites". Such dolerites and diabases with "spilitic pores" are suggested to be intermediate members between diabase and spilite. The sequence gabbro - diabase - spilite is thus a primary differentiation sequence in the ophiolites of the Oman Mountains. What are the possible reasons for the development of this type of fractionation?

From the above descriptions it follows that within the group of hypabyssal rocks, three trends of differentiation become apparent:

1. Increase of silica saturation and appearance of free quartz,

2. beginning of magnetite and ilmenite crystallization,

3. increasing predominance of hornblende (and chlorite) over clinopyroxene.

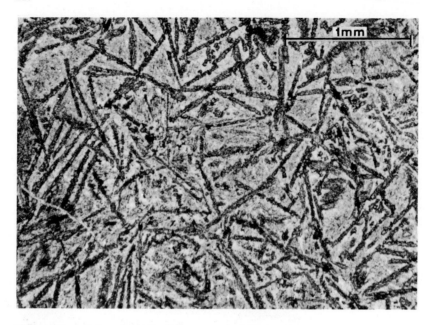

Fig. 14. Subvolcanic dike containing albite, quartz, hornblende and chlorite. The black dots are clusters of ilmeno-magnetite and "leucoxene". Rht 368, (A of Fig. 13), parallel polarizers

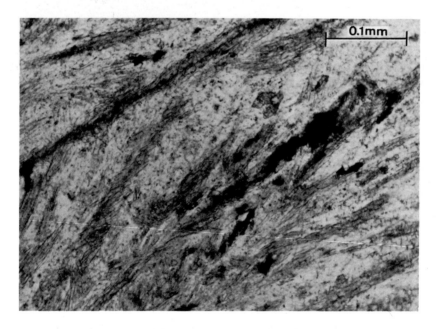

Fig. 15. Same sample as Fig. 14, detail showing the arrangement of hornblende needles and ilmeno-magnetite in the quartz-albite groundmass. Rht 368; parallel polarizers

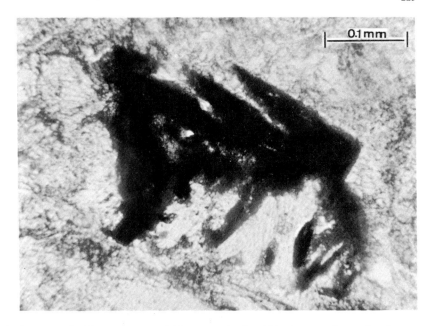

Fig. 16. Skeleton of ilmeno-magnetite, coated with irregular clusters of leucoxene as typically observed in hypabyssal rocks. Diabase, Rht 375; parallel polarizers

The silica increase could be regarded as the result of early separation of olivine. Although this type of fractionation is believed to be effective in the sequences of layered gabbros, there are major difficulties in accepting it as a crucial factor in the differentiation of the shallower igneous rocks. Olivine is virtually absent from the rocks of the subvolcanic feeder zone. If separation were to have taken place at this shallow depth, this would imply that all olivine had been very efficiently separated by gravitational settling. This is difficult to believe, since the hypabyssal zone is formed by a multitude of dikes, many of which have been partly chilled. The chilled parts, however, do not contain olivine. The alternative would be to assume that olivine separation took place at the level of gabbro crystallization where this mineral is actually present. There, however, the silica-enriched differentiation products are entirely missing, which is also difficult to explain.

It is suggested, therefore, that another type of fractionation might be responsible for the striking silica enrichment in the subvolcanic rocks. KUNO (1968) has given many examples that suggest that silica increase in the residual melts of basaltic differentiation series may be caused by the crystallization of magnetite and ilmenite. The reaction series of such sequences of basaltic rocks is characterized by the presence of these ores and the absence of Fe-rich olivine and pyroxenes. This is precisely the case with the members of the Oman Ophiolites. OSBORN (1959 and 1962) has shown that the oxygen pressure P_{O2} in a normal basaltic magma strongly influences the type of fractionation that develops: higher P_{O2} encourages the crystallization of the Ti and Fe ores instead of Ti- and Fe-rich silicates, and thus drives the residual melt towards a relatively SiO_2-rich composition. This is supposed to occur in parallel with the development of the plagioclase sequence anorthite - albite. Thus, the water content, which is an important factor governing oxygen pressure, seems to be of crucial importance in the differentiation of more acid residual liquids.

It is not known whether, in the Oman Ophiolite Suite, the water source is endogenic or whether the basaltic liquids of the feeder system have had a chance to be enriched with ocean water at a subvolcanic level. The fact, however, that increase in water pressure is not limited to the superficial levels of lavas and hypabyssal rocks but is already recognized at the level of gabbro crystallization (re. ubiquity of hornblende), may suggest that the water source was at least partly endogenic.

The separation of the "spilitic" liquids from the hypabyssal dike swarms may be compared to filter flow: the hydrous liquids developed first within the crystalline framework of the ophitic rocks, and this framework is believed to have been rather immobile due to the crystalline connections with the walls of the dikes. Increasing liquid pressure from below (subsequent basaltic liquids) may have been the force that drove the residual fluids through the network of crystals towards the surface. As a result, the differentiated liquids in the interstitial spaces were replaced by less differentiated melt from below. The crystallizing hypabyssal rock became thus a diabase or dolerite without albite and chlorite in the interstitial "pores". Until crystallization had destroyed porosity, the crystallizing hypabyssal rocks acted as a "differentiation filter" that was fed from below with less differentiated basaltic liquids, and that was producing spilitic residual melt towards the top. The hypabyssal rocks that still contain the spilitic residue may be regarded as rocks that were protected from this filter flow replacement. They are the intermediate members between spilite and diabase.

The expulsion of residual fluids from the ophitic rocks may also have been promoted by increase of thermal energy from below or by further crystallization of heavier components (ilmenite, magnetite, hematite, etc.), both of which reduced the specific weight of the liquid and gave it more "buoyancy uplift".

One problem, however, remains unsolved: the spilitic interstitial pores of the diabases and the subvolcanic spilite dikes may be silica-saturated and generally do not contain excess CO_2. The extrusive spilites, however, are usually deficient in free silica and commonly contain excess carbon dioxide. There are many possible explanations for these differences. For instance, as a volatile component, carbon dioxide may diffuse towards the surface with more ease than the other components, and consequently become relatively enriched in the vesicles of the extrusive rocks. Silica, on the other hand, may selectively escape into sea water during crystallization, which would also be in agreement with the observation that spilites are commonly associated with siliceous sediments. As yet, however, the data necessary to confirm these ideas are not available.

Acknowledgement

The author wishes to thank the Koninklijke/Shell Exploratie en Produktie Laboratorium (KSEPL), Rijswijk, The Netherlands, for supporting this study, and Shell Research N.V., The Hague, for kindly giving permission to publish this paper. Sincere thanks are expressed to the colleagues who have contributed to the present study and have critically reviewed the manuscript. The electron-probe microanalyses carried out by Prof. Dr. H. SCHWANDER in Basel (Switzerland) are gratefully acknowledged. Finally, thanks are due to Prof. Dr. P. BEARTH for his critical comments.

References

AMSTUTZ, G.C. (1968): Les laves spilitiques et leurs gîtes minéraux. Geol. Rundsch. 57/3, 936-954.

GREENWOOD, J.E.G.W., LONEY, P.E. (1968): Geology and mineral resources of the Trucial Oman Range. Inst. of Geol. Sci., Overseas Division, London.

KUNO, H. (1968): Differentiation of basalt magmas. In: The Poldervaart Treatise on rocks of basaltic composition. Ed. H.H. Hess and A. Poldervaart, Vol. 2, Interscience.

LEES, G.M. (1928): The geology and tectonics of Oman and parts of south-eastern Arabia. Q. Jour. Geol. Soc. London, 84/4 No. 336, 585-670.

MORTON, D.M. (1959): The geology of Oman. Proc. 5th World Petr. Congr., Sect. 1, Paper no. 14, 227-280.

OSBORN, E.F. (1959): Role of oxygen pressure in the crystallization and differentiation of basaltic magma. Amer. J. Sc. 257, 609-647.

OSBORN, E.F. (1962): Reaction series for subalkaline igneous rocks based on different oxygen pressure conditions. Amer. Miner. 47, 211-226.

REINHARDT, B.M. (1969): On the genesis and emplacement of ophiolites in the Oman Mountains geosyncline. Schweiz. Miner. Petr. Mitt. 49/1, 1-30.

WILSON, H.H. (1969): Late Cretaceous eugeosynclinal sedimentation, gravity tectonics and ophiolite emplacement in Oman Mountains, Southeast Arabia. Amer. Ass. Pet. Geol. Bull. 53/3, 626-671.

Gradation of Tholeiitic Deccan Basalt into Spilite, Bombay, India

R. N. Sukheswala

Abstract

The origin of spilite from tholeiite basalt is discussed; in support, field and laboratory results are presented. It is contended that the formation of spilite from tholeiite involved differentiation with accumulation of Ca and Si and their final removal from the magma system. In this process the overhead pressure of water, both magmatic and marine, played a role. The latter was the result of the subaqueous eruption of lava probably in shallow marine or estuarine waters.

Attention has been drawn to the differentiation of basalts both before and after their extrusion. Chemistry of the three major groups of rocks, viz., tholeiites, spilite and the intertrappeans (mainly hyaloclastites) occurring in the island of Bombay, is discussed. The occurrence of spilite with pillow character in the Deccan traps of India is mentioned for the first time.

Geological Setting

In an earlier paper (SUKHESWALA, 1960), attention was drawn to the existence of albitized basalt in the island of Bombay. This paper contains a discussion of possible conditions resulting in its formation.

The detailed geology of the Bombay Island, dealing mainly with the field relations of its rock types, was published earlier (SUKHESWALA, 1953). It was established that the two lava flows (109, 13) are separated by thick intertrappeans with a north-south strike fault. The older basalt (109) also occurs as sills in the intertrappeans, and is partly albitized. A huge dike of highly chilled basalt (almost tachylitic) of Antop hill represents the final phase in the eruptive cycle. Rhyolite and pyroclastics appear as detached masses in the northern and eastern hill ranges.

Albitized Basalt

The albitized basalt is exposed in the eastern hill ranges (Fig. 1), under the harbor waters (Colaba seaside basalt suggests this) and in road cuts and deep excavations. It is likely that the whole island has a base of this albitized basalt. On the island of Salsette, it has been observed at Thana, and a little further north, in a thick

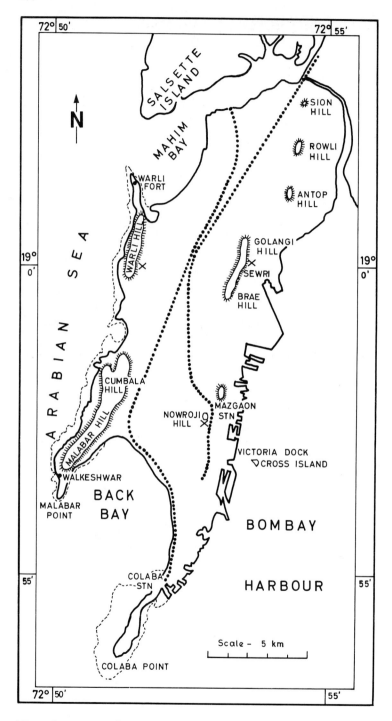

Fig. 1. Map of Bombay Island showing different hill ranges (after C.S. FOX)

Table 1

	109	13	A	D	E	F
Plagioclase	An_{65} $\alpha = 1.559$ $\gamma = 1.571$ labradorite (volcanic)	An_{50-60} $\gamma = 1.561$ labradorite (volcanic)	An_{10}	An_{3-10} $2V = 87°(-)$ low albite	An_{3-8}	An_{5-14} $2V = 88°(-)$ low albite
Pyroxene	$2V = 46.5°(+)$ $\alpha = 1.681$ $\beta = 1.697$ $\gamma = 1.699$ $Ca_{38}Mg_{39}Fe_{23}$	$2V = 47°(+)$ $\alpha = 1.687$ $\gamma = 1.710$		$2V = 42°(+)$		$2V = 48°(+)$ $\alpha = 1.690$ $\beta = 1.697$ $\gamma = 1.715$ $Ca_{39}Mg_{38}Fe_{23}$
Sp. Gr.	2.93	2.74		2.55	2.63	2.77

tunnel section under the Thana creek near Kolshet. Scattered occurrences of albitized basalt north of Bombay in Dahanu, Bulsar, and Surat have been observed by the writer. In all probability, therefore, it extends along the western coast of India. The lavas may have erupted in the Tertiary advancing sea.

The albitized basalt is distinguished in the field from the normal tholeiite basalt by its green color, and in part by the profuse development of secondary minerals - zeolites, calcite, quartz. It is comparatively softer, and sometimes more tuffaceous in appearance. The pillow structure common in spilitic lavas is observed in the Bhoiwada section, but is absent in other parts where the albitic nature of the lava is proved. The Bhoiwada section reveals one additional peculiarity, viz., the intensity of the green color of the albitized basalt and its richness in secondary mineral content diminish from the bottom to the top; the lower green compact albitized basalt with pillow structure grades upward into coarse green crystalline albitized basalt (doleritic), and then into coarse black crystalline normal labradorite basalt (doleritic)(Fig. 2).

Nature of the Pillows

Pillows vary in size from 3 feet to about 10 feet along the longer axis. They are usually ellipsoidal in form, the tops being more or less rounded and the undersides curved or depressed, forming hollows. In places they are highly vesicular, with large and small vesicles

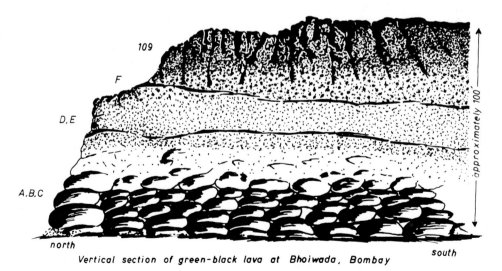

Fig. 2. Green albite basalt (spilite) grading upward into black tholeiite basalt as seen at Bhoiwada

filled with quartz, zeolite and calcite. In parts the pillow structure is not so pronounced; even then the suggestion that the pillow structure has begun to develop cannot be overlooked. The pillows are at times surrounded and separated from each other by a thin black matrix resembling the thin black siliceous bands (Table 2) in albitized basalt. Each pillow has a thin, black glassy skin with a resinous or greasy luster. It appears that the pillow character is related to greater depth of water, for as one goes upward in the sequence, the size of the pillows diminishes and then disappears.

Table 2

	1	2	3	4
SiO_2	50.90	52.28	43.10	81.50
Al_2O_3	20.30	21.73	20.55	4.53
Fe_2O_3	}1.20	}0.11	}4.35	2.49
FeO				2.28
MgO	0.69	0.56	0.79	1.12
CaO	11.20	12.30	25.58	2.36
Na_2O	0.93	0.85	1.50	0.17
K_2O	0.38	0.10	0.28	1.50
H_2O^+	14.32	}12.28	4.06	}3.62
H_2O^-	0.90		0.14	
Total	100.82	100.21	100.35	99.57

Prehnite $2V = 49°(+)$
Laumontite $2V = 45°(-)$, $\alpha = 1.51$, $\gamma = 1.53$
X-ray diffraction patterns confirm the identification of prehnite and laumontite.

1. Laumontite (acicular), Bhoiwada, Bombay.
2. Laumontite (ball-shaped, fibrous), Bhoiwada, Bombay.
3. Prehnite, Bhoiwada, Bombay.
4. Black siliceous bands in spilite, Bhoiwada, Bombay.

Laboratory Data

The albitized basalt in a hand specimen appears very fine-grained to compact without any recognizable crystallization or mineral constituents. A number of slides examined show fine-grained intersertal or intergranular texture with cavities filled by secondary products. The albitized doleritic variety in comparison is completely crystalline, coarse-grained and subophitic. Needle-shaped albite with inclined extinction and refractive index lower than canada balsam appears turbid due to clouding with inclusions of chlorite. In the doleritic variety albite is clear and devoid of clouding. Indistinct radiating or plumose character is noticeable, though only infrequently. In some cases the albite needles are zeolitized or calcitized, thus retaining their shape. In all cases when the original albite was tested (NOBLE, 1965), it was found to be low albite. Pyroxene is quite fresh and often granular. A subophitic relationship with albite is evident in the doleritic type. A cervical form is indistinctly visible in stray crystals. Chlorite appears as a major constituent both in the groundmass and in cavities. The primary nature of chlorite is evident when it stands in ophitic relationship with albite or is interstitial between feldspathic laths. In such rocks pyroxene is completely absent. Accessory minerals include epidote, zeolite in cavities and opaques. The latter appear either as granules or as black dust. Quartz is almost absent or rare in the groundmass. Its presence with zeolite in cavities suggests their late formation.

The skin material of the pillow is very fine-grained with tiny needles of albite, grains of pyroxene, black iron ore and many vesicles filled with zeolite and quartz. The core material is more crystalline, with needle-shaped albite, fresh pyroxene and green chlorite of low as well as high birefringence.

Though calcite, quartz and zeolites occur in considerable quantities as cavity fillings or aggregates, their presence in microsections is almost negligible. The greatest proportion of calcite occurs as aggregates of rhombohedra, a foot or more across. Nail-head and dog-tooth varieties fill cavities with zeolite and quartz. Quartz appears primarily as tiny, transparent crystals in radiating clusters. The most common zeolite, laumontite, occurs as delicate, milky white needles and as white silky, fibrous balls (Table 2). Associated with it are marina-green mammillated prehnite and colorless transparent apophyllite. Tiny cubes of pyrite infrequently occur with quartz.

In a vertical section at Bhoiwada, albitized basalt was sampled at different levels, the assumption being that these divisions stood at different depths of water after eruption. Microscopic examination of samples from different stages showed the following mineral associations:

a) clouded albite + chlorite in subophitic relation + opaques. Rock fine-grained and restricted mainly to the base.

b) clouded albite + fresh augite + chlorite + opaques + cavities filled by chlorite, quartz, zeolite. Rock fine-grained, intergranular of middle stage.

c) fresh clear albite + fresh augite in subophitic relation + opaques. Rock coarse-grained, doleritic, restricted to higher levels.

d) labradorite + augite in subophitic relation + opaques. Rock normal black tholeiite basalt (doleritic) passing upward from (c), and forming the top of the Bhoiwada-Cotton Green section.

The upward passage of green albite-basalt into black labradorite-basalt (109) in Bhoiwada-Cotton Green section with a green-black color boundary may be taken to indicate the upper limit of water in which the magma erupted. The upper black horizon is subaerial in origin.

The Intertrappeans

The albitized basalt alternates with the intertrappeans (about 100 feet thick) as thin and thick sills (2 feet to 20 feet and more) in Bhoiwada, Sewri, Cotton Green and Worli hills. The intertrappeans vary in color and texture, and compositionally approximate palagonite tuff. It is proper to term them stratified hyaloclastites because they are intercalated with black, carbonaceous, fossiliferous clayey material.

Three types are recognized: (I) green black, (II) black, highly carbonaceous, (III) thinly stratified yellowish-brown. It is in the green-black type that the albitized basalt occurs as sills.

Green-black sedimentary (Type I): Stray fossils are found in this zone. Under the microscope the constituents are brown isotropic glass, and green, feebly anisotropic chlorite, calcite, stray crystals of pyroxene, needles of albitic feldspar, quartzo-feldspathic cryptocrystalline groundmass, and angular fragments of albite basalt and felsic rock. The intercalated black material is extremely fine-grained and reacts to light, but is difficult to identify. In the Cotton Green section the black material is purely calcareous and may be termed carbonaceous limestone, though in small patches. Green chlorite and brown palagonite occur in rounded form with interstitial feldspar needles. In some sections calcite is dispersed throughout the groundmass. Overlying this in the Worli hill section is a thick, massive, highly-jointed green rock. It is very fine-grained, but is marked by lamination as in sedimentary rocks. Under very high power, calcite with yellow-brown glass appears to form the matrix, in which are dispersed large patches of albitic feldspar.

Black carbonaceous fossiliferous bed (Type II): It is extremely fine-grained with paper-thin lamellae. Irregularly arranged aggregates of albitic feldspar, occasional pyroxene grains and coarser patches of glass are embedded in a quartzo-feldspathic groundmass. The rock has a patchy occurrence intermingling with green-black sedimentary and is fairly fossiliferous.

Yellowish-brown bed (Type III): Microscopic examination reveals quartzo-feldspathic material in the main. In some sections brown glassy material forms the base with interspersed calcite and rare grains of augite. The rock, though stratified, does not give clear indications of aqueous action.

It is concluded that the above-described intertrappeans are in the main hyaloclastites interspersed with calcareous and clayey fossiliferous sediments. The albite-basalt (pillow lava) alternating with these fossiliferous intertrappeans (SUKHESWALA, 1953) also appears to have erupted under water. Angular fragments of albite-basalt in the green-black sedimentary would suggest the relatively older age of the former. Thus, field and laboratory results point to the fact that the subaqueous eruption of tholeiite basalt (109) gave rise to albite-basalt with pillow structure, as well as to the hyaloclastites (intertrappeans). Two different types of eruptions are envisaged by BONATTI (1967) to have resulted in the formation of pillow lavas and hyaloclastites: (I) quiet fissure eruption of pillow lavas on the sea floor without any significant interaction with sea water due to the formation of a thin insulating crust of glass at the surface of the flows; and (II) extensive physical and chemical interaction between the hot lava and sea water, resulting in shattering and pulverizing of the highly viscous lava, believed to have caused the formation of the hyaloclastites.

Chemistry

a) Chemistry of the tholeiites: Chemical analyses of the three different phases of tholeiite basalt occurring in the Island are presented (in Table 3). Each constituent shows progressive variation, especially MgO and CaO, which show a fall from older to younger. The process of differentiation becomes quite apparent in the gradual enrichment of alkalis, silica and iron, and even water. Chemical differences are significant in view of the fact that the eastern ridge lava (109) is separated from the western ridge lava (13) by thick intertrappeans; the Antop hill basalt (126), which occurs as a dike, is the youngest of the three.

Table 3

	109	13	126
SiO_2	51.54	52.26	54.68
TiO_2	1.13	3.24	1.91
Al_2O_3	13.76	12.74	12.88
Fe_2O_3	2.26	2.21	1.76
FeO	9.25	10.61	10.32
MgO	6.09	3.86	2.74
CaO	10.23	7.46	6.29
Na_2O	2.33	3.33	3.28
K_2O	0.62	1.34	1.64
H_2O+	1.79	2.72	3.03
H_2O-	0.56	0.75	0.81
MnO	0.18	0.12	0.14
P_2O_5	0.13	0.14	0.13
CO_2	0.22	0.00	0.00
Total	100.09	100.78	99.61

109 - Tholeiite basalt (oldest), top of Bhoiwada hill, Sewri cemetery, Bombay Island. Analyst H.B. WIIK.
13 - Tholeiite basalt (younger), top of Malabar hill along Tardeo Road, Bombay Island. Analyst H.B. WIIK.
126 - Tholeiite basalt (youngest), Antop hill, Bombay Island. Analyst H.B. WIIK.

(From SUKHESWALA and POLDERVAART, 1958, p.1480)

These facts suggest that the two basaltic flows and the dike basalt were fed from the same magma reservoir at short or long intervals; during that period the basaltic magma in the chamber was undergoing differentiation. That the interval between the earlier (109) phase of eruption and its successor (13) was a little more protracted is suggested by a somewhat steeper rise and fall in the Harker curves (Fig. 3). With every outburst the rock became more acid and glassy and became densely charged with black opaque iron ore dust. It is likely that the increasingly fluid mobile lava was, on extrusion, limited in its ability to crystallize because of the escape of volatiles, chiefly water. This would explain the comparatively coarse-grained and better crystallized nature of the older lava (109) in comparison to the younger ones, which are quite compact to lithoidal and even tachylitic with conchoidal fracture. Also, the younger basalts (13, 126) are heavily veined with quartzo-feldspathic material (0.2 mm thick), an evidence of the differentiated character of the magma throwing out aqueous siliceous solutions.

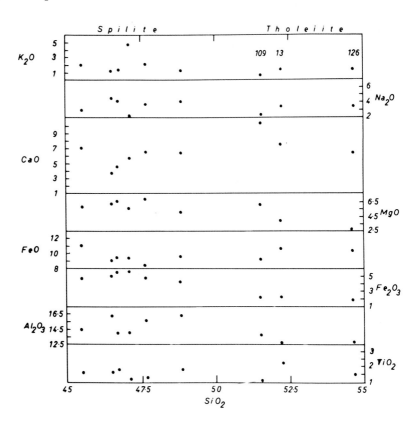

Fig. 3

b) Chemistry of the intertrappeans: In Table 4 the chemical analyses of the stratified intertrappeans are compared with the two different phases of basalt over- and underlying them. The comparison reveals a fairly close resemblance between them, which points to their genetic relationship. In field relations, the stratified yellowish-brown intertrappean (H, Type III) immediately underlies the Malabar-Cumbala-Worli basalt with which it shows chemical identity. The same is the case with the older lava (109) and the Worli green massive bed (G, Type I). Worli green stratified (J, Type I) and Worli black carbonaceous (K, Type II) are considered reworked (aqueous action) sediments because of the sharp fall in SiO_2 and abnormal rise in MgO and CaO, and partly Al_2O_3 with high CO_2.

c) Chemistry of the albitized basalt: To study the albitization process, the albitized segment of the lava was sampled at different levels, based on the assumption that such samples represented the different water levels in which the rocks remained. Comparing the original tholeiite (109) with its albitized segments (Table 5) the following changes seem to have occurred during the process of

Table 4

	109	G	13	H	J	K
SiO_2	51.59	50.00	52.26	52.88	32.93	29.90
TiO_2	1.13	1.50	3.24	1.53	0.83	0.47
Al_2O_3	13.76	10.08	12.74	11.66	12.25	18.02
Fe_2O_3	2.26	8.81	2.21	10.38	10.09	2.77
FeO	9.25	0.73	10.61	0.43	0.15	3.37
MgO	6.09	7.45	3.86	4.85	10.54	4.80
CaO	10.23	7.88	7.46	1.55	13.79	23.61
Na_2O	2.33	2.35	3.33	3.25	0.93	0.30
K_2O	0.62	0.42	1.34	0.40	0.87	1.85
H_2O+	1.79	6.01	2.72	8.52	5.91	2.43
H_2O-	0.56	1.90	0.75	5.60	4.14	0.64
MnO	0.18	–	0.12	–	–	–
P_2O_5	0.13	–	0.14	–	–	–
CO_2	0.22	3.36	0.00	–	7.19	12.54
Total	100.09	100.49	100.78	101.05	99.62	100.70

109 - Tholeiite basalt, top of Bhoiwada hill, Bombay Island.
G - Upper green, massive part of green-black sedimentary, Type I intertrappean, Worli hill, Bombay Island.
13 - Tholeiite basalt, top of Malabar hill, Bombay Island.
H - Yellowish-brown bed, Type III intertrappean, Worli hill, Bombay Island.
J - Green-black sedimentary, Type I intertrappean, Worli hill, Bombay Island.
K - Black, carbonaceous fossiliferous bed, Type II intertrappean, Worli hill, Bombay Island.

albitization: sharp decrease in SiO_2 and CaO; increase in total iron, K_2O, Na_2O, TiO_2 and combined water; MgO and Al_2O_3 remain almost steady. Fe_2O_3/FeO and Fe/Mg ratios increase from tholeiite to albitized basalt. Negligible amounts of CO_2 suggest absence of free calcite in the rock, though it occurs in large quantities in field as segregated masses and in cavities.

Although the specimens of albitized basalt and tholeiite were sampled at short vertical intervals (15-20 feet), the different constituents vary (Table 5), particularly SiO_2 and CaO. When considered from top to bottom (increasing depth of water) there is a suggestion that amongst the albitized types SiO_2 and CaO are on the decrease, while TiO_2, Na_2O and combined water are on the increase. It may be that with increasing overhead pressure of water (magmatic and marine), more SiO_2 and CaO are removed from the magma. Thin black siliceous bands (Table 2) at different horizons in the albitized basalt and crystallized quartz in cavities may have been deposited from sea water, thus being enriched in SiO_2 released with the fall in temperature. Similarly released CaO in the presence of sodium carbonate solutions at moderate temperatures may have formed calcite (ESKOLA, VUORISTO and RANKAMA, 1937; VALLANCE, 1960 and 1965), now so abundantly seen in the albitized portion of the lava. The enrichment of iron, alkalis and water may be ascribed to their retention in the magma crystallizing under a hydrous environment (BATTEY, 1956, p.104).

Among the three tholeiite samples showing marked differentiation, the increase towards iron and alkalis is very regular (Figs. 4,5,6). The albitized basalt samples do show enrichment in these components with respect to tholeiite basalt, but the scatter is rather haphazard. Sample F shows the least amount of enrichment, as it should be, for it immediately grades upwards into the tholeiite. Some trend is, however, discernible from the Ca end to the Na - K join (Fig. 5). Among the tholeitites the enrichment is towards K, along which trend also lies sample F (doleritic albite basalt); while from tholeiite

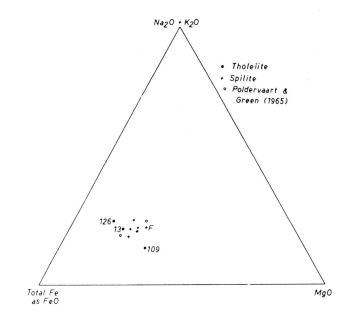

Fig. 4

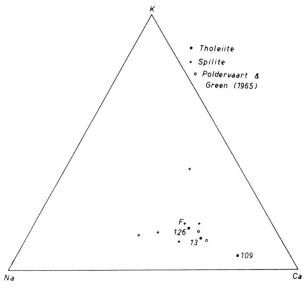

Fig. 5

Table 5

	A	B	C	D	E	F
SiO_2	45.50	46.70	46.50	47.20	48.86	47.66
TiO_2	2.20	2.55	2.20	1.44	2.50	1.50
Al_2O_3	14.30	13.86	16.25	14.04	16.19	15.48
Fe_2O_3	4.66	5.43	4.90	5.53	4.24	4.73
FeO	11.12	9.42	9.14	9.26	9.51	8.36
MgO	5.61	6.34	5.97	5.36	4.99	6.66
CaO	7.09	4.49	3.61	5.56	6.30	6.44
Na_2O	3.00	4.08	4.45	2.18	4.05	3.60
K_2O	1.95	1.33	1.15	4.53	1.15	2.00
H_2O+	3.06	3.62	3.62	2.94	2.62	3.02
H_2O-	0.34	1.66	1.24	0.74	0.34	0.80
P_2O_5	-	-	-	-	-	-
CO_2	0.20	0.40	0.10	0.20	-	0.60
MnO	-	-	-	-	-	-
Total	99.03	99.88	99.13	98.88	100.75	100.85
Fe_2O_3/FeO	0.419	0.576	0.536	0.597	0.445	0.565
qu	-	-	-	-	-	-
or	11.12	7.78	6.67	26.69	7.23	11.68
ab	25.15	34.06	37.73	18.34	34.58	30.39
an	20.02	15.57	17.79	15.01	22.24	20.29
cor	-	-	1.22	-	-	-
wo	6.38	2.78	-	5.34	3.83	4.87
en	2.90	1.70	-	3.10	2.00	3.30
fs	3.43	0.92	-	1.98	1.72	1.19
en	0.60	5.90	4.80	0.60	4.10	-
of	0.53	3.83	2.90	0.40	2.64	-
fo	7.42	5.18	7.00	6.72	4.90	9.38
fa	6.73	3.88	4.69	6.12	3.67	6.12
hm	-	-	-	-	-	-
mt	6.73	7.89	7.19	7.89	6.03	6.73
il	4.26	4.71	4.26	2.74	4.71	2.89
H_2O	3.60	5.68	4.96	3.88	2.96	4.42
Total	98.87	99.88	99.21	97.68	100.61	99.96

Orthoclase, albite, anorthite recalculated to 100

	A	B	C	D	E	F
or	19.74	13.60	10.73	44.50	11.30	18.74
ab	44.71	59.30	60.64	30.50	54.00	48.72
an	35.55	27.10	28.63	25.00	34.70	32.54

A - Green fine-grained albitized basalt, base of Bhoiwada section, Bombay Island.
B - Core of a pillow, base of Bhoiwada section, Bombay Island.
C - Skin of above pillow.
D - Green fine-grained albitized basalt, middle part of Bhoiwada section, Bombay Island.
E - Green fine-grained albitized basalt, about 20 feet higher than D.
F - Green doleritic albitized basalt (coarse-grained) grading upwards into tholeiite basalt (109), Bhoiwada.
(Analyst: S.F. SETHNA)

Table 5 (cont.)

	109	13	126	1	2	3	I	II
SiO_2	51.54	52.26	54.68	50.83	45.78	49.65	47.62	43.33
TiO_2	1.13	3.24	1.91	2.03	2.63	1.57	3.21	3.10
Al_2O_3	13.76	12.74	12.88	14.07	14.64	16.00	14.74	16.51
Fe_2O_3	2.26	2.21	1.76	2.88	3.16	3.85	7.92	12.98
FeO	9.25	10.61	10.32	9.00	8.73	6.08	2.88	1.08
MgO	6.09	3.86	2.74	6.34	9.39	5.10	5.16	4.00
CaO	10.23	7.46	6.29	10.42	10.74	6.62	7.61	7.31
Na_2O	2.33	3.33	3.28	2.23	2.63	4.29	3.41	2.89
K_2O	0.62	1.34	1.64	0.82	0.95	1.28	1.69	1.15
H_2O+	1.79	2.72	3.03	0.91	0.76	3.49	1.81	3.52
H_2O-	0.56	0.75	0.81	-	-		1.58	2.24
P_2O_5	0.13	0.14	0.13	0.23	0.39	0.26	0.91	0.72
CO_2	0.22	0.00	0.00	-	-	1.63	-	-
MnO	0.18	0.12	0.14	0.18	0.20	0.15	0.14	0.20
Total	100.09	100.78	99.61				99.02	99.37
Fe_2O_3/FeO	0.244	0.208	0.170					
qu	4.42	5.09	8.67				2.32	3.77
or	3.66	7.91	9.69				9.99	6.80
ab	19.70	28.16	27.74				28.85	24.45
an	25.26	15.87	15.58				19.92	25.35
cor	-	-	-				-	1.22
wo	9.74	8.48	6.20				5.07	-
en	5.01	3.68	2.04				4.38	-
fs	4.48	4.80	4.37				-	-
en	10.15	5.93	4.78				8.47	9.96
of	9.10	7.73	10.23				-	-
fo	-	-	-				-	-
fa	-	-	-				-	-
hm	-	-	-				7.92	12.98
mt	3.28	3.20	2.55				-	-
il	2.15	6.15	3.63				5.92	2.24
H_2O	2.35	3.47	3.84				3.39	5.76
Total	100.08	100.77	99.60				99.02	99.37

109 - Black tholeiite doleritic basalt (coarse-grained), top Bhoiwada section, including 0.28 ap and 0.50 cc.
13 - Tholeiite basalt, top of Malabar hill, Bombay Island, including 0.30 ap.
126 - Tholeiite basalt, Antop hill, Bombay Island, including 0.28 ap.
1 - Average tholeiite basalt of NOCKOLDS (1954), No. 96 in VALLANCE (1960).
2 - Average alkali basalt of NOCKOLDS (1954), No. 97 in VALLANCE (1960).
3 - Average of spilites nos. 1 to 92 (No. 93, VALLANCE, 1960).
I - Submarine basalt, POLDERVAART and GREEN (1965), p.1724, Table I, No. 1, including 0.30 Cr_2O_3, 0.07 S, 0.02 O for S, 0.23 tn, 1.99 ap, 0.13 pr, and 0.44 cm.
II - Submarine basalt, POLDERVAART and GREEN (1965), p.1724, Table I, No. 2, including 0.28 Cr_2O_3, 0.08 S, 0.02 O for S, 4.71 tn, 1.57 ap, 0.15 pr, and 0.41 cm.

(109) to albitized basalt it is a straight line with greater enrichment in Na than K, and a gradual decrease in Ca. In Fig. 7 the albite basalt samples plot in the same regions as most of the world spilites (VALLANCE, 1960, p.37). The higher content of K_2O (4.53%) in D is rather inexplicable.

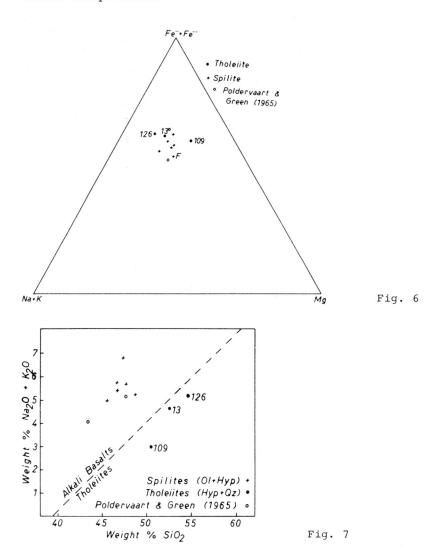

Fig. 6

Fig. 7

Albitized Basalt or Spilite?

YODER (1965-1966) has drawn attention to the following special features of spilites: (I) the spilites lie for the most part in the field of Hawaiian alkali basalts, but they are obviously not alkali basalts. This is true of the Bombay albitized basalt (Fig. 8), with its tholeiite counterpart (109) entering the tholeiite field; (II) the modal feldspar (An_{0-15}) is not in accord with the normative feldspar

(Fig. 7); (III) the apparent stable coexistence of albite, chlorite and augite; (IV) the norms of spilite run the gamut of the normative limits of normal basalts. Specimens with normative Ol + Ne, Ol + Hyp, Hyp + Qz are common; (V) low K_2O and high Na_2O compared with the average basalts; (VI) absence of micas, amphiboles and zeolites.

The albitized basalt of Bombay fulfills the above conditions required for spilites. The author therefore feels justified to call the Bombay albite basalt a spilite.

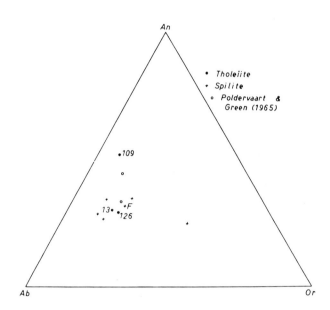

Fig. 8

Origin of Bombay Spilite

The only recorded instance of spilite in India (PICHAMUTHU, 1938) has a geosynclinal environment and is of pre-Cambrian age. The spilite of Bombay is a part of the Deccan trap (Tertiary) sequence. No doubt therefore remains about its magmatic (tholeiite) origin and continental setting.

In the previous discussion it is shown that the tholeiite magma was converted into spilite as a consequence of its eruption in water. That the magma erupted in water in liquid state becomes apparent when we take into account the formation of pillows and the absence of phenocrysts. The subophitic relationship between albite and augite and albite and chlorite suggests the primary nature of albite and chlorite. The highly clouded (BRIDGEWATER, 1966) nature of albite in the lower horizon (greater depth of water) and high Fe_2O_3/FeO ratio (OSBORN, 1962) point to water enrichment in the system. Other points of significance are the greater removal of silica and calcium from the magma system with the greater depth of water. The greater or lesser depth of water has also a role to play in the process of spilitization, i.e., the formation of albite and chlorite with or without fresh pyroxene.

With these points in mind we proceed to decide about the modus operandi of spilite formation out of the original tholeiite magma. A comparison of analyses (Table 5) indicates that the tholeiite magma (109) differentiating at depth within the crust produced results identical to those of the tholeiite magma erupting in water to form spilite, except for the decrease in amount of SiO_2 in the latter. Alkali and total iron contents in spilite in comparison have increased. It can therefore be imagined that the problem of the formation of spilite is the problem in differentiation in a closed system under subaqueous conditions. Two possibilities suggest themselves: (I) highly differentiated tholeiite magma erupted in water to give rise to spilite; or (II) the original tholeiite magma erupted under water and differentiated there in situ to form spilite. In view of the spilite grading upward into undifferentiated tholeiite, the first possibility can be easily ruled out. Alternatively it is reasonable to assume that the tholeiite magma differentiated to spilite only after it erupted in water. It is also apparent (Table 5) that the high water and alkali contents in the spilites are the result of differentiation, and not the result of introduction of sea water into the magma system. This leads us to believe that during the process of differentiation under water there was no escape route for the volatiles nor could any extraneous material enter the magma system. Such a closed system could have been achieved partly by a quickly formed chilled skin and partly by the overhead pressure of sea water on the lava.

The Bombay basalts have differentiated both prior to and after extrusion. The differentiated tholeiites (13, 126) have further fractionated on extrusion to form quartzo-feldspathic veins. Similarly the original tholeiite magma (109) erupted to differentiate under water, but followed a slightly different fractionation trend. The residual mother liquor became enriched in Ca and Si together with water, whilst Na was used up in the system for the crystallization of albite. The residual SiO_2 not being utilized reappeared in the form of black siliceous bands and quartz crystals in cavities in spilites. Excess CaO not forming anorthite was distilled over during differentiation, and later formed calcite and calcic zeolites. This would explain the sharp fall in SiO_2 and CaO. With high internal vapor pressure generated in the magma, the silicate liquidus temperature was also lowered significantly (YODER and TILLEY, 1962, p.453). The accumulation of Ca in the residual melt may therefore be due to a rapid fall in temperature, and consequently rapid crystallization, thus preventing formation of labradorite.

All other constituents vary only slightly, both in the differentiated tholeiite and in the spilites. In spite of irregular scatter, the spilites in Fig. 7 give a clear indication of gradual increase in Na_2O, K_2O and Fe_2O_3 with increasing depth of water. Peculiarly enough CaO shows a very regular and sharp fall with MgO remaining almost steady.

That the differentiation trends among the tholeiites and between tholeiite (109) and its counterpart spilite followed similar patterns is also apparent in the ternary diagrams (Figs. 4,5,6); in these the differentiated tholeiites lie amidst the spilites. It is also significant to observe that F remains closer to 109, and particularly in Fig. 5 it follows the path of the tholeiites. This fact suggests the probable greater alkali enrichment in spilite under a more hydrous environment (or greater depth of water); but when the influence of water on the magma became less the differentiation process followed the normal trend as in tholeiites. There is also an indication (Figs. 4,6) that the tholeiite differentiating to spilite tends to become more alkali-rich than iron-rich.

It is shown that the modal feldspar (An_{0-15}) of Bombay spilite is not in accord with the normative feldspar (An_{25-35}). Neither is the modal feldspar (An_{50-60} in 13) of differentiated tholeiite in accord with its normative feldspar (An_{30}). This, however, may be attributed to the presence of innumerable quartzo-feldspathic veinlets crisscrossing this rock.

The above observations lead us to one conclusion: that to produce spilite from tholeiite there was probably differentiation with enrich-

Fig. 9. Pillows as seen in Bhoiwada-Sewri section

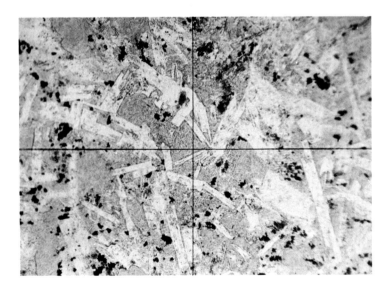

Fig. 10. Albite basalt (spilite), chlorite + albite

ment in Ca and Si, and the subsequent removal of these elements from the magma system. AMSTUTZ (1968) also points to certain differentiation tendencies existing in most spilites and concludes: "the mode of formation may be understood as resulting from a transfer and differentiation of constituents in a separate aqueous phase during primary crystallization".

In the light of the above discussion the hypothesis of a primary spilitic magma appears redundant. Also, in the absence of metasomatic replacement phenomena of the nature of relict labradorite and the pre-

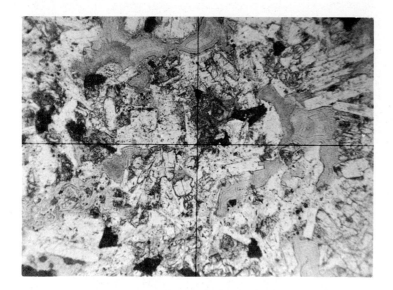

Fig. 11. Albite basalt, doleritic with albite + augite

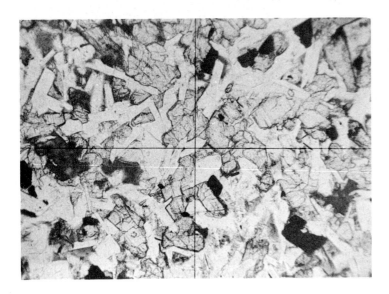

Fig. 12. Black tholeiite, doleritic with labradorite + augite

Fig. 13. Albite with plumose character in spilite

Fig. 14. Green albite basalt (spilite, light colored) grading upward into black tholeiite basalt (dark colored) at Bhoiwada

servation of plumose feldspar, extraneous addition of alkalis, particularly sodium, appears a remote possibility. The metasomatic origin of spilite therefore need not be considered. With subaqueous formation of Bombay spilite and its upward grading into tholeiite as a proven fact, crystallization of tholeiite under a hydrous environment is an obvious choice. In such an event, explanation has to be sought regarding the mineral assemblage of the Bombay spilite, viz., albite + fresh pyroxene + chlorite, and albite + chlorite; for this is the main objection to spilite mineralogy. BATTEY (1956, p.102) overcomes this difficulty by assuming that the "fresh augite persists metastably in a liquid in which high water content and consequent free diffusion has permitted the plagioclase to pass down the curve of continuous reaction to produce albite". The author believes that BATTEY has entered into the heart of the problem when he draws a significant conclusion that the tholeiite magma may approach a spilitic condition to a variable degree, depending upon the wateriness of its environment. Such ideal conditions of dry tholeiite to wet tholeiite to spilite as envisaged by BATTEY (p.105) do seem to obtain in the island of Bombay.

POLDERVAART and GREEN (1965) prepared an alkali-silica diagram of submarine basalts. Although they have not classified any of these basalts as spilites, a number of them fall in the alkali-basalt field. Whether all such basalts are spilites is impossible to know. But analyses nos. I and II (Table 5, POLDERVAART and GREEN, 1965) satisfy to some extent the characteristics of spilite: (I) chemical analyses agree well with Bombay spilite analyses; (II) normative feldspar plots with those of spilites; (III) they fall in the alkali-basalt field; (IV) both of these rocks show serpentine, chlorite (authors have called them highly altered rocks) with one of them having andesine microlites in the groundmass. Since no definite information is given for feldspar, chlorite and serpentine in these rocks, it would be interesting to study their optics.

It has been observed that spilites are common in geosynclinal environments attaining great depths. Also, ENGEL and ENGEL (1963, p.1321) wrote: "to date we have found no evidence of albitization or spilitization of the basalts dredged from the Pacific. These observations are consistent with KORZHINSKY's conclusion that spilites are not developed through the interaction of basaltic lava and sea water, at least at depths of 5,000 m". (Compare also the later papers by ENGEL and ENGEL, 1964a and b; and ENGEL et al. 1965).

In the light of the above conclusions the formation of Bombay spilite in shallow water conditions becomes quite interesting and significant.

Acknowledgement

The author is indebted to Professor M.H. BATTEY of the University Newcastle Upon Tyne, England, and Professor T.G. VALLANCE of the University of Sydney, Australia, for reading the manuscript and offering valuable suggestions. However, for views expressed in the paper the author is solely responsible. Professor VALLANCE also kindly made available X-ray data on chlorite of spilite of Bombay. The author acknowledges with much appreciation the assistance given by his colleague Dr. S.F. SETHNA who carried out chemical analyses of spilite samples, and made helpful suggestions during the preparation of this paper.

References

AMSTUTZ, G.C. (1968): Spilites and spilitic rocks. The Poldervaart Treatise on rocks of basaltic composition. Ed. by H.H. Hess and the late Arie Poldervaart 2, 737-753.

BATTEY, M.H. (1956): The petrogenesis of a spilitic rock series from New Zealand. Geol. Mag. 93, 89-110.

BONATTI, E. (1967): Mechanisms of deep-sea volcanism in the South Pacific. Researches in Geochemistry. 2, 453-491. Ed. by P.H. Abelson. John Wiley & Sons, Inc.

RIDGEWATER, D. (1966): Clouded feldspar (a correspondence). Geol. Mag. 103, 284-285.

ENGEL, C.G., ENGEL, A.E.J. (1963): Basalts dredged from the northeastern Pacific Ocean. Science 140, 1321-1324.

ENGEL, A.E.J., ENGEL, C.G. (1964a): Compositions of basalts from the Mid-Atlantic Ridge. Science 144, 1330-1333.

ENGEL, A.E.J., ENGEL, C.G. (1964b): Igneous rocks of the East Pacific Rise. Science 146, 477-485.

ENGEL, C.G., FISHER, R.L., ENGEL, A.E.J. (1965): Igneous rocks of the Indian Ocean floor. Science 150, 605-610.

ESKOLA, P., VUORISTO, U., RANKAMA, K. (1937): An experimental illustration of the spilite reaction. Bull. Comm. géol. Finlande 119, 61-68.

NOBLE, D.C. (1965): Determination of composition and structural state of plagioclase with the five-axis universal stage. Amer. Mineral. 50, 367-381.

OSBORN, E.F. (1962): Reaction series for subalkaline igneous rocks based on different oxygen pressure conditions. Amer. Mineral. 47, 211-226.

PICHAMUTHU, C.S. (1938): Spilitic rocks from Chitaldrug, Mysore State. Curr. Sci. 7, 55-57.

POLDERVAART, A., GREEN, J. (1965): Chemical analyses of submarine basalts. Amer. Mineral. 50, 1723-1728.

SUKHESWALA, R.N. (1953): Notes on the field occurrence and petrography of the rocks of the Bombay Island, Bombay. Trans. Mining Geol. Met. Inst. India 50, 101-126.

SUKHESWALA, R.N., POLDERVAART, A. (1958): Deccan basalts of the Bombay area, India. Bull. Geol. Soc. Amer. 69, 1475-1494.

SUKHESWALA, R.N. (1960): Albitized basalts of Bombay and the associated intertrappeans: A preliminary survey. Indian Sci. Cong. Assoc., Abstracts, Bombay.

VALLANCE, T.G. (1960): Concerning spilites. Proc. Linnean Soc., New South Wales, Presidential address 85, 8-52.

VALLANCE, T.G. (1965): On the chemistry of pillow lavas and the origin of spilites. Mineral. Mag. 34, 471-481.

WILSON, M.E. (1960): Origin of pillow structure in early pre-Cambrian lavas of Western Quebec. Jour. Geol. 68, 97-102.

YODER, H.S., TILLEY, C.E. (1955-1956): Natural tholeiite basalt - water system. Annual Report of the Director, Geophysical Laboratory 169-171. Carnegie Institution Year Book.

YODER, H.S., TILLEY, C.E. (1962): Origin of basalt magmas: An experimental study of natural and synthetic rock systems. Jour. Petrology 3, 343-530.

YODER, H.S. (1965-1966): Spilites and serpentites. Annual Report of the Director, Geophysical Laboratory, 269-279. Carnegie Institution Year Book.

3. Papers Proposing an Autohydrothermal or Autometamorphic Origin

Vers une Meilleure Connaissance du Problème des Spilites à Partir de Données Nouvelles sur le Cortège Spilito-Keratophyrique Hercynotype

Th. Juteau et G. Rocci

Abstract

The paper deals with the petrographic and chemical study of some spilite-keratophyre associations of the northern part of the Hercynian orogeny of Western Europe.

The first part concentrates on the analysis of the facts, after a brief introduction to the geology of the areas under investigation (Cornwall and Devon; Lahn-Dill, Northern Vosges). The different petrographic facies of each area are examined in detail on the basis of samples taken at typical locations.

The comparison of the results of the three areas makes it possible to derive analogies and differences between each province and to draw a sketch of the distribution of the facies, starting from the center of the provinces under consideration.

Subsequently, the authors derive mineralogical and textural characteristics of the spilites and the keratophyres which represent the facies typical and dominating in the Hercynian association.

The primary nature of the textures is reconfirmed and the spilitic paragenesis is given in detail: albite-chlorite-calcite-(K-feldspar)-(augite)-(quartz)-iron oxides, whereas the absence of olivine is underlined as typical.

The chemical investigation, including 175 analyses, is used to comment on the distribution of the different major elements, as well as on the very important role of H_2O and CO_2. For each facies group important conclusions are derived from the comparison of the analytical results and their chemical properties are thus characterized. A number of diagrams show the general geochemical tendency of the association.

In the second part, the problem of the origin of the spilites and the keratophyres is treated. After a summary of the major observational facts, the primary character of these Hercynian provinces is reconfirmed. Consequently, the existence of a spilitic magma is taken into consideration. The origin of this magma is discussed and the authors defend the hypothesis of a basaltic magma contaminated with H_2O and CO_2 derived to a larger part from the geosynclinal environment. The genesis of the keratophyres and their connections with the spilites is also examined. All through this investigation a magmatism of "folds" (Schirmeck type) to a magmatism of "Graben" (Lahn-Dill type) is made evident. It reflects the importance of paleogeographic and paleostructural conditions on the nature of the production of geosynclinal magmatism.

Introduction

Afin d'apporter notre contribution au problème des spilites, nous nous proposons de montrer que les spilites de différentes régions de la chaîne hercynienne et les roches qui les accompagnent ont en commun un ensemble de caractères pétrographiques, chimiques et de gisement, et que l'on peut définir ainsi une "association spilitique hercynotype" (ROCCI et JUTEAU, 1968) qui caractérise le magmatisme géosynclinal de ce domaine orogénique.

C'est volontairement que nous avons écarté de notre étude les terrains hercyniens trop métamorphisés, en particulier ceux qui affleurent dans le Massif Central et les Alpes (massifs cristallins externes). Nos recherches ont donc porté sur ce qu'il est convenu d'appeler la branche nord de l'orogène hercynien. Cette partie de la chaîne présente l'avantage d'être très peu métamorphisée et de se tenir en dehors de la zone d'influence de l'orogénèse alpine.

Cette branche nord cependant n'affleure pas de façon continue, il s'en faut de beaucoup. Si plus précisément nous recherchons les terrains d'âge dévonien et dinantien contemporains du stade géosynclinal, nous constatons que les granites varisques, les terrains post-hercyniens et les mers actuelles ne nous laissent que peu de surfaces à prospecter en masses isolées et éloignées les unes des autres. Dans ces régions, seules certaines zones montrent un magmatisme géosynclinal important, contemporain de la sédimentation: le Massif ardennais, par exemple, qui occupe une position centrale dans l'édifice, est totalement dépourvu de manifestations magmatiques d'âge dévonien ou dinantien.

Par ailleurs, nous n'avons pas fait figurer dans notre étude certaines portions du segment orogénique. Le Massif Armoricain a été exclu, pour l'instant, car il représente un domaine trop complexe dont l'histoire montre qu'il a été le siège à de nombreuses reprises depuis l'Antécambrien, d'une intense activité orogénique précédée de phénomènes magmatiques. De même nous n'aborderons pas l'étude du volcanisme préorogénique du Morvan et des Vosges du Sud qui font actuellement l'objet de recherches qui seront publiées prochainement.

L'article est divisé en deux parties qui traitent respectivement de la description de quelques associations spilito-kératophyriques et du problème de l'origine des spilites hercynotypes.

En conclusion, nous mettrons l'accent sur les points qui restent obscurs et qui mériteront des études plus approfondies.

I. Revue de Quelques Associations Spilite - Kératophyre Hercynotypes d'Europe Nord-Occidentale

Introduction

Dans une précédente publication (JUTEAU et ROCCI, 1966), nous avions présenté une brève étude géochimique comparative de divers gisements volcaniques et volcano-sédimentaires de la branche nord de l'orogène hercynien. Nous allons préciser cette étude en apportant de nouveaux éléments analytiques.

Reprenant la terminologie de AUBOUIN (1961), nous désignons sous le terme de "branche nord de l'orogène hercynien", le domaine orogénique qui se développe en Europe occidentale sur la bordure méridionale du craton nord-atlantique (continent des VGR) et limité au Sud par la cordillère de l'Europe moyenne essentiellement constituée d'un vieux socle antécambrien.

Malgré de nombreux recouvrements récents, ce domaine affleure de la Cornouaille britannique au Harz allemand et à la Thuringe occidentale en passant par le Massif Armoricain, le Nord du Massif Central français et les Vosges.

Les terrains qui le constituent sont d'âge paléozoïque inférieur à moyen, mais les formations volcaniques sont, sauf exception et notamment en Normandie et en Bretagne, localisées au Dévono-dinantien. Bien que les plissements aient été intenses à l'orogenèse hercynienne, le métamorphisme régional est toujours très léger voire inexistant. Cependant, en plusieurs points, des granites carbonifères recoupent les séries et y développent du métamorphisme de contact.

Dans l'ensemble, la sédimentation dévono-dinantienne qui constitue l'environnement des suites spilite - kératophyre, est de caractère détritique avec des épisodes carbonatés parfois bien marqués (notamment au Dévonien moyen), mais pouvant manquer en certaines régions.

Les laves spilitiques qui vont retenir tout spécialement notre attention, sont très communes et présentent souvent la structure en pillow-lavas. Mais elles sont toujours associées à d'autres roches volcaniques et volcano-sédimentaires en proportions très variables, dont l'importance ne doit jamais être sous-estimée. Nous verrons en effet qu'en aucun cas le problème de la genèse de ces spilites ne peut être dissocié de celui de l'ensemble volcano-sédimentaire. Par ailleurs, nous avons volontairement restreint notre étude au domaine strictement préorogénique dont le volcanisme est dévonien. Les manifestations magmatiques se poursuivent généralement au Dinantien mais avec des caractères sensiblement modifiés.

Afin d'apporter un certain nombre de faits nouveaux, nous avons fait un échantillonnage dans trois régions-types dont nous présentons la description pétrographique et géochimique. Il s'agit des célèbres gisements de Cornwall-Devonshire (Angleterre), du district Lahn-Dill (République Fédérale Allemande) et des Vosges septentrionales (France) (Fig. 1).

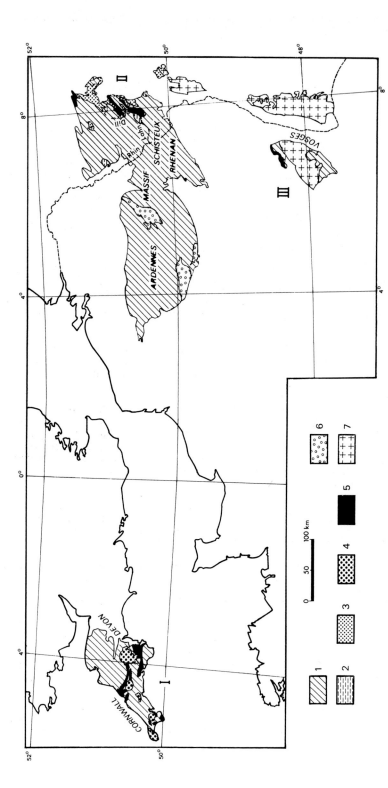

Fig. 1. Répartition dans l'orogêne hercynien des régions étudiées.
I : Cornwall et Devon,
II : Massif schisteux-rhénan,
III: Vosges septentrionales.
1 - Formations essentiellement schisteuses dévono-dinantiennes; 2 - principales masses calcaires dans (1), exagérées; 3 - faciès flysch; 4 - granites hercyniens; 5 - produits du magmatisme géosynclinal (spilites kératophyres, etc.); 6 - formations calédoniennes; 7 - complexes granito-gneissiques (d'après la carte tectonique de l'Europe)

Brève Présentation Géologique des Régions Etudiées

Sud-Ouest de l'Angleterre

Cornwall

Nos recherches ont porté principalement sur la côte nord de Cornwall. En effet, à l'intérieur les affleurements sont très mauvais et sauf dans quelques carrières l'échantillonnage n'est guère possible. En revanche, la côte rocheuse permet d'excellentes observations et des prélèvements de roches fraîches. On sait que dans cette région le Dévonien est marin et complet. Très schématiquement (cf. RAYNER, 1967), on rencontre des formations schisteuses et gréseuses au Dévonien inférieur tandis que le Dévonien moyen est représenté par des calcaires construits passant latéralement à des schistes gris à goniatites qui affleurent largement sur la côte nord de Cornwall. Le Dévonien supérieur est de nouveau schisteux (schistes versicolores)(HOUSE, 1963).

Les masses les plus importantes de roches volcaniques se situent au Dévonien moyen (schistes à goniatites) et au Dévonien supérieur (schistes versicolores). Il s'agit de coulées de pillow-lavas associées à des brèches et tufs, de dykes et de sills de dolérites et dans une bien moindre mesure, de laves plus acides, de type kératophyrique.

Les affleurements étudiés se situent (Fig. 2) :
- à Trevose Head (feuille 335[1] Trevose Head),
- à Padstow, Wadebridge et Pentire Head (feuille 336 Camelford),
- à Penzance (South-West Cornwall),
- à l'Ouest de Plymouth, région de Landrake (feuille 348 Plymouth).

Devonshire

La région du Devon fait suite, à l'Est à celle de Cornwall, les structures étant sensiblement E-W. On retrouve donc les mêmes formations avec toutefois quelques changements de faciès.

Nos recherches dans cette province ont porté essentiellement sur la partie orientale, dans le périmètre Ashburton - Newton Abbot - Torquay - Totnes. Nous avons voulu en effet éviter l'influence métamorphisante du grand batholite granitique de Dartmoor.

Du point de vue stratigraphique, on retrouve au Dévonien inférieur des faciès détritiques plus ou moins fins (Meadfoot et Staddon beds), au Dévonien moyen des calcaires (Torquay limestone) et au Dévonien supérieur des schistes (Saltein cove beds et Ostracod slates).

Les formations volcaniques se situent à divers niveaux du Dévonien moyen - supérieur et constituent les "Asprington series". On trouvera dans l'étude de MIDDLETON (1960) la description précise des affleurements et des types de roches volcaniques rencontrées, dont la position stratigraphique a été déterminée grâce à quelques niveaux repères et notamment un calcaire dit d'Ashburton et des schistes à Ostracodes.

---

[1] Cartes géologiques one inch/one mile du Geological Survey of England and Wales.

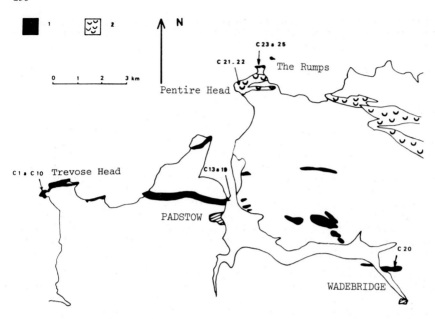

Fig. 2. Emplacement des affleurements étudiés en Cornouailles.
1 - roches intrusives (diabases), spilites massives, tufs; 2 - pillow-lavas spilitiques

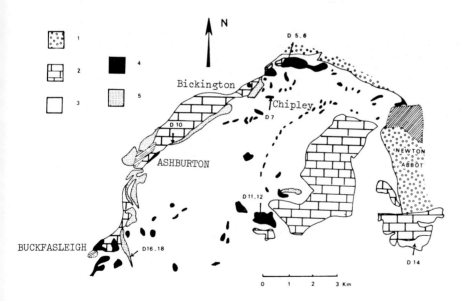

Fig. 3. Emplacement des affleurements étudiés dans le Devon.
1 - couverture secondaire; 2 - calcaires dévoniens; 3 - schistes dévono-dinantiens; 4 - roches magmatiques basiques, intrusives et extrusives; 5 - roches acides, principalement pyroclastiques

Dans cette partie du Devon, on peut distinguer des coulées de laves basiques plus ou moins porphyriques, des sills doléritiques, des pillow-lavas et des tufs kératophyriques.

Les pincipaux affleurements qui ont été visités et échantillonnés sont situés (Fig. 3):
- à l'Ouest de Newton-Abbot (feuille 339 Teignmouth),
- au Sud de Buckfastleigh (feuille 349 Ivybridge),
- dans la région de Torquay (feuille 350 Torquay).

Lahn-Dill

Le deuxième région-type est située à l'Est du massif schisteux rhénan, dans le synclinorium hessois. Il s'agit d'une région classique qui a fait l'objet de très nombreux travaux. Mais comme le matériel dont on trouve des analyses chimiques dans la littérature est fort restreint, il nous est apparu indispensable de refaire un échantillonnage qui a donc porté sur les plus beaux affleurements des vallées de la Lahn et de son affluent la Dill.

Dans cette région dont la stratigraphie est bien connue, l'environnement sédimentaire est très schématiquement le suivant. Le Dévonien inférieur est représenté par des faciès détritiques plus ou moins fins. Au Dévonien moyen, la sédimentation devient plus carbonatée. De véritables récifs s'installent au milieu d'une sédimentation de caractère flysch avec des argilites sombres, des calcaires en plaquettes et des intercalations de tufs polygènes souvent à ciment calcareux (schalsteins). Le faciès flysch se développe nettement au Dévonien supérieur avec prédominance de schistes argileux à goniatites, de lydiennes, de schistes calcareux.

On rencontre des formations volcaniques dès le Dévonien moyen: kératophyres et tufs acides, mais surtout diabases. Le développement le plus considérable se situe au Dévonien supérieur, avec une très large prédominance des diabases de divers types, tantôt grenues en dykes ou en sills, tantôt aphanitiques compactes ou amygdalaires en coulées et en pillow-lavas, tantôt en amas de petites dimensions au milieu des schistes (diabases à olivine), tantôt enfin sous forme de tufs variés. Les spilites sont bien représentées tandis que les kératophyres demeurent toujours très nettement subordonnés.

Les affleurements étudiés sont situés dans la région de Weilburg (feuilles Weilburg[2], Weilmünster et Merenberg); de Dillenburg (feuille Dillenburg); de Wetzlar (feuilles Braunfels et Wetzlar).

Le plupart des affleurements visités sont situés dans les vallées encaissées de la Lahn, de la Dill, de la Weil et de quelques autres rivières. D'assez nombreuses carrières en partie encore en activité ont également fourni d'excellents matériaux.

---

[2] Il s'agit des cartes géologiques au 1/25 000°.

Vosges Septentrionales

La troisième région-type étudiée, située dans les Vosges septentrionales, se situe en bordure immédiate du vieux socle antécambrien des Vosges moyennes; les formations dévono-dinantiennes des vallées de la Bruche et du Rabodeau sont séparées de ce vieux socle par tout un cortège de massifs granitiques, granodioritiques et dioritiques d'âge hercynien (massifs du Champ-du-Feu, de Senones, de Raon l'Etape, etc.) qui les ont transformées en cornéennes à leur contact; au Nord et à l'Ouest, elles disparaissent sous la couverture permo-triasique gréseuse.

La stratigraphie de cet ensemble dévono-dinantien est encore mal connue: la "traînée volcanique du Champ-du-Feu" est la formation la plus méridionale et sans doute la plus ancienne du système (dévonien inférieur?) Au Nord immédiat du massif du Champ-du-Feu, les schistes et arkoses de Champenay ont livré une faune eifélienne typique. Au Nord de cette formation se place le massif éruptif de Schirmeck, qui contient des lentilles de calcaire dolomitique bréchique tout-à-fait analogues à celles qui, à l'Est de Schirmeck, sont associées à des couches à faune givétienne; ceci, joint au fait que le massif de Schirmeck surmonte au Nord les schistes rouges et arkoses de Raon-sur-Plaine datées du Couvinien par RUHLAND (1964) laisse à penser que ce massif représente un faciès particulièrement volcanique du Givétien. Enfin, une flore viséenne a été découverte dans la basse vallée de la Bruche, dans la région de Wisches - Schwartzbach (CORSIN et al., 1960) et à Plaine, près de Champenay (DUBOIS, 1931).

La série sédimentaire dévono-dinantienne est essentiellement schisto-grauwackeuse: schistes et phtanites à radiolaires, grès, arkoses, grauwackes, conglomérats, avec, au dévonien moyen, intercalations de calcaires construits lenticulaires et bréchiques. Cette série est intensément volcanisée et comprend de véritables massifs éruptifs dont le mieux connu actuellement est le massif de Schirmeck, formé d'une association complexe de kératophyres, spilites et diabases diverses, avec un important cortège de pyroclastites, spilitiques et kératophyriques. C'est ce massif, que nous avons décrit en détail il y a quelques années (JUTEAU, 1965; JUTEAU et ROCCI, 1965, 1966), qui nous servira d'exemple pour la description des faciès (cf. carte géologique in JUTEAU et ROCCI, 1966).

Etude Pétrographique de l'Association Magmatique de Chaque Région
──

Cornwall - Devonshire

Définition des faciès

Les faciès éruptifs rencontrés dans cette partie de la chaîne sont disséminés et interstratifiés dans une série sédimentaire dévonienne essentiellement schisteuse, avec intercalations carbonatées. Il s'agit essentiellement de coulées sous-marines spilitiques et kératophyriques, d'intrusions basiques et de pyroclastites volcano-sédimentaires acides et basiques. La nomenclature des faciès que nous avons adoptée est, à quelques détails près, celle de MIDDLETON (1960), qui est une excellente classification descriptive; seuls certains termes ont été modifiés, par souci d'homogénéité avec les autres massifs. Nous distinguerons donc:

Les roches d'origine purement éruptive
Effusives:

   Kératophyres aphanitiques plus ou moins amygdalaires, en minces coulées interstratifiées.

   Spilites microlitiques et amygdalaires, porphyriques ou non, le plus souvent épanchées sous la forme de pillow-lavas.

   Spilites microlitiques intersertales, non amygdalaires, ou spilites en coulées massives.

   Andésites spilitiques porphyriques, microlitiques ou microgrenues, effusives à sub-effusives.

Intrusives:

   Dolérites et diabases microgrenues et grenues en sills, filons-couche, etc.

Les roches d'origine mixte volcano-sédimentaires

   Tufs acides kératophyriques.

   Tufs basiques spilitiques, chloriteux ou chlorito-calcareux.

   Cinérites rubanées.

Description des faciès-type

Roches d'origine purement éruptive effusives

a) Kératophyres: d'après MIDDLETON, ils forment quelques rares coulées de faible épaisseur interstratifiées dans les schistes de la région de Newton Abbot. Ils ont une texture microlitique intersertale et peuvent être plus ou moins amygdalaires.

b) Spilites microlitiques et amygdalaires:

1. Pillow-lavas non porphyriques. Les célèbres pillow-lavas de Chipley (Devonshire) sont le meilleur exemple de ce type. Etroitement moulés les uns sur les autres par l'intermédiaire d'une fine matrice chloriteuse, ils sont formés d'une lave sombre et dure, à petites amygdales blanchâtres.

Au microscope, la lave est finement microlitique, formée de microlites d'albite dans un fond ferrugineux avec calcite et chlorite. Des phénocristaux d'albite antiperthitique sont très disséminés dans cette pâte. Les amygdales, très abondantes, sont en général mixtes et montrent de belles associations sphérolitiques de chlorite, calcite, quartz et calcédoine (pl. I-1).

La matrice des pillows se compose d'une pâte microlitique à deux tailles de plagioclases dans un fond très fin d'oxydes de fer et de chlorite, et de très nombreuses vésicules chloriteuses coalescentes, souvent très allongées; on remarque des débris angulaux de phénocristaux d'albite, et l'absence totale de calcite (pl. I-2).

2. Pillow-lavas porphyriques. Nous prendrons pour type les pillow-lavas de Pentire Head (Cornouailles). Ils sont moins bien conservés que ceux de Chipley et ont subi une forte altération. C'est une lave de couleur brun-rouille, vacuolaire en surface. Au microscope, la lave montre nettement trois tailles de plagioclase: un fond très fin de microlites albitiques enchevêtrés (0.15 mm), une trame lâche de lattes intermédiaires (1 à 1.5 mm) et de gros phénocristaux trapus, très dispersés, de 4 à 5 mm. Le fond de la lave est uniquement formé, entre les microlites

d'albite, de granules de chlorite et d'oxyde de fer. Les amygdales sont très abondantes, souvent coalescentes, à calcite prédominante, à chlorite ou mixtes (pl. I-4).

3. Spilites microlitiques et amygdalaires massives. Sur le terrain, ces roches massives ne montrent pas de débit en pillow-lavas; ce sont visiblement des roches d'épanchement, présentant par ailleurs les mêmes caractères pétrographiques que les pillow-lavas. Les carrières de Landrake (Cornouailles) en sont un bon exemple. On y observe une lave massive verdâtre, montrant de nombreux globules blancs de calcite et de lattes dispersées de plagioclases. En lame mince, la texture est microlitique, porphyrique et amygdalaire, montrant nettement trois tailles de plagioclase, comme les pillows de Pentire Head. Les grand phénocristaux, très dispersés, peuvent atteindre 3 mm. Ils sont fracturés et cicatrisés par la calcite, et contiennent de nombreuses inclusions de pennine, ainsi que des mouches de calcite. Les plagioclases de taille intermédiaire (O.5 mm), se présentent en lattes allongées formant une trame lâche. Les microlites albitiques enfin forment le fond de la roche en un enchevêtrement de fines baguettes avec de très nombreux granules d'opaques et de la pennine. Les amygdales, abondantes et souvent coalescentes, sont surtout remplies de calcite, plus rarement de chlorite ou d'un mélange de deux minéraux.

c) Spilites microlitiques intersertales, non amygdalaires: ce groupe affleure en particulier au lieu-dit "the Rumps" près de Pentire Head (Cornwall); il est représenté par des roches verdâtres, massives, aphanitiques. La texture est microlitique (les lattes d'albite sont inférieures à 0.8 mm) et intersertale, les plagioclases s'enchevêtrent en une trame serrée. Le matériel interstitiel est toujours à calcite et chlorite dominantes, avec sphène, leucoxène, oxydes de fer et parfois pyroxène (augite) poecilitique. La différence essentielle de ce faciès

---

PLANCHE I. Cornouailles-Devonshire
1. Spilite microlitique et amygdalaire - Pillow-lavas de Chipley (Devonshire): écorce de pillow. La lave est finement microlitique, formée de microlites d'albite dans un fond ferrugineux, calciteux et chloriteux; des phénocristaux d'albite perthitique sont très disséminés dans cette pâte. Les amygdales, très abondantes, sont en général mixtes et montrent de belles associations sphérolitiques de chlorite, calcite, quartz et calcédoine. L.N.
2. Spilite microlitique et amygdalaire - Pillow-lavas de Chipley (Devonshire): matrice des pillow-lavas. Elle est formée d'une pâte microlitique fine à deux tailles de plagioclase dans un fond très fin d'oxydes de fer et de chlorite, et de très nombreuses vésicules chloriteuses coalescentes, souvent très allongées; on remarque des débris anguleux de phénocristaux d'albite et l'absence totale de calcite. Le tout est bréchifié avec un ciment chloriteux en grandes traînées. L.N.
3. Spilite microlitique et intersertale, non amygdalaire - Diabase de Wadebridge (Cornouailles). C'est une lave massive et homogène, finement cristallisée, complètement dépourvue d'amygdales. Les microlites d'albite (0,3 mm) sont enchevêtrés et forment une trame dans les mailles de laquelle on trouve la chlorite, la calcite et les oxydes de fer. La calcite cicatrise les nombreuses microfractures. L.P.
4. Spilite microlitique, porphyrique et amygdalaire - Pillow-lavas porphyriques de Pentire Head (Cornouailles). On remarque trois tailles de plagioclase: un fond intersertal très fin albitique (0,15 mm); une trame floue de lattes intermédiaires (1 mm), et de gros phénocristaux trapus, très dispersés, à macles complexes. Le fond est uniquement formé de chlorite et d'opaques. Les amygdales sont très abondantes, souvent coalescentes, en majorité remplies de calcite; d'autres sont à chlorite, ou d'un mélange des deux. L.P.

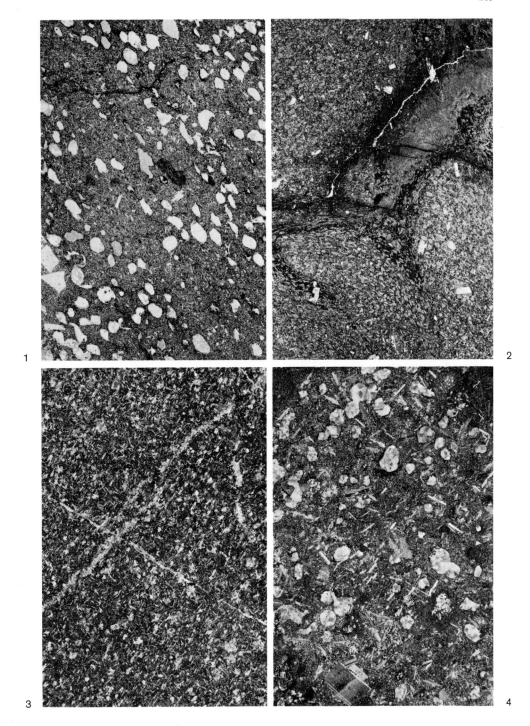

avec les pillow-lavas précédemment décrits est l'absence totale d'amygdales (cf. pl. I-3).

d) Andésites spilitiques porphyriques: ce faciès, intermédiaire entre les roches effusives microlitiques et les diabases grenues intrusives, est bien représenté à Beacon Hill (Devonshire); la roche a l'aspect d'une porphyrite microgrenue. Au microscope, la texture est microgrenue intersertale et porphyrique. Les phénocristaux sont tous plagioclasiques, très fracturés, à nombreuses inclusions chloriteuses. La mésostase, assez confuse et altérée, montre des grains d'augite plus ou moins titanifères, et une association de plagioclase dipyrisé, chlorite, calcite, épidote et opaques (pl. II-1).

Roches d'origine éruptive intrusives

Dolérites et diabases grenues. La région de Padstow (Cornwall) montre plusieurs exemples de ces sills doléritiques et diabasiques intrusifs dans les schistes du Dévonien supérieur. La dolérite de Padstow est une roche sombre à débit en boules caractéristiques. La texture est grenue, grossièrement intersertale. Les plagioclases, en grosses lattes trapues (2-3 mm), à macles complexes, très damouritisés, forment une charpente assez confuse. Ils englobent ou enserrent de grandes baguettes sub-automorphes d'augite titanifère violette pouvant atteindre 5 mm de long. Cette augite est en voie de remplacement par une hornblende brune qui la pseudomorphose. Plagioclases et pyroxènes sont cimentés par des amas de pennine avec un peu de calcite, des opaques avec sphène et leucoxène, de l'albite secondaire limpide et de nombreuses baguettes d'apatite (pl. II-2). Les diabases grenues, comme celles de Padstow ou de Penzance (Cornwall), n'ont pas de pyroxène et sont en général très altérées; le plagioclase originel est souvent difficilement reconnaissable et remplacé par une association complexe de minéraux de basse température. La mésostase est confuse et formée de chlorite, calcite, quartz, calcédoine, actinote, oxydes, etc.

PLANCHE II. Cornouailles-Devonshire (suite)
1. Andésite spilitique porphyrique - Andésite de Knowle, Beacon Hill (Devonshire). La structure est presque microgrenue, intersertale, avec de gros phénocristaux de plagioclase très fracturés, à nombreuses inclusions de chlorite. Le fond est un assemblage de lattes d'albite, de grains de chlorite, calcite, épidote, opaques et augite titanifère. Le dipyre s'y développe en agrégats fibroradiés. L.P.
2. Dolérite - Dolérite de Padstow (Cornouailles). La texture est grenue, grossièrement intersertale. Les plagioclases (2-3 mm) très damouritisés forment une charpente confuse. Ils englobent de grandes baguettes (jusqu'à 5 mm) d'augite titanifère violette, en voie de remplacement par une hornblende brune. Le tout est cimenté par des amas de pennine, avec calcite, sphène leucoxène, apatite, albite. L.P.
3. Brèche spilitique basique (Blair Hill). - Il s'agit d'une brèche d'éléments spilitiques amygdalaires se distinguant mal les uns des autres, accompagnés d'échardes chloriteuses et de traînées de calcite. L.N.
4. Tuf volcano-sédimentaire kératophyrique - Austin's Bridge, Devonshire. Débris et galets étirés de microquartzites, spilites microlitiques et cristaux de quartz dans un ciment finement lité argilo-ferrugineux. L.N.

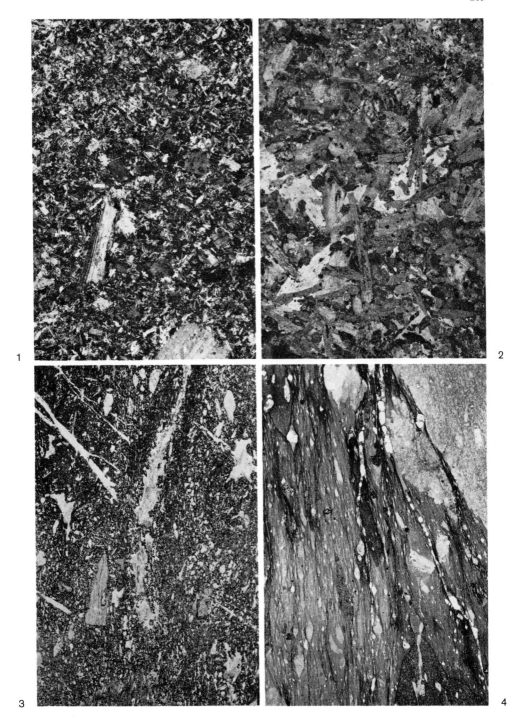

Les roches d'origine mixte volcano-sédimentaires

a) Tufs acides kératophyriques (selon MIDDLETON). Nous citerons comme exemple les tufs de Dartington et de Blair Hill (Devonshire), caractérisés par la présence de débris de feldspath potassique, de quartz (dans les tufs de Darlington seulement) et de fragments andésitiques et spilitiques, dans un ciment calcaro-chloriteux; de nombreux amas chloriteux représentent d'anciennes échardes de verre basique (pl. II-3).

A notre avis, ces tufs sont plus basiques qu'acides, et peuvent être groupés avec les tufs spilitiques, le tout étant équivalent aux "schalsteins" de Lahn-Dill (voir plus loin). L'appellation de "tufs kératophyriques" nous semble donc impropre. C'est notre seul point de désaccord avec la terminologie de MIDDLETON.

b) Tufs basiques spilitiques, chloriteux ou chlorito-calcareux. Ce sont soit des brèches assez grossières de fragments de spilites amygdalaires dans un ciment chloriteux lité, soit des tufs fins formés de débris spilitiques et de minéraux clastiques dans un ciment chloriteux et phylliteux stratifié. Le ciment peut être calcaire, comme à Austin's Bridge (Devonshire)(pl. II-4). Près de cette localité, MIDDLETON signale des formations à bombes scoriacées.

c) Cinérites rubanées. Ce faciès est visible dans la carrière de Trevose Head: ce sont des roches siliceuses aphanitiques très fines, rubanées, composées de quartz, calcite microcristalline, et fins débris de feldspaths.

Conclusion et résumé

Nature des roches

Les roches éruptives dévoniennes en Cornouailles et Devonshire se présentent donc en nombreuses coulées et intrusions isolées les unes des autres. Les coulées de pillow-lavas, assez fréquentes, sont formées de spilites amygdalaires, porphyriques ou non. Les intrusions basiques sont soit des dolérites, soit des diabases grenues plus ou moins altérées. Les roches acides ne sont connues que sous la forme de tufs kératophyriques; enfin, un cortège important de tufs spilitiques et de cinérites accompagne les laves.

Proportions relatives des faciès

Les roches basiques semblent franchement prédominantes sur les roches acides: certes il est malaisé d'estimer le volume des tufs kératophyriques mais, comme nous l'avons dit, les vrais tufs kératophyriques sont rares et les tufs basiques à faciès "schalsteins" sont au contraire très répandus. Parmi les roches basiques, les roches effusives semblent aussi abondantes que les roches intrusives.

Modes de gisement

Les spilites effusives se sont mises en place sous forme d'épanchements sous-marins en coulées de pillow-lavas. Les spilites microlitiques massives non amygdalaires peuvent être considérées comme effusives sous faible couverture empêchant la vésiculation des gaz. C'est vraisemblablement aussi le cas des andésites spilitiques porphyriques. Les diabases grenues et les dolérites se sont mises en place sous la forme de sills et filons-couches généralement bien interstratifiés dans les schistes encaissants. Enfin les tufs kératophyriques et spilitiques témoignent de phénomènes explosifs importants liés à la mise en place sous-marine.

Contexte sédimentaire

La série encaissante est essentiellement siliceuse et schisteuse. Les calcaires sont rares et tout-à-fait subordonnés.

District de Lahn-Dill

Définition des faciès

Comme en Cornouailles et Devonshire, la région de Lahn-Dill se caractérise par un grand nombre de coulées et d'intrusions isolées s'intercalant dans la série schisteuse dévonienne. Les pétrographes allemands ont naturellement subdivisé les faciès avec beaucoup de détail. Nous nous sommes efforcés pour notre part, par souci d'homogénéité et de comparaison avec les autres régions, de les rassembler en un petit nombre de faciès-types caractérisés avant tout par la texture de la roche (donc par le mode de mise en place) et par le chimisme global. Les subdivisions adoptées sont les suivantes:

Les roches d'origine purement éruptive
Effusives:

    Kératophyres à texture trachytique,
    - porphyriques -
    - aphanitiques -

    Spilites microlitiques et amygdalaires, épanchées sous la forme de pillow-lavas

    Spilites microlitiques intersertales non amygdalaires, en coulées massives

Intrusives:

    Diabases spilitiques microgrenues et grenues

    Intrusions picritiques

Les roches d'origine mixte volcano-sédimentaire

    Schalsteins

    Cinérites rubanées

Description des faciès-types

Roches d'origine purement éruptive effusives

a) Kératophyres

Kératophyres porphyriques. L'échantillon-type (sortie de Weilburg) est une roche dure et rougeâtre, montrant des lattes de feldspath rose dans une pâte aphanitique rougeâtre. Au microscope, la texture est microlitique fluidale (type trachytique) et porphyrique. Les microlites d'albite très serrés, en paquets flexueux, forment tout le fond de la roche, avec un peu de quartz interstitiel, des granules d'oxydes de fer et un peu de chlorite jaune. Des plages de calcite ferruginisée sont dispersées dans ce fond. Les phénocristaux feldspathiques sont corrodés et se groupent volontiers en agrégats pouvant atteindre 5 mm. Ils sont complexes, formés en général d'un plagioclase antiperthitique (pl. III-1).

Kératophyres aphanitiques (Wirbelau). Ils sont identiques aux précédents, mais ne présentent pas de phénocristaux. D'autre part ils ne contiennent pas de calcite dans le fond, mais plus de quartz, en agrégats mosaïques et en filonnets (pl. III-2).

b) Pillow-lavas spilitiques amygdalaires

Pillow-lavas non porphyriques. Nous prendrons pour exemple de ce faciès les pillow-lavas de la carrière située à l'Ouest de Ernsthausen (feuille de Weilmünster). La lave est brun-rouille, compacte, à petites amygdales sombres chloriteuses; les pillows bien formés se moulent les uns sur les autres par l'intermédiaire d'une mince matrice chloriteuse. Au microscope, la lave est finement microlitique, les microlites d'albite nageant dans un fond de chlorite parsemé de granules d'oxydes de fer. Les amygdales sont abondantes et toutes chloriteuses, la calcite est pratiquement inexistante et remplit parfois le centre de certaines amygdales chloriteuses; au coeur des pillows, les amygdales à calcite sont plus abondantes. Quant à la matrice, elle est peu différente de la lave de l'écorce des pillows: les amygdales chloriteuses y sont simplement plus abondantes, coalescentes et étirées. Dans d'autres gisements des coulées montrent cependant des amygdales plus "classiques", à calcite prépondérante et omniprésente, chlorite et quartz.

Pillow-lavas porphyriques. Ce faciès, bien représenté au Nord de Löhnberg, diffère du précédent par une texture microlitique plus grossière et une grande quantité de phénocristaux feldspathiques de grande taille (5 mm); ces phénocristaux sont complexes (plagioclase antiperthitique, orthose perthitique) et bourrés d'inclusions de chlorite et calcite. Le fond de la mésostase est chlorito-calcique, ainsi que les amygdales souvent coalescentes, jusqu'à former de grandes traînées irrégulières à travers la roche (pl. III-3).

c) Spilites microlitiques intersertales non amygdalaires. Ce faciès se présente sous la forme de roches massives, verdâtres très finement grenues, activement exploitées dans les carrières de la vallée de la

---

PLANCHE III. Lahn-Dill
1. Kératophyre porphyrique (sortie de Weilburg) - La texture est finement microlitique et fluidale, de type trachytique. De nombreux phénocristaux corrodés de feldspath complexe (plagioclase antiperthitique), souvent groupés en agrégats, flottent dans cette pâte. On remarque l'absence totale de quartz dans les phénocristaux (comparer avec les kératophyres porphyriques "rhyolitiques" de Schirmeck, Planche V, Fig. 2). L.P.
2. Kératophyre aphanitique (Wirbelau) - C'est la même roche que la précédente, mais sans phénocristaux: remarquer la belle texture microlitique fluidale, de type trachytique. L.P.
3. Spilite amygdalaire porphyrique (Löhnberg - Hütte) - Ce sont des pillow-lavas porphyriques. On remarque en effet, en plus du fond microlitique albitique et des amygdales de calcite, de nombreux phénocristaux plus ou moins corrodés de feldspath complexe antiperthitique, bourrés d'inclusions de chlorite et de calcite. L.P.
4. Spilite microlitique intersertale, non amygdalaire - (Philippstein, vallée de la Mott). Nous avons là un exemple caractéristique de ces roches finement cristallines, homogènes et massives, à texture microlitique parfaitement intersertale. Comparer avec Planche I, Fig. 3 et Planche VI, Fig. 1. L.P.

269

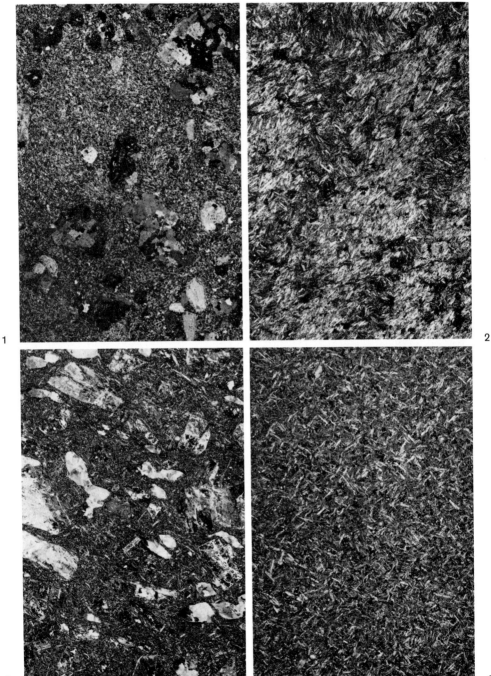

Mott, au Sud de Braunfels. Au microscope, elles montrent une très belle texture microlitique intersertale, la mésostase entre les microlites d'albite étant essentiellement chloriteuse, avec ou sans calcite, et ferrugineuse (oxydes). On remarque l'absence totale d'amygdales. Ce faciès peut être plus ou moins bréchifié et parcouru de filonnets de calcite. On peut rattacher à ce type certaines diabases microgrenues porphyriques altérées (pl. III-4).

Roches d'origine éruptive intrusives

a) Diabases spilitiques microgrenues et grenues à pyroxène. Nous abordons avec ce groupe les roches franchement intrusives. Ces roches sont mésocrates verdâtres, bien cristallines, grenues à microgrenues, où le pyroxène noir se détache bien sur le fond verdâtre. Au microscope la texture est intersertale ou ophitique; l'augite est soit poecilitique en grandes plages englobant les lattes de plagioclases, soit en grains et baguettes occupant les vides de la trame plagioclasique. Le pyroxène est toujours accompagné de chlorite abondante, et d'un peu de calcite, de titano-magnétite et parfois de biotite. Le plagioclase est très damouritisé. La diabase grenue de Weilburg, les diabases prismées de Leun, d'Eyershausen, etc., sont typiques de ce groupe (pl. IV-1 et IV-3).

b) Cumulats ultrabasiques d'origine picritique.[1] Ce faciès, très localisé (carrière de Schwartzstein), se présente comme une belle roche noire serpentinisée, à débit spectaculaire en boules. Au microscope, il s'agit d'une véritable péridotite serpentinisée du type werhlite, à olivine globulaire et augite poecilitique; l'antigorite forme une trame dans laquelle a cristallisé une biotite secondaire, et la magnétite xénomorphe. Des zones interstitielles très altérées représentent peut-être un ancien plagioclase basique très minoritaire. A notre avis, le terme de "picrite" adopté par les Allemands pour désigner cette roche ne convient pas, car il s'agit d'une péridotite serpentinisée (pl. IV-2).

Roches mixtes d'origine volcano-sédimentaire

a) Schalsteins. Ce sont des roches très caractéristiques de ce massif: elles ont un débit schisteux, un aspect noduleux et montrent en coupe

---

[1] Ces intrusions en forme de sills, provenant d'un magma basique riche en phénocristaux ferro-magnésiens, se différencient par gravité en une partie inférieure ultrabasique et une partie supérieure doléritique. Nous respectons l'appellation "picrite" des auteurs allemands, bien qu'en toute rigueur ce terme soit réservé aux laves ultrabasiques.

PLANCHE IV. Lahn-Dill (suite)
1. Diabase grenue à pyroxène (Weilburg) - La texture est grenue, avec une trame de plagioclase très damouritisé à disposition intersertale; l'augite est poecilitique, en grandes plages englobant les lattes de feldspath. Le reste de la roche est constitué de chlorite, zoïsite, épidote, titanomagnétite, albite limpide. L.N.
2. Péridotite serpentinisée (carrière de Schwartzstein) - On remarque de nombreux cristaux globulaires d'olivine résiduelle dans une trame de serpentine. De grands cristaux d'augite poecilitique englobent l'olivine. L.N.
3. Diabase microgrenue (sortie de Leun) - Cette belle diabase, prismée à l'affleurement, montre une texture microgrenue intersertale, avec lattes de plagioclase et d'augite, dans un fond de chlorite interstitielle; on remarque des phénocristaux dispersés de plagioclase très damouritisé. L.N.
4. Schalstein (Oberbiel) - Nombreux débris spilitiques plus ou moins scoriacés dans un ciment calcaro-chloriteux contenant des fragments de plagioclase et de quartz. L.N.

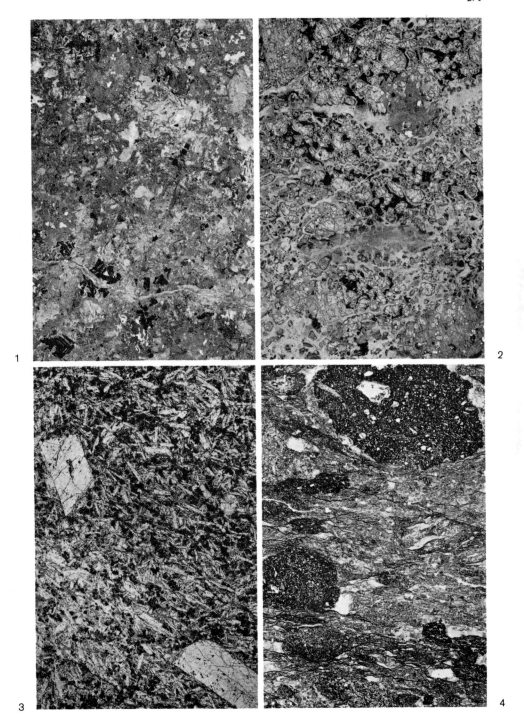

une belle structure pyroclastique. Au microscope, on reconnaît un mélange intime, l'élément spilitique vésiculeux à bords flous se fondant dans un ciment calcaire, ainsi que des débris de cristaux, plagioclases et quartz surtout. La calcite et la chlorite forment le plus souvent de nombreuses vésicules très étirées (pl. IV-4).

b) Cinérites rubanées. Ces formations caractéristique, en bancs décimétriques et à grain fin se rencontrent en de nombreux endroits (près de Leun, Heisterberg, sur la route de Weilburg à Schnitten, etc.), souvent associées aux schalsteins.

Conclusion et résumé

Nature des roches

Les roches éruptives dévoniennes de la région de Lahn-Dill comprennent toute une gamme de roches basiques intrusives, accidentellement ultrabasiques et effusives (pillow-lavas, spilites massives microlitiques), des coulées de kératophyres trachytiques très localisées, et un important cortège de pyroclastites spilitiques à ciment calcaire (schalsteins).

Proportions relatives des faciès

Les roches basiques sont très largement prépondérantes: les kératophyres sont presque négligeables en volume et toujours très localisés. Les picrites ultrabasiques sont des faciès accidentels. Parmi les roches basiques, roches effusives et roches intrusives semblent à peu près équivalentes en quantité. Les schalsteins forment un cortège pyroclastique très important, sans doute au moins aussi important que les coulées de pillow-lavas.

Modes de gisement

Les pillow-lavas spilitiques et les kératophyres représentent des coulées sous-marines, auxquelles il faut ajouter la masse des schalsteins. Les spilites microlitiques massives, non amygdalaires, se sont sans doute épanchées sous faible couverture; les diverses diabases microgrenues et grenues, à pyroxène ou non, sont des intrusions sous forme de sills, filons-couche, etc. qui sont parfois prismés. C'est également le cas de la picrite ultrabasique, qui se présente comme une intrusion très localisée.

Contexte sédimentaire

La série sédimentaire encaissante est essentiellement schisteuse. Cette série s'enrichit en calcaire au niveau des schalsteins, et cet enrichissement est donc directement lié au volcanisme sous-marin.

Vosges Septentrionales: Le Massif de Schirmeck

Définition des faciès

Au point de vue pétrographique, le massif de Schirmeck se caractérise par l'association étroite, sur un territoire relativement restreint, de faciès éruptifs acides et basiques. Dans le détail, les faciès pétrographiques sont très variés, et les anciens auteurs allemands (BÜCKING, 1923) en avaient distingués près d'une trentaine, portant tous des noms différents: le seul groupe des "diabases", par exemple était subdivisé en dix neuf variétés.

On peut aisément ramener ces nombreuses variétés en une dizaine de faciès-types, qui à eux tous représentent bien la pétrographie du massif. La classification que nous avons proposée est la suivante (JUTEAU, 1965; JUTEAU et ROCCI, 1965):

Les roches d'origine purement éruptive
Effusives:

> Les roches acides ou kératophyres, laves effusives hololeucocrates caractérisées par l'absence totale de minéraux ferro-magnésiens autres que la chlorite, peu abondante, et les oxydes de fer, se subdivisent en:
> - kératophyres aphanitiques - sans phénocristaux visibles à l'oeil nu
> - kératophyres porphyriques quartziques - à phénocristaux visibles de quartz automorphe et de plagioclase
> - orthokératophyres - à texture pyroméride sphérolitique
>
> Les roches basiques, comprenant
> - des spilites microlitiques effusives à plagioclase acide (albite), à texture intersertale
> - amygdalaire
> - non amygdalaire
> - des diabases microgrenues porphyriques à gros phénocristaux de plagioclase basique et d'amphibole

Intrusives:

> Les roches basiques intrusives grenues, à plagioclase basique et texture ophitique à sub-ophitique: ce sont les dolérites et ophites.
>
> Les diabases grenues quartziques, roches intermédiaires grenues à texture intersertale, à plagioclase acide et quartz interstitiel.

Les roches d'origine mixte volcano-sédimentaire

> On peut y distinguer:
>
> Des tufs et agglomérats volcaniques à ciment sédimentaire très réduit:
> - acides - ce sont les tufs kératophyriques
> - basiques - ce sont les "schalsteins" des auteurs allemands
>
> Des brèches siliceuses sédimentaires quartzo-pélitiques, à débris volcaniques remaniés acides ou basiques. Nous les avons appelés brèches volcano-sédimentaires siliceuses.

Description des faciès-types

Roches d'origine purement éruptive effusives

a) Kératophyres

Kératophyres aphanitiques. L'échantillon-type, provenant du Nid des Oiseaux, est une roche dure, rougeâtre, à cassure esquilleuse, absolument aphanitique. Au microscope, la texture est microlitique sub-fluidale. Les phénocristaux, très dispersés, sont un peu plus gros que les microlites et toujours feldspathiques (orthose albitisée). Le quartz n'apparaît jamais parmi les phénocristaux. Le fond est formé de microlites mal individualisés d'albite, de grains de quartz très abondants, de chlorite interstitielle et de magnétite. La séricite apparaît en paillettes sur les feldspaths (pl. V-1).

Kératophyres porphyriques quartziques. L'échantillon-type, provenant de la Côte d'Albet, montre dans un fond verdâtre sombre des globules de quartz à éclat gras pouvant atteindre 5 mm de diamètre, de nombreuses petites lattes brillantes de feldspath rose et des mouches vert sombre de chlorite. Au microscope, la texture est microlitique non fluidale. Les phénocristaux, beaucoup plus abondants et plus gros que ceux des kératophyres aphanitiques, se composent de quartz automorphe corrodé, d'albite, d'orthose albitisée et de chlorite en amas pseudomorphosant, avec la magnétite et l'apatite, d'anciens minéraux ferro-magnésiens. Le fond est formé de microlites d'albite dans une trame de quartz spongieux et de chlorite "alvéolaire", ainsi que de grains de magnétite et d'apatite (pl. V-2).

Orthokératophyres. L'échantillon-type, provenant des Evaux, est une roche rose clair, légère, montrant à l'oeil nu des sphérolites feldspathiques roses de 1 à 3 mm de diamètre, dans une pâte aphanitique blanchâtre. Au microscope, les sphérolites sont très denses et formés de fibres de quartz et d'orthose. Le fond est formé de quartz engrené, d'albite et de bouquets de muscovite. Des phénocristaux d'orthose perthitique albitisée peuvent exister. Ce type est caractérisé par la prépondérance du feldspath potassique sur le feldspath sodique (pl. V-3).

b) Spilites et diabases

Spilites microlitiques non amygdalaires. Ce sont des roches homogènes gris verdâtre, finement grenues. Les plagioclases (albite-oligoclase) forment une trame intersertale serrée de microlites rachitiques, bifides, aux contours flous. Dans les mailles de cette trame, on trouve l'augite, l'ouralite, la chlorite, l'épidote, la calcite, le sphène, la magnétite, parfois un peu de quartz. Dans le détail, de nombreuses variétés peuvent être reconnues: spilites à augite plus ou moins ouralitisée, spilite à actinote, spilite à chlorite-épidote, etc. (pl. VI-1).

Spilites amygdalaires. L'échantillon-type montre une structure bréchique caractéristique: des fragments de lave noire très riche en amygdales blanches sont cimentés par une lave spilitique très fine verdâtre et aphanitique. Au microscope, les éléments de la brèche sont formés d'un fond ferrugineux à microlites d'albite, et de très nombreuses amygdales à calcédoine, épidote, pennine, mésotype, albite, hématite. La pâte qui les cimente est formée de très nombreux microlites d'albite à disposition fluidale, d'amygdales à calcite et chlorite et d'un fond de calcite et épidote en fins granules.

PLANCHE V. Schirmeck
1. Kératophyre aphanitique (Nid des Oiseaux) - Remarquer l'extrême finesse des microlites, la rareté et la petitesse des phénocristaux. L.P.
2. Kératophyre porphyrique quartzique (Côte d'Albet) - Nombreux phénocristaux d'albite automorphe, d'orthose et de quartz corrodé, dans une pâte fine non orientée. L.P.
3. Orthokératophyre quartzique à texture sphérolitique (Les Evaux) - Noter les sphérolites de grande taille, jointifs, formés de quartz et d'orthose. En haut, à droite, un phénocristal de quartz corrodé. L.P.
4. Dolérite (SE des Evaux) - On remarquera surtout la texture grenue subophitique. L.P.

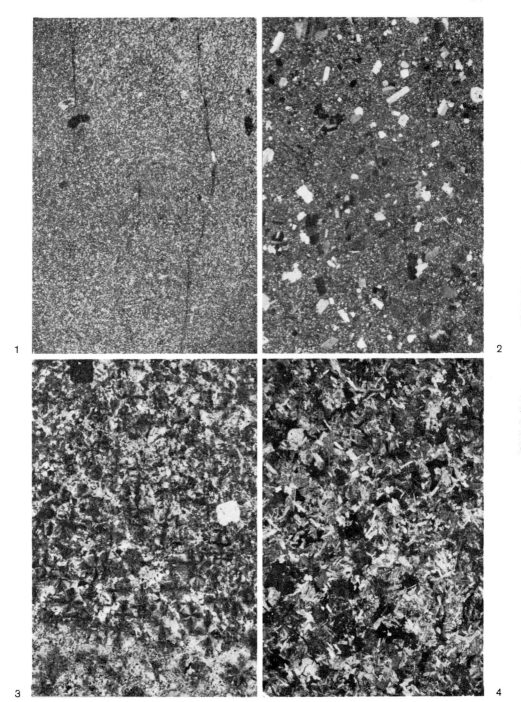

Diabases microgrenues porphyriques. L'échantillon-type montre de nombreux phénocristaux trapus de feldspath rose et des sections rhombiques noires d'amphibole, dans une pâte grise et compacte. Au microscope, les phénocristaux d'andésine (8 mm de long) sont fortement saussuritisés (damourite, chlorite, épidote, calcite, apatite), ceux de hornblende verte automorphe (4-5 mm) sont altérés en chlorite, calcite, épidote et fins agrégats micacés. Le fond microgrenu est essentiellement feldspathique (plagioclases flous à extinction roulante), avec aiguilles de hornblende verte et grains de calcite, épidote et chlorite.

Roches éruptives intrusives

a) Roches basiques grenues. Nous trouvons dans ce groupe des dolérites franches, à texture sub-ophitique, composées de labrador et d'augite, des ophites, à andésine et ouralite fibreuse, et des diabases grenues, à andésine et produits secondaires (chlorite, actinote, bastite, etc.) (pl. V-4).

b) Diabases grenues quartziques. Le faciès-type est une roche rougeâtre, d'aspect grenu, montrant de nombreuses lattes brillantes de plagioclase et de belles aiguilles vert sombre d'amphibole. Au microscope, la texture est grenue intersertale, avec une trame d'albite-oligoclase en grandes lattes automorphes damouritisées, des baguettes d'actinote, des amas de calcite et chlorite, et un quartz interstitiel abondant, souvent en association graphique avec le plagioclase.

Roches pyroclastiques

a) Tufs et agglomérats volcaniques

Tufs kératophyriques. Ce faciès montre, dans un ciment sombre siliceux, de nombreux débris anguleux de feldspaths roses et de quartz détritique. Au microscope, la texture est pyroclastique. Le fond est formé de quartz en mosaïque régulière et de chlorite interstitielle; les débris de feldspath sont fracturés, rubéfiés et chloritisés; les débris de quartz anguleux montrent une auréole de croissance secondaire.

Schalsteins. Ce terme allemand consacré par l'usage désigne de véritables agglomérats de fragments spilitiques moulés les uns sur les autres, pratiquement sans ciment (celui-ci, quand il existe, est soit calcaire, soit chloriteux). Ce groupe rassemble des roches très diverses par leur granulométrie, depuis des tufs fins dont les éléments ne dé-

---

PLANCHE VI. Schirmeck (suite)
1. Spilite microlitique intersertale non amygdalaire (Vallée d'Albet) - La texture est intersertale fine. Les lattes de plagioclase acide sont rachitiques. L.P.
2. Brèche spilitique (Basse Garrat) - Nombreux éléments et débris de spilites microlitiques et amygdalaires, plus ou moins scoriacés, se moulant les uns sur les autres pratiquement sans ciment. L.N.
3. Brèche siliceuse (Grandfontaine) - Débris fins, surtout de quartz, dans un ciment ferrugineux et phylliteux finement lité. L.N.
4. Schalstein fin (Côte de Fréconrupt) - Débris spilitiques et fragments de minéraux dans un ciment chloriteux. L.N.

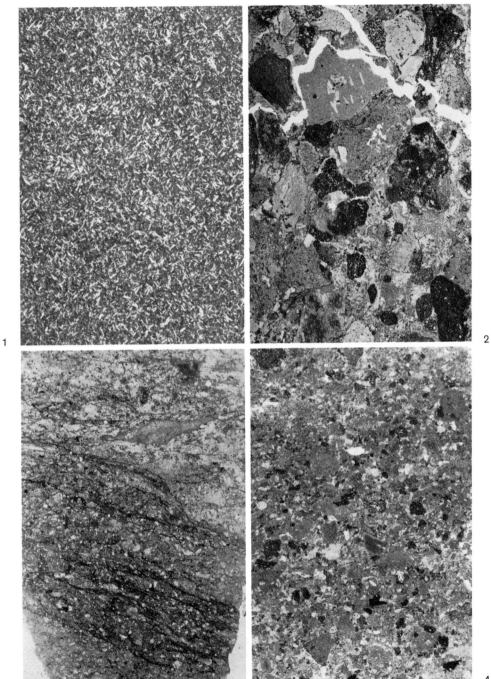

passent pas le millimètre, jusqu'à des brèches à tendance conglomératique dont les plus gros éléments peuvent atteindre 3 à 5 cm (pl. VI-2 et 4).

b) Brèches volcano-sédimentaires siliceuses. Ce sont des roches sombres, très dures et compactes; on y voit de nombreux globules de quartz à éclat gras, des débris de feldspaths roses dans une pâte aphanitique noire siliceuse. Au microscope, la roche est formée de nombreux débris de quartz corrodés et de plagioclases fracturés et recristallisés, dans un ciment essentiellement quartzeux et phylliteux (vermiculites, chlorites); des éléments allongés à microlites et amygdales témoignent de l'apport volcanique (pl. VI-3).

Conclusion et résumé

Nature des roches

Le massif de Schirmeck se compose de laves basique (spilites microlitiques et amygdalaires), associées à des dolérites, ophites et diabases grenues avec ou sans quartz, et de laves acides représentées par des kératophyres aphanitiques et porphyriques, très localement par des ortho-kératophyres. Toutes ces laves s'accompagnent d'un important cortège de brèches magmatiques (spilites amygdalaires) et de roches pyroclastiques (schalsteins, tufs kératophyriques).

Proportions relatives des faciès

Les roches effusives sont largement prépondérantes: les kératophyres d'abord, puis les spilites microlitiques et amygdalaires, ainsi que les pyroclastites associées, c'est-à-dire les schalsteins, très abondants, et dans une moindre mesure les tufs kératophyriques. Les roches grenues basiques (dolérites, ophites, diabases grenues) sont beaucoup plus rares et très localisées (sills, petites intrusions, etc.). Enfin certains faciès sont accidentels, comme les diabases porphyriques et les ortho-kératophyres. Il faut souligner ici la nette prépondérance des faciès kératophyriques sur les faciès spilitiques, ce qui constitue un trait original du massif de Schirmeck par rapport aux autres régions décrites.

Modes de gisement

Nous avons donc essentiellement des coulées volcaniques kératophyriques et spilitiques, accompagnées d'intrusions basiques et diabasiques locales. Le massif de Schirmeck ne montre pas de pillow-lavas: ceci est un autre trait original de ce massif par rapport aux autres régions décrites. Les spilites microlitiques amygdalaires sont pourtant bien représentées, mais toujours sous la forme de brèches magmatiques et de pyroclastites de type "schalsteins".

Contexte sédimentaire

Ce contexte est typiquement siliceux et détritique: grès, arkoses, grauwackes, conglomérats, schistes et phtanites. Les brèches siliceuses volcano-sédimentaires représentent cette sédimentation dans le massif de Schirmeck. Elles contiennent des lentilles de calcaire récifal bréchifié dont l'importance en volume est négligeable.

Comparaison des Trois Régions Etudiées - Répartition des Faciès

Après ces descriptions, il convient maintenant de faire le bilan comparatif des trois régions étudiées. Il est frappant de constater que les produits du magmatisme "initial" sont à peu près les mêmes dans les trois cas, si l'on excepte quelques variantes sur lesquelles nous reviendrons.

Rappel des faits

Si nous considérons d'abord les roches effusives, nous constatons que les spilites sont les faciès les mieux représentés dans les trois régions aussi bien les spilites amygdalaires que les spilites microlitiques intersertales massives. Les kératophyres sont beaucoup moins abondants, sauf dans le massif de Schirmeck, mais existent sous la forme de coulées ou de tufs dans les trois régions.

En ce qui concerne les roches intrusives, nous retrouvons partout des sills et filons-couche de dolérites et de diabases plus ou moins spilitiques grenues et microgrenues. Enfin, le cortège pyroclastique associé est important dans les trois régions, le faciès caractéristique étant le faciès "schalstein", formé de nombreux lapillis et débris de spilites et de fragments de minéraux (quartz, feldspaths, pyroxènes) dans un ciment calcito-chloriteux. C'est cet ensemble de faciès éruptifs communs aux trois régions étudiées que nous avons appelé "association spilite-kératophyre hercynotype" (ROCCI et JUTEAU, 1968) et que l'on retrouve d'un bout à l'autre de la chaîne, non seulement en Cornwall-Devonshire, à Lahn-Dill et dans les Vosges septentrionales, mais aussi en Bretagne, dans le Morvan, dans les Vosges méridionales, dans le Harz, en Thuringe occidentale, etc.

Ayant souligné les analogies, il nous faut montrer aussi les différences qui donnent à chacune des trois régions étudiées son cachet original. Le massif de Schirmeck, en particulier, présente par rapport aux deux autres régions un certain nombre de traits originaux, que nous avons mentionnés au passage: d'abord, il s'agit d'un massif éruptif complexe, où roches basiques et roches acides se côtoient et s'interpénètrent, alors qu'en Cornwall-Devonshire et à Lahn-Dill il s'agit de coulées et d'intrusions isolées, interstratifiées dans les schistes. D'autres massifs de ce type sont connus dans les Vosges septentrionales, et sont actuellement à l'étude (massif de Mayenonautier, par exemple, dans la vallée du Rabodeau): il s'agit donc là d'un trait singulier des terrains dévono-dinantiens des Vosges septentrionales, dans lesquels on peut voir aussi des coulées et des intrusions isolées interstratifiées. Autre trait original, le massif de Schirmeck ne montre pas de coulées spilitiques à débit en pillow-lavas: elles sont remplacées ici par des brèches magmatiques et des pyroclastites (schalsteins), montrant que, lors de l'épanchement des laves, les phénomènes explosifs ont été assez importants pour empêcher la formation de coulées. En Cornwall-Devonshire et à Lahn-Dill, coulées et pyroclastites spilitiques coexistent, vraisemblablement en raison d'une mise en place à plus grande profondeur sous la mer. Il faut signaler aussi la grande abondance des coulées kératophyriques dans le massif de Schirmeck, par rapport aux deux autres régions où ces roches sont franchement minoritaires. Enfin, nous devons mentionner l'existence dans ce massif de diabases quartziques qui, au point de vue chimico-minéralogique, sont intermédiaires entre les kératophyres et les spilites: ce faciès important n'a pas été rencontré dans les autres massifs.

Tous ces caractères tendent à opposer la région des Vosges septentrionales aux deux autres régions, qui ont entre elles beaucoup plus de points communs qu'elles n'en ont avec le massif de Schirmeck: prédominance des roches basiques avec abondance de coulées en pillow-lavas; rareté des kératophyres, toujours très localisés; abondance du cortège pyroclastique, en particulier des pyroclastites basiques à faciès "schalsteins"; mode de gisement en coulées et intrusions isolées et interstratifiées dans la série dévonienne schisteuse. Le district de Lahn-Dill présente toutefois une particularité remarquable, c'est la présence de petites intrusions picritiques, dont la différenciation peut produire localement des cumulats ultrabasiques.

Interprétation

L'omniprésence de l'association "spilite-kératophyre" dans l'orogène hercynien est due à l'implantation de cet orogène sur un vieux socle antécambrien qui affleure abondamment de la Bohême à la Bretagne dans toute la "cordillère de l'Europe moyenne". Ce véritable "écran sialique" entre la partie supérieure du Manteau et les fonds géosynclinaux a d'une part empêché toute participation importante du Sima (on peut constater ici tout le contraste avec les associations de type ophiolitique des chaînes alpines méditerranéennes, où les roches ultrabasiques sont prépondérantes) et d'autre part favorisé l'établissement de réservoirs intracrustaux, donc l'hybridation et la palingenèse d'un matériel sialique. Nous expliquons ainsi l'association constante de roches basiques (spilites) à des roches acides (kératophyres). Mais si cette association est constante, nous avons vu cependant que le rapport en volume entre spilites et kératophyres est très variable: dans les Vosges septentrionales, la proximité immédiate du vieux socle se traduit par l'abondance des kératophyres; dans la région de Lahn-Dill, le seul socle antédévonien connu est le socle calédonien ardennais; par ailleurs aucun massif granitique hercynien n'est visible dans le massif schisteux rhénan: il est donc vraisemblable que l'écran "sialique" y était beaucoup plus réduit, d'où les différences sensibles avec les Vosges septentrionales: rareté des kératophyres et intrusions picritiques indiquant une participation directe du Sima; l'abondance des pillow-lavas indique vraisemblablement une mer plus profonde qu'à Schirmeck. Nous pouvons donc opposer, dans l'association spilite kératophyre hercynotype, un magmatisme de fosse (Lahn-Dill) à un magmatisme de ride (Schirmeck). La Cornouaille et le Devonshire se rapprochent beaucoup plus du type Lahn-Dill, comme nous l'avons vu: mais on n'y connaît pas d'intrusions ultrabasiques[3], et par ailleurs les granites batholitiques hercyniens reparaissent; nous avons donc là un cas intermédiaire, avec un "écran sialique" plus épais qu'à Lahn-Dill (pas d'ultrabasites) mais moins proche qu'à Schirmeck (peu de kératophyres): nous sommes sur le flanc nord du géosynclinal hercynien, en bordure du Continent nord-atlantique.

L'interprétation proposée permet donc de comprendre à la fois les remarquables similitudes que l'on constate d'un massif à l'autre, similitudes dues à l'implantation sialique de l'orogène, sur un vieux socle en grande partie antécambrien, et aussi les différences observées, qui ne font que refléter les variations d'épaisseur et de proximité de cet "écran sialique" sous les sédiments dévoniens.

Il nous reste maintenant à examiner le problème de la spilitisation: en effet les interprétations exposées ci-dessus rendent compte de

---

[3] Rappelons pour mémoire que le massif du Lizard, à l'extrême pointe de la Cornouaille, est considéré à l'heure actuelle comme anté-dévonien, et charrié du Sud vers le Nord sur les terrains dévoniens et carbonifères.

l'association "roches basiques - roches acides", et la variation de la répartition des faciès d'une région à l'autre. Elles n'expliquent pas pourquoi, au lieu de roches volcaniques courantes (basaltes, andésites, rhyolites), ce sont des spilites, des kératophyres, des diabases qui se manifestent.

Avant de discuter de ce problème, nous présenterons d'abord les caractères pétrographiques généraux des spilites et des kératophyres dans les régions étudiées.

Caractères Minéralogiques et Texturaux des Spilites et des Kératophyres dans l'Association Hercynotype

Nous allons dégager maintenant les principaux faits fournis par l'examen pétrographique des roches à "paragénèse spilitique", c'est-à-dire des spilites et des kératophyres, à l'exclusion des roches basiques non spilitiques (dolérites, diabases non spilitiques, ophites, andésites porphyriques, etc.), et des roches pyroclastiques du cortège hercynotype.

Classification des faciès

En reprenant les données du chapitre III, où nous avons présenté les faciès région par région, nous nous proposons de classer les spilites et les kératophyres hercynotypes en fonction de leur texture: en effet une classification minéralogique n'aurait que peu d'intérêt, étant donné la grande uniformité des paragénèses, la considération des textures permet au contraire d'établir une classification descriptive détaillée. La composition minéralogique s'impose en revanche pour définir, dans l'ensemble des roches à "paragénèse spilitique", les deux grands groupes des spilites, roches basiques à minéraux ferro-magnésiens et calciques (chlorite, calcite, épidote, oxydes de fer, etc.) prédominants, et des kératophyres, roches acides leucocrates essentiellement quartzo-feldspathiques. Il convient de prévoir en outre un groupe intermédiaire, pour des roches telles que les diabases quartziques de Schirmeck.

Nous proposons donc le tableau de classification suivant pour les spilites et kératophyres hercynotypes, avec un rapide résumé de la description pétrographique pour chaque type (pour plus de détails, voir chapitre III):

SPILITES

Spilites à texture microlitique (ou spilites effusives):

a) Spilites microlitiques et amygdalaires:
  1. Non porphyriques:
  - Pillow-lavas (CHIPLEY, ERNSTHAUSEN) - Pillows: microlites d'albite dans fond ferrugineux à chlorite et calcite; quelques phénocristaux d'albite antiperthitique; amygdales mixtes à calcite-chlorite-quartz. Matrice: vésicules chloriteux étirés et coalescents, dans fond à microlites d'albite, oxydes de fer et chlorite; débris anguleux de phénocristaux d'albite.

  - Bréchiques (SCHIRMECK) - Eléments: fond ferrugineux opaque à microlites d'albite; nombreuses amygdales à calcédoine, pennine, épidote, albite. Pâte: nombreux microlites d'albite à disposition fluidale dans fond à oxyde de fer, calcite, épidote; amygdales à calcite-chlorite.

2. Porphyriques:
- Pillow-lavas (PENTIRE HEAD, LÖHNBERG) - Trois tailles de plagioclases: fond très fin de microlites d'albite enchevêtrés (0,15 mm), trame lâche de lattes intermédiaires (1 - 1,50 mm) et gros phénocristaux trapus (4-5 mm) et complexes (albite-oligoclase antiperthitique); fond à oxyde de fer, chlorite et calcite. Amygdales abondantes à calcite-chlorite.
- Massives (LANDRAKE) - idem, + filonnets de calcite.

b) Spilites microlitiques et intersertales non amygdalaires:

(THE RUMPS, VALLEE DE LA MOTT, SCHIRMECK). Belle texture microlitique et intersertale régulière; absence totale de phénocristaux et d'amygdales. Deux paragénèses:

- Augite, albite, calcite, chlorite, oxydes de fer.
- Chlorite, albite, calcite, sphène leucoxène, oxydes de fer, éventuellement actinote, épidote, etc.

Spilites à texture microgrenue (ou spilites intrusives, de semi-profondeur):

a) Diabases spilitiques microgrenues, à texture intersertale (PADSTOW, WEILBURG, LEUN, EYERSHAUSEN, SCHIRMECK). Mêmes subdivisions que le groupe précédent:

- Diabases spilitiques à augite, albite, chlorite, calcite, oxydes de fer.
- Diabases spilitiques à chlorite, albite, calcite, sphène leucoxène, oxydes de fer, et éventuellement actinote, épidote, etc.

b) Porphyrites spilitiques microgrenues (BEACON HILL). Texture microgrenue intersertale et porphyrique; phénocristaux d'albite-oligoclase à inclusions chloriteuses; mésostase confuse et altérée à plagioclase dipyrisé, augite titanifère, chlorite, calcite, épidote et hématite.

ROCHES INTERMEDIAIRES

Diabases quartziques de SCHIRMECK. Texture microgrenue à grenue intersertale, avec trame d'albite-oligoclase damouritisé, sur fond d'actinote, calcite, chlorite et quartz interstitiel et micropegmatitique abondant.

KERATOPHYRES

Kératophyres à texture microlitique (ou kératophyres effusifs):

a) Kératophyres microlitiques "trachytiques" (texture fluidale, pas de quartz en phénocristaux):

1. Aphanitiques (WIRBELAU, SCHIRMECK) - Microlites d'albite en paquets flexueux, dans fond réduit à quartz, chlorite, hématite.
2. Porphyriques (WEILBURG) - Idem, avec phénocristaux corrodés et complexes (albite, oligoclase antiperthitique); plages de calcite ferruginisée dans le fond.

b) Kératophyres microlitiques "dacitiques" (fond vitreux, phénocristaux de quartz): SCHIRMECK (kératophyres porphyriques quartziques). Texture microlitique non fluidale à microlites d'albite dans ancien fond vitreux (quartz, chlorite, magnétite, apatite). Phénocristaux de quartz automorphe corrodés, d'albite et d'orthose albitisée. Amas chloriteux.

Kératophyres à texture microgrenue (ou kératophyres intrusifs):

Pas d'exemples jusqu'à maintenant.

Conclusions de l'étude pétrographique

a) Caractère primaire des textures. Toutes les textures examinées nous ont semblé indiscutablement primaires: elles dérivent de la cristallisation directe d'un magma, aussi bien pour les spilites que pour les kératophyres. L'examen approfondi des relations entre les minéraux, en particulier la forme, l'agencement et la limpidité des lattes d'albite, aussi bien dans les roches microlitiques que microgrenues, ne laissent aucun doute à ce sujet. Nous n'avons pas trouvé de reliques de plagioclase plus basique dans les microlites ou les lattes d'albite. Seuls les phénocristaux de certains faciès porphyriques montrent, comme nous l'avons signalé, une histoire plus complexe: il s'agit dans tous les cas d'une substitution entre Na et K: albite antiperthitique, orthose albitisée, etc.

b) Remarques sur la "paragénèse spilitique". Un certain nombre de faits concernant la composition minéralogique des spilites-kératophyres nous semblent particulièrement importants:

1. Absence totale de l'olivine. Dans le groupe des spilites, et a fortiori dans celui des kératophyres, aucune trace d'olivine, fraîche ou pseudomorphosée, n'a pu être identifiée. Ceci est particulièrement important car, on le sait, ce minéral est fréquent dans les spilites alpines. Il s'agit donc là d'une particularité des spilites hercynotypes qui demande à être expliquée (cf. 2e partie).

2. Le cas de l'augite. Le pyroxène des spilites, quand il existe dans la paragénèse, est toujours de l'augite, souvent plus ou moins titanifère; sa présence n'est cependant pas obligatoire: les spilites amygdalaires (pillow-lavas) en sont complètement dépourvues. Il apparaît dans les spilites microlitiques intersertales, soit en petits grains entre les lattes d'albite, soit en plages poeciliticiques: dans les deux cas, c'est un minéral manifestement primaire, nullement "résiduel", ayant cristallisé avec l'albite et la chlorite, un peu avant (grains) ou un peu après (plages poeciliticiques). Dans les roches microgrenues, l'augite est fréquente, avec ces deux mêmes aspects. Donc:

- l'augite peut faire partie de la "paragénèse spilitique". Elle cristallise avec l'albite, la chlorite, la calcite, etc.

- elle ne se trouve jamais dans les roches d'épanchement à amygdales (pillows, etc.), mais dans les roches microlitiques intersertales (cristallisation sous faible couverture), ou microgrenues: donc avant la vésiculation des gaz, c'est-à-dire sous forte pression des fluides (H_2O, CO_2).

3. Le cas de l'orthose. L'orthose n'est pas un minéral exceptionnel des spilites: on ne l'a peut-être pas assez souligné: si l'albite est certes le minéral caractéristique des spilites, l'orthose l'accompagne

souvent: non seulement la teneur en K_2O des spilites hercynotypes est souvent supérieure à celle des basaltes normaux, mais on trouve aussi de véritables "spilites potassiques" à microlites d'orthose. L'expression minéralogique du potassium n'est pas toujours évidente; dans la grande majorité des cas, l'albite semble admettre le potassium dans son réseau et l'orthose ne s'exprime pas. Ceci étant, les weilburgites de LEHMANN (1949, 1952) ne sont donc pas, dans l'hercynotype, une exception locale, mais un faciès qu'on peut considérer comme typique, et plus répandu qu'on ne pourrait le penser de prime abord.

c) Ces remarques étant faites, nous pouvons donc caractériser la "paragénèse spilitique" par l'association albite-chlorite-calcite-(orthose)- (augite)-(quartz) et oxydes de fer. En somme, un petit nombre de minéraux et une grande variété de textures, dont les deux plus typiques sont la texture microlitique amygdalaire et la texture intersertale, microlitique ou microgrenue.

Etude Chimique

Présentation des tableaux d'analyses

Toutes les analyses chimiques présentées dans cet article ont été effectuées au Centre de Recherches Pétrographiques et Géochimiques de Nancy. Les plus anciennes concernent le massif de Schirmeck et ont déjà été publiées en 1966; il y manquait toutefois une donnée importante, qui est la perte au feu (pratiquement, P.F. = $H_2O + CO_2$); cette lacune a été comblée et le lecteur trouvera donc ici ces analyses maintenant complétées. Le dosage complémentaire du CO_2 n'a pas été jugé utile car, sauf pour le n° 8301 (ancien n° 36), qui représente une ancienne spilite amygdalaire secondairement calcitisée, la calcite n'est jamais assez abondante pour que la teneur en CO_2 puisse influer d'une manière sensible sur les teneurs des autres éléments.

Pour la Cornouailles, le Devonshire et la région de Lahn-Dill, nous avons fait analyser tous les faciès-types prélevés dans ces massifs au cours de nos études de terrain en 1965 et 1966: toutes ces analyses sont nouvelles et inédites. Le dosage complémentaire du CO_2 n'a été effectué, pour des raisons de temps, que sur la moitié de ces échantillons: on trouvera donc dans les tableaux soit la teneur globale en $H_2O + CO_2$ sous la forme de "perte au feu" (P.F.), soit les teneurs respectives en H_2O et en CO_2 lorsque le dosage complémentaire a été fait.

Les tableaux d'analyses sont présentés par régions, et les analyses regroupées en un certain nombre de faciès-types définis au chapitre III. Les nombres respectifs d'analyses sont les suivants: 21 pour la Cornouailles, 20 pour le Devonshire, 43 pour Lahn-Dill et 91 pour Schirmeck: soit au total 175 analyses nouvelles.

Ce stock d'analyses nouvelles, toutes effectuées par le même laboratoire entre 1964 et 1968, nous semble suffisamment représentatif pour que nous ne reproduisions pas ici les analyses chimiques peu nombreuses et dispersées que l'on peut trouver dans la littérature consacrée aux régions étudiées: pour le massif de Schirmeck, ce sont quelques analyses antérieures à 1918, publiées par BÜCKING en 1923; pour le Devonshire, nous avons trouvé une analyse de kératophyre (GILLULY, 1935) et sept analyses de diabases (LEHMANN, 1965); pour la région de Lahn-Dill, nous disposons de 11 analyses de diabases et 6 de kératophyres (LEHMANN, 1952 et 1965); au total, 24 analyses postérieures à 1950: on voit à quel point il était indispensable, avant même de discuter du problème de la spilitisation, de mieux connaître la pétrochimie de l'association spilite-kératophyre hercynotype.

Le lecteur intéressé dispose donc maintenant de près de 200 analyses chimiques postérieures à 1950: étant donné la variété des faciès et la grande variabilité des teneurs, ce n'est pas encore suffisant (en particulier pour une étude statistique sérieuse), ni parfaitement représentatif; certains faciès sont manquants, comme ces minces coulées kératophyriques du Devonshire signalées par MIDDLETON, que nous n'avons pas retrouvées sur le terrain et dont nous n'avons pas d'analyse chimique.

Compte tenu de ces réserves, l'ensemble des 175 analyses présentées ici jointes aux descriptions pétrographiques du chapitre III donnent pour la première fois un tableau pétrochimique assez précis de l'association hercynotype.

Commentaire des tableaux

Les spilites

La première impression qui se dégage, à la lecture des tableaux d'analyses des spilites, amygdalaires ou non, est celle d'une grande variabilité des teneurs, pour tous les éléments majeurs: le tableau 1, qui donne les maxima et minima atteints pour chaque élément, est très significatif à cet égard: la silice, par exemple, varie entre 29,60 et 67,20%, le CaO entre 0,70 et 20,30, etc. Cette variabilité des teneurs est un caractère essentiel des spilites, elle est bien supérieure à celle des basaltes et montre qu'il serait illusoire - comme certains ont tenté de le faire - de vouloir définir les spilites par leur chimisme seul.

a) Les spilites amygdalaires

1. Les pillow-lavas (Cornouailles, Devonshire, Lahn-Dill) se caractérisent par une très forte perte au feu, jamais inférieure à 5%, souvent supérieure à 10; en moyenne: 8,80 pour les pillows de Lahn-Dill et 10,35 pour ceux de Cornouailles et Devonshire: de telles teneurs sont dues évidemment à l'abondance des amygdales de calcite, et la teneur en CO_2 suit fidèlement les variations de la teneur en CaO. La teneur en H_2O n'est cependant jamais négligeable et montre que la chlorite contribue pour une bonne part à cette perte au feu élevée. D'un échantillon à l'autre, les variations sont considérables, même sur un seul affleurement, comme ces trois échantillons des carrières de Landrake (Devonshire):

	D 2	D 3	D 4(1)
SiO_2	41,70	37,50	31,40
CaO	1,97	8,12	20,30
CO_2	0,20	5,59	12,77
H_2O	5,99	4,09	2,34

Dans le premier échantillon à amygdales calciques rares, les teneurs en CO_2 et CaO sont très faibles et la chlorite seule fournit la quasi totalité de la perte au feu (près de 6% de H_2O); le deuxième échantillon est beaucoup plus riche en amygdales de calcite, le plus souvent mixtes (calcite-chlorite): les teneurs en CaO et CO_2 montent brusquement, CO_2 et H_2O sont en gros à égalité; on remarque la diminution corrélative de SiO_2. Enfin le troisième échantillon, à amygdales de calcite très abondantes et coalescentes, contient 20,30% de CaO et 12,77% de CO_2, soit 33% de calcite, un tiers du poids de la roche! La chlorite devient subordonnée (2,34% de H_2O) et la chute corrélative de SiO_2 s'accentue. A ce dernier stade, les teneurs des autres éléments (cf. tableau XVII) sont évidemment gravement perturbées par la teneur anormale en calcite: outre la silice, le fer, le magnésium, l'alumine diminuent fortement; par contre, et c'est tout à fait remarquable, les teneurs en Na_2O et TiO_2 augmentent, et celle en K_2O reste stable.

Après avoir souligné l'importance de la perte au feu, considérons maintenant les teneurs des éléments majeurs:

- les moyennes du tableau II montrent nettement que les ferro-magnésiens (Fe_2O_3-MnO-MgO et TiO_2) et l'alumine (Al_2O_3) ont des teneurs comparables à celles des basaltes, Fe_2O_3 étant un peu en excès et MgO nettement en retrait; après correction pour tenir compte de la perte au feu, la somme Fe_2O_3 + MgO est exactement comparable à celle des basaltes. Les tableaux montrent que ce sont ces mêmes éléments dont les teneurs sont les plus régulières et varient le moins. Ce groupe d'éléments "invariants" est à opposer au groupe SiO_2-CaO-Na_2O-K_2O, éléments dont les fluctuations peuvent être considérables.

- SiO_2 est le plus régulier de ces éléments: sa teneur est inversement proportionnelle à celle de $CaCO_3$; dans les échantillons pauvres en calcite, elle varie peu autour de 45%, soit un peu au-dessous de celle des basaltes (compte tenu de la perte au feu); elle décroît régulièrement quand $CaCO_3$ augmente, et peut descendre au-dessous de 30% dans les échantillons les plus calcitisés.

- la teneur en CaO, comme nous l'avons vu, est éminemment variable, et va pratiquement de 0 à 20%: elle dépend de la proportion de calcite dans la roche, ce qui revient à dire que pratiquement tout le calcium des spilites amygdalaires est calcitisé. En moyenne, nous constatons cependant (tableau II) que la teneur en CaO est très proche de celle des basaltes (8,40% pour Lahn-Dill et 10,02% pour Cornouailles-Devonshire). Si l'élément CaO se rapproche donc du groupe des ferro-magnésiens par sa teneur moyenne proche de celle des basaltes, il s'en distingue par sa très grande variabilité, qui est d'ailleurs un autre trait caractéristique des spilites.

- les alcalins Na_2O et K_2O ont un comportement complexe: en moyenne d'abord, il est clair que la teneur en Na_2O est franchement supérieure à celle des basaltes, malgré la perte au feu importante qui accentue encore ce caractère; la teneur en potassium est au moins égale (Angleterre) et même nettement supérieure (Lahn-Dill) à celle des basaltes, malgré la perte au feu: ceci mérite d'être souligné, car on exalte classiquement le rôle de Na_2O dans les spilites, en tenant K_2O pour négligeable: il apparaît en fait que les alcalins Na_2O et K_2O voient leur teneur augmenter dans les spilites, Na_2O augmentant cependant relativement plus que K_2O, le rapport Na_2O/K_2O devenant bien plus élevé que celui des basaltes. Dans le détail, on remarque de spectaculaires renversements de ce rapport dans certains échantillons, aussi bien dans le Devonshire qu'à Lahn-Dill:

	D 15(2)	LH 9(b)	LH 9(c)	LH 19(1)
Na_2O	0,95	0,10	0,23	3,11
K_2O	1,95	6,24	7,49	3,71

On peut parler de véritables spilites potassiques. Les échantillons LH (9-b) et LH 9(c), en particulier, sont remarquables par leur teneur presque nulle en Na_2O et la valeur élevée de K_2O. Ces anomalies sont cependant purement locales, et exceptionnelles. Elles sont cependant signalées par de nombreux auteurs (cf. entre autres, LEHMANN, 1952; NAREBSKI, 1964), et sont une preuve de plus de l'hétérogénéité des spilites dans le détail.

2. Les spilites amygdalaires bréchiques de Schirmeck. Nous ne disposons que de deux analyses de ces spilites, qui sont nettement différentes des précédentes:

- SiO_2 est nettement plus élevé (53,50%) et supérieur aux basaltes,
- corrélativement, CaO (5,05%) et P.F. (3,20%) sont nettement plus bas,
- nous retrouvons un groupe d'"invariants": Al_2O_3, Fe_2O_3 et MgO, dont les teneurs sont exactement comparables à celles des basaltes,
- Na_2O (5,52%) est particulièrement élevé, et K_2O (0,79%) particulièrement faible,
- TiO_2 (1,23%) est beaucoup moins élevé que dans les pillows.

b) Les spilites microlitiques intersertales non amygdalaires

Nous pouvons faire sur ces spilites les mêmes remarques générales que sur le groupe précédent. La disparition des amygdales entraîne une diminution de P.F. (qui reste cependant importante, avec prédominance de H_2O sur CO_2) et de CaO. Nous retrouvons le groupe des "invariants": Al_2O_3, Fe_2O_3 et MgO, dont les teneurs restent comparables à celles des basaltes. La teneur en SiO_2 est beaucoup moins fluctuante, maintenant que $CaCO_3$ reste subordonné, et varie peu autour de 45% (Angleterre) ou 50% (Allemagne); enfin, les spilites de Schirmeck se singularisent de la même façon (SiO_2 plus élevé, P.F., TiO_2 et CaO nettement plus faibles); leur moyenne diffère peu de celle des spilites amygdalaires, ce qui traduit simplement le fait que les spilites amygdalaires de Schirmeck sont surtout chloriteuses.

c) Les diabases spilitiques microgrenues et grenues

Ce groupe se caractérise par la présence fréquente de pyroxène (augite), minéral quasiment inexistant dans les spilites franchement effusives, et par l'existence de tous les types intermédiaires avec ce groupe des dolérites: par rapport aux spilites effusives microlitiques, les spilites microgrenues et grenues présentent des caractères spilitiques "atténués":

- perte au feu moins importante, autour de 3-4%,
- corrélativement, remontée de la teneur en CaO (qui entre dans la composition de l'augite et du plagioclase) et de SiO_2, qui se stabilise autour de 48%,
- la teneur en Fe_2O_3 a tendance à augmenter, et celle en MgO à diminuer,
- le rapport Na_2O/K_2O reste élevé. La somme des alcalins est toutefois nettement plus faible que dans les spilites microlitiques, sauf à Schirmeck (cf. diabase albitique n° 8314).

Dans ce dernier massif, nous avons un bon exemple du passage dolérite - spilite:

	Dolérite n° 8312	Dolérite albitique n° 8313	Diabase albitique (spilite) n° 8314
SiO_2	47,30	48,00	52,60
CaO	10,39	3,50	2,02
Na_2O	2,88	3,45	6,36
K_2O	1,25	0,84	1,06
P.F.	3,77	6,66	3,60

La dolérite n° 8312 contient du pyroxène et de l'andésine; sa composition est celle d'un basalte normal (à part P.F., un peu élevée); dans la dolérite albitique, la teneur en CaO s'effondre, alors que SiO_2, Na_2O et P.F. s'accroissent: ces modifications correspondent à la diminution du pyroxène, à la disparition de l'andésine, remplacée par l'albite et à l'augmentation de la chlorite. Enfin dans la diabase albitique, ces tendances s'accentuent: la teneur en Na_2O double presque, et la somme Na_2O + K_2O atteint 7,42; CaO diminue encore (2,42%), SiO_2 augmente nettement (52,60%). La perte au feu reste modérée en l'absence de calcite: cette modération de la perte au feu est typique des spilites de Schirmeck.

d) Conclusions sur les spilites

En résumé, les principaux caractères chimiques des spilites hercynotypes sont les suivants:

- Une grande variabilité des teneurs en éléments majeurs (tableau I). Cette variabilité est bien supérieure à celle des basaltes et rend illusoire toute tentative de vouloir cerner une "composition spilitique",

- une forte perte au feu, directement liée aux proportions de chlorite (H_2O) et de calcite (CO_2) dans la roche. Les pillow-lavas ont une perte au feu considérable, de l'ordre de 8 à 10%, en raison de leurs nombreuses amygdales à calcite (surtout) et à chlorite. Cette teneur tombe entre 5 et 8% dans les spilites non amygdalaires, et autour de 3% dans les diabases spilitiques plus ou moins grenues. A Schirmeck, la perte au feu est modérée, de l'ordre de 3% quel que soit le type,

- une teneur en CaO typiquement capricieuse, dépendant de la teneur en calcite de la roche. A ce sujet, il faut ici détruire un mythe, celui des spilites pauvres en calcium (FONTEILLES, 1968): nous avons vu que certaines spilites peuvent contenir plus de 20% de CaO, alors que d'autres en sont pratiquement dépourvues. La proposition correcte est que les spilites sont pauvres en calcium non calcitisé: si l'on soustrait en effet le CaO combiné à la calcite du CaO total de la roche, il reste en moyenne 2 à 3% de CaO non calcitisé: en ce sens, mais en ce sens seulement, on peut dire que les spilites sont pauvres en calcium (les spilites de Schirmeck sont pratiquement dépourvues de calcite et ont donc des teneurs en CaO très basses, d'où la confusion de FONTEILLES, qui a voulu généraliser ce qui n'est qu'un cas particulier),

- la teneur en SiO_2 est de l'ordre de 45% (sauf à Schirmeck) dans les spilites dépourvues de calcite. Cette teneur diminue corrélativement à l'augmentation de la teneur en $CaCO_3$, et peut descendre au-dessous de 30% dans les pillows les plus amygdalaires: le groupe (SiO_2 - $CaCO_3$) est typiquement antithétique,

- les alcalins sont "exaltés", la somme Na_2O + K_2O se situant entre 5 et 8%. Par rapport aux basaltes, Na_2O augmente fortement, K_2O reste comparable ou augmente légèrement. Il faut insister sur le fait que la valeur élevée du rapport Na_2O/K_2O n'est pas due à une baisse de la teneur en K_2O, comme on le sous entend généralement: bien au contraire, la teneur en K_2O tend en moyenne à augmenter par rapport à celle des basaltes mais dans des proportions beaucoup moins spectaculaires que Na_2O. L'importance de K_2O se manifeste encore dans l'existence locale de spilites potassiques, dans lesquelles le rapport Na_2O/K_2O s'inverse complètement,

- il existe un groupe d'éléments dont les variations sont relativement modérées et dont les teneurs moyennes sont proches de celles des basaltes: il s'agit de Al_2O_3, qui est sans doute l'élément le plus stable et dont les teneurs moyennes sont exactement comparables à celles des basaltes[4]; de Fe_2O_3 dont les teneurs ont tendance à être un peu supérieures à celles des basaltes; de MgO, dont les teneurs sont un peu (parfois nettement) inférieures à celles des basaltes, et de TiO_2, dont les valeurs sont soit typiquement "océaniques" à Lahn-Dill et en Cornouailles-Devonshire (2 à 3%), soit typiquement "continentales" à Schirmeck (1,20%),

- il existe une sorte de polarité de la spilitisation dont les effets, à leur maximum dans les pillow-lavas effusifs, s'atténuent progressivement dans les roches de faible profondeur, puis les roches microgrenues et grenues intrusives: cette "atténuation" se traduit par une diminution de la perte au feu, une diminution de Na_2O, une stabilisation de SiO_2 et de CaO: le développement du pyroxène, l'évolution du plagioclase vers l'andésine ménagent des transitions avec le groupe des dolérites,

- enfin, les spilites de Schirmeck se distinguent des autres par leur haute teneur en silice (53-54%), leur perte au feu modérée (de l'ordre de 3%), leur faible teneur en CaO (4-5%) et en TiO_2 (1,20%), une plus forte teneur en alcalins et un rapport Na_2O/K_2O plus élevé.

Les kératophyres

Les kératophyres forment un groupe beaucoup plus homogène que celui des spilites. Les variations des teneurs sont assez modérées et les moyennes du tableau III sont parfaitement représentatives. On y distingue aisément trois groupes:

a) Les kératophyres siliceux de Schirmeck, caractérisés par une très haute valeur de SiO_2 (plus de 71% en moyenne). Remarquons d'abord qu'il n'existe pas de différence chimique sensible entre les kératophyres aphanitiques et les kératophyres porphyriques quartziques; les deux types ont une teneur assez faible en Al_2O_3 (12-13%) et en Fe_2O_3 (4-5%), une teneur très faible en MgO et en CaO (autour de 1%), des teneurs en alcalins comparables aux spilites (4 à 6% pour Na_2O, 1 à 2% pour K_2O) une faible teneur en TiO_2 (0,30% à 0,40%) et une perte au feu très modérée, autour de 1,80%.

b) Les kératophyres trachytiques de la Lahn. Ils se caractérisent par une teneur nettement moins élevée en silice (61-63%), plus élevée en

---

[4]Cette remarque montre tout l'intérêt du diagramme (Al/3 - K, Al/3 - Na) préconisé par de la ROCHE; cf., dans ce même volume, l'article de la ROCHE, ROCCI, JUTEAU.

alumine (15-16%) et une très haute teneur en alcalins ($Na_2O + K_2O = 10$, contre 6 à Schirmeck). Dans le détail, les kératophyres aphanitiques de Wirbelau sont deux fois plus riches en Fe_2O_3 que les kératophyres porphyriques de Weilburg. Ils sont en revanche un peu moins sodiques et moins riches en H_2O et CO_2.

c) Les orthokératophyres de Schirmeck sont des roches rares et très localisées, comme l'étaient les spilites potassiques de la Lahn et du Devonshire. Ce sont des roches hautement siliceuses (76%), moyennement alumineuses (12,55%) et très alcalines ($Na_2O + K_2O = 8,07\%$, dont 5,42% de K_2O et 2,65% de Na_2O). Les autres éléments sont réduits à l'état de traces.

d) Conclusions sur les kératophyres. Nous sommes bien documentés sur les kératophyres de Schirmeck, mais il faudrait pouvoir disposer de plus d'analyses des kératophyres de la Lahn et surtout connaître le chimisme des kératophyres du Devonshire. En attendant, on peut distinguer clairement deux grands types de kératophyres:
- les kératophyres siliceux à tendance dacitique, type Schirmeck, hypersiliceux et sodiques, avec un rapport Na_2O/K_2O élevé,
- les kératophyres à tendance trachytique, type Lahn-Dill, hyper-alcalins, avec un rapport Na_2O/K_2O un peu supérieur à 1.

A côté de ces deux grandes tendances, il faut ajouter la présence occasionnelle des orthokératophyres, roches hyper-siliceuses et hyper-alcalines, dans lesquelles le rapport Na_2O/K_2O s'inverse, comme dans les spilites potassiques.

Les diabases quartziques

Ce groupe, qui n'existe qu'à Schirmeck, est très intéressant, car ses caractères chimiques sont intermédiaires entre ceux des spilites et ceux des kératophyres, comme en témoigne le tableau suivant:

	Spilites (1)	Diabases quartziques	Kératophyres (2)
SiO_2	54,05	61,90	71,56
Al_2O_3	15,50	13,75	12,97
Fe_2O_3	11,26	9,54	5,12
MnO	0,21	0,16	0,07
MgO	4,92	2,70	0,94
CaO	4,66	1,51	0,46
Na_2O	4,92	4,66	4,72
K_2O	1,04	1,30	1,35
TiO_2	1,25	0,79	0,37
P.F.	2,97	2,72	1,81

(1) moyenne de toutes les spilites microlitiques de Schirmeck
(2) moyenne de tous les kératophyres de Schirmeck

La teneur en SiO_2 est exactement intermédiaire entre les deux groupes, ainsi que celles en Al_2O_3 et en TiO_2. La teneur en Fe_2O_3 (9,54%)

apparente ce groupe à celui des spilites; mais l'effondrement de la teneur en CaO rappelle les kératophyres. Il est remarquable de constater que les alcalins restent sensiblement constants dans les trois groupes, avec un rapport Na_2O/K_2O identique (forte prédominance sodique): ceci confère indiscutablement au groupe intermédiaire des diabases quartziques un caractère franchement spilitique. Le massif de Schirmeck est le seul à présenter ainsi une évolution continue et progressive des spilites aux kératophyres: dans les autres massifs, il y a un "trou" entre les deux groupes (voir plus loin les diagrammes de NIGGLI). Nous verrons plus loin ce qu'il faut penser de cette particularité du massif de Schirmeck.

Les autres faciès

Nous commenterons maintenant rapidement le chimisme des autres faciès, c'est-à-dire des faciès non spilitiques du cortège hercynotype.

a) Les roches basiques de semi-profondeur. Ce sont essentiellement des dolérites, des diabases plus ou moins transformées, des ophites, etc. Ce groupe est assez hétérogène. Les moins transformées de ces roches ont une composition basaltique typique, avec toutefois une perte au feu plus importante (de l'ordre de 3%). Le rapport Na_2O/K_2O est généralement plus élevé que dans les basaltes. On remarque une très faible teneur en TiO_2 à Schirmeck, alors qu'elle est forte en Cornouailles, Lahn-Dill étant intermédiaire.

b) Les porphyrites. Dans chaque massif, il existe quelques gisements très localisés de roches porphyriques microlitiques ou microgrenues. A Schirmeck, ce sont des diabases microgrenues porphyriques à hornblende (n° 8218, 19, 20 et 21); dans le Devonshire, ce sont les microgabbros porphyriques à pyroxène de Beacon Hill (D-11 et 12), dans la Lahn, nous avons trouvé une diabase microgrenue porphyrique (LH-14). Ces roches ont toutes la particularité d'être très alumineuses (Al_2O_3 = 17 à 18%), ceci étant dû à l'abondance de phénocristaux de plagioclases plus ou moins basiques, et pauvres en calcium (sauf la D-11(1)). Celles de Schirmeck sont comme toujours les plus siliceuses et ont une teneur élevée en potassium (Na_2O/K_2O un peu supérieur à 1). Le microgabbro de Beacon Hill se caractérise par une faible teneur en Silice (46%) et un rapport Na_2O/K_2O élevé, identique à celui des spilites. La perte au feu est assez uniforme, entre 3 et 4%.

c) Les roches picritiques. Nous avons signalé quelques petites intrusions ultrabasiques très localisée dans le district de Lahn-Dill. Dans la carrière de Schwartzstein, il s'agit d'une werhlite serpentinisée plutôt que d'une "picrite". L'analyse chimique (DL-6) confirme ce diagnostic: les teneurs en silice (38,15%), Fe_2O_3 (14,94%) et MgO (28,45%) sont celles d'une péridotite; les teneurs en Al_2O_3 (4,86%) et CaO (2,37%) trahissent la présence d'un peu d'anorthite, que nous avons reconnue sous le microscope (minéral accessoire interstitiel très altéré). La perte au feu importante (10,38%) est due évidemment à la serpentine abondante.

d) Le cortège pyroclastique. En ce qui concerne le cortège pyroclastique, notre documentation est malheureusement inégale suivant les régions, le massif de Schirmeck étant de loin le mieux connu à cet égard.

MASSIF DE SCHIRMECK

1. Le chimisme du groupe volcano-sédimentaire siliceux (tableau IX) reflète le mélange de microquartzites et d'éléments phylliteux avec des débris kératophyriques, dans un ciment siliceux. Les proportions de ce mélange sont variables, d'où les teneurs en silice allant de 52 à plus de 80%. La fraction sédimentaire est très généralement prépondérante.

2. Le groupe volcano-pyroclastique, à éléments sédimentaires en faible proportion, comporte:

- un type acide: les tufs de kératophyres porphyriques quartziques, dont la composition est peu différente de celle des kératophyres s.s. (tableau X),

- un type basique: les "schalsteins", où prédominent largement les débris de spilites vacuolaires, et dont le chimisme est proche de celui des spilites amygdalaires s.s. (tableau XI), avec toutefois plus de silice, car le ciment est siliceux.

LAHN-DILL

Un certain nombre de "schalsteins" de la vallée de la Lahn ont été analysés (tableau XVI): il s'agit ici de "schalsteins calciques", avec mélange en toutes proportions de débris spilitiques et d'un ciment calcaire. On voit toute la différence avec les "schalsteins" siliceux de Schirmeck.

CORNOUAILLES - DEVONSHIRE

La série de Wadebridge (tableau XXI) est à notre avis à rapprocher des "schalsteins" de la Lahn, avec un mélange spilite-calcite reconnu en lame mince et bien apparent dans le chimisme (éch. D-14 à D-18). MIDDLETON préfère le terme de "tufs kératophyriques", mais nous n'avons pas pu trouver de vrais tufs kératophyriques sur le terrain.

Les cinérites de Pentire Head (éch. C-4 et C-8(b)) ont une composition andésitique, avec une très faible teneur en CaO (2%).

Diagrammes chimiques

Pour caractériser les tendances géochimiques de l'association hercynotype, nous avons utilisé d'une part quelques diagrammes classiques universellement employés (paramètres de NIGGLI, triangle Na-K-Ca, diagramme de calco-alcalinité de JUNG), d'autre part les diagrammes chimico-minéralogiques de de la ROCHE (1964).

Les diagrammes chimico-minéralogiques de de la ROCHE: individualisation de la lignée spilitique.

La présentation et le commentaire détaillé de ces diagrammes, appliqués à l'association hercynotype, se trouve dans ce même volume (de la ROCHE, ROCCI, JUTEAU), aussi ne retiendrons-nous ici que les conclusions qui s'en dégagent:

1. Le diagramme (Al/3 - K, Al/3 - Na) montre clairement que la lignée spilite-kératophyre est originale: elle s'enracine dans le domaine des basaltes et s'écarte nettement des grandes lignées classiques: basalte-dacite-rhyolite ou basalte alcalin-trachyte.

2. Le diagramme (Si/3 - (K+Na+2 Ca/3), K - (Na+Ca)) montre deux grandes tendances dans l'hercynotype: une tendance "tholéiitique" à spilites pauvres en calcium et kératophyres très siliceux (type SCHIRMECK) et une

tendance "alcaline", à spilites carbonatées (pillow-lavas) et kératophyres très alcalins (type LAHN-DILL).

Les diagrammes de de la ROCHE ont donc le grand intérêt et le grand avantage sur les diagrammes classiques, de faire ressortir d'une manière très claire l'individualité de l'association spilite-kératophyre par rapport aux lignées magmatiques classiques, et de mettre en évidence, à l'intérieur d'une même province (ici, la province hercynienne), les diverses tendances géochimiques de cette association en fonction de l'implantation paléogéographique des zones étudiées.

Les diagrammes classiques

a) Paramètres de NIGGLI (Figs. 4, 5, 6 et 7)

- Schirmeck: "Si" varie de 80 à 550; les courbes sont régulières, bien étalées, les diabases quartziques assurant la liaison entre les spilites et les kératophyres, lesquels s'étalent beaucoup plus loin que ceux des autres régions. Nous attribuons cet effet de continuité, d'étalement et de silicification générale des faciès, que nous avons constatée à

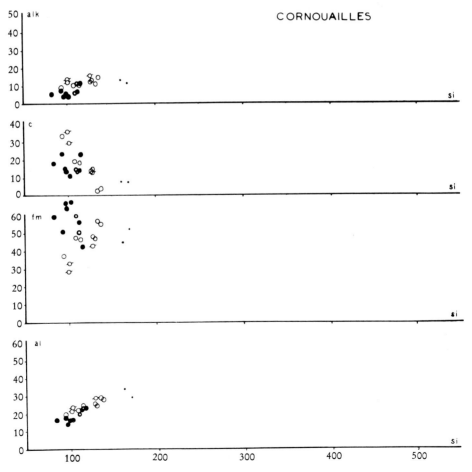

Fig. 4. Série de Cornouailles: représentation des paramètres de NIGGLI

maintes reprises, à une silicification progressive du magma spilitique avant l'épanchement,

- Lahn-Dill: "Si" varie de 70 (pillow-lavas amygdalaires) à 350 (kératophyres); on remarque un "trou" entre les spilites et les kératophyres,

- Cornouailles-Devonshire: c'est la même type que Lahn-Dill, auquel il manque les kératophyres.

b) Triangles Na - K - Ca (Figs. 10 et 11)

Ces triangles schématisent l'opposition entre Schirmeck et les autres massifs. A Schirmeck, les deux rapports Na/K et Na/Ca sont élevés, d'où le groupement général des points à proximité du pôle sodique. La faible teneur générale en calcium de la série de Schirmeck apparaît ici clairement. Les deux autres massifs montrent un rapport Na/K élevé, avec toutefois un certain étalement vers le potassium (kératophyres de la Lahn), et un rapport Na/Ca généralement faible, en particulier dans les pillow-lavas amygdalaires.

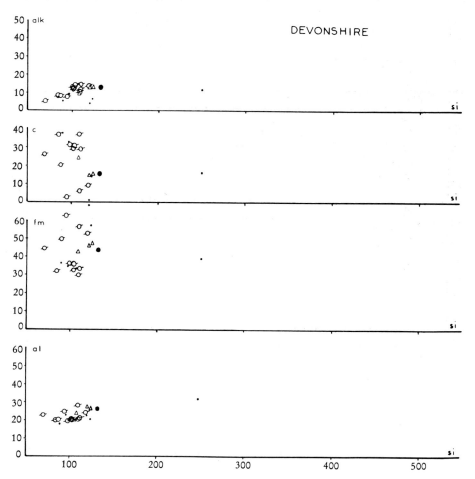

Fig. 5. Série de Devonshire: représentation des paramètres de NIGGLI

c) Diagrammes de calco-alcalinité de JUNG

- La série de Schirmeck (Fig. 9) s'aligne bien, suivant une courbe intermédiaire d'allure plutôt alcaline. L'indice "i" de calco alcalinité est égal à 49,1, ce qui correspond au type alcalin (i < 51).

- Les autres massifs (Fig. 8) ont une distribution plus confuse: les pillow-lavas amygdalaires en particulier perturbent sensiblement l'alignement des points. La courbe moyenne (non tracée sur le diagramme) et l'emplacement des kératophyres indiquent une série plus alcaline encore que celle de Schirmeck. L'indice "i" est de l'ordre de 48.

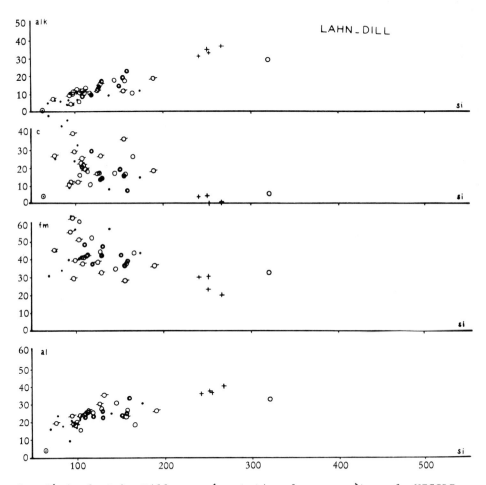

Fig. 6. Série de Lahn-Dill: représentation des paramètres de NIGGLI

296

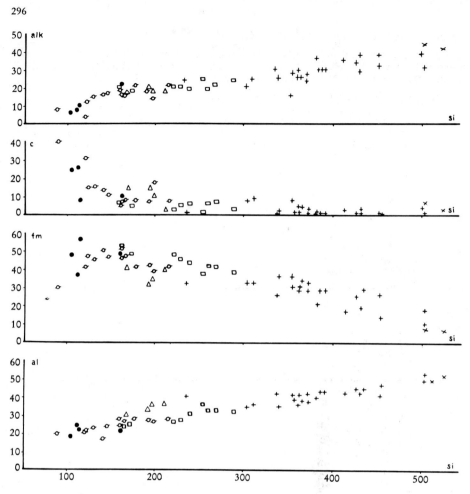

Fig. 7. Série de Schirmeck: représentation des paramètres de NIGGLI

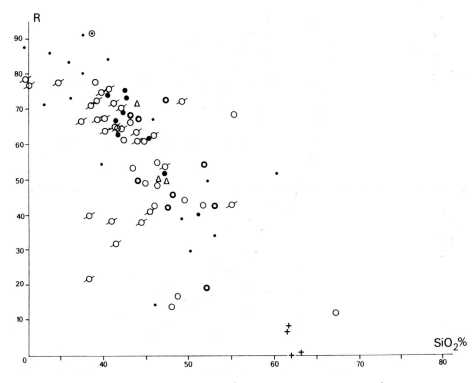

Fig. 8. Diagramme de calco-alcalinité de JUNG, pour les séries de Cornouailles, Devonshire et de Lahn-Dill

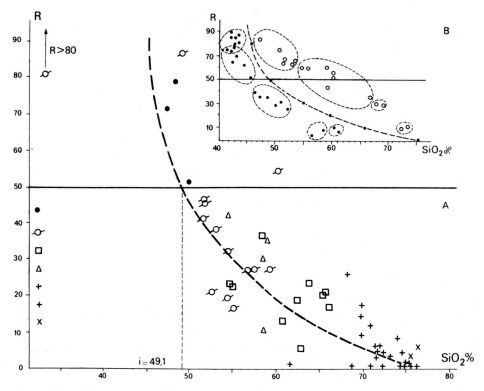

Fig. 9. Diagramme de calco-alcalinité de JUNG, pour la série de Schirmeck

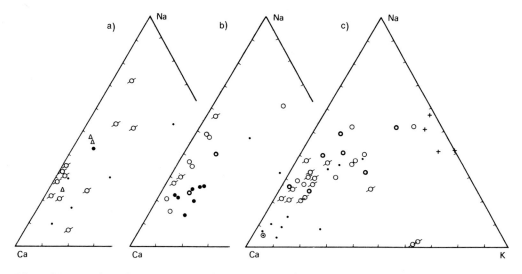

Fig. 10. Triangles Na-K-Ca (en nombre d'atomes) - a) Devonshire; b) Cornouailles; c) Lahn-Dill

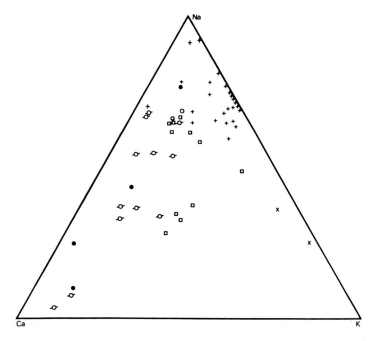

Fig. 11. Triangle Na-K-Ca (en nombre d'atomes) - pour la série de Schirmeck

Tableau 1. Maxima et minima atteints par les éléments majeurs dans les spilites effusives microlitiques

	Spilites Amygdalaires						Spilites non Amygdalaires					
	Lahn-Dill		Cornwall-Devon		Schirmeck		Lahn-Dill		Cornwall-Devon		Schirmeck	
	M	m	M	m	M	m	M	m	M	m	M	m
SiO_2	55.00	34.60	45.80	29.60	56.90	50.10	67.20	43.00	48.60	39.50	60.20	49.10
Al_2O_3	19.40	12.75	18.40	12.80	16.10	15.60	18.00	10.65	18.00	14.10	16.25	13.55
Fe_2O_3	15.50	8.28	17.39	8.95	10.97	10.70	14.25	6.76	13.71	11.51	12.37	10.48
MnO	0.21	0.09	0.30	0.15	0.21	0.19	0.22	0.09	0.21	0.09	0.30	0.16
MgO	12.33	1.97	6.09	2.20	7.36	3.40	11.47	0.87	7.34	4.70	6.69	1.92
CaO	15.46	4.20	20.30	1.97	5.97	2.67	8.33	0.97	13.30	0.70	11.88	1.20
Na_2O	4.32	0.10	5.77	0.95	6.71	4.33	4.61	2.55	4.90	3.34	5.70	0.63
K_2O	7.49	0.31	2.32	0.17	1.20	0.38	2.82	0.46	1.27	0.36	2.81	0.33
TiO_2	3.03	0.60	4.09	2.49	1.45	1.01	2.96	0.68	2.61	1.11	1.48	1.05
P.F.	13.10	5.55	15.11	5.03	3.69	2.72	10.35	2.92	10.73	4.56	3.89	0.93

M: maximum; m: minimum

Tableau 2. Moyennes des analyses chimiques des spilites effusives microlitiques. Comparaison avec des moyennes de basaltes (MANSON, 1967).

	Spilites Amygdalaires			Spilites non Amygdalaires			Basalte Océanique	Basalte Continental
	Lahn-Dill	Cornwall-Devon	Schirmeck	Lahn-Dill	Cornwall-Devon	Schirmeck		
SiO_2	43.36	39.38	53.50	50.17	45.17	54.65	48.5	49.9
Al_2O_3	15.68	15.24	15.85	14.67	15.93	15.14	15.0	16.2
Fe_2O_3	11.14	12.79	10.84	10.72	12.95	11.67	11.6	10.8
MnO	0.15	0.20	0.20	0.15	0.17	0.21	0.17	0.17
MgO	5.39	3.76	5.38	4.89	5.68	4.45	7.2	6.3
CaO	8.40	10.02	5.05	5.35	5.46	4.22	10.5	9.8
Na_2O	3.15	4.18	5.52	3.84	4.11	4.32	2.5	2.8
K_2O	1.95	0.74	0.79	1.50	0.56	1.29	0.8	1.1
TiO_2	2.27	3.08	1.23	2.15	1.91	1.28	2.6	1.6
P.F.	8.80	10.35	3.20	4.90	7.44	2.73	0.8	1.0

Tableau 3. Moyennes des analyses chimiques pour chaque type de kératophyre.

	Schirmeck		Lahn	Schirmeck
	Kératophyres siliceux "dacitiques"		Kératophyres alcalins "trachytiques"	Orthokératophyres
	aphanitiques	porphyriques		
SiO_2	71.79	71.38	62.17	76.00
Al_2O_3	13.43	12.49	15.67	12.55
Fe_2O_3	4.65	5.58	7.79	1.08
MnO	0.06	0.07	0.16	0.03
MgO	0.90	0.98	0.36	0.00
CaO	0.36	0.55	0.43	0.41
Na_2O	4.80	4.64	5.49	2.65
K_2O	1.44	1.26	4.86	5.42
TiO_2	0.34	0.40	0.32	0.10
P.F.	1.78	1.85	2.36	0.93

Tableau 4. Schirmeck - Analyses chimiques et paramètres de NIGGLI des kératophyres aphanitiques

Echantillon N°.	8101	8102	8103	8104	8105	8106	8107	8108
SiO_2	61,30	68,30	68,60	69,80	70,90	70,80	71,80	72,10
Al_2O_3	18,20	13,20	15,20	13,50	12,10	13,60	13,40	13,15
Fe_2O_3	6,76	6,97	4,90	7,05	6,09	4,71	5,95	5,17
MnO	0,20	0,11	0,05	0,09	0,09	0,06	0,10	0,06
MgO	2,46	1,46	1,27	1,22	1,21	1,26	0,83	1,33
CaO	tr.	1,85	-	1,32	0,68	-	0,39	-
Na_2O	5,12	4,88	6,17	2,89	4,43	4,47	4,57	5,61
K_2O	1,78	0,19	0,34	0,19	0,48	1,65	1,77	1,23
TiO_2	0,57	0,58	0,36	0,53	0,42	0,35	0,41	0,93
Perte au Feu	3,34	2,39	2,00	1,99	2,52	2,73	1,61	1,45
Total	99,73	99,93	98,89	98,58	98,92	99,63	100,83	101,03
Paramètres de NIGGLI								
al	42,00	35,00	43,00	41,00	38,00	42,40	40,00	38,90
fm	34,00	34,00	27,10	37,00	34,00	29,00	30,00	29,80
c	0	9,00	0	7,00	4,00	0	2,00	0
alk	24,00	22,00	29,90	15,00	24,00	28,50	28,00	31,30
si	239,00	312,00	329,80	358,00	375,00	374,30	368,00	361,80
mg	0,40	0,29	0,34	0,26	0,28	0,35	0,21	0,34
k	0,17	0,01	0,07	0,04	0,05	0,39	0,19	0,25

Echantillon N°.	8109	8110	8111	8112	8113	8114	8115	8116
SiO_2	72,10	73,40	74,10	74,40	74,70	74,80	75,40	76,20
Al_2O_3	12,85	12,50	13,00	12,40	12,30	13,30	13,25	12,90
Fe_2O_3	2,86	4,44	2,61	3,46	5,70	3,08	2,77	1,97
MnO	0,02	0,02	0,03	0,02	0,04	0,02	0,08	-
MgO	1,08	0,44	0,83	0,44	0,29	-	0,24	-
CaO	0,39	0,24	0,64	-	0,24	-	tr.	-
Na_2O	5,43	4,69	4,78	5,45	4,32	5,07	3,89	4,99
K_2O	2,71	1,77	2,51	2,13	0,88	2,08	1,50	1,83
TiO_2	0,13	0,27	0,22	-	-	0,34	0,11	0,17
Perte au Feu	1,19	1,26	1,25	0,86	1,71	0,57	2,64	1,04
Total	98,76	99,03	99,97	99,31	100,54	99,26	99,88	99,10
Paramètres de NIGGLI								
al	40,30	43,00	43,00	42,40	43,00	47,70	52,00	50,40
fm	20,20	24,00	18,00	19,00	28,00	14,20	17,00	9,80
c	2,30	1,00	4,00	0	1,00	0	0	0
alk	37,40	32,00	35,00	38,70	28,00	38,10	31,00	39,90
si	384,00	425,00	416,00	432,10	438,00	454,80	506,00	505,50
mg	0,42	0,16	0,38	0,20	0,08	0	0,14	0
k	0,49	0,19	0,26	0,41	0,11	0,42	0,20	0,39

Tableau 5. Schirmeck - Analyses chimiques et paramètres de NIGGLI des kératophyres porphyriques quartziques et des orthokératophyres

Echantillon N°.	Kératophyres porphyriques quartziques							Orthokératophyres	
	8201	8202	8203	8204	8205	8206	8207	8208	8209
SiO_2	69,80	69,80	70,00	71,60	71,60	71,80	75,10	75,60	76,40
Al_2O_3	12,10	12,40	12,40	12,90	13,05	13,20	11,40	12,50	12,60
Fe_2O_3	7,05	5,91	5,75	5,80	5,78	4,81	4,00	0,98	1,19
MnO	0,11	0,09	0,07	0,08	0,03	0,09	0,05	0,03	0,02
MgO	1,37	0,92	0,97	1,03	0,63	1,08	0,88	-	-
CaO	0,63	1,02	1,17	0,49	0,39	0,19	-	0,29	0,54
Na_2O	4,21	4,74	4,24	4,57	4,71	5,35	4,69	2,12	3,17
K_2O	1,66	1,17	1,17	2,16	1,17	0,24	1,28	5,93	4,91
TiO_2	0,50	0,45	0,46	0,42	0,40	0,38	0,17	0,12	0,09
Perte au Feu	1,97	2,25	2,45	1,60	1,87	1,69	1,13	1,08	0,79
Total	99,40	98,75	98,68	100,65	99,63	98,83	98,70	98,65	99,71

Paramètres de NIGGLI

al	34,90	37,20	37,90	38,10	41,00	42,00	40,90	51,50	49,10
fm	36,60	30,00	31,00	30,00	29,00	28,00	26,50	5,30	6,00
c	3,30	5,60	6,50	2,60	2,00	1,00	0	2,20	3,80
alk	25,20	27,20	25,20	29,20	28,00	29,00	32,60	40,90	41,10
si	342,50	355,30	363,20	359,70	384,00	387,00	456,20	529,60	505,80
mg	0,28	0,23	0,25	0,26	0,17	0,29	0,30	0	0
k	0,41	0,28	0,17	0,48	0,13	0,02	0,31	1,30	1,00

Tableau 6. Schirmeck - Analyses chimiques et paramètres de NIGGLI des diabases grenues quartziques et des diabases microgrenues porphyriques

Echantillon N°.	Diabases grenues quartziques								Diabases microgrenues porphyriques			
	8210	8211	8212	8213	8214	8215	8216	8217	8218	8219	8220	8221
SiO_2	54,80	58,70	60,80	62,40	62,90	63,90	65,60	66,10	54,70	58,60	58,80	59,00
Al_2O_3	13,80	15,25	12,80	13,10	14,80	13,65	13,60	13,05	16,57	18,10	17,30	17,50
Fe_2O_3	>13,00	7,05	9,71	9,99	9,44	10,04	9,04	8,04	3,79	7,08	6,27	6,76
FeO									3,87			
MnO	0,18	0,08	0,23	0,18	0,09	0,21	0,15	0,14	0,13	0,15	0,10	0,16
MgO	3,58	3,11	3,84	3,12	1,36	2,43	2,38	1,81	4,80	3,10	3,85	3,16
CaO	1,84	3,11	1,00	1,36	0,38	1,70	1,45	1,27	4,30	2,90	0,76	3,89
Na_2O	5,28	2,92	4,90	4,92	5,83	4,53	4,57	4,36	3,04	3,54	3,57	3,65
K_2O	0,92	2,43	1,89	0,97	0,68	0,97	0,82	1,71	2,89	2,95	2,94	3,26
TiO_2	-	1,35	1,01	1,04	1,26	0,93	-	0,75	0,89	0,73	0,85	0,76
Perte au Feu	3,10	2,57	3,84	2,45	2,68	2,51	2,71	1,94	5,12	3,01	4,85	2,60
Total	100,67	100,67	99,52	99,53	99,42	100,87	101,21	99,17	100,10	100,16	99,24	100,74
Paramètres de NIGGLI												
al	26,00	33,60	27,40	28,70	35,60	30,60	32,00	33,80	30,00	36,30	37,10	34,00
fm	49,00	37,30	48,30	45,80	37,60	43,30	41,90	39,10	41,00	34,50	40,70	33,00
c	6,00	12,50	3,90	5,40	1,70	6,90	6,20	3,50	14,00	10,70	3,00	14,00
alk	19,00	16,40	20,20	20,00	24,90	19,10	19,80	23,40	15,00	18,10	19,30	19,00
si	177,00	219,80	221,40	231,50	256,30	243,40	261,80	291,50	169,00	199,90	214,20	197,00
mg	0,35	0,47	0,43	0,38	0,22	0,32	0,34	0,31	0,54	0,46	0,52	0,47
k	0,10	0,71	0,29	0,23	0,14	0,25	0,21	0,41	0,37	0,71	0,71	0,36

Tableau 7. Schirmeck - Analyses chimiques et paramètres de NIGGLI des spilites microlitiques non amygdalaires

Echantillon N°.	Spilites microlitiques non amygdalaires										
	8301	8302	8303	8304	8305	8306	8307	8308	8309	8310	8311
SiO_2	32,90	49,10	51,40	51,80	53,20	54,30	54,50	55,10	57,50	59,40	60,20
Al_2O_3	12,10	15,80	16,25	15,80	14,75	16,15	15,20	15,30	14,50	13,55	14,10
Fe_2O_3	5,85	12,37	10,48	12,16	11,37	12,37	12,11	12,09	11,78	10,86	11,09
MnO	0,16	0,30	0,20	0,21	0,26	0,17	0,17	0,21	0,19	0,22	0,16
MgO	4,49	4,67	6,69	5,86	5,70	4,83	4,13	5,87	2,58	1,92	2,27
CaO	>13,00	11,88	4,99	5,66	4,01	1,59	2,77	1,20	2,38	2,30	5,44
Na_2O	1,32	0,63	4,17	4,48	5,70	5,26	4,28	5,01	4,72	5,67	3,31
K_2O	1,88	1,12	2,81	1,92	0,67	1,11	1,36	0,91	1,46	0,33	1,23
TiO_2	0,68	1,13	1,32	1,33	1,45	1,35	1,48	1,23	1,24	1,05	1,27
Perte au Feu	19,76	2,56	2,95	1,48	3,23	3,32	2,70	3,62	2,58	3,89	0,93
Total		99,56	101,26	100,70	100,34	100,45	98,70	100,54	98,93	99,19	100,00

Paramètres de NIGGLI

al	21,00	23,00	18,30	24,00	23,90	28,20	27,60	26,50	28,80	29,11	27,70
fm	32,00	42,00	50,30	46,00	47,50	49,50	47,60	53,30	43,60	40,90	39,60
c	40,00	32,00	15,00	16,00	11,80	5,07	9,20	3,80	8,70	9,00	19,40
alk	7,00	3,00	16,20	14,00	16,40	17,30	15,40	16,10	18,60	20,90	13,30
si	95,00	124,00	142,90	133,00	146,00	160,90	167,80	162,30	194,30	216,40	200,20
mg	0,60	0,42	0,56	0,48	0,49	0,43	0,40	0,49	0,30	0,26	0,29
k	0,46	0,54	0,62	0,21	0,14	0,24	0,35	0,21	0,34	0,07	0,39

Tableau 8. Schirmeck - Analyses chimiques et paramètres de NIGGLI des dolérites et diabases albitiques, d'une ophite et des spilites amygdalaires

	Dolérite s.s.	Dolérite albitique	Diabase albitique	Ophite	Spilites amygdalaires	
Echantillon N°.	8312	8313	8314	8315	8316	8317
SiO_2	47,30	48,00	52,60	48,30	50,10	56,90
Al_2O_3	18,30	16,25	16,80	15,80	16,10	15,60
Fe_2O_3	9,43	9,94	11,32	9,51	10,97	10,70
MnO	0,19	0,20	0,20	0,22	0,21	0,19
MgO	5,77	11,06	5,06	9,80	7,36	3,40
CaO	10,39	3,50	2,02	10,69	5,97	2,67
Na_2O	2,88	3,45	6,36	1,42	4,33	6,71
K_2O	1,25	0,84	1,06	1,37	1,20	0,38
TiO_2	0,96	-	-	0,82	1,01	1,45
Perte au Feu	3,77	6,66	3,60	1,91	3,69	2,72
Total	100,14	100,73	100,08	99,84	100,94	100,72
Paramètres de NIGGLI						
al	25,88	23,10	21,36	20,44	23,50	28,60
fm	38,40	58,47	50,60	48,44	48,00	41,40
c	27,00	9,07	6,72	25,26	16,00	8,90
alk	8,56	9,36	21,22	5,61	12,20	21,00
si	111,10	115,80	162,75	106,13	124,30	177,10
mg	0,54	0,69	0,46	0,66	0,57	0,38
k	0,45	0,28	0,20	0,92	0,31	0,07

Tableau 9. Schirmeck - Analyses chimiques des brèches siliceuses volcano-sédimentaires. Comparaison avec les kératophyres quartziques (N°. 8203)

Echantillon N°.	9101	9102	9103	9104	9105	9106	9107	9108	9109	9110	9111	8203
SiO_2	52,90	53,80	59,20	65,00	66,20	67,30	67,50	68,00	68,70	69,50	70,00	70,00
Al_2O_3	14,10	14,80	13,90	15,70	16,80	11,75	12,20	14,90	13,70	13,80	12,65	12,40
Fe_2O_3	13,67	7,72	10,89	8,68	4,84	8,97	6,02	5,24	7,30	6,90	6,15	5,75
MnO	0,24	0,09	0,19	0,10	0,04	0,08	0,11	0,05	0,10	0,07	0,04	0,07
MgO	4,85	9,59	3,71	1,21	2,22	3,15	1,43	1,13	1,91	0,58	1,09	0,97
CaO	6,53	6,49	2,32	1,12	2,07	1,72	2,96	2,17	-	1,02	1,63	1,17
Na_2O	4,95	1,96	4,01	2,87	3,16	2,02	3,53	4,45	5,15	5,77	4,01	4,24
K_2O	0,89	2,80	2,87	1,85	4,44	2,86	1,33	1,58	0,98	0,34	2,08	1,17
TiO_2	1,29	0,54	1,09	0,50	0,36	0,72	0,45	0,27	0,54	0,52	0,40	0,46
Perte au Feu	0,94	1,54	0,98	2,44	1,18	1,32	0,86	1,07	1,88	2,05	0,78	2,45
Total	100,36	99,33	99,16	99,47	101,31	99,89	100,52	98,86	100,26	100,55	98,83	98,68

Echantillon N°.	9113	9114	9115	9116	9117	9118	9119	9120	9121	9122	9123	8203
SiO_2	70,70	70,80	72,90	73,40	73,40	74,10	75,50	76,80	76,90	77,40	>79,00	70,00
Al_2O_3	13,00	14,35	13,70	12,60	12,25	11,25	13,40	10,60	11,00	<10,00	<10,00	12,40
Fe_2O_3	5,20	3,25	5,05	3,96	3,53	4,59	2,28	4,60	3,40	4,58	3,46	5,75
MnO	0,09	0,04	0,09	0,08	0,04	0,03	0,03	0,05	0,04	0,05	0,05	0,07
MgO	1,47	1,28	1,09	0,54	0,98	1,43	0,59	0,93	0,78	0,68	0,89	0,97
CaO	0,73	0,98	0,79	0,49	0,39	1,67	tr.	0,84	1,57	-	0,89	1,17
Na_2O	3,73	4,93	4,32	4,46	4,21	4,74	4,95	3,90	1,77	3,99	1,48	4,24
K_2O	2,45	2,02	1,93	2,27	2,05	1,13	2,77	1,48	3,00	0,19	2,17	1,17
TiO_2	0,45	0,25	0,29	0,13	0,21	-	0,14	-	0,19	0,20	0,32	0,46
Perte au Feu	1,70	1,33	0,90	0,87	1,93	1,23		1,06	1,44	1,44	1,05	2,45
Total	99,52	99,23	101,06	98,80	98,99	100,45		100,56	100,09			98,68

Tableau 10. Schirmeck - Analyses chimiques des tufs kératophyriques.
Comparaison avec les kératophyres quartziques (N°. 8203)

Echantillon N°.	9201	9202	9203	9204	9205	9206	8203
SiO_2	66,70	71,80	73,50	74,20	76,80	77,70	70,00
Al_2O_3	13,70	12,30	11,60	12,30	<10,00	9,85	12,40
Fe_2O_3	7,62	6,82	5,41	3,63	7,03	4,48	5,75
MnO	0,09	0,03	0,02	0,03	0,03	0,06	0,07
MgO	2,00	0,78	0,49	0,98	0,83	0,59	0,97
CaO	0,73	-	0,44	0,19	tr.	0,09	1,17
Na_2O	5,57	4,38	4,53	5,05	2,65	3,94	4,24
K_2O	0,53	0,73	1,08	0,49	1,23	0,49	1,17
TiO_2	0,56	0,36	0,40	0,14	0,68	0,14	0,46
Perte au Feu	2,21	2,46	1,51	1,87	1,57	1,43	2,45
Total	99,71	99,66	98,98	98,88		98,77	98,68

Tableau 11. Schirmeck - Analyses chimiques des schalsteins. Comparaison avec les spilites amygdalaires (N°. 8316)

Echantillon N°.	9301	9302	9303	9304	9305	9306	9307	9308	9309	8316
SiO_2	51,10	51,80	52,60	54,70	54,90	57,70	60,00	61,00	61,10	50,10
Al_2O_3	16,35	16,10	15,10	14,20	14,50	13,95	14,50	15,30	14,30	16,10
Fe_2O_3	10,14	11,39	12,10	11,64	10,49	11,41	10,81	10,14	10,14	10,97
MnO	0,22	0,17	0,18	0,18	0,20	0,15	0,11	0,16	0,22	0,21
MgO	7,61	5,26	5,76	4,03	5,10	3,78	4,26	2,54	2,57	7,36
CaO	7,01	1,56	1,82	3,99	3,33	3,78	1,74	1,86	3,54	5,97
Na_2O	3,61	4,60	5,28	3,89	5,39	4,31	4,26	5,19	1,89	4,33
K_2O	1,65	1,47	1,05	1,75	0,85	0,86	1,40	0,78	3,15	1,20
TiO_2	0,87	–	1,24	1,21	1,11	1,07	1,10	–	0,95	1,01
Perte au Feu	2,41	5,07	3,95	4,95	4,58	4,11	2,97	1,99	2,95	3,69
Total	101,03	98,93	99,08	100,54	100,45	101,12	101,15	100,08	100,81	100,94

Tableau 12. Lahn-Dill - Pillow-lavas spilitiques amygdalaires

	LH4b(1)	LH9b	LH9c	LH19(1)	LH33	DL1(a)	DL1(b)
SiO_2	49,40	38,40	40,90	55,00	42,10	40,80	41,10
Al_2O_3	12,75	16,80	16,25	12,80	14,30	14,20	14,70
Fe_2O_3	8,28	15,50	14,74	9,66	12,60	9,89	10,67
MnO	0,12	0,12	0,11	0,18	0,19	0,11	0,09
MgO	1,97	7,53	6,23	2,18	12,33	3,37	5,62
CaO	10,72	4,20	4,71	5,06	5,34	15,46	11,38
Na_2O	3,00	0,10	0,23	3,11	1,66	4,22	4,02
K_2O	1,30	6,24	7,49	3,71	0,69	0,94	0,58
TiO_2	2,85	3,03	2,92	0,60	1,89	1,34	1,16
CO_2	7,95			5,28	1,22		
H_2O	3,23			2,40	7,54		
PF		7,55	5,78			10,06	10,71
Total	101,57	99,47	99,36	99,98	99,86	100,39	100,03
Paramètres de NIGGLI							
al	23,44	23,82	23,83	25,96	18,99	19,80	20,78
fm	28,97	55,49	51,10	36,78	63,46	29,79	39,67
c	35,90	10,84	12,58	18,69	12,91	39,28	29,30
alk	11,67	9,83	12,47	18,54	4,62	11,10	10,23
si	154,44	92,56	101,99	189,68	95,06	96,75	98,77
mg	0,31	0,49	0,45	0,30	0,65	0,40	0,51
k	0,22	0,97	0,95	0,44	0,21	0,12	0,08

Tableau 12 (suite)

	LH24(a)	LH24(b)	LH24(c)	LH24(d)	LH24(e)	LH28(1)	LH28(2)
SiO_2	44,00	34,60	42,30	43,90	47,20	45,80	41,50
Al_2O_3	17,70	15,00	16,80	17,20	19,40	15,80	15,85
Fe_2O_3	10,80	13,40	10,30	10,08	9,44	9,16	11,48
MnO	0,12	0,17	0,17	0,15	0,12	0,20	0,21
MgO	5,73	7,05	5,72	5,87	4,96	2,98	3,91
CaO	7,19	11,86	8,46	8,32	6,19	9,26	9,47
Na_2O	4,32	3,17	3,82	4,07	4,20	4,31	3,83
K_2O	0,37	0,31	0,87	0,83	1,23	1,37	1,41
TiO_2	2,64	1,96	2,48	2,63	2,41	3,02	2,92
CO_2	1,30	6,82	3,35	2,69	0,20	5,05	4,61
H_2O PF	5,71	6,28	5,54	5,61	5,35	3,96	4,96
Total	99,88	100,62	99,81	101,37	100,70	100,91	100,15
Paramètres de NIGGLI							
al	26,47	19,36	24,92	25,29	30,41	25,98	24,07
fm	42,70	45,58	41,48	41,23	38,97	32,17	37,83
c	19,58	27,88	22,86	22,29	17,67	27,73	26,19
alk	11,23	7,16	10,72	11,17	12,92	14,10	11,89
si	111,87	75,93	106,69	109,77	125,81	128,04	107,15
mg	0,51	0,50	0,52	0,53	0,50	0,38	0,40
k	0,05	0,06	0,13	0,11	0,16	0,17	0,19

LH-4b(1): pillows assez écrasés près de Weilburg (route Weilburg-Schnitten) - LH-9: pillows, embranchement routes Essershausen et Laimbach; b: écorce; c: coeur - LH-19(1): spilite amygdalaire près de Philippstein (vallée de la Mott) - LH-33: spilite amygdalaire de Wetzlar (au-dessus de Naunheim); DL-1: entrée de Herborn, vallée de la Dill; a: spilite amygdalaire massive; b: pillow - LH-24(a,b,c,d,e): pillows porphyriques de Löhnberg - Hütte (route de Niedershausen) - LH-28 (1 et 2): spilite amygdalaire assez écrasée, 1 km au Nord de Weinbach (route de Weinbach à la vallée de Essershausen).

Tableau 13. Lahn-Dill - Spilites microlitiques intersertales non amygdalaires

	LH6	LH14(2)	LH18	LH19(2)	LH19(3)	LH20	LH31(1)	DL2
SiO_2	49,50	67,20	46,20	45,60	43,30	51,50	55,10	43,00
Al_2O_3	18,00	11,70	16,30	16,70	16,90	15,00	10,65	12,10
Fe_2O_3	9,33	7,30	13,47	14,25	11,80	12,30	6,76	10,57
MnO	0,14	0,09	0,16	0,18	0,22	0,21	0,10	0,13
MgO	3,09	0,87	3,87	6,37	4,84	2,26	6,34	11,47
CaO	5,52	0,97	5,81	3,99	6,82	5,09	8,33	6,27
Na_2O	4,61	4,57	3,69	3,42	3,90	4,52	3,47	2,55
K_2O	2,42	2,82	1,24	2,04	2,07	2,35	0,46	0,62
TiO_2	2,95	0,68	2,28	2,90	2,96	1,66	1,62	2,24
CO_2		1,12	3,77	0,49	1,48		3,94	
H_2O		1,80	4,80	5,14	3,93		4,09	
PF	4,76					3,84		10,35
Total	100,32	99,12	101,59	101,08	98,22	98,73	100,86	
Paramètres de NIGGLI								
al	30,90	32,77	26,47	25,12	25,72	26,78	18,69	16,97
fm	34,29	32,64	44,29	52,15	42,17	38,83	43,76	60,18
c	17,26	4,94	17,18	10,93	18,91	16,55	26,63	16,01
alk	17,53	29,63	12,04	11,79	13,18	17,83	10,89	6,82
si	144,48	319,99	127,55	116,60	112,06	156,32	164,44	102,52
mg	0,39	0,19	0,36	0,46	0,44	0,26	0,64	0,68
k	0,25	0,28	0,18	0,28	0,25	0,25	0,08	0,13

LH-6: diabase fine de Freihenfels (route de Essershausen) - LH-14(1): diabase fine de Wirbelau - LH-18: diabase fine, carrière entre Ernsthausen et Altenkirchen, juste après la vallée de la Mott - LH-19(2): diabase fine, carrière 500 m avant Philippstein (vallée de la Mott) - LH-20: diabase fine de Niederbiel (route de Leun à Niederbiel) - LH-31(1): diabase fine de Kraftsolms (route de Kraftsolms à Kröffelbach) - DL-2: diabase de Dillenburg, vallée de la Dill (derrière dépôt locomotives).

Tableau 14. Lahn-Dill

	Diabases spilitiques microgrenue et grenues à pyroxène - picrite							Diabase microgrenue porphyrique altérée
	LH1	LH11	LH12(b)	LH12(d)	DL7	DL7(1)	DL6	LH14(l)
SiO$_2$	47,80	44,40	52,00	53,10	47,77	48,40	38,15	52,00
Al$_2$O$_3$	16,00	15,50	13,90	15,20	15,75	14,60	4,86	17,90
Fe$_2$O$_3$	10,61	14,60	12,55	11,90	10,02	9,14	14,94	11,73
MnO	0,14	0,18	0,20	0,18	0,25	0,10	0,17	0,10
MgO	4,60	5,93	3,47	2,63	5,05	7,41	28,45	2,55
CaO	11,19	7,86	6,13	5,27	5,40	5,16	2,37	2,02
Na$_2$O	3,90	2,84	4,54	6,24	4,30	4,69	0,14	5,67
K$_2$O	0,48	1,20	0,82	1,13	3,29	1,54	0,08	2,88
TiO$_2$	1,88	3,21	2,01	2,04	2,40	1,97	0,58	1,61
CO$_2$			1,07	0,65	0,17	1,72	0,21	1,26
H$_2$O			2,98	2,58	5,35	5,02	10,38	3,30
PF	3,50	3,51						
Total	100,10	99,23	99,67	100,92	99,75	99,75	100,33	101,02
Paramètres de NIGGLI								
al	23,26	22,21	23,73	26,01	25,21	22,78	4,78	32,17
fm	37,01	48,71	42,92	37,88	42,00	47,89	90,64	38,82
c	29,63	20,51	19,06	16,42	15,74	14,66	4,25	6,61
alk	10,08	8,56	14,27	19,66	17,03	14,64	0,31	22,38
si	118,14	108,15	150,95	154,49	129,99	128,40	63,87	158,89
mg	0,46	0,44	0,35	0,30	0,49	0,61	0,78	0,30
k	0,07	0,21	0,10	0,10	0,33	0,17	0,27	0,25

LH-1: diabase de Weilburg - LH-11: diabase prismée de Leun (route de Weilburg) - LH-12(b,d): carrières de Heisterberg, diabase fine - DL-7 et 7(l): diabase prismée d'Eyershausen (vallée de la Dill) - DL-6: péridotite feldspathique (wehrlite), carrière de Schwartzstein, vallée de la Dill - LH-14(l): diabase microgrenue porphyrique altérée de Wirbelau.

Tableau 15. Lahn-Dill - Kératophyres

	LH5(1)	LH5(2)	LH15'(a)	LH15'(b)
SiO_2	61,70	63,30	62,10	61,60
Al_2O_3	15,70	16,20	15,20	15,60
Fe_2O_3	5,30	5,75	10,05	10,06
MnO	0,19	0,22	0,04	0,18
MgO	1,01	0,43	Traces	Traces
CaO	0,92	0,07	Traces	0,74
Na_2O	6,07	6,53	4,58	4,78
K_2O	4,34	4,09	5,76	5,27
TiO_2	0,19	0,21	0,39	0,49
CO_2	2,74	0,99	0,26	0,39
H_2O PF	1,25	1,67	1,17	0,98
Total	99,41	99,46	99,55	100,09
Paramètres de NIGGLI				
al	37,67	40,24	36,31	35,76
fm	23,04	21,72	30,75	30,00
c	4,02	0,31	0,00	3,09
alk	35,26	37,71	32,93	31,14
si	251,67	267,33	252,22	240,09
mg	0,26	0,12	0,00	0,00
k	0,32	0,29	0,45	0,42

LH-5(1 et 2): kératophyres de Weilburg, faciès porphyrique; LH-15'(a et b): kératophyres de Wirbelau, faciès aphanitique.

Tableau 16. Lahn-Dill - Schalsteins

	LH3	LH3(2)	LH7(a)	LH7(b)	LH10(1)	LH10(2)	LH16(1)	LH21(a)	LH21(b)
SiO_2	33,10	30,70	37,70	40,70	39,90	49,10	52,30	33,60	37,80
Al_2O_3	16,90	11,90	6,73	11,35	15,30	15,05	15,60	11,90	13,35
Fe_2O_3	13,90	8,93	8,86	12,92	16,48	12,12	10,33	11,64	13,72
MnO	0,11	0,30	0,16	0,13	0,18	0,12	0,22	0,31	0,24
MgO	5,63	4,15	6,44	8,68	10,23	7,47	3,40	3,07	2,91
CaO	10,03	20,16	18,01	9,16	3,51	2,63	4,09	16,38	11,15
Na_2O	1,10	0,90	1,23	0,87	2,17	2,63	2,93	1,87	1,71
K_2O	2,94	1,93	0,45	0,87	0,79	1,50	1,24	0,79	1,07
TiO_2	3,39	2,75	2,02	3,01	3,70	3,43	1,22	2,76	3,26
CO_2	6,98	14,76			0,53	0,29	4,83		
H_2O	5,86	4,19			7,66	5,98	3,98		
PF			17,94	12,68				16,85	14,19
Total	99,94	100,67	99,54	100,37	100,45	100,32	100,14	99,17	99,30

Paramètres de NIGGLI

	LH3	LH3(2)	LH7(a)	LH7(b)	LH10(1)	LH10(2)	LH16(1)	LH21(a)	LH21(b)
al	23,34	15,95	9,61	16,40	20,82	24,88	30,36	17,30	21,22
fm	44,52	30,02	39,92	56,05	64,45	57,34	43,12	33,60	40,16
c	25,23	49,22	46,86	24,10	8,70	7,92	14,50	43,37	32,29
alk	6,90	4,79	3,58	3,43	6,02	9,84	12,00	5,71	6,31
si	77,71	69,96	91,56	99,98	92,31	138,03	173,06	83,04	102,17
mg	0,44	0,47	0,58	0,57	0,55	0,54	0,39	0,33	0,29
k	0,63	0,58	0,19	0,39	0,19	0,27	0,21	0,21	0,29

LH-3 et 3(2): schalsteins au lieu-dit "Am Gänsberg", route Weilburg-Schnitten; LH-7 (a et b): schalsteins, carrière à l'entrée de Essershausen; LH-10 (1 et 2): schalsteins, sortie de Braunfels, route de Leun (carrière); LH-16(1): schalsteins, carrière après le pont sur la Lahn menant à Gröveneck; LH-21 (a et b): schalsteins, carrière à la sortie de Oberbiel.

Tableau 17. Cornouailles-Devonshire. Spilites microlitiques et amygdalaires

	D7(1)	D7(m)	D8	D9	D2	D3	D4(1)	D5	D6	D15	D15(2)	C21	C21(2)	C22
SiO$_2$	42,10	38,40	39,70	40,40	41,70	37,50	31,40	40,50	44,60	40,20	29,60	39,50	38,90	45,80
Al$_2$O$_3$	14,35	17,80	13,70	14,90	18,40	15,45	13,25	14,20	15,95	12,80	17,50	15,60	14,50	17,50
Fe$_2$O$_3$	10,76	21,18	12,26	12,02	16,60	16,07	9,84	12,47	16,80	10,55	17,39	10,90	8,95	11,72
MnO	0,25	0,20	0,27	0,24	0,19	0,18	0,18	0,15	0,16	0,19	0,30	0,15	0,16	0,17
MgO	3,00	6,32	3,52	3,72	6,05	6,09	3,01	2,32	4,79	2,20	4,08	3,19	2,66	4,21
CaO	10,49	1,07	12,00	10,77	1,97	8,12	20,30	11,41	3,22	13,01	10,65	10,86	13,18	4,25
Na$_2$O	5,56	3,51	4,58	5,11	3,18	3,57	3,60	4,23	4,51	3,58	0,95	4,79	4,97	5,77
K$_2$O	0,22	0,32	0,17	0,18	1,03	0,32	0,30	2,32	0,85	0,90	1,95	0,58	0,48	0,28
TiO$_2$	2,82	4,21	2,74	3,09	3,80	2,98	3,01	3,14	3,46	3,02	4,09	2,63	2,49	2,77
CO$_2$	8,27	0,34	9,01	0,82	0,20	5,59	12,77	6,86	1,28	11,78	8,28			
H$_2$O	2,05	5,90	2,76	9,42	5,99	4,09	2,34	2,15	3,75	1,74	4,74			
PF												11,15	12,65	6,96
Total	99,87	99,25	100,71	100,67	99,11	99,96	100,00	99,75	99,37	99,97	99,53	99,35	98,94	99,37

Paramètres de NIGGLI

	D7(1)	D7(m)	D8	D9	D2	D3	D4(1)	D5	D6	D15	D15(2)	C21	C21(2)	C22
al	22,22	25,69	20,06	21,82	28,21	21,24	20,81	21,35	24,91	20,41	23,78	23,58	22,09	28,70
fm	33,65	62,65	36,61	36,84	56,53	49,87	32,17	33,13	52,89	30,83	44,85	33,63	28,07	42,52
c	29,59	2,81	32,01	28,73	5,50	20,33	37,19	31,25	9,16	37,79	26,36	29,90	36,57	12,69
alk	14,53	8,83	11,30	12,59	9,73	8,55	9,81	14,25	13,03	10,95	4,99	12,86	13,25	16,07
si	110,83	94,22	98,85	100,59	108,72	87,64	83,85	103,54	118,42	109,00	68,38	101,53	100,76	127,72
mg	0,35	0,37	0,35	0,37	0,41	0,42	0,37	0,26	0,36	0,29	0,31	0,36	0,36	0,41
k	0,02	0,05	0,02	0,02	0,17	0,05	0,05	0,26	0,11	0,14	0,57	0,07	0,05	0,03

D-7, 8 et 9: pillows de Chipley (Devon); D7(m): matrice; D-2, 3 et 4(1): spilite amygdalaire massive, carrières de Landrake (Cornwall); D-5 et 6: spilite amygdalaire, Ingsson Hill (NE de Bickington, Devon); D-15 et D-15(2): spilite amygdalaire, Austin's Bridge près de Buckfastleigh, route de Totnès; C-21, 21(2) et 22: pillows porphyriques assez déformés, Pentire Head, Cornwall.

Tableau 18. Cornouailles-Devonshire - Spilites microlitiques intersertales non amygdalaires

	C9(2)	C10	C20	C20(m)	C23	C24	C25
SiO_2	48,00	39,50	48,60	46,60	45,30	42,90	45,30
Al_2O_3	18,00	14,10	16,65	15,10	16,50	15,70	15,45
Fe_2O_3	12,80	11,51	13,71	12,35	13,33	13,44	13,53
MnO	0,09	0,17	0,12	0,21	0,20	0,21	0,21
MgO	7,34	4,77	6,22	5,15	5,58	5,98	4,70
CaO	0,70	13,30	1,03	4,75	6,94	7,04	4,43
Na_2O	3,34	3,48	4,90	4,61	4,00	4,12	4,34
K_2O	1,27	0,49	0,42	0,45	0,52	0,41	0,36
TiO_2	1,62	1,11	1,49	1,50	2,58	2,44	2,61
PF	5,83	10,73	5,61	9,49	4,56	7,28	8,56
Total	98,99	99,16	98,75	100,22	99,51	99,52	99,49
Paramètres de NIGGLI							
al	29,35	19,67	27,49	24,75	24,33	22,94	25,50
fm	57,35	37,79	55,34	47,83	46,47	47,76	48,75
c	2,07	33,80	3,09	14,18	18,64	18,73	13,31
alk	11,20	8,73	14,06	13,23	10,53	10,55	12,42
si	133,08	93,70	136,43	129,85	113,58	106,57	127,10
mg	0,53	0,44	0,47	0,45	0,45	0,46	0,40
k	0,20	0,08	0,05	0,06	0,07	0,06	0,05

C-9(2): spilite de Trevose Head; C-10: idem avec augite; C-20, C-20(m): spilites de Wadebridge, Cornwall; C-23, 24, 25: diabases de "The Rumps", Pentire Head, à reliques d'augite.

Tableau 19. Cornouailles-Devonshire - Dolérites et diabases grenues

	D13	COO	C13	C14(1)	C15(1)	C15(2)	C18(a)	C19(1)	C19(2)	C19(3)
SiO_2	47,50	44,40	43,60	45,80	40,90	43,10	43,00	41,90	42,80	42,10
Al_2O_3	16,50	16,00	13,25	15,54	13,40	13,75	10,10	11,80	11,75	14,40
Fe_2O_3	13,50	15,73	12,97	11,63	15,40	14,33	14,16	13,08	14,39	13,22
MnO	0,18	0,30	0,17	0,17	0,21	0,23	0,17	0,16	0,16	0,22
MgO	3,82	5,26	9,59	5,29	11,67	8,57	>13,00	11,58	11,84	7,76
CaO	5,71	5,16	5,38	8,48	8,32	10,17	6,32	5,48	4,55	4,98
Na_2O	4,74	4,15	1,87	3,57	2,48	3,00	1,13	2,00	1,32	2,05
K_2O	0,72	1,06	0,73	1,81	0,52	0,82	1,03	0,80	0,82	0,98
TiO_2	2,95	3,86	2,37	2,67	2,53	3,06	1,81	2,23	2,15	2,76
CO_2	0,20									
H_2O	2,91									
PF		3,44	8,65	4,61	4,29	3,11	5,60	10,85	8,88	10,50
Total	98,73	99,36	98,58	99,57	99,72	100,14		99,88	98,66	98,97

Paramètres de NIGGLI

al	26,31	23,78	19,44	23,06	16,16	17,50	13,28	16,30	16,34	22,19
fm	43,40	50,38	60,50	42,38	59,95	51,49	67,66	64,15	67,85	56,97
c	16,58	13,97	14,37	22,92	18,28	23,58	15,13	13,79	11,53	13,98
alk	13,68	11,85	5,67	11,63	5,60	7,41	3,91	5,74	4,25	6,83
si	128,80	112,19	108,75	115,54	83,87	93,27	96,13	98,41	101,24	110,33
mg	0,35	0,39	0,59	0,47	0,59	0,54	0,64	0,63	0,61	0,53
k	0,09	0,14	0,14	0,25	0,12	0,15	0,37	0,20	0,29	0,23

D-13: dolérite de Blackhead (Newton Abbot, Devon), à augite poécilitique; C-OO: diabase à actinote de Penzance (Cornwall); C-13: diabase de Padstow; C-14(1), 15(1), 15(2), 18(a): dolérite de Padstow, à augite, hornblende brune, chlorite et calcite; C-19(1): dolérite de Padstow, bordure à chlorite-biotite-calcite.

Tableau 20. Cornouailles-Devonshire - Microgabbros porphyriques à pyroxène

	D11(1)	D11(2)	D12
SiO_2	44,30	47,80	46,70
Al_2O_3	17,20	17,50	17,80
Fe_2O_3	11,90	11,84	12,10
MnO	0,17	0,31	0,21
MgO	5,61	5,82	5,67
CaO	9,52	5,10	5,20
Na_2O	3,33	4,77	4,71
K_2O	0,57	0,42	0,52
TiO_2	2,58	2,57	2,62
CO_2	0,54	0,38	0,40
H_2O	4,24	4,12	4,26
Total	99,96	100,63	100,19
Paramètres de NIGGLI			
al	24,44	26,72	27,06
fm	42,24	46,40	45,89
c	24,64	14,18	14,40
alk	8,66	12,68	12,63
si	107,03	124,10	120,70
mg	0,48	0,48	0,47
k	0,10	0,05	0,06

D-11 et 12: microgabbros porphyriques à augite-chlorite de Beacon Hill.

Tableau 21. Cornouailles-Devonshire - Tufs et pyroclastites

	D14	D14(b)	D15(3)	D16	D18	C4	C8(b)
SiO_2	46,10	46,20	36,30	36,50	60,60	50,40	53,10
Al_2O_3	13,00	15,10	12,80	15,20	13,30	18,00	15,60
Fe_2O_3	12,10	14,30	13,49	8,96	7,74	11,46	10,01
MnO	0,10	0,15	0,29	0,33	0,28	0,17	0,15
MgO	8,31	10,98	2,77	3,98	2,34	3,65	5,93
CaO	5,01	0,36	14,34	11,37	3,75	2,12	2,08
Na_2O	2,21	1,41	1,62	2,57	2,25	2,45	3,01
K_2O	0,31	0,77	1,28	1,74	1,31	2,68	1,15
TiO_2	2,21	2,29	3,00	1,28	1,45	0,60	0,74
CO_2	3,71	0,54	11,33	14,77	3,94		
H_2O	5,91	7,95	3,29	2,18	2,18		
PF						7,51	7,30
Total	98,97	100,05	100,51	98,88	99,14	99,04	99,07
Paramètres de NIGGLI							
al	20,68	23,10	18,92	23,72	32,05	33,98	29,05
fm	58,48	71,06	36,47	34,40	39,13	45,62	52,33
c	14,51	1,00	38,60	32,32	16,46	7,29	7,05
alk	6,31	4,82	5,99	9,54	12,34	13,10	11,54
si	124,67	120,16	91,21	96,84	248,29	161,77	168,13
mg	0,57	0,60	0,28	0,46	0,36	0,38	0,53
k	0,08	0,26	0,34	0,30	0,27	0,41	0,20

D-14: tufs de Blair Hill (Devon; D-14(b): tufs au carrefour des routes de Compton et Ipplepen; D-15(3) et 16: tufs calcareux d'Austin's Bridge, près de Buckfastleigh, route de Totnès (Devon); D-18: idem, sans calcite; C-4 et 8(b): cinérites rubanées de Trevose Head (Cornwall).

II. Le Problème de l'Origine des Spilites et des Kératophyres Hercynotypes

Aborder le problème de l'origine des roches du magmatisme préorogénique hercynotype revient à tenter de comprendre les phénomènes de spilitisation. Nous utiliserons ce terme sous une acception à la fois très rigoureuse et cependant plus large qu'il n'en est généralement fait usage. En effet, nous avons insisté à plusieurs reprises sur la nécessité de ne jamais séparer les spilites s.s. des albitophyres acides (ou kératophyres) qui leur sont génétiquement liées. Donc nous entendons apporter notre contribution au problème de la spilitisation et de la "kératophyrisation", terme peu élégant que nous n'utiliserons pas davantage, ayant ainsi prévenu le lecteur de l'usage que nous ferons désormais de son homologue.

Nous allons aussi brièvement que possible résumer les faits d'observation rapportés dans la première partie et qui nous paraissent fondamentaux pour le choix des hypothèses les plus plausibles que l'on peut en déduire.

a) Un premier fait d'observation qui nous semble incontestable est celui de la présence dans la plupart des roches étudiées de textures primaires. Ceci est particulièrement évident dans les laves où les textures microlitiques parfois fluidales, souvent porphyriques et fréquemment intersertales ne se différencient en rien des textures habituelles des roches volcaniques coenotypiques. Seul l'examen attentif de la nature des minéraux permet d'y voir une grande originalité.

Par conséquent s'il y avait eu remplacement des constituants primaires de la roche volcanique "normale" après complète consolidation, il faudrait admettre que ce remplacement ait été particulièrement sélectif et soucieux de respecter jusque dans les moindres détails les arrangements primitifs les plus délicats. Sur ce point nous partageons l'opinion de AMSTUTZ (1968, p.946).

Quoique moins spectaculaires, les textures des roches grenues présentent les mêmes caractéristiques. Nous en déduisons que les constituants des roches hercynotypes ont, dans leur très grande majorité (le cas de la calcite et du quartz peut être discuté) cristallisé sous la forme que nous observons.

b) Une seconde remarque d'une grande importance concerne la généralisation de la spilitisation des roches volcaniques. Aucune relique de roches volcaniques normales telles que basalte, trachyte ou rhyolite n'est observable. Et pourtant les conditions de gisement de ces laves sont très variées: coulées massives, nappes de pyroclastites, coulées de pillow-lavas, etc.

En revanche, certaines roches grenues ou microgrenues ne sont pas spilitiques. Elles sont toutes de nature intrusive.

c) Si l'on regarde maintenant du côté de la minéralogie, on peut faire les constatations suivantes. Sans revenir sur le caractère de basse température des associations spilitiques qui est bien connu, nous remarquons que sauf dans le cas très particulier signalé dans le district de Lahn-Dill (petites intrusions de péridotites) l'olivine, minéral banal des laves basiques, est totalement absente. Il se peut que dans certains cas l'olivine ait été entièrement pseudomorphosée comme le souligne TANE (1967) pour les spilites du Pelvoux et que ce minéral ne soit plus reconnaissable. En revanche l'augite persiste dans un certain nombre de roches. La présence de spilites à pyroxène qui est reconnue dans de nombreuses régions pose le problème de la stabilité de ce minéral dans les conditions de la spilitisation. Il s'agit, cela va de soi, d'un minéral primaire faisant partie, nous l'avons vu, d'une association géométrique (texture) également primaire. C'est également un minéral de haute température. Or il côtoie des minéraux de basse température: chlorite, albite, calcite, etc. Ce déséquilibre évident a une origine pour l'instant assez peu claire mais qui ne doit pas manquer d'entraîner de sérieuses conséquences.

d) Un autre fait minéralogique non négligeable concerne les phénocristaux feldspathiques de certaines laves et roches microgrenues. Ces minéraux sont souvent complexes car ils présentent des transformations pouvant faire intervenir plusieurs phases. Dans les orthokératophyres, par exemple, le plagioclase basique est remplacé partiellement par l'orthose qui subit à son tour une albitisation incomplète.

e) Nous avons vu que les roches amygdalaires sont riches en calcite et que les vésicules sont également remplies en moindre proportion de chlorite, quartz et zéolites. Ce fait, qui peut paraître banal, n'est pas sans intérêt. Il se traduit par une grande variabilité des spilites quant à la teneur en CaO et SiO_2.

f) Lorsqu'on passe sur le plan de la géochimie des associations spilites-kératophyres hercynotypes, nous avons vu qu'il était très important de souligner leur caractère original. Contrairement à ce qui est souvent avancé, nous avons pu montrer (de la ROCHE et al. 1969) que le magma "spilitique" n'avait la composition ni d'un basalte alcalin ni d'un basalte tholéïtique. Ce caractère peut ne pas apparaître si l'on n'envisage que les spilites s.s. mais il est incontestable lorsqu'on examine l'ensemble des roches magmatiques de l'association hercynotype.

g) Enfin, pour terminer là cet inventaire des faits les plus saillants que nous retiendrons, nous rappellerons que par rapport aux roches magmatiques des suites basaltiques normales Fe, Mg et Al ne varient pratiquement pas tandis que les spilites s'enrichissent en CO_2, H_2O et Na_2O et que la teneur en silice décroît quand celle en $CaCO_3$ augmente.

Les conséquences logiques de tous ces faits nous portent à envisager la nature essentiellement primaire des spilites et kératophyres hercynotypes (théorie VII de AMSTUTZ, 1968) sans pour autant rejeter certaines recristallisations métasomatiques postérieures à la consolidation (silicification, calcitisations, etc.), transformations localisées et secondaires clairement reconnaissables sous le microscope.

Existence d'un magma de nature spilitique

Nous avons montré que dans les régions étudiées, les roches montrent des textures primaires qui ne peuvent qu'être le résultat d'une cristallisation directe d'un matériel magmatique de composition originale. Tout autre processus faisant intervenir des modifications aussi profondes que la transformation (parfois partielle) d'une paragénèse de haute température en une paragénèse de basse température, après consolidation, devrait modifier les textures. De plus, ces modifications devraient affecter préférentiellement les laves les plus fragmentées en préservant tout ou partie des laves massives comme on peut l'observer au cours d'un métamorphisme. Or il n'en est rien, toutes les laves ayant acquis une composition spilito-kératophyrique. Nous ne pouvons donc accepter l'hypothèse d'une métasomatose secondaire récemment proposée pour le massif de Schirmeck par FONTEILLES (1968) qui ne repose que sur des considérations chimiques, d'ailleurs discutables. Nous admettons donc que les spilites-kératophyres hercynotypes sont de cristallisation directe. En ce sens on peut dire qu'elles sont d'origine primaire.

Cependant, il est clair que des phénomènes postérieurs à la consolidation ont pu intervenir. Il s'agit essentiellement de silicifications, de calcitisation et parfois de potassification. On peut expliquer ces transformations qui sont toujours partielles par une métasomatose de percolation (KORZHINSKI, 1963), étant bien entendu que ce phénomène est pour nous secondaire et tardif, et affecte des roches déjà spilitisées: nous rejetons formellement l'intervention de ce phénomène dans la genèse des spilites, car elle est en contradiction flagrante avec les faits d'observation (textures primaires, etc.).

Origine du magma spilitique

L'hypothèse de la cristallisation directe d'un matériau fondu sous la forme de spilites et de kératophyres - mécanisme simple tenant compte de tous les faits observés - ne nous renseigne pas sur l'origine de ce magma.

Considérons d'abord le problème du magma spilitique, sans tenir compte des kératophyres. Deux possibilités viennent à l'esprit:

1. Magma de nature particulière se trouvant en réserve dans l'écorce terrestre, sans rapport avec les magmas basaltiques, et s'épanchant occasionnellement,

2. magma basaltique banal, transformé en magma spilitique avant ou pendant l'émission (nous avons vu qu'il ne saurait être question de spilitisation postérieure à la cristallisation).

L'hypothèse d'un magma de nature particulière a eu beaucoup de partisans et en a encore, depuis DEWEY et FLETT (1911). Dans le cas qui nous préoccupe, où toutes les roches de la suite sont transformées, on pourrait à la rigueur le concevoir, encore que la présence de dolérites, de diabases non spilitiques et parfois de péridotites nécessiterait l'intervention en un même lieu d'un magma spilitique et - plus discrètement - d'un magma basaltique. Mais dans le cas des ophiolites alpinotypes, où l'association des roches spilitiques et des roches basaltiques normales est la règle, on devrait admettre la coexistence systématique en un même lieu et au même moment de deux magmas proches mais cependant distincts. Nous rejetons donc cette hypothèse comme inutilement compliquée et trop peu satisfaisante, car elle n'explique ni l'origine de ce magma spilitique, ni la présence des kératophyres.

Il nous reste donc à examiner l'hypothèse d'un magma basaltique normal transformé avant ou pendant l'émission. Etant donné qu'il est bien connu - nous l'avons confirmé dans cette étude - que l'eau et les volatils (notamment CO_2) jouent un rôle considérable dans les associations spilito-kératophyriques, on peut envisager la transformation du magma basaltique par l'eau et le gaz carbonique. Deux processus peuvent intervenir, qui ne sont d'ailleurs pas incompatibles:

1. Action de l'eau et du CO_2 juvéniles sur leur propre magma (altération auto-hydrothermale de VALLANCE, 1960; auto-métamorphisme, etc.). Ce processus est fort plausible car, à la différence du volcanisme aérien où la séparation des phases s'effectue au moment de l'émission (les gaz se perdent dans l'atmosphère et les laves dégazées s'épanchent puis cristallisent), le volcanisme sous-marin empêche généralement ce dégazage: la séparation des phases ne s'effectue pas. Ceci est particulièrement visible dans les spilites amygdalaires, où la vésiculation des gaz s'est effectuée: l'évolution normale s'est bloquée à ce stade dès l'émission des laves en raison de l'énorme pression hydrostatique exercée par des centaines, voire des milliers de mètres d'eau de mer.

Ce processus ne saurait cependant expliquer à lui tout seul la spilitisation: en effet, tous les pillow-lavas devraient, par voie de conséquence, être spilitiques: on sait qu'il n'en est rien et que bien au contraire la plupart des pillow-lavas récents sous-océaniques ne sont pas spilitiques (CANN, 1969).

2. Le deuxième processus est un processus d'échange avec le milieu environnant, et en particulier l'assimilation par le magma basaltique de quantités considérables d'eau et de CO_2 emprunté au milieu géosynclinal. L'assimilation d'eau de mer piégée (eau connée des sédiments, réservoirs et nappes superficielles, etc.) que le magma basaltique est amené à traverser lors de son ascension est un mécanisme qui a été évoqué par plusieurs auteurs et notamment en France par TANE (1967). Ce processus nous paraît bien rendre compte des faits observés, en particulier:

- L'assimilation massive d'eau et de CO_2 abaisse la température du magma basaltique et favorise ainsi le développement de la "paragénèse spilitique",

- elle augmente la proportion, donc la pression des fluides, ce qui entraîne un brassage énergique du magma et une redistribution plus ou moins désordonnée des éléments les plus légers, soit les alcalins, le calcium, la silice. Ceci explique la grande variabilité de la composition chimique des spilites, et la mobilité constatée du groupe d'éléments cités ci-dessus,

- cette assimilation peut être plus ou moins importante, et même ne pas exister, suivant le trajet emprunté par les laves: on peut donc ainsi avoir côte à côte des spilites, des basaltes et des roches intermédiaires, suivant le degré d'assimilation d'eau de mer.

En ce qui concerne le mécanisme de cette assimilation, plusieurs faits importants nous suggèrent qu'il s'est effectué à partir de l'établissement de réservoirs de magma basaltique proche de la surface. Le premier fait d'observation est l'absence totale d'olivine (fraîche ou pseudomorphosée) dans les roches basiques de l'association, ce qui suggère fortement une décantation efficace de ce minéral à la suite de stations prolongées du magma dans des réservoirs intra-crustaux. A ce sujet, le contraste est frappant avec les spilites alpinotypes du Taurus (1) et de Chypre (2), où les reliques d'olivine calcitisée sont extrêmement

---

(1) Th. JUTEAU, thèse en cours.
(2) H. LAPIERRE, thèse en 1972. Soutenue à l'Université de Nancy.

abondantes: il est évident que dans ce dernier cas le magma basaltique est monté rapidement et directement, sans que la gravité ait eu le temps d'exercer ses effets.

Un deuxième fait important est l'existence des kératophyres, qui posent le problème de leur origine: nous avons vu que leurs textures primaires sont indubitables et qu'il existe donc un "magma kératophyrique": quelle est la relation entre celui-ci et le magma spilitique? Les fait observés à Schirmeck, où les kératophyres sont particulièrement abondants, hololeucocrates et essentiellement quartzofeldspathiques, et où d'autre part la liaison est continue entre les spilites et les kératophyres, laissent à penser qu'ils sont nés par fusion anatectique et que des hybridations anatectiques ont eu lieu entre le magma basique spilitisé et la croûte sialique; là encore, l'existence de réservoirs intra-crustaux s'impose.

Les kératophyres de la Lahn, par contre, semblent être issus plutôt de la différenciation magmatique; ils sont très localisés et de nature trachytique; d'autre part, un "trou" les sépare des spilites dans les diagrammes d'évolution; il ne semble donc pas qu'il y ait eu un phénomène d'hybridation, mais une évolution par différenciation magmatique, calquée sur le type basalte alcalin - trachyte.

En résumé, nous proposons donc le schéma suivant pour expliquer la genèse des spilites-kératophyres hercynotypes: montée lente d'un magma basaltique normal à travers l'épaisse couche sialique; en milieu géosynclinal et près des fonds sous-marins, contamination de ce magma basique par d'importantes quantités de H_2O et CO_2 empruntées au milieu géosynclinal environnant; dès ce stade, deux types d'évolution s'individualisent: le type Schirmeck, en milieu géosynclinal très siliceux, contaminé essentiellement par H_2O, qui donnera des spilites siliceuses et peu calciques, et le type Lahn-Dill, en milieu géosynclinal beaucoup plus calcaire, qui donnera des spilites peu siliceuses et calciques. A ce niveau s'établissent des réservoirs intra-crustaux, dans lesquels s'effectue la décantation totale de l'olivine. Le magma basaltique est alors devenu un magma spilitique, dont l'évolution dépend toujours étroitement du milieu environnant: à Schirmeck, il y a contamination sialique progressive et continue par hybridation anatectique, qui fournira toute la gamme des diabases quartziques et des kératophyres siliceux à hyper-siliceux; dans le district de Lahn-Dill, l'évolution est plus rapide: l'hybridation n'a pas le temps de se faire et la différenciation magmatique classique conduit à des kératophyres trachytiques hyper-alcalins.

Nos renseignements sur les kératophyres de Cornouailles-Devonshire sont trop fragmentaires pour que nous puissions définir le type d'évolution de ce secteur: il semble toutefois se rapprocher beaucoup plus du type Lahn-Dill que du type Schirmeck.

Les considérations développées ci-dessus ne sont que des hypothèses de travail et nous prions le lecteur de les considérer comme telles: elles ont été élaborées en tenant compte du maximum de faits d'observation. Elles offrent l'intérêt de proposer pour la première fois un schéma général d'évolution concernant la genèse des spilites et des kératophyres hercynotypes: on a trop souvent négligé ces derniers dans l'étude de la spilitisation et des spilites hercyniennes. Bien des questions restent en suspens, en particulier le bilan chimique précis de la spilitisation: les spilites hercynotypes sont parfois appauvris en calcium (à Schirmeck), presque toujours en magnésium (c'est général): ces éléments ont-ils été cédés par le magma spilitique au milieu environnant, en échange de H_2O et CO_2? La nature du magma basaltique originel reste encore à préciser: nous manquons de données suffisamment nombreuses sur les roches basiques non spilitiques, précisément en raison de leur rareté dans l'hercynotype: était-ce un basalte du type alcalin à olivine, ou un basalte du type tholéïtique à deux pyroxènes? Nous ne le savons pas: dans le dernier cas bien sûr, la décantation de l'olivine dont nous avons parlé se serait faite bien avant, dans les zones profondes souscrustales, et non en cours de montée. La composition chimique précise des minéraux des spilites et des kératophyres est encore une inconnue: nous avons entrepris cette étude qui est actuellement en cours. On le voit, le champ des recherches est encore vaste.

Conclusion

L'association spilite-kératophyre de la branche nord de l'orogène hercynien d'Europe occidentale est interprétée comme d'origine primaire à la suite de la consolidation d'un magma basaltique dont la tendance chimique est variable suivant la position paléogéographique et paléostructurale, comme il est montré par ailleurs (de la ROCHE, JUTEAU, ROCCI, dans ce même volume), magma basaltique préalablement transformé et contaminé par la traversée de réservoirs d'eau marine piégée en milieu géosynclinal.

Nous avons mis en évidence deux types d'évolution de la série hercynotype, dans laquelle on peut opposer un "magmatisme de ride" (type Schirmeck) à un "magmatisme de fosse" (type Lahn-Dill): sous une forme lapidaire, c'est l'opposition entre "spilites siliceuses" et "spilites calciques", dont l'existence dépend étroitement de la nature du milieu géosynclinal environnant. A Schirmeck, les spilites sont pauvres en calcium et en CO_2 et riches en chlorite; des phénomènes d'hybridation avec le sial environnant conduisent à toute une gamme de faciès de plus en plus siliceux, la liaison spilite-kératophyre étant progressive et continue. A Lahn-Dill, les spilites sont riches en calcite et la différenciation magmatique conduit à des kératophyres trachytiques hyperalcalins; nous attribuons ces différences à la présence d'un socle sialique moins épais (petites intrusions ultrabasiques, absence de réactivation du sial) et à un milieu géosynclinal indiscutablement plus calcaire (rappelons la différence entre les "schalsteins siliceux" de Schirmeck et les "schalsteins calcaires" de Lahn-Dill). Quant à la Cornouailles et au Devonshire, leur type d'évolution est encore assez ambigu, en raison du manque d'informations sur leurs kératophyres: les tendances générales sont plutôt celles du type Lahn-Dill. Nous avons pleinement conscience que ce schéma ne peut avoir valeur générale, connaissant les problèmes souvant bien différents qui se posent dans d'autres domaines orogéniques: nous avons d'ailleurs opposé récemment deux types d'associations souvent confondus (ROCCI et JUTEAU, 1968). Nous savons que demeurent bien des points obscurs concernant les spilites et la spilitisation. C'est ainsi que l'on observe (cf. ROCCI et LAPIERRE, 1968) souvent dans les ophiolites tous les intermédiaires entre basaltes normaux, basaltes transformés et spilites. Ces transformations devront être étudiées soigneusement et nous nous y employons actuellement. On peut également parler du problème des varioles albitiques, si répandues dans les spilites alpines, présentes également dans les spilites antécambriennes de Bretagne (AUVRAY, 1967), et totalement absentes dans le domaine que nous venons d'examiner.

Bref, nous n'excluons pas, pour finir, que certaines spilites puissent avoir une autre origine que celle que nous avons proposée pour les spilites hercyniennes. Mais nous demeurons persuadés que chaque domaine géotectonique particulier possède son magmatisme préorogénique original, que des études approfondies doivent permettre de caractériser.

References

AMSTUTZ, G.C. (1968): Les laves spilitiques et leurs gîtes minéraux. Geol. Rundsch., Bd. 57, H. 3, 936-954.

AUBOUIN, J. (1961): Propos sur les géosynclinaux. Bull. Soc. géol. Fr. (7), III, 621-702.

AUVRAY, B. (1967): La série volcanique de la pointe de la Heussaye (Erquy, Côtes du Nord). Paris: Thèse 3e Cycle.

BÜCKING, H. (1923): Beiträge zur Geologie des oberen Breuschtals in den Vogesen. Mitt. geol. Land. Els. Lothr., t. XII, 1-364.

CANN (1969): Spilites from the Carlsberg Ridge, Indian Ocean. Journ. of Petr. 10, 1, 1-19.

CORSIN, P., DUBOIS, G. (1933): Description de la flore dinantienne de Champenay. Bull. du Serv. de la Carte géol. d'Als. Lorr., T. II, fasc. 1, 1-33.

CORSIN, P. et al. (1960): Sur l'âge viséen inférieur des schistes de Schwartzbach (vallée de la Bruche) dans les Vosges du Nord. Bull. Serv. Carte Géol. Als.-Lor., 13, 4, 163-164.

DEWEY, H., FLETT, J.S. (1911): On some British pillow-lavas and the rocks associated with them. Geol. Mag., dec. V, vol. 8, No. V, 202-209 et No. VI, 241-248.

DOUBINGER, J. (Mlle), RUHLAND, M. (1963): Découverte d'une faune de chitinozoaires d'âge dévonien au Treh (région du Markstein, Vosges méridionales). C. R. Acad. Sc., T. 256, 2894-2896 (micropaléoontologie).

FONTEILLES, M. (1968): Séries volcaniques paléozoïques de la Bruche (Vosges) et de la Brévenne (Massif Central français). Etude de la spilitisation. Bull. B.R.G.M., 2e sér., II, 3, 1-54.

GILLULY, J. (1935): Keratophyres of eastern Oregon and the spilit problem. Amer. Journ. Sci., 5th ser., vol. XXIX, No. 171, 225-252 et No. 172, 336-352.

HOUSE, M.R. (1963): Devonian ammonoid successions and facies in Devon and Cornwall. Quat. Journ. Geol. Soc. Lond., No. 473, vol. 119, I.

JUTEAU, Th. (1965): Etude pétrographique et géochimique des roches volcaniques dévono-dinantiennes du massif de Schirmeck (Vosges septentrionales). Nancy: Thèse 3e Cycle.

JUTEAU, Th., ROCCI, G. (1965): Contribution à l'étude pétrographique du massif volcanique dévonien de Schirmeck (Bas-Rhin). Bull. Serv. Carte Géol. Als.-Lor., Strasbourg, t. 18, fasc. 3, 145-176.

JUTEAU, Th., ROCCI, G. (1966): Etude chimique du massif volcanique dévonien de Schirmeck (Vosges septentrionales); évolution d'une série spilite-kératophyre. Sciences de la Terre, XI, n° 1, 68-104.

KORZHINSKIJ, O.S. (1963): Das Spilitproblem und die Trans-Vaporisationhypothese im Lichte neuer ozeanologischer und vulkanologischer Ergebnisse. Ber. Geol. Ges. DDR, Sonderh. 1, 89-95.

la ROCHE (de), H. (1964): Sur l'expression graphique des relations entre la composition chimique et la composition minéralogique des roches cristallines. Sciences de la Terre, IX, 3, 293-337.

la ROCHE (de), H., JUTEAU, Th., ROCCI, G. (1969): Essai de caractérisation chimique des associations spilitiques. (Spilite volume).

LEHMANN, E. (1949): Das Keratophyr-Weilburgit Problem. Heidelb. Beitr. z. Min. u. Petr., 2, 1-166.

LEHMANN, E. (1952): Diskussionsbemerkung zum Thema "Weilburgit" und "Schalstein". Z. Deutsch. geol. Ges., Hannover, 1952, 104, 255-256.

LEHMANN, E. (1965): Diabasgesteine SW - Englands und damit verbundene Probleme. Z. Deutsch. geol. Ges., Hannover, Bd. 115, 1, 228-276.

MIDDLETON, G.V. (1960): Spilitic rocks in South-East Devonshire. Geol. Mag., vol. XCVII, No. 3, 192-207.

NAREBSKI, W. (1964): Petrochemistry of pillow-lavas of the Kaczawa Mountains and some general petrogenetical problems of spilites. Prace Muzeum Ziemi 7, 69-205.

RAYNER, D.H. (1967): The stratigraphy of the British Isles. Cambridge, Un. Press.

ROCCI, G., JUTEAU, Th. (1968): Spilite-kératophyres et ophiolites. Influence de la traversée d'un socle sialique sur le magmatisme initial. Geol. en Mijnbouw, 47, 5, 330-339.

ROCCI, G., LAPIERRE, H. (1969): Etude comparative des diverses manifestations du volcanisme préorogénique au Sud de Chypre. Schweiz. Miner. Petr. Mitt. 49, H. 1, 31-46.

TANE, J.L. (1967): Contribution à l'étude du phénomène de spilitisation. Trav. Lab. Géol. Grenoble, 43, 187-192.

VALLANCE, T.G. (1960): Concerning spilites. Proc. Linn. Soc. N.S. Wales, IXXXV, Part 1, 8-52.

VALLANCE, T.G. (1965): On the chemistry of pillow lavas and the origin of spilites. Mineralog. Mag. 34, 471-481.

Spilites of the Lucanian Apennine (Southern Italy)

P. Spadea

Abstract

Field and petrographical descriptions of some volcanic rocks occurring at two outcrops in the Lucanian Apennine (Southern Italy) are reported. The volcanic rocks, belonging to the ophiolite complex of Alpine age present in this area, are represented chiefly by pillow lavas, among which several types may be distinguished structurally, by autoclastic lavas and volcanoclastic rocks. A Tithonian age is inferred for the pillow lavas associated with a polygenic sedimentary breccia. The mineralogy and fabric of the volcanic rocks are characteristically spilitic. Besides spilites, diabases, of probable hypabyssal occurrence, are also recorded. Some data about the spilite-forming minerals (particularly the chlorites) and four chemical analyses of spilites are reported. The variations in mineral and chemical composition exhibited by spilites are discussed and interpreted as chiefly dependent on the primary features of the extruded lavas. Furthermore, differences between spilites and some intrusive diabases and a gabbro occurring within the same area and in a similar tectonic position are pointed out. A causal relationship between spilitization and a volcanic occurrence is inferred, together with a probable primary origin of the actual essential petrographical features of the spilites. Effects of metamorphic alteration in some volcanic rocks, doubtfully referred to as medium grade burial metamorphism, and in associated sedimentary rocks are pointed out. The extent and importance of this alteration with regard to the petrogenesis of the spilites is, however, very uncertain.

Introduction

Basic and ultrabasic volcanic and intrusive rocks, which make up a complex association known as ophiolites or "green rocks", of Alpine age are widespread in the allochtonous geosynclinal terrains of southern Lucania and northern Calabria. The igneous rocks are mostly tectonically associated with Lower Cretaceous formations (Frido and Crete Nere Formations, VEZZANI, 1968; OGNIBEN, 1969); in a few localities extrusive rocks are found interbedded, with primary contacts, in radiolarites, cherts and limestones of Upper Jurassic to Lower Cretaceous age (BOUSQUET, 1961, 1963; SPADEA, 1968; VEZZANI, 1968).

A comprehensive geological description of the Lucanian-Calabrian ophiolites has recently been given by OGNIBEN (1969) and VEZZANI (1968). A division of the ophiolites into four groups, according to their geographical and tectonic position, is proposed in the works cited, each group including rocks with variable occurrence (basic lavas and hypabyssal or intrusive rocks, serpentinites), but characterized by the prevalance of one type of rock over the others. Various features related to diagenesis and/or metamorphism are also found in the associated igneous and sedimentary rocks of each group.

The knowledge of the petrography of the Lucanian-Calabrian ophiolites is still quite fragmentary. A list of previous works dealing with the petrography of these rocks is limited to QUITZOW (1935a, b), who gave detailed microscopic descriptions of the different igneous rocks, and COTECCHIA (1959) who reported some chemical data about the intrusive rocks. Brief petrographical descriptions of the lavas are also given by BOUSQUET (1961, 1963). In a recent paper by LO GIUDICE (1968) a detailed petrographical and chemical study of the serpentinites and diabases of S. Severino Lucano is reported.

The present paper deals with the field and petrographic studies of basic volcanic rocks exposed at two localities (M. Tumbarino and Contrada Iazzicelli) which display clearly recognizable features of the responsible eruption mechanism (pillowy, autoclastic and pyroclastic structures) and which in part clearly reveal their relationship with primary associated sedimentary rocks. Almost all of the rocks studied revealed spilitic character. Besides spilites some diabases, occurring at Masseria Carlomagno near the M. Tumbarino exposure as small masses, also were examined.

Field Observations

At Mount Tumbarino (coordinates: $3°$ 49' 5" Long. E; $40°$ 00' 29" Lat. N) volcanic rocks are exposed with an areal extent of about 0.2 sq km and a maximum thickness of 60 meters; they overlie sedimentary beds predominantly made up of shale and laminated siltite, quartzarenite and limestone.

The volcanic rocks are represented chiefly by pillow lavas and by brecciated (autoclastic) lavas. Also present are lavas which show a vaguely spheroidal to massive structure. In addition, volcanoclastic rocks with heterogeneous fragments occur as small intercalated layers between the lavas. Because of the tectonic movements which affected the igneous mass, primary contacts between all of these bodies are rarely found, thus making it difficult to infer the stratigraphic sequence of the volcanites and, in part, to interpret the observed structural features, particularly of pillow lavas.

Several types of pillow lavas have been distinguished on the basis of megascopic features, i.e., size and zoning of the pillows, ratio between pillow and interstitial matrix (chiefly made up of chlorites), color and structure of the rock:

1. porphyritic pillow lavas, with pillows of 1-1.5 m in diameter, gray-green color, poor in chloritic matrix and showing slight zoning at the periphery of the pillows;

2. oligophyric green pillow lavas, with smaller pillows (a few decimeters in diameter), rich in chloritic matrix and with a homogeneous texture;

3. aphyric reddish-black pillow lavas, showing conspicuous textural zoning: the pillows, about half a meter in diameter, consist of a central core of brecciated aphyric lava and several peripheral layers, each a few centimeters thick, with variolitic texture;

4. aphyric reddish-black pillow lavas, made up of small pillows (10-15 cm in diameter) with a matrix poor in chlorite.

Significant structural features of these pillow lavas also include the
frequent occurrence of cooling fractures and amygdales. Both are most
fully developed in types 3 and 4 and seem to indicate, together with
the other textural features, a more rapid gas escape and chilling rate
in comparison with the other types; the degree of oxidation of iron,
indicated by the abundance of hematite, also seems to be related to
this process.

The autoclastic lavas are texturally similar to the aphyric pillow
lavas with small pillows (type 4) and sometimes grade into them; their
distinctive character is a small-scale brecciation which divides the
rock into fragments a few centimeters to 1-2 decimeters in size. The
calcite cement is randomly distributed as white, well-crystallized
intersecting veins which give the rock the appearance of an ophicalcite.

The volcanoclastic rocks are made up of aphyric and porphyritic lava
fragments of various sizes, of crystals and of scarce green cement;
they include volcanic breccias and tuffs. Probable tuffites, showing
thin alternating layers of volcanic fragments, roughly sorted, and
of cherty-chloritic material, are found in association with the tuffs
and breccias. At the contact with the underlying sedimentary formation
a monogenic breccia of small lava fragments, containing abundant
amounts of calcite-pumpellyite cement and which seems to be of tectonic
origin, has been sampled.

At Contrada Iazzicelli (coordinates: 3° 40' 45" Long. E; 40° 00' 8"
Lat. N) volcanic rocks are intermittently exposed for a few square
meters within sedimentary beds of shales and laminated marls similar
to the sedimentary beds associated with the volcanites of M. Tumbarino.
At the immediate lower contact a two-meter-thick calcarenite bed
occurs.

At the lower part of the exposure a polygenic breccia made up of
rounded to angular fragments of gray, rosy and reddish limestones and
of reddish-black aphyric lava 30 cm to 50 cm in diameter, occurs. The
matrix is composed of minute limestone and lava fragments included
in a groundmass of brick-red limestone; the same limestone also appears
as thin and short pseudoveins cross-cutting the matrix and the larger
limestone fragments. A network of white calcite veins is developed
everywhere. Locally the lower breccia contains lava fragments with
a pillow-like appearance, sometimes showing elliptical or irregularly
curved outlines, which seem to have been plastically introduced into
the limestone blocks. Above the breccia some isolated bodies of
reddish-black fractured lava, showing a moderately developed pillow
structure, are present; rare limestone fragments occur again between
the pillows, and calcite veins are also widespread.

An extrusive origin of the volcanic rocks described is easily inferred
from their structure; as to the polygenic breccia, one is unable to
determine from field observations whether its origin is explosive or
sedimentary (see below).

Sedimentary Rocks and Age of the Volcanites
───

The sedimentary rocks occurring at the immediate area of contact with
the volcanites or primarily associated with them were studied in order
to recognize possible metamorphic influences from lavas and the degree
of alteration of the sedimentary rocks in comparison with the extrusive
ones. The fossil content was also examined to provide chronological

reference points for the volcanic episode and to establish the geologic relation between the associated igneous and sedimentary rocks.

At Mount Tumbarino the following rock types have been recognized: laminated marl and marly micritic limestone, sometimes fossil-bearing with Heterohelix sp. and Hedbergella sp.; biomicritic limestone, partly recrystallized, containing a microfaunal assemblage predominantly of Pithonella ovalis Kaufmann and Stomiosphaera sphaerica Kaufmann together with radiolaria remnants, Heterohelix sp., Hedbergella sp. and Rotalipora sp.; fine-grained quartzarenite and subarkose, the latter with a sparse matrix of recrystallized sericite-chlorite; quartz-feldspar-sericite siltite, showing moderate schistosity. The least degree of alteration shown by all of the rocks studied consists of textural transformations which are responsible for their slight schistosity. The newly formed minerals are represented primarily by authigenic carbonate and albite, both clear and undeformed, and by sericite and chlorite (probably derived from the original clay minerals). The greatest alteration is shown by the siltite, which therefore was classified as a slate. No effect of thermal metamorphism, on the other hand, is recognizable in any of the samples examined. A Lower Cretaceous age can be assigned to the fossiliferous limestones.

The calcarenite exposed at Contrada Iazzicelli below the polygenic breccia can be correlated, apart from the lithological difference, with the fossiliferous limestones of M. Tamburino because the former contains abundant remains of Pithonella ovalis Kaufmann and Stomiosphaera sphaerica Kaufmann. The rock also contains numerous authigenic albite crystals.

In the sedimentary component of the polygenic breccia found at Contrada Iazzicelli, three types of limestone have been distinguished, chiefly through their fossil content. The first type, which makes up the largest fragments of the breccia and locally a kind of cement around lava pillows, is a micrite largely void of fossils, but sometimes containing Calpionella sp. and Stomiosphaera sp., with variable content of opaque minerals (probably hematite and limonite). The second one, which is found only in the finely brecciated matrix as minute fragments together with the former limestone type is a fossiliferous micrite containing abundant remains of pelagic pelecypods. Authigenic albite crystals and recrystallized carbonate patches with rhombohedral outlines occur frequently in both limestones. Finally the brick-red cement of the finely brecciated matrix and the pseudoveins within the larger limestone fragments are formed from a fossiliferous micrite rich in hematite and limonite, in which is occasionally found calcitized chert concretions, and in which, in order of abundance, the following microfossils were found: radiolaria, Calpionella sp., Stomiosphaera sp., ostrachoda, globochaetae, sponges, gastropoda, together with rare Tintinnopsella sp., Trocholina sp. and Cornuspira sp. A Tithonian age can be assigned to this limestone because of the presence of calpionellae and globochaetae; a similar age can also be assumed for the fossil-poor limestone. The micrite with pelagic pelecypods shells can doubtfully be referred to the Lower-Middle Jurassic period.

The problem of dating the polygenic breccia and consequently the volcanic episode, is closely connected with the interpretation of its mechanism of formation. Significant data were obtained by examining a great number of samples collected from different parts of the breccia. The observed relationships, particularly those between the lavas and limestones and between the various limestone types, revealed mutual changes in both components (chiefly deep weathering and calcitization of the lavas, enrichment in iron ore, probably connected with the volcanic activity, and perhaps in albite within the limestones) and the later origin of the brick-red fossiliferous limestone with respect to the other components of the breccia. The presence of a sedimentary fossiliferous cement, therefore, excludes the possibility of an explosive genesis. Moreover, the structure of the lavas associated with and overlying the breccia suggests a period of relatively quiet volcanic activity. In addition some features observed in the breccia are remindful of the hard-ground levels described from many localities (VOIGT, 1959; MALARODA, 1962; PICCOLI, 1967). A sedimentary origin synchronous with the volcanic extrusion may thus be assumed for the breccia: the deposition of the breccia probably immediately followed (perhaps as a consequence of the volcanic activity[1]) that of the micritic limestone in which lavas appear plastically intruded. In view of this interpretation, the age of the volcanic episode can be assumed to be the same as that of the fossiliferous brick-red limestone cement. In any case, the volcanic extrusion cannot be younger than Tithonian.

The tectonic superposition of lavas and associated breccia on younger terrains (Lower Cretaceous p.p.) is therefore proved.

Petrography

The essential mineral assemblage of all volcanic rocks studied is relatively simple and uniform. Two fundamental assemblages are found:

1. albite-chlorite-titanite ± pyroxene, epidote and calcite;
2. albite-chlorite-titanite-hematite ± calcite.

The high content of colored components and the predominance of low-temperature minerals, particularly of fresh albite, permit the assignment of most of the rocks to the unique type of spilites (see, e.g. VUAGNAT's definition in STRECKEISEN, 1967). In view of the megascopic and textural features exhibited by all of the rocks under study (see below), the term spilite as defined by AMSTUTZ (1968) is also applicable. In addition to the spilites, altered rocks which occur in the field as masses of unknown origin, probably hypabyssal, were reported, which according to VUAGNAT's nomenclature (in STRECK-EISEN, 1967) can be classified as diabases. These rocks are made up of a mineral assemblage similar to the first one reported above and differ from the spilites chiefly in that they contain an altered plagioclase, completely decomposed into albite and saussurite, whose original calcic composition, unlike in spilites, is still evident.

Apatite, epidote, quartz, amphiboles, white mica, pumpellyite, serpentine and pyrite occur as minor and only locally abundant components of the rocks studied.

With regard to texture, strongly variable features can be observed, even in rocks which in the field display a similar structure and which occur within the same volcanic body, as for instance in zoned pillow lavas and in some autoclastic lavas.

---

[1] A causal relationship between volcanic activity and hard-grounds is denied by PICCOLI (1967).

The recognized textures, both aphyric and porphyritic, named primarily according to VUAGNAT[2] include the following types, when arranged in order of decreasing size of the plagioclase crystals: subophitic, intersertal, divergent intersertal (=diktytaxitic), arborescent and pilotaxitic. In the most finely grained rocks, in which plagioclase is partly irresolvable and shows a decidedly fibrous habit, only aphyric spherulitic textures are present; these include subvariolitic (continuous spherulitic according to VUAGNAT) and microvariolitic[2], the latter including a cellular variety (characterized by radiating aggregates showing polygonal outlines).

The porphyritic rocks are oligophyric to mesophyric: phenocrysts do not exceed 20% in volume and are mostly or exclusively represented by albite or saussuritized plagioclase; relicts of olivine phenocrysts and small pyroxene phenocrysts may also occur. In the porphyritic pillow lavas albite makes up two distinct phenocryst generations.

The petrographic features of the structurally different volcanic rocks distinguished in the field are reported below. The mineral composition of a few chemically analyzed samples and the calculation of a "Niggli Variante", are also given.

Unzoned or Slightly Zoned Pillow Lavas

The pillow lavas types 1 and 2 described on page 332 are made up respectively of porphyritic chlorite-pyroxene-epidote and chlorite-titanite spilites with a divergent intersertal to arborescent and pilotaxitic groundmass. The volumetric mineral composition of one (Table 2, analysis 2) pyroxene-bearing spilite without calcite, is: 1.9% quartz; 48.8% albite (An_2); 7.9% white mica; 15.0% pyroxene (augite with composition $Wo_{40}En_{40}Hy_{20}$); 7.2% epidote; 15.5% chlorite; 3.4% titanite and 0.3% apatite. The calcite content is 0.4 wt. %. Amygdales and veinlets, made up of albite-chlorite-calcite or epidote $\pm$ chlorite, are rare in these rocks. A slight zoning is exhibited by the largest pillows as the result of diminishing grain-size and fluidic orientation of the plagioclase microlites and second-generation phenocrysts towards the periphery.

The aphyric pillow lavas (type 4, page 332) consist of chlorite-hematite-titanite spilite with a divergent intersertal texture.

A uniform type of chlorite spilite, very rich in carbonates and hematite and with intersertal texture, is shown by the pillow lava and the lava fragments of the breccia from Contrada Iazzicelli. The high carbonate content, which can approach 1/3 of the rock volume and which occurs in the form of amygdales or interstitial patches and of partial replacements of plagioclase, seems to be chiefly related to the diagenetic transformations favored by the intimate association with limestones. It is not, however, excluded that some calcite, together with hematite, which in part appears to be of late origin, was formed during deuteric crystallization.

---

[2] The subophitic texture is distinguished from the intersertal one on the basis of the size of the pyroxene crystals and the disposition (felty or not) of the plagioclase laths. The variolitic texture, as defined by JOHANNSEN (1939), is characterized by spherulites ranging in size from some mm to a few cm. In the rocks under study the spherulites do not exceed one millimeter in diameter.

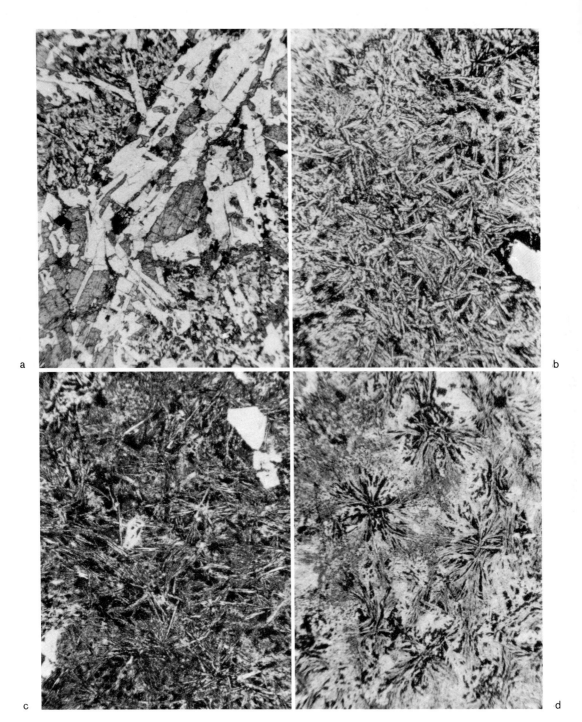

PLATE I. Textures of spilites of the Lucanian Apennine. Photomicrographs, plain light.
a) Augite-chlorite spilite with subophitic texture; fragment of a monogenic volcanic breccia, M. Tumbarino. x 25. b) Chlorite-hematite spilite with divergent intersertal texture; autoclastic lava (analysis 4), M. Tumbarino. x 25. c) Chlorite-augite spilite with arborescent texture; massive lava (analysis 3), M. Tumbarino. x 30. d) Chlorite-hematite spilite with subvariolitic texture; from the intermediate zone of a zoned pillow, M. Tumbarino. x 25

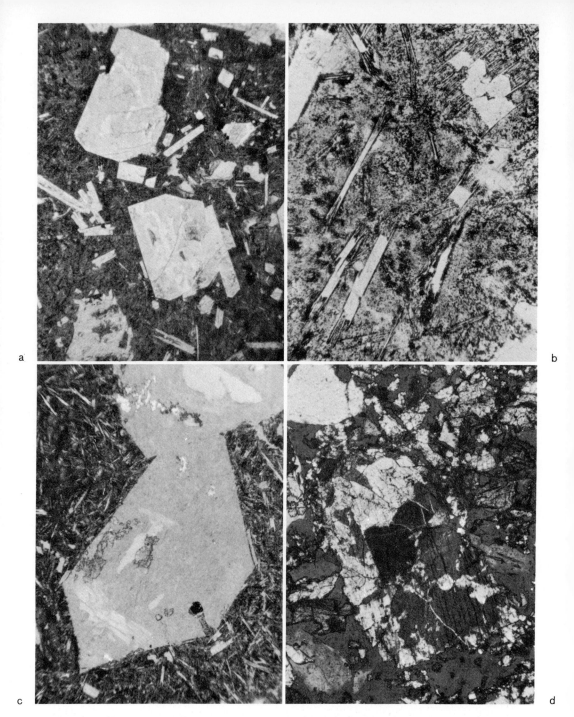

PLATE II. Textures and microstructures of spilites of the Lucanian Apennine.
Photomicrographs a,b,c, plain light; d, with crossed Nicols.
a and b) Porphyritic chlorite-epidote spilite; pillow-lava (with pillows of middle size), M. Tumbarino. Two main generations of albite phenocrysts and rare relict olivine phenocrysts in an aphanitic, vaguely arborescent textured matrix. Second generation albite phenocrysts and microlites show acicular overgrowths of the same feldspar and skeletal forms b). a) x 12; b) x 60. c) Porphyritic augite-chlorite spilite; pillow-lava (analysis 2), M. Tumbarino. Phenocrysts of olivine as pseudomorphs with chlorite. x 25. d) Altered lithic tuff with spilite fragments, M. Tumbarino. Parallel replacement of augite (extinguished, at the centre of the figure) at the margins and along a fracture by pumpellyite (illuminated). x 60

The chemical analysis of a sample, showing a minimum carbonate content of 8.1 wt. % revealed the following composition not including calcite: 4.7% quartz; 51.0% albite (An_0); 35.6% chlorite and clay minerals; 7.3% hematite; 1.0% rutile and 0.4% apatite.

Zoned Pillow Lavas

The textural zoning of the type 3 pillow lavas found at M. Tumbarino results from a number of concentric shells in which the texture ranges, from the core towards the periphery, from divergent intersertal to subvariolitic and microvariolitic. The mineral association consists of albite, chlorite, calcite, titanite and hematite, with moderate changes in the ratios between the named minerals either concurrent with the textural zoning or within different adjacent portions of a single shell. The chlorites are most abundant at the extreme periphery of the pillows as matrix interposed between the spherulites. In the intermediate subvariolitic shells hematite and titanite, chiefly concentrated at the core of the spherulites, show antithetic relationships in their abundance, characterized by roughly radial sectors in which either hematite or titanite respectively prevails. Other primary features related to chilling of the lava such as vesicles and cooling fractures, are still recognizable as calcite-chlorite amygdales and veinlets composed of calcite and skeletal hematite.

Autoclastic Lavas

These consist of chlorite-hematite-titanite spilites showing divergent intersertal or arborescent textures. For one analyzed sample (analysis 4) the following mineral composition has been calculated: 2.2% quartz; 64.5% albite (An_0); 24.1% chlorite, serpentine and clay minerals; 3.3% titanite; 5.0% hematite; 0.3% rutile and 0.6% apatite. Calcite and calcite-chlorite veins and veinlets as well as amygdales are frequent. At the point of contact with underlying slates, some samples of autoclastic lava show a poorly developed schistosity, attributable to parallel chlorite-muscovite-calcite layers containing oriented muscovite flakes. The fabric and mineralogical transformations are comparable to those observed in siltites and similarly can chiefly be referred to dynamometamorphism.

Massive Lavas

Massive lavas include spilites, often pyroxene-bearing, and showing various textures. The most frequent type is a chlorite-pyroxene-titanite spilite with arborescent texture. Its mineral composition has been calculated on the basis of one analyzed sample (analysis 3): 4.5% quartz; 49.0% albite; 25.5% chlorite; 8.7% augite; 4.2% titanite; 0.6% apatite; 5.0% clay minerals. Calcite (very scarce, with 0.2 wt. %) and epidote are occasionally found as amygdales. A porphyritic chlorite-titanite spilite with a divergent intersertal groundmass is also frequent among the massive lavas. A third, less represented type of rock is a pyroxene-chlorite-calcite spilite showing subophitic texture and generally containing abundant calcite and/or chlorite amygdales.

Rocks of Unknown Occurrence

This group of rocks includes, besides some spilites exposed at M. Tumbarino which in the field do not show any significant features as to their origin, two diabase varieties from a small exposure near Masseria Carlomagno. One variety consists of an augite-chlorite-epidote diabase and the other of a porphyritic chlorite diabase containing few plagioclase phenocrysts. The groundmass of both diabases shows a relict subophitic texture: the effects of a weak cataclasis, followed by the subsequent deposition of epidote for the former rock and of fresh albite as thin veins in the fractures for the latter, are recognizable.

Volcanoclastic Rocks

These include monogenic volcanic breccias, tuffs, tuffites and a fine-grained breccia of probable cataclastic origin. Excepting the last type of rock, the rock fragments show the same variability in texture and mineral composition observed in the lavas, the only difference being that the former contain very little or no hematite. The crystal fragments are represented by albite, augite and rarely hornblende. Among the tuffs, two types may be distinguished chiefly on the basis of the composition of the cement. The first one, generally free of pyroxene, contains abundant chlorite accompanied by a little calcite and epidote as the cement. Other tuffs contain instead epidote and chlorite in similar proportions, together with some titanite, albite, hematite and pumpellyite as a cement which partially replaces the fragments. These rocks show a generally more advanced degree of alteration, probably due to metamorphism, than the remaining pyroclastic rocks as well as the extrusive ones.

In the above mentioned fine-grained, probably tectonic breccia, the fragments are made up of micro-variolitic-cellular-textured or almost isotropic lava containing abundant and irregularly diffused hematite. The cement is present as irregular veins made up of large crystals of calcite in the center and of fine-grained acicular pumpellyite at the rims.

Remarks on Minerals

Some optical and X-ray data on the minerals of the rocks studied are reported here together with some observations concerning their distribution and frequence.

Albite is represented by almost pure, low-temperature forms with $\varepsilon + (100) = 13° - 16°$ and $2V_\gamma = 77° - 81°$ and comprises one to three or four crystal generations in the lavas. A late generation is often found filling in veinlets and amygdales. The microlites sometimes show curved or skeletal forms with plumose or forked ends and are observed both in the groundmass and as fibrous rims around albite phenocrysts. Albite twinning is widespread; also albite-Karlsbad (only in diabases), pericline, Baveno and albite-Ala twins are recognized.

The larger albite crystals in spilites, though exhibiting almost perfect crystal forms, often contain well-crystallized calcite as well as epidote and chlorite as irregular patches and veins, the latter clearly filling in small fractures. The small albite crystals and microlites also contain chlorite, chiefly positioned along the contact planes between albite-twinning lamellae. Only the late albite is totally free of inclusions.

The pyroxene found in some spilitic lavas and pyroclastic rocks consists uniformly of normal, weakly-zoned augite with $2V_\gamma = 51° - 46°$, $c/\gamma = 40° - 44°$ and $\underline{n}_\beta = 1.693 - 1.697 \pm 0.001$. Unzoned crystals have $2V_\gamma = 48° - 49°$ and $\underline{n}_\beta = 1.694$, which indicates, according to HESS's diagram (in TRÖGER, 1959), a composition $Wo_{41}En_{41}Hy_{18}$. A trace of titanium is probably present in the zoned crystals, judging from color and dispersion.

In pyroxene diabases augite is predominant also; however, titanaugite, showing moderate pleochroism and $2V_\gamma = 54°$, occurs at the periphery of zoned crystals.

In spilitic lavas the augite is replaced, at various stages, by chlorite and calcite. In some monogenic volcanic breccias and in the more transformed, previously described tuffs, augite shows an incipient complex alteration which could be related to at least two different processes. The products of this alteration include nonfibrous uralite and aegirine-augite, the latter accompanied by pistacite, which replace augite along fractures and at the crystal periphery, or occur as scattered small patches. Secondary amphibole is represented by moderately colored green or reddish-brown hornblende (showing respectively $2V\alpha = 77°$ and $74° - 75°$). Reddish-brown hornblende is also found as isolated crystals often fringed by acicular actinolite. Aegirine-augite with moderate aegirine content, as deduced by optical properties (marked pleochroism: α = grass green, β = yellowish green, γ = greenish yellow, $2V_\gamma = 85°-90°$, $c/\alpha = 19°-13°$) occurs, like hornblende, in parallel growth with augite. When amphibole and aegirine-augite are found together, the former clearly appears to have been formed before aegirine-augite. The epidote associated with aegirine-augite does not show any parallel orientation and is also found in continuity with the cementing epidote.
cementing epidote.

With regard to the occurrence of augite, peculiar textures are present in the groundmass of porphyritic spilites, where pyroxene developed as radiated, branching or pectinate aggregates of small acicules between the feldspar laths. Similar textures were described by BATTEY (1956) as resulting from crystallization in the presence of volatiles.

Chlorite, which is by far the most abundant colored component of the rocks studied, is represented by at least four optically distinguishable varieties, often coexisting with each other and showing different textures: a) almost isotropic green chlorites with maximum refractive indices between 1.617 and 1.621; b) a pale green optically negative variety with anomalous gray-blue interference colors and $\underline{n}_\beta = 1.607$; c) an almost colorless, optically positive variety with anomalous brass-colored interference colors and $\underline{n}_\beta = 1.610$; d) a green-yellow chlorite showing appreciable pleochroism, a rather high negative birefrigence and $\underline{n}_\beta = 1.62 - 1.63$.

The average composition of the chlorites from six different rocks, as determined by X-ray study according to the method reported by BRINDLEY (1961), is given in Table 1. Notwithstanding the variable optical properties of the chlorites in the samples analyzed, only small differences are found as far as the ratios between Al, Mg and Fe" are concerned. A variation in the relative amounts of Fe" and Fe"', which could not be detected by the method used, may perhaps account for some

Table 1. X-ray data and composition of chlorites in spilites and diabases of the Lucanian Apennine (SPADEA, 1968)

Sample	$\underline{d}$(001) (Å)	$\underline{b}$ (Å)	Composition	Classification (HEY, 1954)	Types distinguished optically (see text)[+]
1	14.26	9.281	$(Mg_{3.1}Fe_{1.9}Al_1)(Si_3Al_1)O_{10}(OH)_8$	pycnochlorite	a, d, (b)
2	14.28	9.269	$(Mg_{3.3}Fe_{1.7}Al_1)(Si_3Al_1)O_{10}(OH)_8$	diabantite-pycnochlorite	b, (a)
3	14.20	9.277	$(Mg_{2.9}Fe_{1.9}Al_{1.2})(Si_{2.8}Al_{1.2})O_{10}(OH)_8$	ripidolite	b, a
4	14.27	9.289	$(Mg_{2.9}Fe_{2.1}Al_1)(Si_3Al_1)O_{10}(OH)_8$	pycnochlorite	a, d, (b)
5	14.18	9.280	$(Mg_3Fe_{1.9}Al_{1.1})(Si_{2.9}Al_{1.1})O_{10}(OH)_8$	pycnochlorite	c, (b)
6	14.22	9.276	$(Mg_{2.9}Fe_{1.8}Al_{1.3})(Si_{2.7}Al_{1.3})O_{10}(OH)_8$	ripidolite	b, (c), (d)

[+] less abundant types in brackets
1: pyroxene-bearing spilite; massive lava (?) of M. Tumbarino.
2: porphyritic spilite (unoxidized); from a large pillow of M. Tumbarino.
3: hematite-bearing spilite (oxidized); autoclastic lava of M. Tumbarino.
4: hematite-bearing spilite; from the variolitic border of a zoned pillow of M. Tumbarino.
5, 6: porphyritic diabases (hypabyssal?), near Masseria Carlomagno.

of the optical differences observed: in fact, it is known that the presence of Fe"' largely influences the optical properties (HEY, 1954; ALBEE, 1962; TRÖGER, 1967). With regard to the occurrence of chlorites, a complex paragenesis can be inferred from their widespread diffusion into different parts of the rock fabric. Only to a small extent are the chlorites recognizable as secondary minerals after pyroxene and olivine or as products of late vein- and amygdale-filling processes. They are mostly diffused as fine-grained mesostases between albite laths and are also often enriched at the periphery of pillows in both zoned and unzoned pillow lavas. This textural development of chlorites, which is frequently observed in spilites, has been interpreted as indicative of a primary origin of the chlorites (VUAGNAT, 1946, 1949; BATTEY, 1956; BAILEY and McCALLIEN, 1960; VALLANCE, 1965). The composition of chlorites in the rocks studied, particularly the high Al content, is also in agreement with that of chlorites regarded as a primary mineral (BATTEY, 1956).

Among the opaque or semiopaque minerals, hematite, ilmenite (in some pyroxene-bearing spilites and in diabases), fine-grained titanite and rutile have been recognized under the microscope and by X-ray study. Hematite seems to be primarily of late origin with respect to the other spilite components; fine-grained titanite is diffused in a manner similar to that of most of the chlorites and seems to have formed directly as such. It is, however, also found in leucoxenized ilmenite.

Epidotes usually range from clinozoisite to pistacite, the latter variety being predominant. Zoisite (pseudozoisite with $2V_\gamma = 10°$) occurs as veins in one pyroxene diabase.

Pumpellyite occurring in volcanoclastic rocks is represented by a moderately iron-rich variety showing strong pleochroism ($\alpha = \gamma =$ colorless, $\beta =$ emerald green) and dispersion, with $2V_\gamma = 57°-71°$. The presence of pumpellyite in the previously described, fine-grained breccia has been determined by X-ray study of a pure powder sample. The optical and X-ray data agree with those of COOMBS (1955).

Chemical Composition
_____

The chemical composition of three lavas of M. Tumbarino and of the pillow lava of Contrada Iazzicelli is given in Table 2. The analyzed rocks coming from M. Tumbarino are representative of the prevailing rock types with minimum calcite content (the calcite present as veins in sample 4 was mechanically eliminated before crushing). The high Na_2O content and the very low K_2O/Na_2O ratio should be noticed; these agree with the chemical character often found, though not exclusively, in spilitic rocks (BAILEY and McCALLIEN, 1960). Other peculiar chemical features of the rocks analyzed are the low CaO content, either total or combined in silicates, the variable Fe_2O_3/FeO ratio and the high Al_2O_3 and H_2O^+ content. The reported CaO values probably do not represent the real Ca-content of the rocks as a whole, given the criteria followed in choosing the analyzed sample (AMSTUTZ, 1968). Anyway it is impossible to evaluate the real CaO content, even if calcite is considered, because of the random and discrete distribution of calcite in lavas and, in the case of the pillow lava from Contrada Iazzicelli, because of its possible external origin. The high Fe_2O_3 content of two rocks (analyses 2 and 3), though uncommon in spilitic rocks, is not exceptional (cf. data reported by VALLANCE, 1965; BAILEY and McCALLIEN, 1960).

Table 2. Chemical composition of spilites of the Lucanian Apennine (SPADEA, 1968)

Analysis	1	2	3	4
SiO_2	46.95	52.35	53.40	53.05
TiO_2	1.24	1.46	2.01	1.82
P_2O_5	0.18	0.17	0.25	0.28
Al_2O_3	19.34	15.20	15.59	17.34
Fe_2O_3	9.15	0.18	0.41	6.80
FeO	1.26	8.55	8.80	3.85
MnO	0.14	0.16	0.38	0.18
MgO	3.15	5.41	5.83	3.73
CaO	4.50	6.41	4.30	1.96
Na_2O	5.00	5.11	5.17	6.73
K_2O	1.13	0.89	0.05	0.31
H_2O^-	0.30	0.10	0.09	0.06
H_2O^+	4.37	3.39	3.49	3.48
CO_2	3.56	0.16	0.09	0.52
Total	100.27	99.54	99.86	100.11

1 - chlorite-calcite-hematite spilite; pillow lava of Contrada Iazzicelli.
2 - porphyritic chlorite-augite-epidote spilite; unzoned pillow lava of M. Tumbarino.
3 - chlorite-augite spilite; massive lava (?) of M. Tumbarino.
4 - chlorite-hematite spilite; autoclastic lava of M. Tumbarino.

The high Al_2O_3 and water content, particularly in two rocks (analyses 1 and 4), must be primarily related to an advanced degree of diagenetic alteration and possibly of weathering.

None of the rock analyses can therefore reflect the original composition of extruded lavas. Those which are likely to approach more closely the original composition are those of the two pyroxene- and epidote-bearing rocks (analyses 2 and 3). A comparison of chemical and mineralogical composition (the latter reported in the preceding pages) of the rocks analyzed, reveals that the presence of pyroxene and epidote coincides with a decrease in the content or absence of clay minerals, which is reflected by an excess of Al relative to the alteration of feldspar, and also with a lower degree of oxidation of the iron. In the pyroxene-free rocks (analyses 1 and 4), which contain Ca only as carbonate, the feldspar is also partly altered (into sericite and undetermined clay minerals), and iron shows a high degree of oxidation. These features can chiefly be referred to different primary processes of deuteric alteration and crystallization, but diagenesis also may be regarded as a possible cause of alteration. These processes could have been favored in the rocks exposed at C. Iazzicelli by the intimate association between lavas and limestones.

In order to frame the reported spilitic rocks within the complex of basic igneous rocks of the same cycle of igneous activity in the Lucanian Apennine, the chemical composition of some diabases recently studied by LO GIUDICE (1968) and of an altered hornblende gabbro, analyzed by the writer, is reported in Table 4; the mineral composition of each rock analyzed is indicated in the same table. The diabases occur as a sheet-like hypabyssal mass and as a dyke. The gabbro makes up a small body of intrusive appearance. A normal basaltic and gabbroic composition is inferred, although the rocks are altered. Some contrasting chemical features with respect to spilites are apparent, particularly a lower SiO_2 content and higher CaO and Na_2O contents (except for one rock, analysis 5, in which fresh albite veinlets were found). In spite of the scarcity of reported data, it can be assumed that the different chemical and mineralogical characteristics of the group of rocks analyzed represent primary features, because they are found in rocks which were subjected to the same post-magmatic geological events. The spilitic character therefore seems to be chiefly related to an extrusive occurrence.

Table 3. Chemical composition of nonspilitic basic igneous rocks of the Lucanian Apennine

Analysis	5	6	7	8	9
SiO_2	49.13	42.52	38.75	44.90	47.85
TiO_2	1.78	1.15	1.20	1.43	1.03
P_2O_5	0.28	0.13	0.20	0.21	1.10
Al_2O_3	15.81	17.71	19.71	17.46	16.74
Fe_2O_3	2.57	1.67	-	1.93	1.92
FeO	6.61	5.89	9.91	6.46	7.71
MnO	0.18	0.21	0.21	0.14	0.16
MgO	8.56	10.88	11.68	9.47	8.82
CaO	7.15	11.89	8.81	10.09	7.11
Na_2O	4.06	0.74	1.47	1.74	3.55
K_2O	0.36	1.37	0.09	1.73	0.28
H_2O^-	0.46	0.21	0.40	0.13	0.02
H_2O^+	3.82	6.11	7.61	4.74	3.95
CO_2	-	-	-	-	0.14
Total	100.77	100.48	100.04	100.43	100.38

5,6,7: lower (5), middle (6) and upper (7) part of a sheet-like intrusive diabase body near S. Severino Lucano. Mode, 5: albite, augite, chlorite, hornblende, epidote, sericite, quartz, ilmenite, hematite, titanite, apatite; 6: epidote, augite, chlorite, sericite, actinolite, albite, ilmenite, hematite, apatite; 7: chlorite, epidote, albite, actinolite, ilmenite, sericite, apatite (LO GIUDICE, 1968).

8: diabase dyke near S. Severino Lucano. Mode: augite, chlorite, albite, sericite, hornblende, epidote, actinolite, titanite, calcite, magnetite, apatite, ilmenite (LO GIUDICE, 1968).

9: hornblende gabbro near S. Costantino Albanese. Mode: saussuritized plagioclase, hornblende, chlorite, albite, magnetite, titanite, calcite, hematite (Analysis by P. SPADEA).

Conclusions

The presence of spilites among the products of the igneous geosynclinal activity of Alpine age in the Lucanian Apennine has been ascertained by the study of some volcanic rocks exposed at two outcrops. These volcanic rocks are chiefly represented by pillow lavas which attest, together with some associated pyroclastic and sedimentary rocks (tuffs, tuffites, polygenic breccia with fossiliferous limestone cement) to a submarine volcanic activity. Various and often well-preserved structures shown in the field, coupled with the textural features of the rocks, made it possible to differentiate between different volcanic rock types, as follows: massive lavas, unzoned pillow lavas, autoclastic lavas, and zoned pillow lavas. Different cooling-rates, increasing from massive lavas to autoclastic and zoned pillow lavas, can be inferred; concurrently an increase in the degree of oxidation of the lavas, which chiefly reveals itself through their hematite content, which very likely is a primary feature, can be recognized in most cases.

With regard to the mineral composition, a variable but distinct distribution of augite, epidote and calcite, as well as of hematite, as noted above, is shown in the rocks studied. As a result oligomineralic assemblages prevail, in which, besides albite, which is the most abundant component in all of the rocks examined - the following mineral assemblages are found: augite-chlorite-epidote, chlorite-epidote-calcite, chlorite-calcite, chlorite-calcite-hematite. Among the accessory components titanite is ubiquitous and abundant. Augite occurs only in unoxidized lavas, in the coarser grained rocks and in porphyritic rocks as a groundmass component and in some tuffs; it is the only high-temperature mineral and for the most part has undergone alteration, to variable degrees, to chlorite, epidote and/or calcite or to uralite and/or aegirine-augite. Epidote has a distribution similar to that of augite. In oxidized lavas, which usually are finer-grained and aphyric, only chlorite and calcite are found in place of augite and epidote. As a possible origin for some of the chlorites, the glass-component could be assumed, given the textural features observed and the abundance of them in the matrix interposed between pillows.

Generally, the mineralogical features observed in the rocks studied, and as a consequence the chemical features, seem to reflect chiefly the manner of extrusion and the different physical character of the lavas, rather than a more or less advanced degree of post-magmatic alteration. The presence of albite, which is a peculiar and common characteristic of the rocks studied, must be considered individually: though the greatest part of albite is found as apparently primary and unique feldspar, some albite also occurs in amygdales and veins and in unusual abundance in the sedimentary rocks associated with lavas, as an undeformed authigenic mineral. There is, therefore, evidence of the great stability of albite, apart from limited effects of weathering or diagenesis (particularly calcitization). It must be noted, however, that the presence of fresh albite represents a striking difference between the spilites and some of the intrusive and hypabyssal rocks (diabases and gabbro) found in the same tectonic position, suggesting a causal relationship between complete albitization and a volcanic occurrence. In this respect and in accordance with the various theories advanced to explain the spilite problem (as reported and summarized by TURNER and VERHOOGEN, 1960; AMSTUTZ, 1968), either a primary origin during or after consolidation of the lavas, or a metasomatic origin through chemical exchange with sea water, can be assumed for the albite.

With regard to the general problem of spilitization, the hypothesis of a primary origin of albite, and of chlorite as well, seems to be chiefly connected with the possible existence of a special spilite magma which on extrusion differs in physical character and perhaps chemical character from normal basalts; tholeiites, to which spilites are often connected, are included as well (WATERS, 1955; BATTEY, 1956). It must be remembered that YODER and TILLEY (1962, p.468) concluded on the basis of experimental data that a basalt magma, even if saturated with sea water, would, under deep sea conditions, solidify in the same manner as under subaerial conditions and, particularly, that hydrous phases could not crystallize out in the presence of a liquid phase.

Whether changes could have come about through regional or burial metamorphism has been considered for the volcanic as well as the associated sedimentary rocks. The following observations can be considered significant in this respect:

a) the slight metamorphism exhibited by sedimentary rocks of appropriate composition and fabric (siltites and marls) and exceptionally by lavas;

b) the presence in some strongly altered tuffs and in a probably tectonic breccia of the association epidote-chlorite and/or pumpellyite-calcite as newly formed minerals. Besides to deuteric alteration, these could also be attributed to a metamorphism intermediate in conditions between that of a zeolitic facies and a greenschist facies, as defined by COOMBS et al. (1959), or corresponding to the chlorite-pumpellyite facies proposed by SEKI (1961). No textural differences were observed between the spilite rock fragments and the texturally similar lavas; this feature is in agreement with that found in rocks altered by burial metamorphism (WINKLER, 1967);

c) if a metamorphic origin for the mineral association of the above-named tuffs and breccia is admitted, a metamorphic origin for the albite could also be expected. In this respect the relict character of the textures shown by the spilites included in the tuffs should be mentioned.

The extent, importance and role of metamorphism as a cause of development of the actual petrographic features of all of the volcanic rocks studied remains, however, very uncertain.

References

ALBEE, A.L. (1962): Relationship between the mineral association, chemical composition and physical properties of the chlorite series. Am. Mineral. 47, 851-870, 5 fig., Menasha.

AMSTUTZ, G.C. (1968): Spilites and spilitic rocks. In: Basalts, H.H. Hess and A. Poldervaart, ed., vol. 2, 737-753, I fig., I pl., Interscience Publishers.

BAILEY, E.B., McCALLIEN, W.J. (1960): Some aspects of the Steinmann trinity: mainly chemical. Quart. Journ. Geol. Soc. London 116, 365-395, II fig., London.

BATTEY, M.H. (1956): The petrogenesis of a spilitic rock series from New Zealand. Geol. Mag. 93, 89-110, 4 fig., I pl., Hertford.

BOUSQUET, J.C. (1961): Nouvelles données sur les diabases-porphyrites de la région de Sangineto (Calabre). Bull. Soc. Géol. France, 7ème ser. 3, 370-378, 4 fig., I pl., Paris.

BOUSQUET, J.C. (1963): Age de la série des diabases-porphyrites (roches vertes dy flysch calabro-lucanien: Italie méridionale). Bull. Soc. Géol. France, 7ème ser. 4 (1962), 712-718, 3 fig., I pl., Paris.

BRINDLEY, G.W. (1961): Chlorite minerals. In: The X-ray identification and crystal structure of clay minerals, G. Brown, ed., 242-296, 9 fig., 2 pl., Miner. Soc., London.

COOMBS, D.S. (1955): The pumpellyite mineral series. Am. Miner. 55, 113-135, 9 fig., Menasha.

COOMBS, D.S., ELLIS, A.J., FYFE, W.S., TAYLOR, A.M. (1959): The zeolite facies. With comments on the interpretation of hydrothermal syntheses. Geochim. Cosmochim. Acta 17, 53-107, 18 fig., London.

COTECCHIA, V. (1959): Argille scagliose ofiolitifere della valle del Frido a nord del Monte Pollino. Boll. Soc. Geol. Ital. 77 (1958), 205-245, 20 fig., 3 pl., Roma.

HEY, M.H. (1954): A new review of the chlorites. Miner. Mag. 30, 277-292, 4 fig., London.

JOHANNSEN, A. (1939): A descriptive petrography of the igneous rocks, vol. I, 2nd ed., 318, 145 fig., The University of Chicago Press, Chicago.

LO GIUDICE, A. (1968): La massa ofiolitica di S. Severino Lucano (Potenza). Atti Acc. Gioenia Sc. Nat. Catania, s. 6^a 20, 173-188, I fig., 2 pl., Catania.

MALARODA, R. (1962): Gli "hard grounds" al limite tra Cretaceo ed Eocene nei Lessini occidentali. Mem. Soc. Geol. Ital. 3, 111-147, 9 fig., 6 pl., Roma.

OGNIBEN, L. (1969): Schema introduttivo alla geologia del confine calabro-lucano. Mem. Soc. Geol. Ital. (in press).

PICCOLI, G. (1967): Studio geologico del vulcanismo paleogenico veneto. Mem. Ist. Geol. Miner. Univ. Padova 26, 100, 16 fig., 5 pl., Padova.

QUITZOW, H.W. (1935a): Der Deckenbau des Kalabrischen Massivs und seiner Randgebiete. Abh. Ges. Wiss. Göttingen, Mat. Phys. Kl. 3, n. 13, 63-179, 36 fig., 6 pl., Göttingen.

QUITZOW, H.W. (1935b): Diabas-porphyrite und Glaukophangesteine in der Trias von Nord-Kalabrien. Nachr. Ges. Göttingen, Mat. Phys. Kl., s. I, n. 9, 83-118, Göttingen.

SEKI, Y. (1961): Pumpellyite in low-grade metamorphism. Journ. Petrol. 2, 407-423, 6 pl., Oxford.

SPADEA, P. (1968): Pillow lavas nei terreni alloctoni dell'Appennino lucano. Atti Acc. Gioenia Sc. Nat. Catania, s. 5, 20, 105-142, 2 fig., 5 tav., Catania.

STRECKEISEN, A.L. (1967): Classification and nomenclature of igneous rocks. N. Jb. Miner. Abh. 107, 144-214, 22 fig., Stuttgart.

TRÖGER, W.E. (1959): Optische Bestimmung der gesteinsbildenden Minerale. Teil I, 147, Stuttgart: Schweizerbart'sche Verlagsbuchhandlung.

TRÖGER, W.E. (1967): Optische Bestimmung der gesteinsbildenden Minerale. Teil II, 822, 259 fig., Stuttgart: Schweizerbart'sche Verlagsbuchhandlung.

TURNER, F.J., VERHOOGEN, J. (1960): Igneous and metamorphic petrology. 694, 117 fig., New York: McGraw-Hill.

VALLANCE, T.G. (1965): On the chemistry of pillow lavas and the origin of spilites. Miner. Mag. 34, Tilley Volume, 471-481, 2 fig., London.

VEZZANI, L. (1968): Rapporti tra ofioliti e formazioni sedimentarie nell'area compresa tra Viggianello, Francavilla sul Sinni, Terranova del Pollino e S. Lorenzo Belizzi. Atti Acc. Gioenia Sc. Nat. Catania, s. 6, 19 Suppl. Sc. Geol. 109-142, 15 fig., 6 pl., Catania.

VOIGT, E. (1959): Die ökologische Bedeutung der Hartgründe ("hardgrounds") in der oberen Kreide. Pal. Zeitsch. 33, 129-147, I fig., 4 pl., Stuttgart.

VUAGNAT, M. (1946): Sur quelques diabases suisses. Contribution à l'étude du problème des spilites et des pillow lavas. Schw. Miner. Petr. Mitt. 26, 116-228, 29 fig., Zürich.

VUAGNAT, M. (1949): Variolites et spilites. Comparaison entre quelques pillow lavas britanniques et alpines. Arch. Sci. (Soc. Phys. Hist. Nat. Genève) 2, 223-236, Genève.

WATERS, A.C. (1955): Volcanic rocks and the tectonic cycle. Geol. Soc. America, Special Paper, n. 62, 704-707, I pl., Baltimore.

WINKLER, H.G.F. (1967): Petrogenesis of metamorphic rocks. 2nd ed., 237, 53 fig., Berlin-Heidelberg-New York: Springer.

YODER, H.S., TILLEY, C.E. (1962): Origin of basalt magmas: An experimental study of natural and synthetic rocks systems. Journ. Petrol. 3, 342-532, 50 fig., II pl., Oxford.

Quelques Observations Nouvelles Relatives à la Genèse des Laves Spilitiques

J.-L. Tane

Abstract

The following subjects will be discussed in this article:
- The telescoping effect in the case of a juvenile fluid phase encountering a dense phase of peridotite, considered as an inevitable explanation of the genesis of the spilites.

- The relationship between lamprophyric rocks and spilitic rocks, the former being considered the dike-counterpart of the latter.

- The process of granitization interpreted as a possible reproduction of the spilitization mechanism on a capillary scale.

Enoncé Pétrographique du Problème

Quel que soit l'endroit où il se manifeste, le volcanisme spilitique semble présenter un caractère immuable quant à la constitution pétrographique de la lave à laquelle il donne naissance. En effet, celle-ci consiste toujours en une paragenèse hydrothermale (albite, chlorite, calcite, hématite, quartz, épidote) au sein de laquelle on distingue, avec plus ou moins de facilité d'ailleurs, de nombreux témoins de la pré-existence de l'olivine.

Ces deux constatations sont désormais classiques et elles permettent d'assimiler le phénomène de spilitisation à une intense et brutale intervention d'eau sur une lave dont les premiers cristaux formés (olivine) laissent présumer au contraire une teneur initiale en eau tout à fait normale.

Nous nous proposons dans cette note de contribuer à mieux connaître ce phénomène et nous procéderons pour ce faire en trois étapes.

Dans la première, nous tenterons de le situer dans le temps au moyen de considérations purement pétrographiques.

Dans la seconde, nous essaierons de replacer le résultat ainsi obtenu dans un contexte géologique plus général.

Dans la troisième étape enfin, nous nous attacherons à cerner plus particulièrement, et au moyen d'un exemple local, le mécanisme proprement dit de ce phénomène de spilitisation.

La Situation du Phénomène dans le Temps

Puisque les laves spilitiques présentent des reliques d'olivine, ce minéral a nécessairement cristallisé dans un premier temps. Cela signifie par conséquent que, momentanément au moins, les conditions de stabilité de l'olivine ont été satisfaites et qu'ensuite seulement, elles ne l'ont plus été.

Si l'on observe alors ces reliques d'olivine, on constate le plus souvent qu'elles n'ont conservé de l'olivine que les formes, mais qu'elles appartiennent du point de vue de leur constitution minéralogique à l'une ou l'autre, ou même simultanément à plusieurs, des espèces hydrothermales que nous avons mentionnées un peu plus haut.

Il s'agit donc de pseudomorphoses d'olivine en minéraux hydrothermaux et par conséquent des conditions hydrothermales, incompatibles avec la stabilité de l'olivine, ont succédé aux conditions magmatiques initiales qui avaient déclenché sa cristallisation.

Dans ces observations, aujourd'hui bien connues, les géologues ont trouvé l'inspiration de plusieurs hypothèses. Nous ne voulons pas les rappeler toutes ici d'autant que cet inventaire a été dressé récemment (AMSTUTZ, 1968) mais nous évoquerons cependant ce problème de manière à ne pas approuver sans justification, la préférence qui est actuellement donnée à l'hypothèse dite "primaire" (au sens pétrogénétique du terme).

La première de ces hypothèses consistait à imaginer que la lave arrivait vers la surface sous la forme normale d'un basalte, mais que celui-ci s'épanchant généralement dans une mer, il subissait alors à son contact le phénomène hydrothermal de la spilitisation.

Effectivement, la majeure partie des laves spilitiques connues affleurent dans un contexte géologique qui implique un épanchement sous-marin. Néanmoins lorsque des émissions basaltiques se produisent de nos jours sous une mer, on ne constate pas qu'il se produise à son contact un véritable phénomène de spilitisation.

En outre, et cela aussi condamne assez sérieusement l'hypothèse d'un basalte initial, toutes les émissions de laves spilitiques anciennes ne se sont pas produites sous les eaux, certaines ont vu le jour à l'air libre.

De sorte que la coïncidence entre volcanisme spilitique et milieu marin reste un état de fait fréquent et qui, en première approximation, paraît supposer une intervention de l'eau de mer dans le phénomène de spilitisation encore que manifestement, ce phénomène ne peut se résumer à un pur et simple épanchement de lave basaltique dans la mer elle-même.

Du point de vue pétrographique d'ailleurs, rien n'indique que des constituants basaltiques autres que l'olivine aient eux-mêmes laissé quelques reliques dans les échantillons de laves spilitiques. Or, un basalte n'est pas fait que d'olivine, et si dans les pseudomorphoses on peut éventuellement confondre avec elle certains pyroxènes, il est bien difficile en revanche d'identifier les ruines de feldspaths basaltiques (labrador, andésine) qui, en toute logique auraient dû cristalliser en même temps que l'olivine et les pyroxènes.

Certes, on a imaginé quelquefois que l'ensemble albite-calcite résultait de la déstabilisation complète et systématique d'un labrador préexistant, mais n'est-ce pas ouvrir la porte à des interprétations un peu trop libres que d'accepter en pétrologie le principe d'une trans-

formation dont l'état initial n'aurait laissé précisément aucune trace
matérielle qui puisse réellement témoigner de son existence passée.

Au surplus, et cela a été souligné il y a déjà longtemps (VUAGNAT, 1946)
si l'albite et la calcite étaient nées d'un labrador préexistant, leur
association présenterait au moins localement un contour global compatible de par sa forme avec celui d'un plagioclase initial. Or la
réalité ne montrant même pas cet indice, il ne subsiste finalement
aucune raison matérielle d'attribuer à la calcite et à l'albite le
caractère de constituants secondaires formés aux dépens d'un minéral
préexistant.

Nous les considérons par conséquent comme primaires.[1]

Le Contexte Géologique du Phénomène de Spilitisation

Si le volcanisme spilitique peut se manifester dans des contextes géologiques assez variés, il montre cependant (comme nous l'avons signalé
plus haut) une préférence très marquée pour les zones géosynclinales
ou pour la bordure de ces zones géosynclinales.

Et puisqu'il implique une intervention d'eau tout à fait exceptionnelle
il était logique, a priori, de voir dans ces fosses géosynclinales le
réservoir susceptible de la lui fournir.

Comme nous l'avons rappelé ci-dessus, le phénomène ne peut se résumer
cependant à l'épanchement d'une lave initialement basaltique dans une
mer géosynclinale. Mais de là à rejeter l'idée d'une intervention de
l'eau de mer il y a peut-être un pas à ne pas franchir car cela re-
viendrait à négliger l'observation que nous venons de rappeler: cette
préférence du volcanisme spilitique pour les zones géosynclinales.

C'est au contraire pour en tenir le plus grande compte que nous avions
envisagé précédemment (TANE, 1967) une hypothèse selon laquelle le
phénomène de spilitisation aurait pu consister en une intervention in-
directe de l'eau de mer. Pour les raisons que nous indiquons un peu
plus loin nous abandonnons aujourd'hui cette hypothèse dont nous
rappelons brièvement ci-dessous, et pour mémoire, le principe dont
elle s'inspirait:

"Si l'on admet que toute émission volcanique ne peut se faire qu'à la
faveur de fissures s'ouvrant progressivement, et de haut en bas, à
l'intérieur de l'écorce terrestre, les épanchements volcaniques sous-
marins sont nécessairement précédés par un mouvement de descente de
l'eau de mer à l'intérieur de ces fissures. Il pourrait donc se créer
en profondeur de véritables réservoirs d'eau de mer et à la limite
ceux-ci finiraient par télescoper les réservoirs proprement magmatiques
(Fig. 1).

Mais la différence des densités respectives et l'imperfection de la
miscibilité aurait alors pour conséquence que l'eau demeurerait en
suspension au-dessus du liquide magmatique. Dans la mesure où le
contact avec la mer aurait été entre temps rompu (fissures refermées)
cet état pourrait momentanément demeurer relativement stable, et ne
cesser qu'avec la réouverture des fissures, c'est-à-dire avec la
naissance d'une activité volcanique.

---

[1] L'hypothèse de la transformation d'un basalte initial par métamorphisme appellerait les mêmes objections.

L'interaction eau-magma d'une part et l'interaction eau-socle sus-jacent d'autre part consisterait naturellement en une migration vers la phase H_2O des constituants les plus solubles, c'est-à-dire SiO_2 et Na_2O principalement.

De sorte que le rétablissement des communications avec la surface (c'est-à-dire la réouverture d'anciennes fissures ou l'ouverture de nouvelles fissures) déclencherait une activité volcanique dont la première production serait une lave très fluide, riche en eau et en constituants solubles. Que ce soit alors sur le fond de la mer ou à l'air libre, la lave pourrait acquérir de la même façon le caractère hydrothermal des roches spilitiques. Et ce ne serait qu'après évacuation préalable de cette phase particulièrement fluide que pourrait à son tour monter vers la surface la phase magmatique beaucoup plus dense restée jusqu'alors en profondeur".

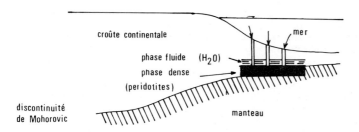

Fig. 1

Le mécanisme initial évoqué dans cette hypothèse ne nous semble aujourd'hui plus guère soutenable: en effet, et du simple point de vue de la dynamique des fluides, il est difficile d'admettre que H_2O dont la densité est égale à 1 puisse s'enfoncer profondément dans une écorce terrestre dont la densité est de l'ordre de 3,3. L'eau ne peut donc décrire une trajectoire importante en profondeur, suivant la verticale d'un lieu donné, que si son mouvement est décrit de bas en haut le long de cette verticale. Cela revient à dire que sur un intervalle de temps (t_1, t_2) de dimensions géologiques, le volume total de l'hydrosphère ne peut aller qu'en augmentant. Cette idée est d'ailleurs conforme au point de vue actuel de nombreux géochimistes et nous-mêmes l'avons adoptée récemment dans une communication relative au problème du granite (TANE, 1970).

Tout en conservant donc le principe que le volcanisme spilitique résulte du télescopage en profondeur d'une phase dense péridotique et d'une phase fluide essentiellement aqueuse (ces deux phases restant peu miscibles), nous admettons donc que l'origine première de la phase fluide est nécessairement juvénile et que, par conséquent, le volcanisme spilitique (au même titre que la granitisation) compte parmi les phénomènes géologiques qui apportent de l'eau à l'hydrosphère.

Mise à part cette importante nuance quant à l'origine de l'eau, l'imparfaite miscibilité des deux phases fluide et dense demeure à notre avis une chose certaine, facile à vérifier expérimentalement et dont la répercussion peut apparaître dans les observations de terrains.

Celles-ci fournissent à cet effet deux enseignements significatifs: elles montrent tout d'abord que cette phase magmatique plus dense ne parvient pas elle-même dans tous les cas à se hisser jusqu'à la surface de la Terre; elles montrent aussi que lorsqu'elle y parvient, le matériau qui se trouve alors déversé a une composition de péridotites.

A cette remarque correspond vraisemblablement la distinction établie récemment, pour le bassin méditerranéen tout au moins, entre volcanisme hercynotype et volcanisme alpinotype (ROCCI et JUTEAU, 1968). L'un se caractérisant par des émissions de laves spilitiques mais par l'absence quasi totale de péridotites, l'autre se caractérisant au contraire par la production des deux types de roches, avec une large prédominance des péridotites.

A la lumière de cette distinction nous proposons dans le paragraphe qui suit, d'apporter une contribution locale à la connaissance du phénomène de spilitisation.

Contribution Locale à la Connaisance du Phénomène de Spilitisation[2]

Le caractère local de cette contribution provient de ce que les observations que nous allons rapporter ont été faites dans une région relativement petite des Alpes françaises mais particulièrement bien située vis-à-vis du problème des laves spilitiques: il s'agit du Massif du Pelvoux.

Du point de vue géologique, en effet, ce massif est compris entre le môle hercynien du Massif Central français et la zone alpine interne proprement dite. Dans cette dernière, l'ensemble ophiolitique est représenté de manière complète par des émissions d'âge alpin (spilite, gabbros, péridotites) alors que rien de semblable ne s'observe dans le Massif Central où il faut atteindre l'ère tertiaire pour que des

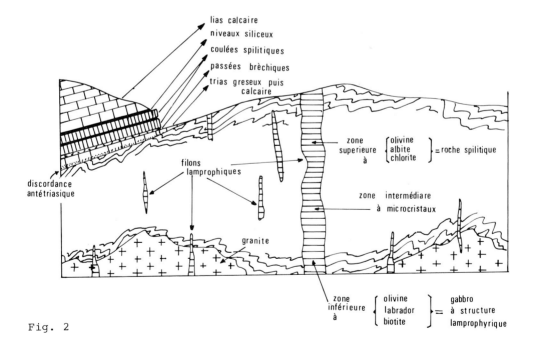

Fig. 2

---

[2]La figure 2 représente d'une manière schématique et synthétique l'essentiel de ce paragraphe.

coulées (souvent aériennes) de laves principalement basaltiques viennent s'épancher sur le socle granitique et cristallophyllien de l'Hercynien.

Entre ces deux extrêmes, le Massif du Pelvoux se présente comme un intermédiaire non seulement sur le plan proprement géographique, mais sur le plan de la volcanologie également.

Sur un socle hercynien encore fait de granites et de roches métamorphiques, un volcanisme limité à des coulées spilitiques s'est en effet manifesté au Trias.

En l'absence de venues péridotitiques qui auraient pu succéder éventuellement à ces émissions de laves spilitiques, il devenait très intéressant d'examiner le remplissage des cheminées volcaniques.

Car le matériau qui s'est arrêté à ce stade filonien présente du seul point de vue théorique un intérêt particulier: il matérialise la limite pétrographique entre ce qui a réussi à sortir du volcan et ce qui est resté en profondeur. De sorte que dans notre hypothèse, il représente la roche de transition entre la lave fluide et spilitique parvenue jusqu'à la surface et la lave dense et péridotitique demeurée plus bas.

Il est significatif à ce propos que les cheminées des volcans spilitiques du Pelvoux se trouvent remplies d'un matériau qui, lui-même, a plus souvent la composition d'un lamprophyre que celle d'un spilite, néanmoins, le passage de l'un à l'autre de ces deux faciès peut s'observer de manière continue pour peu que l'on procède à un échantillonnage systématique depuis le bas jusque vers le haut des cheminées (TANE, 1967). Le relief étant ici assez escarpé, un tel échantillonnage est, sinon aisé, du moins possible et nous avons pu, de cette manière, vérifier nos observations sur plusieurs cheminées.

Nous sommes parvenus ainsi à un schéma moyen qui peut se résumer comme suit: dans la partie basse des cheminées, l'olivine est pseudomorphosée en des minéraux très variables, le plus souvent une phyllite brune, quelquefois aussi un feldspath ou même du quartz. Autour d'elle et respectant ses contours, cristallisent des biotites ou des plagioclases basiques (labrador) de grande taille. Ces derniers s'altèrent localement en séricite mais ils ne paraissent nullement se transformer en albite et calcite.

Dans la partie haute des cheminées, où l'on ne voit plus trace de labrador, on voit au contraire proliférer l'albite non plus sous forme de grands cristaux mais sous forme de microlites.

A ses côtés cristallise alors la chlorite, soit directement, soit aux dépens d'une olivine préexistante.

L'intermédiaire entre ces deux types extrêmes est une roche presque aphanitique prouvant ainsi qu'au niveau où on la récolte, aucune des espèces minérales précédemment citées ne trouve les conditions de sa stabilité au point de cristalliser facilement.

Quant à la teneur en olivine pseudomorphosée, elle paraît plus abondante dans ces cheminées lamprophyriques qu'elle ne l'est habituellement dans un basalte. Au surplus, cette olivine se présente ici en cristaux de très grande taille. Devant cet état de fait, il semble bien par conséquent que juste au-dessous de ce matériau lamprophyrique demeuré dans les cheminées à la manière d'un bouchon, seul un magma à potentiel péridotitique ait pu intervenir.

Dans cette optique, les lamprophyres représentent un matériau quantitativement peu important par rapport à la lave spilitique vomie en surface et par rapport à la masse péridotitique invisible, mais que nous supposons présente au-dessous d'eux et en grande quantité.

Il y aurait donc de ce seul point de vue quantitatif, un parallélisme possible, semble-t-il, entre ces lamprophyres filoniens du volcanisme spilitique du Pelvoux et les gabbros de la zone interne alpine, qui participent là de façon modeste à un volcanisme ophiolitique complet. On ne peut s'empêcher de remarquer d'ailleurs que le labrador et le pyroxène étant présents à la base des cheminées lamprophyriques du Pelvoux, ce rapprochement avec les gabbros des cortèges ophiolitiques complets paraît assez bien se confirmer.

Mais il reste à propos des lamprophyres eux-mêmes une assez grosse énigme. Car si, dans le Massif du Pelvoux, ils matérialisent effectivement le stade filonien terminal du volcanisme spilitique, il n'en va manifestement pas de même en ce qui concerne d'autres régions de France et en particulier celles dont l'histoire orogénique est telle qu'on a pu les qualifier sans ambiguïté possible de massifs hercyniens (Massif Central, Vosges, Bretagne).

Dans le Massif Central par exemple, où les filons lamprophyriques ne sont pas rares, on ne voit pas jusqu'à ce jour à quelles émissions spilitiques ils pourraient faire suite. On ne peut là que constater leur mise en place dans des massifs de granites et du point de vue chronologique, la phase lamprophyrique fait donc suite à la phase granitique.

Mais précisément, on peut se demander si, dans une certaine mesure, le phénomène de granitisation n'aurait pas la même signification pétrogénétique que celui de la spilitisation.

Nous avons été conduits à propos de ce dernier, à imaginer le télescopage en profondeur d'une masse d'eau d'origine juvénile venant se superposer au sein de l'écorce terrestre à une masse magmatique à potentiel péridotique.

Cet état intermédiaire étant admis, les phénomènes qui lui succèderont ne seront pas tout à fait identiques selon que l'établissement des communications avec la surface se fera facilement ou difficilement, c'est-à-dire selon que les fissures s'ouvriront largement ou selon qu'elles ne s'ouvriront pas, ou tout au moins pas complètement.

Si un champ de fissures se met en place et permet à l'ensemble eau + magma de s'élever, on pourra avoir une phase d'activité volcanique elle-même décomposable en trois temps.

Dans le premier temps, c'est essentiellement la phase fluide que sera évacuée; disposant de cheminées relativement larges, elle pourra entraîner mécaniquement certains constituants solides de la phase dense, c'est-à-dire des cristaux d'olivine. Mais elle entraînera aussi, et en solution cette fois, les constituants chimiques empruntés par elle aux deux matériaux qu'elle aura contactés en profondeur (phase dense d'une part et socle sus-jacent d'autre part). Ainsi que nous l'avons rappelé plus haut, il s'agira alors de constituants particulièrement solubles, c'est-à-dire SiO_2, Na_2O, Al_2O_3, préférentiellement à des constituants moins solubles qui ne monteront donc pas aussi aisément avec la phase fluide.

Parvenue à la surface elle entraînera une série de précipitations chimiques au fur et à mesure que sa température diminuera et que les conditions de solubilité des différents solutés cesseront d'être satisfaites. L'observation des gisements de laves spilitiques montre, à cet effet, que le corps principal de ces laves est constitué de pseudomorphoses d'olivine, d'albite et de chlorite. Au-delà, le dernier constituant restant en solution (SiO_2) précipite à son tour sous la forme d'amas siliceux qui peuvent, le cas échéant, contenir des radiolaires. Enfin, et conformément à ce que nous avons dit plus haut, un excès d'H_2O ira augmenter le volume de l'hydrosphère.

Dans le second temps, c'est la phase de transition entre phase fluide et phase dense qui va à son tour s'élever. Elle est volumétriquement peu importante et elle ne s'épanchera à la surface que dans la mesure où la phase dense y viendra elle-même au troisième temps (volcanisme alpinotype). Dans le cas contraire (volcanisme hercynotype) elle avortera dans le volcan et s'arrêtera donc au niveau des cheminées (filons lamprophyriques).

Inversement, si la phase aqueuse présente en profondeur, au contact de la phase péridotitique, ne bénéficie pas de la réouverture d'un champ de fissures lui permettant de remonter facilement vers la surface, elle se trouvera captive. Compte tenu de sa densité, elle aura néanmoins tendance à remonter et, ne disposant pas de fissures franches, elle ne pourra le faire que de manière plus ou moins capillaire[3] à travers le socle sus-jacent. Elle pourra donc solubiliser certains constituants de celui-ci au même titre que certains constituants de la phase dense sous-jacente et véhiculer vers le haut l'ensemble de ces solutés. Mais sa progression étant ici relativement difficile, c'est pratiquement au même endroit qu'elle laissera précipiter (par refroidissement) les constituants devenus insolubles, de sorte que les phyllites, les feldspaths et le quartz ne s'étageront pas à des niveaux très différents comme dans le cas du volcanisme spilitique, mais se trouveront au contraire groupés ensemble pour constituer une seule et même roche. Il s'agirait donc en l'occurrence de granite qui apparaîtrait ainsi comme un cas limite du volcanisme hercynotype et qui relèverait, à l'échelle capillaire, d'un processus de mise en place assez bien calqué cependant sur celui des laves spilitiques.

On peut du reste remarquer que beaucoup d'apophyses des granites intrusifs consistent en filons d'aplites d'abord et en filons de quartz ensuite, ce qui reproduit, dans une certaine mesure la zonation observée dans un volcan spilitique, où les phyllites sont concentrées dans les cheminées (zones basses) l'albite dans le corps des coulées (zones moyennes) et la silice au-dessus des coulées (zones élevées).

En appliquant cette hypothèse au Massif du Pelvoux nous considérons donc que la phase fluide a été évacuée en deux temps successifs, tout d'abord sous forme de granite, et ensuite sous forme de laves spilitiques. Nous assimilons évidemment l'instant de leur séparation au passage d'un état dépourvu de champ de fractures à un état pourvu de champ de fractures.[4]

Plus généralement et suivant la distinction reconnue entre volcanisme hercynotype et volcanisme alpinotype, il semble donc que le premier cité implique des surpressions tectoniques capables d'évacuer la phase

---

[3] Ce qui interdit généralement tout entraînement mécanique lointain analogue à celui que nous avons imaginé pour l'olivine dans le cas du volcanisme spilitique.

[4] Ce que correspond ici, et du point de vue le plus général, au passage de l'orogenèse hercynienne à l'orogenèse alpine.

fluide exclusivement, tandis que dans le cas du second, celles-ci seraient de taille à évacuer également la phase dense (tout au moins en partie).

Il est remarquable enfin que dans le Massif Central français plus que dans les Vosges et beaucoup plus que dans le Pelvoux, l'évacuation de la phase fluide se soit produite à peu près intégralement sous forme capillaire (granites). La fracturation fini-hercynienne y aurait déclenché ensuite la mise en place de la phase intermédiaire (remplissage lamprophyrique de nombreux filons) mais l'évacuation de la phase dense n'aurait été possible là qu'à la faveur d'actions tectoniques tertiaires (épanchements basaltiques consécutifs à la surrection des Alpes).

Conclusion

L'hypthèse qui vient d'être proposée repose essentiellement sur la superposition en profondeur de deux phases très peu miscibles: l'une dense et d'origine ignée (péridotites), l'autre très fluide (H_2O essentiellement) et dont l'origine doit être également juvénile.

Il est remarquable à ce propos que l'intervention de cette phase aqueuse dans le processus de la spilitisation est bien reconnue de tous les spécialistes. Mais, si ceux-ci ont abandonné l'idée de P. TERMIER, c'est-à-dire l'idée d'un épanchement de basalte dans la mer, beaucoup d'entre eux invoquent la transformation d'un basalte par l'eau des sédiments non encore indurés que ce basalte traverse nécessairement au cours de sa montée vers la surface.

A notre avis, cette explication se heurte aux mêmes objections que celles de P. TERMIER car la profondeur qui correspond aux sédiments non indurés (et donc effectivement riches en eau) équivaut pour un volcan basaltique à un niveau déjà très élevé dans les cheminées et où, par conséquent, des feldspaths de type labrador ou andésine auraient déjà cristallisé autour de l'olivine.

Au surplus, si l'on s'en réfère à la figure 2 ci-dessus, où la zone supérieure du plus important des filons lamprophyriques est remplie d'un matériau déjà spilitique, on est bien conduit à admettre que ce n'est pas l'eau des sédiments qui a pu jouer. Car à ce niveau là en effet, le matériau filonien n'a encore traversé aucune formation sédimentaire, il n'a cheminé qu'à travers des granites et des gneiss, qui à l'époque des éruptions volcaniques, c'est-à-dire au Trias, étaient déjà bien des granites et des gneiss et non des sédiments.

Par conséquent, l'idée d'une intervention de la phase aqueuse à un niveau plus profond nous paraît s'imposer logiquement et contrairement à l'idée que nous avons avancée antérieurement (TANE, 1967) son origine initiale ne peut vraisemblablement pas être marine, mais bien au contraire, juvénile.

Il est d'ailleurs remarquable, et non seulement dans la région du Pelvoux, que postérieurement aux émissions de laves spilitiques, la sédimentation marine cesse d'être essentiellement siliceuse pour devenir essentiellement carbonatée.

On peut se demander si elle n'est pas due à une ascension massive vers la surface de la terre d'une phase juvénile riche en CO_2 et susceptible par conséquent d'entraîner les constituants CaO, MgO et FeO que la seule phase H_2O n'avait elle-même pu entraîner.

C'est là un problème que nous examinerons ultérieurement et sur lequel le Massif du Pelvoux et plus encore le Massif Central fournissent d'ores et déjà un précieux indice, car certains des filons lamprophyres qu'ils contiennent présentent, même à des niveaux très profonds, une altérations généralisée en carbonates.

Bibliographie Sommaire

AMSTUTZ, C.G. (1968): Les laves spilitiques et leurs gîtes minéraux. Geol. Rundschau 57.3, 936-954.

IYAMA, J.T. (1961): Etude préliminaire de la solubilité d'un basalte dans l'eau à haute température. Bull. soc. fr. Min. Crist. 84, 2, 187.

ROCCI, G., JUTEAU, T. (1968): Spilite-kératophyres et ophiolites. Influence de la traversée d'un socle sialique sur le magmatisme initial. Geol. en Mijnbouw 47.5, 330-339.

TANE, J.L. (1967): Contribution à l'étude du phénomène de spilitisation. Géol. Alpine Grenoble 43, 187-192.

TANE, J.L. (1970): Sur l'origine du granite et le problème de la formation des constituants. C.R. Acad. Sci., t. 270, 1559-1562.

VUAGNAT, M. (1946): Sur quelques diabases suisses. Contribution à l'étude des spilites et des pillow-lavas. Schweiz. Min. und Petr. Mitt. 26, 116-298.

ZWART, H.J. (1967): The duality of orogenic belts. Geol. en Mijnbow, 283-309.

Comments on Spilitization of the Permian Eruptive Rocks of the Choč Nappe in the West Carpathians, Slovakia

J. Vozár

Abstract

The paper deals with the position and character of the Permian embryonal magmatism in one of the West-Carpathian nappes. Products of the volcanism, i.e. melaphyres, were affected by autometamorphic spilitization, and locally by Alpine epizonal dynamometamorphism.

The Alpine geosyncline in the area of the Central West Carpathians is characterized by the development of the following series: the mantle series (Tatrides), Sub-Tatrides and Gemerides. The series of Sub-Tatrides is formed by the Krížna nappe, the Choč nappe and some higher nappes. Among the nappes cited, only the Choč nappe is built, in its basal part, of Upper Carboniferous and Permian rocks. The younger Paleozoic of the Choč nappe represents the first constituent of the developing Neoidic geosyncline. The Permian period is characterized by shallow-water clastic sediments and by the opening of embryonal faults. These faults offered ways for the ascent of the magmatic masses represented predominantly by the products of basic magma (melaphyres) with some indications of local differentiation characterized by the rise of melaphyre-porphyrites and porphyrites.

The Permian volcanism in the Choč nappe was submarine, pulsating in character, and its products (mainly melaphyres) were accompanied by tuffs, tuffites, and tuffaceous sediments. The melaphyres may be divided into fine-grained, medium-grained, porphyritic, porphyritic-amygdaloidal, and amygdaloidal varieties on the basis of their structural-textural features. The subsurface dike and sill bodies are homogeneous, consisting mainly of porphyritic or medium-grained varieties.

The Permian volcanism of the Choč nappe in the Central West Carpathians is pre-initial in relation to the Alpine geosyncline, and properly it is the embryonal magmatism in the sense of DE ROEVER (1959). The subsurface bodies in the underlying Carboniferous and in older Permian sequences, and effusions accompanied by volcanoclastic sediments only in the Permian, represent the products of this magmatic activity. The vein as well as the effusive bodies of the melaphyres occasionally shows spilitization phenomena.

Spilitization includes albitization, chloritization, epidotization. These alterations are of autometamorphic character in the broadest sense of the word, i.e., they represent processes occurring during the last stages of the solidification as well as those occurring later in the post-magmatic stage in the hydrothermal phase. More than 90 silicate analyses of these spilitized melaphyres show the following contents of some oxides:

SiO_2	44.0	- 58.0 %
TiO_2	0.9	- 3.47%
Fe_2O_3	2.3	- 5.7 %
FeO	1.2	- 5.5 %
CaO	3.26	- 11.4 %
Na_2O	2.6	- 5.49%
K_2O	0.23	- 2.16%
H_2O	0.39	- 8.15%

The contents of oxides are highly variable. Generally, analyses with increased alkali contents predominate: Na_2O usually exceeds 3.45%, CaO is usually lower than 9.15%. In most analyses, SiO_2 exceeds 48%, TiO_2 1.24%, Fe_2O_3 exceeds 3.45%; and FeO is lower than 3.15%.

Three analyses of the strongly epidotized samples (more than 90% alteration) show CaO contents from 18.84 to 22.80%. Their planimetric analyses show, in addition to predominating epidote, an increased content of calcite, and then chlorite, pumpellyite, chalcopyrite, pyrite, magnetite, hematite (VOZÁR, 1973).

The above mentioned alterations produced the following mineral assemblages:

1. albite + chlorite
2. albite + chlorite + epidote
3. albite + chlorite + epidote + calcite

These assemblages were identified mainly according to the work of VALLANCE (1960).

Apart from the above mentioned alterations there occurs the pumpellyite-prehnite-quartz association which was studied earlier by VRÁNA and VOZÁR (1969). Both of these authors propose the genesis of this association (facies) in relation to the overthrusting of the Choč nappe, spatially bound to the megastructure of the West Carpathians. The complex of younger Paleozoics including the Permian volcanics has been overlain by a thick sequence of complex rocks which leads to the assumption of a deep geosynclinal sinking. To this are added the effects of thrusting of the Choč nappe and of the resulting dislocation metamorphism. This can be seen especially in the lower part of the younger Paleozoic rocks which form the base for the Choč nappe and which have played an important role in its thrusting. The mineral association of the pumpellyite-prehnite-quartz facies has been found in the near-surface vein bodies of porphyries and melaphyres within the Carboniferous sequence as well as in a few veins and effusive bodies within the lower part of the Permian. These are occurrences which lie near the thrust line of the Choč nappe. In the remaining occurrences of eruptive rocks - the younger vein and effusive bodies in the lower and predominantly in the middle and upper part of the Permian sequence - the above mentioned metamorphic facies has not been found.

As has already been stated (VOZÁR, 1967, 1968), the Permian volcanic rocks, mainly of the fine-grained variety, can be compared with some other occurrences in the Alps, especially the spilites (AMSTUTZ, 1954) described from Glarner Freiberg, belonging to the Permian of the Mürtschen nappe.

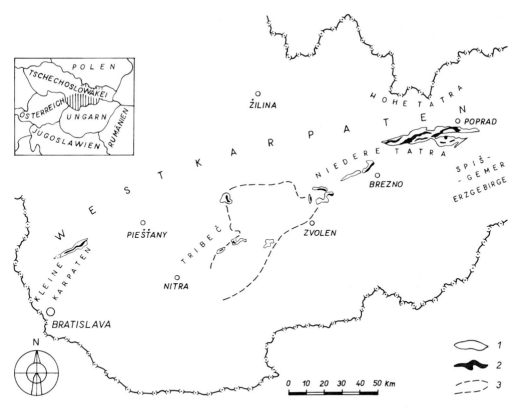

Map of the occurrence of the Upper-Paleozoic vulcano-sedimentary complex in the Choč nappe of the West Carpathians. 1. Upper-Paleozoic sediments; 2. the volcanic rocks (melaphyre, melaphyre-porphyrite, porphyrite); 3. assumed extent of the Upper-Paleozoic vulcano-sedimentary complex of the Choč nappe below the Neogene

References

AMSTUTZ, G.C. (1954): Geologie und Petrographie der Ergußgesteine im Verrucano des Glarner Freiberges. Vulk. Immanuel Friedlaender, No. 3, 150, Zürich.

DE ROEVER, W.P. (1959): Schwach alkalischer frühgeosynklinaler Vulkanismus im Perm der Insel Timor. Geol. Rundsch. 48, 179-184.

VALLANCE, T.G. (1960): Concerning spilites. Linn. Soc. N.S. Wales Proc. 85, pt. 1, 8-52.

VOZÁR, J. (1967): Der permische Vulkanismus in der Choč-Decke (Westkarpaten). GP, Zpr. 42, 79-83, Bratislava.

VOZÁR, J. (1968): Der permo-mesozoische Vulkanismus in den Westkarpaten und Ostalpen. Published on the occasion of the XXIII Intern. Geol. Congress. GP, Zpr. 44-45, 149-162, Bratislava.

VOZÁR, J. (1973): Chemismus der permischen Vulkanite der Choč-Einheit in den Westkarpaten. Science of the Earth, Geologica VII, 1-105, SAV Bratislava.

VRÁNA, St., VOZÁR, J. (1969): Minerálna asociacia pumpellyit-prehnit-
 kremennej fácie z Nízkych Tatier (Západné Karpaty). GP, Správy 49,
 91-100, Bratislava.

# 4. Papers Proposing a Secondary, Diagenetic or Metamorphic Origin

Spilites as Weakly Metamorphosed Tholeiites

M. H. Battey

Abstract

A review of recent work on the spilite problem leads to the conclusions 1. that the original magma was of tholeiitic type, 2. that the production of clinopyroxene by subsolidus reactions does not provide a general explanation of spilite mineralogy and texture, 3. that reaction with sea water is limited to palagonitization of glassy pillow lava selvedges, and that modified selvedge compositions are not relevant to the production of spilitic composition in the rocks as a whole, and 4. that weak metamorphism of tholeiitic basalt, in an open system, in the presence of saline connate waters, provides the best explanation of the overwhelming majority of spilites.

Definition

All discussions of the spilite question still require one to begin with a definition of the kind of spilite that is meant. VALLANCE's (1960) compilation and display of 92 chemical analyses shows that, unless some selection is exercised, any primary characteristics that the group may have are hopelessly lost in the secondary changes to which this composition range is so susceptible.

YODER (1967) wisely confined his study to "apparently unmetamorphosed", calcite-free clinopyroxene-albite-chlorite assemblages. If we may add to this that it should have a well-preserved igneous texture, we have a good definition of the most promising kind of spilite to study. Good examples are described by BATTEY (1956). In composition this rock will be a sodic basalt, but its definition is essentially mineralogical, not chemical. It is not merely a sodic basalt, but an albite-basalt.

When selected according to this definition, spilites seem to possess many characteristics allying them with the tholeiitic basalts and dolerites of the continents, but they share several chemical features that distinguish them from other members of this group. These features include high contents of water, soda and ferric iron, and low contents of potash (leaving aside poenites and SARGENT's (1917) potash-spilite), of lime and of magnesia. The chemical trends from tholeiitic dolerite to spilite, which seem to be nearly, but not quite, continuous, have been exhibited by BATTEY (1956, Table 5).

The question that arises is when and how did the albite and chlorite form, and the resulting high Na and other chemical differences arise?

Reality of the Chemical Differences

There is a tendency at present to minimize the characteristically high Na content of spilites. Thus YODER (1967) states that spilites, apart from a high water content, are not very different from basalts in composition. The high Na content found in many "could be attributed in part to metasomatism, yet may be a normal feature of the parental material". Beyond this, he does not discuss the high Na content, though his general thesis is that external metasomatising agents are not required.

AMSTUTZ (1968) has suggested that the supposed Na-enrichment of spilites is a sampling error, and that if duly weighted analyses of all members of the rock association were taken, we should arrive at the composition of an "ordinary" basalt.

With the present tendency to move the spilites from the igneous to the metamorphic realm, interest is focused on the stability field of the mineral assemblage rather than on the origin of the bulk composition. Thus it is possible to have a discussion of the relation of spilites to the zeolite facies in which the fact that spilites are Na-rich is not even mentioned!

That point of view is not taken here. Whatever may be true of the average composition of all associated rocks, the spilitic rock with whose origin we are concerned is a chemically distinctive rock. This is what first brought it to notice, and it has been established by many later analyses of typical, apparently fresh samples.

Primary Phases

Protospilite is accepted, on the basis of the affinities mentioned above, as being a magma of tholeiitic basalt composition which was almost completely liquid, as demonstrated by selvedges chilled originally to glass, and the common paucity of phenocrysts. Experimental studies have failed to show any way in which clinopyroxene, albite and chlorite could precipitate together or successively at the liquidus of such a melt. The primary phases would be clinopyroxene and calcic plagioclase. As in many tholeiites, olivine is absent or in small amount.

In spilites there are sufficient records of calcic plagioclase preserved in selvedges or as unreplaced cores of crystals, to show that the original plagioclase probably was calcic.

It is only by invoking some unproved influence (previously suggested to be that of water) upon the melt at the time of crystallization that albite and chlorite can be supposed to be primary minerals. Rejecting, as a result of recent experiments, this unspecified influence of water, we must conclude that the primary minerals have been altered after crystallization.

Processes of Alteration

Many possible processes of alteration have been suggested, and many indeed have affected this or that group of spilites. Effects due to metamorphism, carbonation, saussuritization, hydrogrossular formation (which we might call rodingitization), zeolitization, palagonitization and albitization have in the past been stitched together to form a Frankenstein's monster called the "Spilite Problem". The process that shifts about in the order listed above is albitization, which is held by different authors to take place at different times, from initial contact with sea water at the time of eruption to the long-subsequent period of burial metamorphism at depths of 15 000 to 20 000 feet.

A useful approach would seem to be to examine these different processes separately and see to what extent they may be independent of one another and how closely some of them may be linked.

Some of the suggested processes will be considered below in the following order:

1. subsolidus reactions under high water pressure
2. reaction with sea water during cooling
3. zeolite facies metamorphism
4. palagonitization

Subsolidus Reactions in Wet Magma

From the early experiments performed at high pressures of water vapor (YODER, 1952), it was found that chlorite was stable to temperatures of 680°C. It was thought then (e.g. LEHMANN, 1965) that it might later be found to be stable at the liquidus. More recent experiments (YODER, 1967) have shown that, though chlorite is not found on the liquidus, clinopyroxene and chlorite will form together in the solid state, from forsterite + anorthite by the reaction

$$CaAl_2Si_2O_8 + 3Mg_2SiO_4 + 4H_2O = CaMgSi_2O_6 + Mg_5Al_2Si_3O_{10}(OH)_8$$

or from enstatite + anorthite by

$$CaAl_2Si_2O_8 + 6MgSiO_3 + 4H_2O = CaMgSi_2O_6 + Mg_5Al_2Si_3O_{10}(OH)_8 + 3SiO_2$$

These reactions yield clinopyroxene and chlorite below about 650°C to 700°C at 1 to 2 kb water pressure.

On this basis, YODER suggests that spilites are either autometasomatic products of an earlier basaltic or gabbroic mineral assemblage brought about by retention of water in the magma, or normal basalts that have been metamorphosed at depth to the spilite mineral assemblage. For some reason, not specified, he thinks the second alternative is uncommon, though because of the environment of spilites in geosynclinal deposits later uplifted with folding, one might think it would be very common.

The difficulty with either alternative, and indeed with the whole idea of subsolidus formation of clinopyroxene and chlorite from an assemblage including olivine and anorthite, is that of texture. The texture of many spilites can be closely matched amongst tholeiitic basalts and may show features of great delicacy, such as curving fronds of pyroxene and regular interlacing growths of titaniferous magnetite. The chlorite commonly occurs in intersertal patches and ocelli like those typical of the latest-solidifying constituents of basic lavas and dikes. It is difficult to believe that these textures were formed by a solid-state transformation such as that proposed by YODER. If textures mean anything, the clinopyroxene of spilites, at least in many cases, is a phase precipitated on the liquidus. The form of the albite laths is often also strongly indicative of a phase separating on the liquidus.

If it is assumed that the textural evidence is reliable on this point, it follows that any change in the composition of the pyroxene or plagioclase has been brought about by a strictly pseudomorphous process and the chlorite is unlikely to be a partial product of a reaction between earlier anorthite molecules and olivine or enstatite. It seems more probable that the pyroxene has persisted metastably since it precipitated at the liquidus.

If YODER's scheme were acceptable, however, it would still be necessary to account for the retention of water to promote the reactions, the high Na content of the rock and other smaller chemical differences from common tholeiites.

Autometasomatism by retention of water requires that, amongst spilitic lava flows, the mineral transformations should take place during what must be rapid cooling on the sea bed at pressures well below 1 kb. Opposing this are the many examples of submarine pillow lavas that are not spilitic, and the fact that the spilite mineral assemblage is known very rarely, if at all, from young lavas that have not been under considerable later cover (with the long time span that this implies).

The case is, of course, different for intrusive spilites and for the central parts of thick flows (NIGGLI, 1952), where the pressure and cooling rate might permit autometasomatism. Some of these examples ought perhaps to be treated separately from the pillow lavas and quickly cooled masses.

The high Na content and other compositional features different from basalts appear to demand the transfer of elements beyond the distances involved in YODER's proposed reactions. YODER himself does not deal with this point, as he assumes that the bulk composition falls within the spectrum of basalt. On that basis its expression as spilite is simply the result of crystallization in a watery environment, much as suggested (without any specified reaction such as that now demonstrated) by BATTEY (1956). It is now felt, however, that the bulk composition demands an explanation not provided by this mechanism.

Reaction with Sea Water

This old hypothesis is disproved by the existence of vast amounts of nonspilitic basalts, including pillow lavas, erupted at both great and shallow depths in the oceans, and by nonspilitic intrusions emplaced in marine sediments relatively soon after their deposition (e.g. quartz-dolerites of the Midland Valley of Scotland). The inclusion of saline waters amongst the reactants remains, however, the only hypothesis offering a specific source of Na to account for the characteristic compositional feature of spilites.

Zeolite Facies Metamorphism

Since the recognition by COOMBS (1954) of the mineralogical changes produced by low-grade burial metamorphism of tuffaceous sandstones, it has been clear that this zeolite facies metamorphism must have a very important bearing on the origin of spilites. The following points are particularly significant: 1. apart from pyroxene, which must be regarded as persisting metastably, the mineral assemblage albite + chlorite with associated epidote, prehnite, pumpellyite, and sphene comprises minerals found to be stable under zeolite facies conditions. Less common associates like hydrogrossular and babingtonite fit into this picture. Augite was found to persist in the first-described zeolite facies assemblages. 2. As has been mentioned above, spilites are sparsely represented amongst lavas that have not been subsequently buried, and in general, spilites are not found in young (Tertiary and later) formations. 3. Preservation of original textures of tuffaceous sediments is a notable feature of this low-grade metamorphism. The pseudomorphous replacement of volcanic glass shards by zeolites has been described by COOMBS (1954), and calcic plagioclase crystals appear to be directly replaced by albite.

These are strong arguments for believing that processes at least closely related to those of zeolite facies metamorphism have been responsible for the spilite mineralogy.

The bulk composition of spilites must be borne in mind when applying to them the concept of low-grade metamorphism. The whole rock body has apparently been modified from the composition of other basalts, so that the reacting materials must have constituted an open system. This is supported by the common occurrence of the epidote, prehnite, pumpellyite etc. in former cavities in the rock, both as amygdales and veinlets. It is not possible to say anything about the relative pressures except that it seems likely that P_{H_2O} would probably be about equal to P_{load}; hydrous phases occur in both rock and former cavities, but Ca-bearing phases occur mainly in the former cavities and not in the rock. The solubility of Ca was greater in the rock than in the fluid-filled spaces, and this state of affairs persisted for a long time. We are not dealing with isochemical changes. In this system some source for the Na of the albite is required. Connate saline waters, acting over a long time under conditions of mild reheating, might provide this.

With these additional considerations in mind, it seems that metamorphism of tholeiitic basalt under zeolite facies conditions provides the best explanation of the origin of spilites.

Palagonitization

The early observations of SLAVIK (1929) drew attention to the curious Si-poor, Fe- and Mg-rich compositions of pillow lava selvedges in rocks described by him as spilitic. This observation was foreshadowed by the very basic compositions of analyzed spilites from Devonport Workhouse Quarry (FLETT, 1907, p.97), Chipley (FLETT, 1913, p.56) and Amlwch, Anglesey (GREENLY, 1919, p.55) which may be explained by the inclusion of some selvedge material in the samples. Such selvedges have been studied particularly by VUAGNAT (1949a and b) who found them to be composed largely of chlorite. The present writer has analyzed chemically and by X-ray diffraction a selvedge of former glass with well-preserved perlitic cracks that is now composed almost wholly of extremely fine-grained biotite.

These selvedges have been regarded as indicating local metasomatic movement of material connected with change of the pillow lava into spilite. Allied to this idea is that of radial differentiation within individual pillows.

Recent work on palagonite formation from basalt glass on the present sea bed near Hawaii (MOORE, 1966) has thrown new light on the situation and suggests that this aspect of spilite chemistry can be divorced from the spilite problem and treated separately. MOORE shows that the chemical changes include reduction of Na by about 2/3 and of Ca by 9/10 with an increase of K by 2 to 4 times. He points out that ancient submarine pillow lavas must have been subjected to the same changes. Upon subsequent burial the palagonite may crystallize to chlorite or biotite.

MOORE also suggests that the radial chemical differences reported in some lava pillows may be due to an extension of the same process. This possibility, though undeveloped, should certainly be kept in mind.

The recognition of the role of sea water in palagonitization of cold basaltic glass, as a development separate from albitization, removes one extraneous element from the problem of spilitic alteration of basalts, a result for which any worker in this field must be grateful.

Relationships to Keratophyres

The keratophyres have never posed as great a problem as the spilites, because the high alkali content could be explained by fractional crystallization. The writer believes that alkali migration under conditions of burial metamorphism, with concentration of K in originally glassy areas and of Na in originally crystalline parts accounts for most of their compositional features as compared with ordinary rhyolites (BATTEY, 1955). They may also, of course, have lost their little original Ca and gained some Na in the same manner as spilites.

There are certainly intermediate members between spilite and keratophyre that lie compositionally on variation curves consistent with an origin of the felsic from the mafic magma by fractional crystallization. Any questions about this relationship are separate from the spilite problem itself and need not be discussed here.

Summary

Though experiments show that clinopyroxene, albite and chlorite can develop in the solid state from anorthite + forsterite or anorthite + enstatite, the required water pressures and cooling time for this will not have been available for many spilites, while the textures of the rocks indicate clinopyroxene has persisted metastably from the liquidus. The reactions may be possible inside deep-seated sills or thick flows.

Reaction with sea water during cooling does not cause spilitization of basalt.

The overwhelming majority of spilites are pre-Tertiary in age and have been buried under later strata. Their mineralogy is consistent with zeolite facies metamorphism, but the differences in bulk composition from basalt imply an open system and some external source of Na during the process.

Chloritic or biotitic compositions of pillow selvedges are ascribed to palagonitization by reaction of the original glass with sea water, in the cold, over thousands of years. Possibly radial change in composition within pillows is related to this.

These views conform with the explanation of the compositions of associated keratophyres by local alkali metasomatism of original rhyolitic rocks.

References

AMSTUTZ, G.C. (1968): Spilites and spilitic rocks in Hess, H.H. and Poldervaart, A., Basalts: The Poldervaart Treatise on rocks of basaltic composition, Vol. 2, 737-753.

BATTEY, M.H. (1955): Alkali metasomatism and the petrology of some keratophyres. Geol. Mag. 92, 104-126.

BATTEY, M.H. (1956): The petrogenesis of a spilitic rock series from New Zealand. Geol. Mag. 93, 89-110.

COOMBS, D.S. (1954): The nature and alteration of some Triassic sediments from Southland, New Zealand. Trans. roy. Soc. N.Z. 82, 65-109.

DICKINSON, W.R.

FLETT, J.S. (1907): The geology of the country around Plymouth and Liskeard. Mem. geol. Surv. Gt. Britain.

FLETT, J.S. (1913): The geology of the country around Newton Abbot. Mem. geol. Surv. Gt. Britain.

GREENLY, E. (1919): Geology of Anglesey, Vol. I. Mem. geol. Surv. Gt. Britain.

LEHMANN, E. (1965): Non-metasomatic chlorite in igneous rocks. Geol. Mag. 102, 24-35.

MOORE, J.G. (1966): Rate of palagonitization of submarine basalt adjacent to Hawaii. U.S. geol. Surv. Prof. Paper 550-D, D163-D171.

NIGGLI, P. (1952): The chemistry of the Keweenawan lavas. Amer. J. Sci. Bowen Volume, 381-412.

SARGENT, H.C. (1917): On a spilitic facies of Lower Carboniferous lava-flows in Derbyshire. Quart. J. geol. Soc. Lond., 73, 17-23.

SLAVIK, F. (1929): Les "pillow-lavas" algonkiennes de la Bohême. C.R. XIV Congr. geol. internationale, 1926, fasc. 4, Madrid, 1387-1395.

VALLANCE, T.G. (1960): Concerning spilites. Proc. Linnean Soc. N.S.W. 85, 8-52.

VUAGNAT, M. (1949a): Sur les pillow-lavas dalradiennes de la péninsule de Tayvallich (Argyllshire). Schweiz. Min. petrogr. Mitt. Band 29, H. 2, 524-536.

VUAGNAT, M. (1949b): Variolites et spilites: comparaison entre quelques pillow-lavas britanniques et alpines. Archives des Sciences (Société de Physique et d'Histoire naturelle de Genève), vol. 2, fasc. 2, 223-236.

YODER, H.S. (1952): The $MgO-Al_2O_3-SiO_2-H_2O$ system and related metamorphic facies. Amer. J. Sci. Bowen Volume, 569-627.

YODER, H.S. (1967): Spilites and serpentinites. Carnegie Institution Year Book 65, 269-279.

On the Mineral Facies of Spilitic Rocks and Their Genesis

D. S. Coombs

Abstract

A consideration of stability fields of spilite minerals precludes a primary, magmatic origin for these minerals. Furthermore the water content of a typical spilite is not readily interpreted as being primary magmatic water originally contained in the same flow or sill. Chemical and mineralogical features of spilites are shown by burial metamorphic sequences of volcanogenic sediments. In terms of mineral facies, spilites occasionally may be ascribed to the zeolite facies, and more commonly to the prehnite-pumpellyite, pumpellyite-actinolite schist, and greenschist facies. The prehnite-pumpellyite facies is particularly commonly represented, especially in New Zealand. In general, the mineral facies of spilites conforms to that of the associated sediments, bearing in mind possible differences in μ_{CO_2}, μ_{H_2O}, μ_{O_2} between volcanic rock and sediment and the fact that chemically mature sediments deficient in volcanogenic materials are commonly relatively insensitive to metamorphism at low grades and are often described as "unmetamorphosed". An important exception arises in the case of exotic masses in tectonic mélanges. The common albite-chlorite-calcite assemblage of many spilites and metasediments is not critically diagnostic of any specific low-grade mineral facies. These arguments do not preclude the possibility that some spilites obtained their present mineralogy during the initial cooling phase, perhaps under a blanket of other volcanic and sedimentary materials under abyssal conditions. Nevertheless the fact that typical spilites do not preserve a mineralogical record of progressively falling temperature, as in some clearer cases of deuteric alteration, helps make this origin improbable. Most or all spilites can be interpreted as low-grade metabasalts.

Introduction

In a recent review, AMSTUTZ (1967) concludes that spilites are volcanic rocks of widely variable mineralogical and chemical composition. The main minerals are albite, chlorite, epidote, calcite, and iron oxides, sometimes with abundant pyroxene. Actinolite sometimes occurs but is commonly regarded as metamorphic. One might add that pumpellyite and to a less extent prehnite are by no means uncommon and that tiny granules of sphene are almost ubiquitous. "This composition and its variability may be the only real property that spilites do not have in common with basalts" (AMSTUTZ, 1967, p.749). Many spilitic rocks are mineralogically far from homogeneous. AMSTUTZ suggests that heterogeneity results from a transfer and differentiation of constituents in a separate aqueous phase during primary crystallization. Similar views have been put forward by DONNELLY (1966). Placing the origin of the mineralogy at an only slightly later stage during the cooling of the spilites, is the autolysis theory of DEWEY and FLETT (1911). In contrast, in a study of alteration of Ordovician submarine basic lavas in New South Wales, SMITH (1968) finds variants in the one outcrop rich in albite-chlorite-pyroxene on the one hand and pumpellyite and/or epidote plus quartz on the other hand. He attributes this to segregation under low-grade burial-metamorphic conditions affecting country rock sediments and volcanics alike. Such alteration follows deep sedimentary or tectonic burial and hence moderate rise in temperature, perhaps long after the effusion and complete cooling of the spilite parent lavas.

Agreeing that the mineral assemblage of spilites can be produced by metamorphism, AMSTUTZ states (op. cit., p.741): "Metamorphosed spilites are probably abundant, but the first problem to be solved is that of the nature of unmetamorphosed spilites". In more radical fashion, the question may be posed: "Do nonmetamorphic spilites exist?"

The mineral assemblage of spilites, keratophyres, and albite dolerites might conceivably be generated under the following conditions:

1. primary magmatic,

2. following separation of an aqueous phase, but still during primary crystallization (i.e. the spilite minerals do not replace pre-existing minerals),

3. by autolysis, i.e. hydrothermal alteration of pre-existing minerals while the parent volcanic rock is cooling,

4. by burial, regional, or hydrothermal metamorphism involving some rise in temperature subsequent to the initial cooling.

In this paper mineralogical and certain other evidence bearing on these possibilities will be examined. Most of the arguments will be centered on spilites, but can also be applied to the commonly associated keratophyres and albite dolerites.

Stability Fields of Spilite Minerals

Some available data on mineral stabilities relevant to the origin of spilites are summarized in Fig. 1. Assuming a maximum P_{H_2O} of a little over 1 kb for the crystallization of basalts under deepest oceanic conditions (i.e. at a depth of about 10 km), it is seen that laumontite, prehnite, pumpellyite plus quartz, clinozoisite plus quartz, and the chlorites, all have maximum stability temperatures well below temperatures of complete crystallization of the varied suite of basalts investigated by YODER and TILLEY (1962). While the upper stability limit of pistacitic epidote plus quartz may be a little to the right of the epidote curve plotted (HOLDAWAY, 1972), and the full complexity of pumpellyite relations remains to be explored, the general conclusions to be drawn from Fig. 1 are unlikely to be modified.[1]
In complex natural systems the phases concerned will tend to disappear by reactions taking place at temperatures below the maximum

---

[1] The pumpellyite + quartz boundary shown in Fig. 1 is that of LANDIS and ROGERS (1968) and represents the P-T conditions under which natural pumpellyite was destroyed in the presence of quartz in runs of 7 to 14 days. Reversal was not demonstrated and the equilibrium boundary must occur at a lower temperature (cf. HINRICHSEN and SCHÜRMANN, 1969).

stability limits, and in all cases decomposition will take place at lower temperatures than those shown if the fluid phase is other than pure water under the same pressure as the solid phases. Unless the observed mineralogy is the result of non-equilibrium quench processes, it is clear that basaltic lavas, even those of the deep oceans, will be completely crystalline long before cooling reaches the point where characteristic spilite mineralogy can develop. This prediction is supported by the numerous examples of fresh, unspilitized basalts recovered from deep oceanic sites and by the data of YODER and TILLEY (1962, p.468). A primary magmatic origin for spilite mineralogy appears to be precluded.

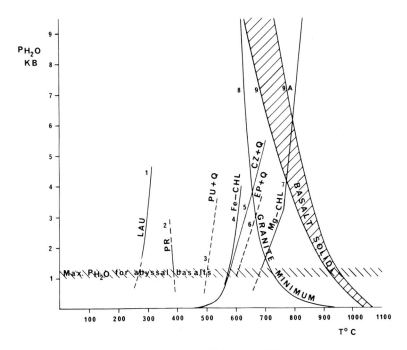

Fig. 1. Phase equilibria relevant to the spilite problem.
Curves 1-7: Upper temperature limits of stability for minerals indicated.
1. Laumontite (LIOU, 1971a). 2. Prehnite (STRENS, 1968). 3. Pumpellyite + quartz (LANDIS and ROGERS, 1968). 4. Fe-chlorites (TURNOCK, 1960). 5. Clinozoisite + quartz (HOLDAWAY, 1966). 6. Epidote + quartz (minimum temperature; HOLDAWAY, 1966). 7. Mg-chlorite (FAWCETT and YODER, 1966). Curve 8: The granite minimum melting liquid (TUTTLE and BOWEN, 1958). Curves 9-9A: Upper- and lower-temperature extremes of basaltic solidi investigated by YODER and TILLEY (1962).

Note: For more recent data on the stability relations of prehnite, pumpellyite and epidote, see LIOU (1971b), HINRICHSEN and SCHÜRMANN (1969) and HOLDAWAY (1972)

The Role of Magmatic Water in Spilitization

According to HAMILTON et al. (1964) the solubility of water in typical basalt and andesite liquids at 1 kb, 1100°C, is approximately 3% and 4.5%, respectively. At pressures corresponding to normal oceanic depths of 3,800 meters, the maximum solubilities would be roughly 1.5% and 2.5%, respectively, although these water contents apparently were not attained in the oceanic basalts erupted at depths as great as 5 km, investigated by MOORE (1965). The composition of the vapor phase that escapes from basalt magma while silicate crystallization proceeds, or which forms vesicles as pressure is released, presumably approximates that of the volcanic gases which escape during basaltic eruptions. Numerous analyses of these gases, together with the lack of solid phases in the vesicles of historic lavas and of fresh basalts dredged from abyssal depths, indicate that this vapor phase contains little dissolved solid matter. This vapor phase is not a promising medium, either in quantity or in composition, for the pervasive <u>primary</u> crystallization of the hydroxyl-bearing and carbonate minerals of spilites, nor for any significant metasomatism of the spilite, diabase, or adjacent country rock. Such water as remains after vesiculation might of course partake in hydration reactions during cooling if cooling is sufficiently protracted, but the amount of such water is very small, much less than the 3 to 5% water commonly shown in spilite analyses (VALLANCE, 1960). It appears that spilite mineralogy, pervasive through a lava flow or sill, is not to be ascribed to the agency of magmatic water initially contained in that flow or sill.

The Proposed Spilite Reaction of YODER (1967)

YODER had recently presented evidence for a reaction:

forsterite + anorthite + water = diopside + chlorite

which proceeds to the right at temperatures below about 650°C at 1 kb P_{H_2O}, or 700° at 2 kb P_{H_2O}. He suggests that this reaction is critical to the formation of the augite-chlorite assemblage found in many spilites. A maximum P_{H_2O} of about 1 kb for deep-sea basalts implies that this reaction would be confined to subsolidus conditions. Two arguments can be levelled against YODER's proposal as a common subsolidus reaction in spilites:

1. The textures of augite-bearing spilites and of albite dolerites appear to be incompatible with the proposed reaction and in fact are typically igneous in character. We may here include as typically igneous the variolitic textures and "bow-tie" sheaves of pyroxene fibers found in some fresh oceanic basalts such as those pictured by MUIR and TILLEY (1966). These textures closely match those of many spilites and indicate primary magmatic (including quench) crystallization of augite as well as feldspar.

2. YODER's proposed reaction implies that at temperatures below the reaction curve, either of the three-phase assemblages - diopside-chlorite-forsterite or diopside-chlorite-anorthite - should be stable. Neither of these assemblages is known in spilitic rocks. Furthermore although in some spilites augite appears to be almost ideally fresh, it often shows at least partial replacement by assemblages such as chlorite ± sphene ± pumpellyite, as has been noted in the case of spilites from Northern Maine (COOMBS et al. 1970).

Augite was thus unstable relative to chlorite - and to chlorite-calcium aluminosilicate-bearing assemblages - during the production of the diagnostic spilitic mineralogy. It is interpreted by the present writer as relict. This writer would attribute the ubiquitous chlorite of spilitic rocks in part to alteration of basaltic glass (including glassy inclusions in plagioclase), in part to the post-magmatic infilling of vesicles, and in part to the alteration of pyroxene and other mafic minerals especially olivine (cf. MELSON and VAN ANDEL, 1966).

Spilitic Mineralogy in Volcanogenic Sediments

The writer has pointed out (COOMBS, 1954) that mineral assemblages and chemical compositions approaching those of spilites have been produced by burial metamorphic processes in andesitic volcanic gray-wackes of the North Range Group in the Taringatura district of southern New Zealand. In this case, massive volcanic rocks do not occur and possibilities of spilitization of the sediments during primary cooling are not applicable. Close analogies with spilite mineralogy and chemistry are also provided by the Taveyanne Sandstone of the Helvetic nappes in the western Alps. MARTINI (1968), in a detailed account of this formation as it occurs in Haute-Savoie, shows that the faciès moucheté is characterized by patches segregated on the millimeter or occasionally centimeter scale consisting essentially of albite + laumontite, in a matrix dominated by albite and chlorite and containing relict clinopyroxene. Prehnite, pumpellyite, or epidote take the place of laumontite in higher-grade variants. Other minerals in these altered andesitic graywackes include dusty sphene and rare quartz. The faciès verte contains little or no calcium aluminosilicate and the essential authigenic mineral assemblage is albite-chlorite-sphene. Individual meta-andesitic or metabasaltic lithic fragments in the Taveyanne Sandstone, as in the volcanic gray-wackes of the North Range Group, are of keratophyric or spilitic aspect and mineralogy, though demonstrably of andesitic or basaltic origin as is proved by unaltered relicts such as in the calcitized margins of sandstone beds described by MARTINI.

These "spilitic" features of the Taveyanne and North Range sediments can only be regarded as low-grade metamorphic, or alternatively as of late-diagenetic origin, depending on the definition adopted for these intergradational processes (cf. COOMBS, 1962, p.213-214). Presumably basaltic and andesitic lava flows may be affected in the same way as the sediments by low-grade metamorphic processes.

The Mineral Facies of Spilitic Rocks

Evidence has been cited above to support the writer's belief that pyroxenes of spilitic rocks (and similarly occasional high-temperature plagioclase) are to be regarded as relict pyrogenic phases. The phases responsible for the distinctive mineralogy of spilites are low-albite and chlorite, commonly in association with one or more of the following: calcite, epidote, pumpellyite, prehnite, sphene, actinolite, hematite. These are minerals typically developed in low-grade mineral facies[2], specifically the prehnite-pumpellyite facies, the pumpellyite-actinolite schist facies, and the greenschist facies. Occasional reports of wairakite (DONNELLY, 1962) and of laumontite in spilitic rocks, and perhaps the laumontite-analcime-thomsonite assemblage of certain albite metagabbros of the Otama Complex, New Zealand (COOMBS et al. 1959) suggest that the zeolite facies may sometimes be represented. Reconstitution under physical conditions appropriate to the prehnite-pumpellyite facies may well be the commonest situation. However some spilitic assemblages which lack calcium aluminosilicates, such as albite-chlorite-calcite $\pm$ hematite, sphene, quartz, are non-diagnostic as to the specific low-grade mineral facies to which the assemblage should be ascribed.

Critical assemblages of the prehnite-pumpellyite facies are shown in Fig. 2. Epidote is not shown on this diagram; the epidote of spilitic rocks is characteristically rich in ferric iron. Assemblages containing alkali-free phases more aluminous than pumpellyite are not well known in this facies, and are not shown. More complete analyses of prehnite-pumpellyite and contiguous facies assemblages, as seen in an Al_2O_3-CaO-$(Fe,Mg)O$ projection, and as a function of μH_2O, μCO_2 are given by COOMBS et al. (1970). Variants of the spilites such as those of the Tasman Geosyncline, New South Wales, described by SMITH (1968),

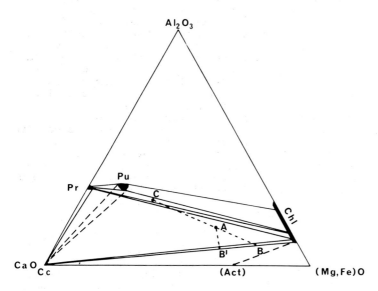

Fig. 2. Some partial mineral assemblages of the prehnite-pumpellyite facies. Al_2O_3: Residual alumina after formation of micas, albite and adularia. CaO: Residual CaO after formation of sphene and apatite. A: Typical basaltic composition. B,C: Spilitic variants formed by complementary segregation of components of pumpellyite, prehnite and/or epidote and of chlorite-calcite. B': Chlorite-calcite assemblage formed by addition of CO_2 and Na_2O or K_2O, forming micas, albite or adularia at the expense of other aluminous phases

---

[2] For a recent review of low grade mineral facies, see SEKI (1969). While the present writer would insert a lawsonite-albite-chlorite facies at slightly lower pressures than the glaucophane schist facies, and feels that the temperatures indicated in SEKI's Fig. 3 may be underestimates, he is otherwise in general agreement with the conclusions outlined.

and attributed by him to low temperature metamorphic segregation processes from basaltic compositions, may be represented on Fig. 2 by points near the line BAC joining pumpellyite-, prehnite- and/or epidote-rich assemblages (C) to chlorite-calcite assemblages (B). Point A represents a typical basaltic or andesitic composition, and Point B^1 represents a calcite-chlorite-(albite) assemblage which might be developed from A where μCO_2 and μNa^+ or μK^+ are sufficiently high to convert the Ca component of potential Ca-Al silicates to calcite, and the liberated Al to micas, albite, or adularia.

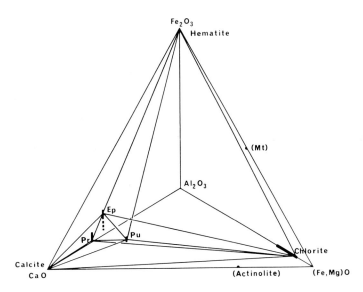

Fig. 3. Some mineral assemblages in the Al_2O_3-poor portion of the tetrahedron Al_2O_3-CaO-(Fe,Mg)O-Al_2O_3 for the prehnite-pumpellyite facies. Ep: epidote; Pr: prehnite; Pu: pumpellyite; Mt: magnetite

Many spilitic rocks with the assemblages albite-chlorite-calcite and albite-epidote-chlorite-calcite show a reddish pigmentation which is generally ascribed to dusty hematite. It is sometimes found, for example in the Torlesse Supergroup of New Zealand, that such spilitic metabasalts lack prehnite, which however may be present in associated graywacke-suite sediments. This is not to be ascribed to difference of mineral facies, but to a composition richer in ferric iron, as may be demonstrated in the Fe_2O_3-Al_2O_3-CaO-(Fe,Mg)O tetrahedron (Fig. 3; cf. LANDIS, 1969). On the Fe_2O_3-Al_2O_3-CaO face, it is seen that the assemblages hematite-epidote and hematite-epidote-calcite virtually preclude prehnite. Within the tetrahedron, the assemblages hematite-epidote-chlorite-calcite and hematite-epidote-chlorite-pumpellyite, found in relatively highly oxidized volcanics, are to be contrasted with isofacial assemblages epidote-chlorite-pumpellyite-calcite and prehnite-pumpellyite-chlorite-calcite typical of associated volcanogenic sediments in which more highly reducing conditions are maintained by the almost ubiquitous presence of partially graphitized carbon (LANDIS, 1969). Vertical bars in Fig. 3 represent Al_2O_3-Fe_2O_3 substitution in prehnite, epidote and pumpellyite.

Corresponding assemblages of the pumpellyite-actinolite schist facies are shown in Fig. 4. Here the critical assemblage is pumpellyite-actinolite-chlorite, and prehnite is restricted to relatively uncommon whole-rock compositions before disappearing completely. For higher values of μ_{CO_2} relative to μ_{H_2O} (COOMBS et al. 1970), the diagnostic assemblage may be suppressed and replaced by calcite-chlorite plus aluminosilicates.

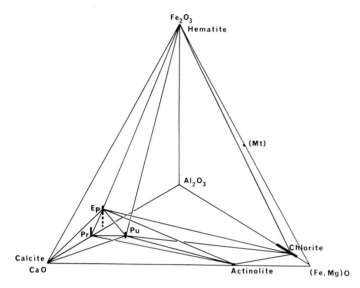

Fig. 4. Some mineral assemblages in the Al_2O_3-poor portion of the tetrahedron Al_2O_3-CaO-(Fe,Mg)O-Al_2O_3 for the pumpellyite-actinolite schist facies. Abbreviations as for Fig. 3

Production of Low Grade Mineral Facies During Cooling

Clearly a cooling lava flow must pass through a range of temperatures (but not necessarily through a range of μ_{H_2O}, μ_{CO_2}) appropriate for the production of the low grade mineral facies mentioned above, as well as the higher grade greenschist and amphibolite facies. Is the spilitic assemblage produced during this cooling? Many recent studies of deep oceanic basalts demonstrate that these rocks, although cooled under a great depth of water, have not reacted with sea water and have preserved basalt, not spilite, mineralogy. Spilites have recently been described from midocean ridge systems (e.g. MELSON and VAN ANDEL, 1966; CANN, 1969), but in these cases it is shown that the rocks concerned are metamorphic in the sense that their present mineralogy and chemistry has been produced under low grade metamorphic conditions of the greenschist or contiguous facies, apparently involving mild reheating after initial cooling and burial in the volcanic pile. Pumpellyite is rare or absent in these cases. Similarly, WALKER (1951, 1960) has shown that zeolitic amygdales in the plateau basalts of Antrim and Iceland have been formed <u>after</u> the accumulation of at least part of the lava piles in which they are found, and that mineral zones are superimposed on, and slightly oblique to, the volcanic stratigraphy. Nevertheless RAAM et al. (1969) have recently described the formation of pumpellyite associated with albite, chlorite, prehnite, biotite, analcime, clay minerals etc. in vugs and deuterically altered syenite differentiates in the Prospect alkali dolerite intrusion, New South

Wales, which is believed to have been emplaced under a cover of only about 200 meters. In this case there appears to be no serious possibility of temperature rise after cooling, and the pumpellyite presumably formed deuterically during primary cooling. It is to be noted however that even in such a comparatively thick and correspondingly slowly cooled intrusive sheet, the quantity of late-stage spilite-like mineral phases is small. Furthermore a range of secondary minerals is reported that is suggestive of crystallization through a range of falling temperatures.

It is conceivable that in a thick pile of pillow lavas emplaced <u>within</u> wet sediments or pyroclasts, cooling might be slow enough, and μH_2O might be maintained sufficiently high, to allow pervasive spilitic alteration. Whether or not this ever happens, relatively thick sequences of pillow lavas, pillow breccias and hyaloclasites are known which are extensively palagonitized but <u>not</u> spilitized, e.g. the Eocene Waiarekan and Deborah Volcanic Formations of North-East Otago, New Zealand. Furthermore, spilitization accompanying falling temperature according to an autolysis-type model might be expected to result in a sequence of minerals indicative of the plurifacial conditions inevitably encountered during such cooling. In fact, the mineralogy of spilites is usually simple, and apart from relict pyroxene or post-orogenic zeolite-filled joints, can normally be ascribed to a single low grade mineral facies.

Relationships of the Mineral Facies of Spilites to That of Their Country Rocks

It is crucial to the solution of the spilite problem that the mineral facies of spilitic suites be compared with that of country rock sediments. Too often it has been claimed without adequate evidence that the country rocks are unmetamorphosed. In fact, most spilites and albite dolerites occur in geosynclinal piles in which they have been buried by sedimentary or tectonic processes to depths of many kilometers, depths at which temperatures appropriate to the zeolite, prehnite-pumpellyite, or higher grade mineral facies cannot have been avoided. It cannot be overemphasized in this contect that by normal field and textural criteria, sediments subjected to these conditions may appear nonmetamorphic, yet mineralogically they may be very extensively reconstituted (e.g. COOMBS, 1954; CHAPPELL, 1968; MARTINI and VUAGNAT, 1965). In the writer's experience cleavage is rare in zeolite facies rocks, though rather common in the finer grained sediments of the prehnite-pumpellyite facies and usually conspicuous in the pumpellyite-actinolite schist facies. Workers on spilites will recognize that incipient cleavage is common at least in the finer grained sediments associated with spilites. This in itself is suggestive, albeit non-definitive, that typical spilitic suites have been exposed to physical conditions approximating those of the prehnite-pumpellyite or contiguous facies. CANN (1969) pointed out that in the case of environments of high heat flow such as midocean ridges, the requisite temperatures can be reached at surprisingly modest depths.

A major difficulty arises from the fact that the readily diagnostic mineral assemblages of the low-grade facies are more or less restricted to rocks of metavolcanic composition rather than to shales and limestones. In the Helvetic nappes of the Alps, MARTINI and VUAGNAT (1965) demonstrated that the Taveyanne Sandstone has mineral assemblages ranging from the zeolite facies to the prehnite-pumpellyite facies. Recent observations of COOMBS, MARTINI and VUAGNAT (in preparation) extend this progression to the pumpellyite-actinolite-schist facies. Yet the associated argillaceous rocks and limestones, relatively insensitive to the corresponding range of pressures and temperatures, are by most criteria unmetamorphosed. One may well suspect that the spilites occurring in the Pre-alps have also been exposed to P-T conditions capable of producing low grade metamorphism (cf. VUAGNAT, this volume, for a re-appraisal of Alpine spilites).

It may further be suggested that relatively thin volcanic formations associated with abundant calcareous and especially dolomitic sediments, may tend to be exposed to conditions of high μCO_2, resulting from reactions such as kaolinite + dolomite = calcite + chlorite + CO_2 + H_2O. This must favor the development of the calcite-chlorite-albite type of assemblage in the volcanics during low-grade metamorphism, whereas thicker volcanic piles, or flows and sills interbedded with thick volcanic graywackes, will tend to be reconstituted under conditions of lower μCO_2 relative to μH_2O, allowing the development of the diagnostic hydrated calcium aluminosilicates.

A few examples of spilites and spilitic rocks that appear to be in metamorphic harmony with the surrounding rocks include:

Torlesse Supergroup, New Zealand. Numerous examples, mostly prehnite-pumpellyite facies.

Northland, New Zealand (Waipapa Group). Prehnite-pumpellyite facies (MAYER, 1969).

Mossburn, New Zealand (Livingstone Volcanics). The spilites and albite dolerites described by REED (1950) and YODER (1967) are often prehnite- and pumpellyite-bearing and are associated with prehnite-pumpellyite facies tuffaceous sediments (observations of L.M. FORCE, pers. comm., and the writer), as well as serpentinites. An important reservation in this ophiolitic suite arises from the fact that the Livingstone Volcanics and correlatives in part form part of a tectonic mélange. Examples of volcanics from the same general belt that are out of metamorphic equilibrium with juxtaposed sediments will be described by the writer elsewhere.

Taieri Reef, East Otago, New Zealand. Pumpellyite-actinolite schist facies (ROBINSON, 1958; and writer's observations).

Northern Maine. Prehnite-pumpellyite, and pumpellyite-actinolite schist facies (COOMBS et al. 1970).

Lyndhurst district, New South Wales. Prehnite-pumpellyite facies and in part greenschist facies (SMITH, 1968, 1969).

Builth volcanics, Wales. Prehnite-pumpellyite facies, at least in part (NICHOLLS, 1959; RAAM et al. 1969; writer's observations).

At higher grades of metamorphism, the mineralogy of metabasalts commonly varies systematically with the metamorphic grade of the country rocks whether it be of greenschist, blueschist, or amphibolite facies. It is here contended that the mineralogies of many or all spilites except where they occur as exotic blocks in tectonic mélanges, also conform to the metamorphic facies of the country rocks, most commonly in this writer's experience, the prehnite-pumpellyite facies. As a result of unfavorable bulk chemistry and of unfavorable values of μCO_2, μH_2O during alteration, <u>diagnostic</u> mineralogy required for unequivocal recognition of this <u>fact is not</u>, however, always developed.

Conclusions

Spilites belong to the prehnite-pumpellyite and contiguous low-grade mineral facies. This may be developed by burial, regional, or hydrothermal low-grade metamorphism long after the consolidation and complete cooling of parent basaltic lavas. Where associated sediments are chemically sensitive and show diagnostic mineral assemblages, they commonly conform to those of the spilites, bearing in mind that factors such as μH_2O, μCO_2, μO_2 may well have varied from country rock to lava. There is a high probability in such cases that the spilites and the sediments derive their mineralogy from the same metamorphic process.

Critical conditions for the prehnite-pumpellyite and contiguous facies may prevail transiently during the cooling of submarine basaltic lavas. The duration of such conditions may be greatest in the case of lavas emplaced under thick sedimentary or volcanic blankets in deep-sea situations. A major question pertaining to the spilite problem as it exists today is: Have in fact any spilites been produced during this primary cooling episode, and if so, how can they be distinguished from spilitized basalts? Spilites of which this writer has experience all appear to be explicable in terms of post-consolidational metamorphism.

Acknowledgements

The writer has benefitted from the opportunity to discuss ideas presented in this paper with American colleagues, and with Professor M. VUAGNAT, while he held Visiting Professorships at Yale University and the University of Geneva respectively during the academic year 1967-68. The writer takes full responsibility for the views expressed. He is grateful to the New Zealand University Grants Committee and NSF Grants G18700 and GA 754 for financial support for field work in New Zealand.

References

AMSTUTZ, G.C. (1967): Spilites and spilitic rocks. In: Basalts: The Poldervaart Treatise on rocks of basaltic composition. H.H. Hess and A. Poldervaart, eds., Interscience, New York, 737-754.

CANN, J.R. (1969): Spilites from the Carlsberg Ridge, Indian Ocean. J. Petrology 10, 1-19.

CHAPPELL, B.W. (1968): Volcanic graywackes from the Upper Devonian Baldwin Formation, Tamworth-Barraba district, New South Wales. J. Geol. Soc. Aust. 15, 87-102.

COOMBS, D.S. (1954): The nature and alteration of some Triassic sediments from Southland, New Zealand. Trans. Roy. Soc. New Zealand 82, 65-109.

COOMBS, D.S. (1962): Some recent work on the lower grades of metamorphism. Aust. J. Sci. 24, 203-215.

COOMBS, D.S., ELLIS, A.J., FYFE, W.S., TAYLOR, A.M. (1959): The zeolite facies, with comments on the interpretation of hydrothermal syntheses. Geochim. Cosmochim. Acta 17, 53-107.

COOMBS, D.S., HORODYSKI, R.J., NAYLOR, R.S. (1970): Occurrence of prehnite-pumpellyite facies metamorphism in northern Maine. Am. J. Sci. 268, 142-156.

DEWEY, H., FLETT, J.S. (1911): On some British pillow-lavas and the rocks associated with them. Geol. Mag. Lond. 563, 202-209, 241-243.

DONNELLY, T.W. (1962): Wairakite in West Indian spilitic rocks. Am. Miner. 47, 794-802.

DONNELLY, T.W. (1966): Geology of St. Thomas and St. John, U.S. Virgin Islands. Mem. Geol. Soc. Amer. 98, 85-176.

FAWCETT, J.J., YODER, H.S. (1966): Phase relationships of chlorites in the system $MgO-Al_2O_3-SiO_2-H_2O$. Am. Miner. 51, 353-380.

HAMILTON, D.L., BURNHAM, C.W., OSBORN, E.F. (1964): The solubility of water and effects of oxygen fugacity and water content on crystallization in mafic magmas. J. Petrology 5, 21-39.

HINRICHSEN, T.H., SCHÜRMANN, K. (1969): Untersuchungen zur Stabilität von Pumpellyit. Neues Jahrb. Mineral. Monatsh. 10, 441-445.

HOLDAWAY, M.J. (1966): Hydrothermal stability of clinozoisite plus quartz. Am. J. Sci. 264, 643-667.

HOLDAWAY, M.J. (1972): Thermal stability of Al-Fe epidote as a function of fO_2 and Fe content. Contr. Mineral. and Petrol. 37, 307-340.

LANDIS, C.A. (1969): Upper Permian rocks of South Island, New Zealand: lithology, stratigraphy, structure, metamorphism and tectonics. Ph.D. Thesis, University of Otago.

LANDIS, C.A., ROGERS, J. (1968): Some experimental data on the stability of pumpellyite. Am. Miner. 53, 1038-1041.

LIOU, J.G. (1971a): P-T stabilities of laumontite, wairakite, lawsonite and related minerals in the system $CaAl_2Si_2O_8-SiO_2-H_2O$. J. Petrology 12, 379-411.

LIOU, J.G. (1971b): Synthesis and stability relations of prehnite, $Ca_2Al_2Si_3O_{10}(OH)_2$. Am. Miner. 56, 507-531.

MARTINI, J. (1968): Etude pétrographique des Grès de Taveyanne entre Arve et Giffre (Haute-Savoie, France). Bull. suisse Min. Pétr. 48, 539-654.

MARTINI, J., VUAGNAT, M. (1965): Présence du faciès à zéolites dans la formation des "grès" de Taveyanne (Alpes franco-suisses). Bull. suisse Min. Pétr. 45, 281-293.

MAYER, W. (1969): Petrology of the Waipapa Group, near Auckland, New Zealand. New Zealand. J. Geol. Geophys. 12, 412-435.

MELSON, W.G., VAN ANDEL, T.H. (1966): Metamorphism in the mid-Atlantic Ridge, 22° N latitude. Marine Geol. 4, 165-186.

MOORE, J.G. (1965): Petrology of deep-sea basalt near Hawaii. Am. J. Sci. 263, 40-52.

MUIR, I.D., TILLEY, C.E. (1966): Basalts from the northern part of the mid-Atlantic Ridge. II The Atlantis collections near 30° N. J. Petrology 7, 193-201.

NICHOLLS, G.D. (1959): Autometasomatism in the Lower Spilites of the Builth Volcanic Series. Quart. J. Geol. Soc. (London) 114, 137-162.

RAAM, A., O'REILLY, S.Y., VERNON, R.H. (1969): Pumpellyite of deuteric origin. Am. Miner. 54, 320-324.

REED, J.J. (1950): Spilites, serpentinites and associated rocks of the Mossburn district, Southland. Trans. Roy. Soc. New Zealand 78, 106-126.

ROBINSON, P. (1958): The structural and metamorphic geology of the Brighton-Taieri Mouth area, East Otago, New Zealand. M. Sc. thesis, University of Otago.

SEKI, Y. (1969): Facies series in low grade metamorphism. J. Geol. Soc. Japan 75, 255-266.

SMITH, R.E. (1968): Redistribution of major elements in the alteration of some basic lavas during metamorphism. J. Petrology 9, 191-219.

SMITH, R.E. (1969): Zones of progressive regional burial metamorphism in part of the Tasman Geosyncline, Eastern Australia. J. Petrology 10, 144-163.

STRENS, R.G.J. (1968): Reconnaissance of the prehnite stability field. Mineralog. Mag. 36, 864-867.

TURNOCK, A.C. (1960): The stability of iron chlorites. Carnegie Inst. Washington Year Book 59, 98-103.

TUTTLE, O.F., BOWEN, N.L. (1958): Origin of granite in the light of experimental studies in the system $NaAlSi_3O_8 \cdot KAlSi_3O_8-SiO_2-H_2O$. Mem. Geol. Soc. Amer. 74.

VALLANCE, T.G. (1960): Concerning spilites. Proc. Linn. Soc. New South Wales 85, 8-52.

WALKER, G.P.L. (1951): The amygdale minerals in the Tertiary lavas of Ireland. I. The distribution of chabazite habits and zeolites in the Garron plateau area, County Antrim. Mineralog. Mag. 29, 773-791.

WALKER, G.P.L. (1960): Zeolite zones and dike distribution in relation to the structure of the basalts of Eastern Iceland. J. Geol. 68, 515-528.

YODER, H.S. (1967): Spilites and serpentinites. Carnegie Inst. Washington Year Book 65, 269-279.

YODER, H.S., TILLEY, C.E. (1962): Origin of basalt magmas: an experimental study of natural and synthetic rock systems. J. Petrology 3, 342-532.

The Pillow Lavas of Sakhalin and the Kurile Islands and Their Significance for the Solution of the Spilite Problems

V. N. Shilov

Abstract

The paper contains data on age-distribution of various pillow lavas of the Sakhalin and the Kurile Islands and on the peculiarities of their petrographic and chemical compositions. The presence of marine fossils within the rocks underlying the pillow lavas and the peculiar structure of these volcanic rocks testify to their undersea origin. The Sakhalin pillow lavas are subdivided into five groups: Jurassic-Early Cretaceous, Late Cretaceous, Early Miocene, Middle Miocene and Late Miocene-Pliocene. The same formations of the Kurile Islands are subdivided into three groups: Late Cretaceous, Early Miocene and Pliocene. The correlation of the petrographic and chemical compositions of the pillow lavas of different ages confirms the viewpoint of those students who state that the peculiarities of mineralogical and chemical compositions of the spilites are not caused by the occurrence of independent spilitic magmas, but are the consequence of the transformation of common volcanic rocks under the conditions peculiar to folded geosynclinal areas.

Resumé

The volcanogenic formations of Sakhalin and the Kurile Islands are of great interest in connection with the problem of spilite origin. In these formations, one observes almost a complete transition from recent lavas to very old lavas that have been subjected to metamorphism. Because many authors connect spilite formation with submarine lava outflow, we confine this report to a discussion of pillow lavas only. The subaqueous character of the latter is confirmed not only by their peculiar structure, but also by the discovery of numerous remnants of marine organisms within the normal sedimentary deposits, overlying and underlying the volcanogenic formations. Only in the Jurassic-Early Cretaceous group do we consider, besides the above, diabases that are genetically connected with the pillow lavas.

The most recent (Late Miocene-Pliocene) pillow lavas of Sakhalin are distributed primarily in the area of the Lamanon massif. Elsewhere, they are known along the coast of the Kril'on peninsula and in some other districts.

The pillow lavas within the Lamanon massif are included among the deposits of the Maruyam series and are found throughout its succession in the form of variably thick layers. At the bass of the pillow lavas and within the normal sedimentary deposits interbedded with them, great numbers of organic remains are found, which not only show that the lava outflow occurred in submarine conditions, but also permit estimation of the depth of the basin, at the bottom of which the volcanic processes took place. The sea was the deepest in the Late Miocene time, when the sediments of the first horizon of the Maruyam series were deposited. At first its depth reached 100 m, which is confirmed by the characteristic fauna of the first horizon (ZHIDKOVA and SHILOV, 1969): Acila (Truncacila) cobboldiae (Sow.), Nuculana majamraphensis Khom., N. (Sacella) chinaensis Ilyina, Yoldia thraciaeformis (Stor.), Y. temblorensis And. et Mart., Y. scapha Yok., Laternula (Aelga) besshoensis (Yok.), Thyasira bisecta (Conr.), Th. disjuncta (Gabb.) var. ochotica L. Krisht., Solemya tokunagai Yok., Mya majanatschensis Ilyina, Macoma optiva (Yok.), Buccinum lencostoma Lischk. var. sachalinensis Yok., Naptunea lirata (Mart.), Turritella ocoyana Conr.

Significant shallowing of the Maruyam basin occurred in the Pliocene age, which is confirmed by the discovery of the following mollusks: Acila (Truncacila) inisignis (Gould), Yoldia (Cnesterium) supraoregona Khom., Anadara (Anadara) trilineata Conr., An (Anadara) devincta (Conr.), Glycymeris yessoensis (Sow.), Chlamys swiftii (Bern.) Pecten (Fortipecten) takahashii Yok, Astarte alaskensis Dall., Venericardia ferruginea (Clessian), Taras (Felanielaa) usta (Gould), Clinocardium californiense Desh., Chione (Securella) securis (Shum).

At the end of the Pliocene age, the formation of pillow lavas stopped because the volcanic eruptions became subaerial. The composition of the Lamanon massif pillow lavas was discussed earlier (SHILOV and KALISHEVICH, 1958). The rocks under consideration are basaltic in composition. Because of quick cooling in submarine conditions they are often partly glassy and have microporphyritic textures. In different parts of the pillows the mineralogical composition and the groundmass texture are different. The marginal areas of the rocks are of vitrophyric texture, and volcanic glass constitutes 73-87% of the volume (n = 1.572 - 1.575). The fine laths of plagioclase and mafic minerals (1-6%), represented predominantly by olivine and rarely by clinopyroxene, are present in only small amounts (12-15%). In the middle and central areas of the pillows, the glass volume decreases to 50-58%, and the texture becomes hyalopilitic. The quantity of plagioclase phenocrysts increases to 32-37%, and that of the mafic minerals (almost exclusively by pseudomorphs) to 7-15%.

Plagioclase (An_{67-71}), distinguished by considerable freshness, is practically unchanged in composition in different parts of the pillows. Olivine, predominant among the mafic minerals, is transitional in composition from forsterite to chrysolite and in most cases is replaced by an iddingsite-like mineral and/or is chloritized. Also pyroxene is almost completely chloritized. From small remnants, we may suppose that it is similar to augite. The degree of alteration of the mafic minerals and the rocks in general is greatest in the intermediate areas of the pillows and decreases towards their center. But the rocks appear freshest in the marginal (vitreous) parts ot the pillows. It is interesting, that besides the greater alteration of the middle parts of the pillows, it is to these areas that original variolitic texture is confined.

On the Kril'on peninsula the pillow lavas occur in Late Miocene-Early Pliocene deposits, belonging to the Aniva series (GOLOVINSKY, 1963), which has the following fauna: Acila sp., Serripes gronlandicus (Chem.), Chione (Seculla) securis (Shum), Tullina pulchra Slod., Macoma echabiensis Slod., M. optiva (Yok.), Mactra ex. gr. polynyma (Stim.), Panope sp., Neptunea sp., Buccinum sp., Turritella sp., Natica sp., Polinices sp., Dentalium sp.

The above fauna shows relatively shallow sedimentation conditions.

The pillow lavas are confined to the middle horizon of the upper subseries of the Aniva series. In composition they are andesite-basalts. The inner segments of the "pillows" are porphyritic rocks with hyalopilitic textured groundmass. The phenocrysts make up about 30% of the rock volume. They are formed primarily by plagioclase (labradorite) and by single grains of augite and olivine. The groundmass of the rock is composed of pure, slightly greenish glass, in which clear plagioclase microlites are included.

The Middle Miocene pillow lavas within the region under consideration are found on the west coast of the southwestern part of Northern Sakhalin and in the Moneron Island. In all of these places the volcanogenic formations are included in the Chekhov series.

On Cape Markevich within the deposits of the Chekhov series many fossils are found: Pecten (Chlamys) cf. branneri Arn., Mytilus (Mytilocondia) expansis Arn., Modiolus tetragonalis Slod., Modiolaria kryshtafovitschi Sim., Phaciodes (Lucinoma) ex. gr. acutilineatus (Conrad), Saxidomus sachalinensis (Sim.), Tellina chibana Yok., Macoma ex. gr. simizuensis L. Krisht., Sanguinolaria sachlinensis Laut., Mya markevitschi Laut., Podosesmus sp., Cardita sp., Cardium sp., Thracia sp., Saxicava sp., Acmaea markevitschi Laut., Crepidula markevitschi L. Krisht., Turritella cf. tokunagai Yok., Priscofusus sp., Trochus sp., Polinices sp., Balanus sp.

This fauna indicates that the pillow lavas were formed in shallow-water marine conditions, changed periodically into continental conditions. This is confirmed by the presence of coal-bearing deposits and subaerial volcanogenic formations in the upper parts of the Chekhov series.

The inner parts of the "pillows" in the region of Cape Markevich are composed of olivine basalt, which are porphyritic rocks with an intersertal groundmass and a weakly porous structure. Phenocrysts comprise up to 60% of the rock volume. Of this, plagioclase (basic labradorite) comprises up to 75%, augite - 20% and olivine/chlorite-serpentine pseudomorphs (in different cases) 5%. The groundmass, excluding the clinopyroxene grains, is intensively altered. The pores are filled with chlorite.

On the east coast of South Sakhalin the pillow lavas are commonly found in the lower part of the Chekhov series. Within the latter volcanomictic formations predominate, being often psephitic. In the upper part of the series subaerial lava flows occur.

In the lower part of the Chekhov series, one finds remains of sea fishes and in the upper part, the remains of freshwater mollusks (Metanella sp. and Viviparus sp.), as well as of large fossilized trees. The formation of the pillow lavas occurred in shallow (a few tens of m deep) marine conditions.

The pillow lavas of the Chekhov series of the east coast of the South Sakhalin are andesitic in composition. They are porphyritic rocks, composed of basic plagioclase phenocrysts (labradorite and bytownite), pyroxene and olivine, replaced by secondary minerals. The groundmass of the andesite-basalts is composed of plagioclase microlites with small grains of pyroxene and ore mineral, inclosed in chloritized volcanic glass. The texture of the groundmass changes, according to the glass content, from hyalopilitic to intersertal.

In the Moneron Island (EROKHOV et al. 1968), situated in the Japan Sea near the southwestern extremity of Sakhalin Island, Middle Miocene pillow lavas are widely distributed and are underlain by normal sedimentary and pyroclastic rocks of the Nevel'sk series containing numerous organic remnants: Acila (Acila) ex. gr. divaricata submirabilis (Mak.), Sacella calkinsi (Moor), S. sp. ex. gr. S. Ochsneri (And. et Mart.), S. cf. taphria (Dall.), Yoldia nitida Slod., Lima (Acesta) cf. sachalinensis Slod., Delectopecten pedroanus (Trask), Venericardia cf. yokoyamai (Slod.), Thyasira disjuncta var. nipponica (Yabe et Nomura), Macoma cf. optiva (Yok.), M. astori (Dall), M. cf. sejugata (Yok.), M. simizuensis L. Krisht., Mya ex. gr. diskersoni Clark, Laternula (Aelga) besshoensis sachalinensis Slod., Buccinum ex. gr. kurodai Kanehara, Turritella ocoyana Conrad, T. cf. montereyana Wiedey.

These organisms are indicative of the comparatively deep environment of sedimentation, not, however, deeper than shelf conditions. That is, these conditions preceded the beginning of the volcanic outflow. Remains of Lima sp. Delectopecten sp. and others were found in the base of the lowest flow of the pillow lavas.

The pillow lavas of Moneron Island are basaltic in composition. Phenocrysts constitute up to 50-55% of the rocks of the inner part of the "pillows" among which plagioclase (bytownite) makes up 35-40%, olivine (iddingsite and chlorite pseudomorphs), 5% and augite, 5-7%. The groundmass is composed of partly chloritized volcanic glass (up to 40%), comparatively rare (up to 7%) plagioclase crystals (labradorite), clinopyroxene grains (up to 3%) and magnetite. The marginal "pillow" parts are composed mainly of volcanic glass (n = 1.579), in which rare small crystals of plagioclase (basic labradorite), grains of olivine and magnetite are distributed.

The Early Miocene pillow lavas in Sakhalin are located in the southwest area of Northern Sakhalin (near Cape Hoingo) and in the western part of the Schmidt Peninsula. In the district of Cape Hoingo the pillow lavas are confined to the basement of the Arakay series. Organic remains are not found within the volcanogenic formation. The pillow lavas here are deposited on normal sedimentary deposits containing marine fauna.

The rocks comprising the inner parts of the "pillows" in the region of Cape Hoingo are basaltic. They are porphyritic rocks with phenocrysts almost solely of basic plagioclase (labradorite-bytownite). The groundmass is composed of plagioclase microlites, small pyroxene grains, ore mineral and chloritized volcanic glass. The texture of the groundmass in most cases is hyalopilitic.

On the Schmidt Peninsula the pillow lavas are found among the deposits of the Michigar series. Along the Vodopadnaya River, within deposits of this series, RATNOVSKY found numerous molluscan fossils: Nuculana (Sacella) washingtonensis (Weaver), Yoldia (Kallayoldia) matschigaria L. Krisht., Pecten (Chlamus), matchigarensis Mak., Cardita matschigarica Khom., Taras gastelloi Evseev, Laevicardium esutoruensis L. Krisht., C. kirkinskaya L. Krisht., Liocima furtiva Yok., Tellina makarovi L. Krisht., T. cf. townsendensis Clark, Cultellus sachalinensis L. Krisht., Mactra (Spisula) vaqisana Laut., Mya grewingki Mak., Cryptomya praebussoensis Laut., Thracia schmidti L. Krisht., Th. condoni Dall, Fusinus sp., Polinices sp., Neptunea sp., indicating the relatively shallow marine character of the deposits containing them.

The pillow lavas of the Schmidt Peninsula are of basaltic composition and are distinguished by slightly increased alkalinity. They are microporphyritic porous rocks with a hyalopilitic groundmass. The phenocrysts make up nearly 25% of the rock and are comprised of plagioclase (labradorite) and mafic minerals, replaced by chlorite and carbonate. Nearly 80% of the phenocrysts are of plagioclase. The groundmass is composed of plagioclase microlites and pyroxene grains, enclosed in chloritized glass, contaminated by ore dust.

Late Cretaceous pillow lavas are found in the East-Sakhalin Mountains and on Schmidt Peninsula. In the East-Sakhalin Mountains, in the deposits directly underlying the volcanic formations, assigned by SHUVAEV (1969) to the Nerpich'a series (supposedly Early-Campian stage), the following faunal remains are found: Inoceramus naumanni Yok., Gandryceras denmarense Whiteaves, G. aff. sachalinensis Schmidt, Hypophylloceras cf. subramosum Shim. In Late Campanian and Early-Maastrichtian deposits of the East-Sakhalin Mountains, containing the pillow lavas, the following faunal remains are found: Inoceramus schmidti Mich., Patella (Helcion) gigantea Schmidti var. depressa Schmidt, Anomia sp., Cenosphaera sp., Porodiscus sp. These fossils indicate a foreshore-marine environment at the commencement of the volcanic process.

The pillow lavas are represented by spilites. These are greenish-gray rocks, commonly without phenocrysts, containing amygdales composed of calcite and chlorite. Irregularly distributed fine laths of plagioclase (albitized), fine grains of augite and ore mineral are enclosed in a groundmass composed of highly chloritized volcanic glass. The texture of the rocks is spilitic and rarely approaches pilotaxitic or intersertal.

On Schmidt Peninsula Late Campanian-Early Maastrichian pillow lavas form part of the Mariyskaya series, distributed only in the northwestern part of the peninsula. According to SHUVAEV (1969), the normal sedimentary deposits of the series contain rare remains of radiolarians, represented by the following forms: Cenospaera sp., Porodiscus sp., Stylodictya sp.

The pillow lavas of the Schmidt Peninsula are composed of spilites which are aphyric rocks with an amygdaloidal texture. In the decomposed groundmass of these rocks, which is semi-opaque, we distinguish only fine laths of albitized plagioclase. Amygdales contained within them are filled with carbonate and zeolite. The marginal areas of some pillows are composed of nearly fresh basalt and andesite, which are porphyritic rocks, made up of small numbers of coarse laths of labradorite, grains of augite and ore mineral, enclosed in brown palagonitic volcanic glass. The latter often has perlitic texture.

Additional old (pre-Late Cretaceous) volcanogenic formations of the spilitic type are found in the Tonino-Anivskom Peninsula and in the East-Sakhalin Mountains. In the former they are assigned to the middle segment of the Novikovskaya series and are dated as Early-Cretaceous. Among the deposits of the Novikovskaya series, TARA-SEVICH found a small number of fossils, represented by the following forms: Pterotrigonia hokkaidoana Iehara, Trigonia sp., Sonnerata sp., indicating relatively deep waters in the corresponding sea basin.

In addition to the pillow lavas formed at the sea bottom, conformable bodies of igneous rocks are widely distributed within the deposits of the Novikovskaya series. These bodies intruded into the poorly consolidated bottom sediments. In both cases the rocks are spilitic, usually of aphyric and, rarely of oligophyric structure. They also are often amygdaloidal. Sometimes, due to the process of ferrugination, the rocks have assumed a cherry-brown color.

In oligophyric differentiates of the spilites, the phenocrysts are represented by crystals of albitized plagioclase, the composition of which is difficult to determine because of the extensive development of the secondary alteration. The groundmass and aphyric differentiates of the spilites consist of thin elongated laths of plagioclase (albite-oligoclase), the interstices between which are filled with the products of devitrification of basic glass, namely sparse aggregates of chlorite, ore mineral grains (titanomagnetite) and monoclinic pyroxene phenocrysts. Of the secondary minerals, chlorite-leucoxene and carbonate are present. The amygdales in the rocks are usually filled with calcite and chlorite and rarely with quartz, chalcedony and albite. Quartz and chlorite sometimes form thin veinlets, which rarely contain epidote.

In the East-Sakhalin Mountains, the volcanogenic formations are placed in the Ostrinskaya series. The age of these deposits is Jurassic. In the normal sedimentary deposits of the Ostrinskaya series, SAVITSKY found remains of the hexactinellid corals: Stylina sachalinica Krasnov, Convexastrea fukazawaensis Eguchi, Thamnasteria vereschagini Krasnov, Calamophyllia flabellum Blainv., Diplocoehia sp., and also great numbers of radiolarians of the genus Dictyomitra. The presence of these remains in association with argillo-silicic deposits indicates the abyssal character of the Ostrinskiy basin.

Conformable layers in the deposits of the Ostrinskaya series are composed of spilites, and the cross-cutting sheets are of diabase. Pillow structure is not characteristic of the volcanogenic formations under consideration.

Spilites are compact massive rocks often containing numerous amygdales, filled with chlorite, carbonate or prehnite. Their texture is aphyric or oligophyric, the texture of the groundmass being often spilitic, rarely approaching the pilotaxitic, intersertal, hyalopilitic or sideranitic. The rocks consist of albitized plagioclase, decomposed volcanic glass and ore mineral. Plagioclase has the form of acicular laths, being intensively sericitized and replaced by argillaceous minerals. The volcanic glass is greatly decomposed, being generally a cloudy, often chloritized mass, which fills the interstices between the laths of plagioclase. Rarely, fine grains of clinopyroxene (titan-augite) are found. Besides this the whole groundmass of the rock is often carbonized and ferruginized.

Diabases are greenish-gray and dark gray massive rocks, composed of basic plagioclase and pyroxene. The interstices between the two are rarely filled with altered volcanic glass, which is commonly replaced by chlorite and contains fine grains of the ore minerals (magnetite, rarely ilmenite and others). The plagioclase is considerably altered (by altitization, saussuritization and often sericitization). Pyroxene is commonly represented by augite or titan-augite. Sometimes relicts of serpentinized olivine are found. The texture of these rocks is in most cases diabasic-ophitic.

In the Kurile Islands the oldest deposits are Late Cretaceous. Pillow lavas are found here both in Late Cretaceous and Neogene deposits. They are most widely distributed among the Pliocene deposits in the islands of Kunashir, Iturup, Urup and others. In Kunashir the pillow lavas belong to the Golovninskaya series in the southwestern part of the island. In the Golovninskaya series here, PR'ALUKHINA found numerous fossils, including the following forms: Venericardia borealis Conrad, V. crebricostata (Krause), Clinocardium cilliatum (Fabr.), C. aff. angustum (Yok.), Serripes sp., Liocyma fluctuosa (Gould), Macoma calcarea (Chem.), Panope (Panomya) arctica (Lam.), Hiatella arctica (L.), H. pholadis (L.), Mya truncata (L.), M. cf. arenaria (L.), Spisula sp., Echinarachnius cf. parma (Lam.), Epitonium sp.

In Iturup pillow lavas are often found among the deposits of the Parusnaya series. The west coast of the island in the region of the Gornaya River is the most characteristic area of occurrence. North of this area along the Osenn'aya and Blagodatnaya Rivers, SERGEEVA found the following faunal remains: Serripes gronlandicus (Chem.), Clinocardium cf. citiatum (Fab.), Macoma aff. calcarea (Chem.), Panope (Panomya) cf. arctica (Lam.), Limatula sp. (cf. L. pilvaensis Laut.), Natica sp., and at Cape Przhevalsky (Geologica SSSR, 1964) Pecten cf. akitanus Yok., P. (Chlamys) cf. brenneri Arn., P. (Chlamys) tanassevitschi Khom., P. (Chlamys) turpiculus Yok., Macoma calcarea (Chem.), Clinocardium sp. cf. C. californiense (Desh.), Mya japonica Jay, Mya cf. truncata (L.), Cardita matitukensis Slod.

In Urup Pliocene pillow lavas are known among the deposits of the Natalinskaya series (PISKUNOV, 1966). Faunal remains are not found there. The fauna listed above indicates the shallowness of the sea enclosing the deposits. Pliocene pillow lavas are possibly present in other islands, particularly in Paramushir and Shumshu, where they belong to the Okeanskaya series.

The Pliocene pillow lavas of the Kurile Islands are basaltic and andesito-basaltic in composition. Generally they are porous, dark gray porphyritic rocks. The phenocryst content generally varies from 15-30%, the majority being plagioclasic, composed of labradorite or bytownite. Magnesian orthopyroxene, augite and olivine (replaced by secondary minerals) occur rarely. The groundmass of the rocks has a variable degree of crystallinity: in marginal areas of the pillows it is holo- or hypohyaline, in the inner parts hyalopilitic and intersertal, sometimes even pilotaxitic. The refractive index of the volcanic glass generally varies between 1.550 - 1.575. The characteristic feature of the rocks is the widespread hydration of the volcanic glass of the groundmass with the formation of minerals of the palagonite-chlorophaeite group.

The Early Miocene pillow lavas of the Kurile Islands are known to date only in Kunashir and Paramushir. In Kunashir, they belong to the Kunashir series, and in the latter, to the Vasilievskaya series. These are the oldest series of the respective islands. In Paramushir the fossils, found in the oldest parts of the Neogene section, are represented, according to SERGEEV (1966), by the following forms, indicative of a relatively abyssal environment of deposition: Nuculana (Sacella) crassatelloides Laut., Yoldia (Portlandella) tokunagai Yok., Malletia inermis (Yok.), M. longa L. Krisht., Limatyla pilvoensis Laut., Mytilus miocenicum L. Krisht., Laternula altarata L. Krisht., Periploma mactra L. Krisht.

The Early Miocene pillow lavas of the Kurile Islands, represented by basalts and andesito-basalts, are dark gray and greenish-gray massive, porous porphyritic rocks often having amygdaloidal structure. The phenocrysts comprise from 20 to 50% of the rock and are represented mainly by plagioclase (labradorite, bytownite) and rarely by augite. The groundmass possesses vitrophyric texture and is highly chloritized. Thin albite veinlets and also epidote, calcite and pyrite segregations are often found here.

Late Cretaceous pillow lavas are known only in some islands of the Small Kurile Chain. They are best represented in Shikotan Island, where they are included in the Matakatan series. Fossils are found in the overlying Malokuril'skaya series. They include: Inoceramus shikotaensis Nagao et Mat., I. aff. orientalis Sok., I. tisimacense Nagao et Mat., Gaudryceras sp. (Geologia SSSR, 1964), indicating the shallow depth of the sea basin.

The pillow lavas are comprised of amygdaloidal basalts, porphyritic rocks, with plagioclase (labradorite) and pyroxene (predominantly augite) phenocrysts. Plagioclase is predominant among the phenocrysts. The texture of the groundmass is hyalopilitic, intersertal or microlitic; it often contains large phenocrysts of ore mineral. The amygdales are filled with carbonate, zeolite and chlorite. The same minerals sometimes form small phenocrysts and veinlets in the groundmass.

The chemical compositions of the volcanogenic rocks of Sakhalin described above are given in Table 1, and of the Kurile Islands, in Table 2. In these tables the analyses are grouped according to age. For each of these groups the average content of lime and alkalis, as well as the losses during ignition, is calculated, these being the constituents whose behavior is especially important when considering the processes involved in spilite formation. The ratios Na_2O/CaO, Na_2O+K_2O/CaO and Na_2O/K_2O were determined for each analysis;

Table 1. Chemical composition of spilites, diabases and pillow lavas of Sakhalin

	Jurassic - Early Cretaceous								
	1	2	3	4	5	6	7	8	Avg.
SiO_2	47.81	48.25	48.72	49.61	49.73	49.99	50.86	53.16	
TiO_2	0.89	0.09	1.42	0.84	0.85	0.86	0.75	0.33	
Al_2O_3	16.63	16.57	17.55	15.99	15.53	16.20	15.66	17.69	
Fe_2O_3	10.85	5.24	4.72	3.95	1.80	12.19	4.09	1.17	
FeO	1.97	8.18	9.69	6.32	7.57	3.81	7.15	6.97	
MnO	0.22	0.14	0.14	0.31	0.21	0.20	0.17	0.26	
MgO	3.66	6.76	1.51	4.40	6.10	2.68	5.83	5.99	
CaO	9.38	5.66	6.93	9.98	6.68	5.55	6.08	4.27	6.82
Na_2O	3.92	4.45	3.77	5.39	4.66	4.91	4.04	5.39	4.57
K_2O	1.23	0.60	1.20	0.36	1.20	0.77	1.56	1.20	0.89
P_2O_5	1.29	0.13	0.20	0.12	0.09	0.34	0.10	0.12	
S	0.27	0.30	0.02	0.01	0.11	0.01	0.18	0.05	
loss on ignition	1.47	3.13	3.66	2.79	3.45	1.92	2.78	2.49	2.71
H_2O	0.84	0.19	0.17	0.24	1.56	0.09	0.42	0.61	
total	100.43	99.69	99.70	100.31	99.54	99.52	99.67	99.70	
Na_2O/CaO	0.42	0.79	0.54	0.54	0.70	0.89	0.66	1.26	0.73
Na_2O+K_2O/CaO	0.55	0.89	0.72	0.58	0.88	1.03	0.92	1.55	0.89
K_2O/Na_2O	0.31	0.13	0.32	0.07	0.26	0.16	0.39	0.22	0.23

1 - sample 2613, spilite, Korsakov district;
2 - sample 2826b, diabase, Cape Delil-de-la Croyer;
3 - sample 2836, spilite, Lunsky range;
4 - sample 2768a, diabase, Rim. River;
5 - sample 3155, spilite, Dolinsk district, hole 5 op, interval 2596-2598 m
6 - sample 2837b, spilite, Lunsky range;
7 - sample 2765a, spilite, Mishkins River;
8 - sample 2767, diabase, Pursh-Pursh River

in addition, for each age group the average value of these relations was calculated. Because the Early Miocene lavas of Schmidt Peninsula are characterized by increased alkalinity, the average values of the oxide contents and the ratios between them have been calculated for the corresponding age group without taking them into consideration and the results are given in brackets in Table 1. The same has been done in calculating the mean ignition losses from the Late Cretaceous and Pliocene groups of the Kurile pillow lavas. Furthermore one analysis, in which the absolute value of ignition loss was markedly different from the others was also excluded.

The tables show clearly the basic (basaltic and andesito-basaltic) composition of all of the pillow lavas. Only two analyses of Late Cretaceous lavas of Schmidt Peninsula and one of a Pliocene rock from Kunashir Island belong to andesites.

In spite of the similar basicity of the rocks under consideration, the individual oxide contents differ in the different age groups.

Table 1 (continued)

	Late Cretaceous						
	9	10	11	12	13	14	Average
SiO_2	49.06	48.60	51.72	53.10	55.97	56.06	
TiO_2	1.95	2.10	0.92	0.50	0.50	0.16	
Al_2O_3	16.65	14.93	17.04	16.50	14.84	14.80	
Fe_2O_3	11.46	5.98	8.90	4.81	4.50	3.52	
FeO	1.29	3.16	2.49	4.86	2.96	2.35	
MnO	0.39	0.33	0.19	0.16	0.06	0.07	
MgO	1.80	5.03	2.83	6.26	4.03	3.71	
CaO	5.37	7.06	5.23	7.01	6.66	5.44	6.13
Na_2O	5.79	4.45	4.02	4.11	4.04	3.91	4.39
K_2O	1.76	2.83	2.41	0.87	3.01	4.06	2.49
P_2O_5	0.49	0.25	0.14	0.08	0.27	0.20	
S	0.18	0.24	0.16	0.11	0.01	0.22	
loss on ignition	2.76	4.46	2.17	1.97	1.92	3.82	2.85
H_2O	0.84	0.41	2.29	-	1.32	2.37	
total	99.79	99.85	100.51	100.34	99.99	100.69	
Na_2O/CaO	1.08	0.63	0.77	0.59	0.61	0.72	0.73
Na_2O+K_2O/CaO	1.41	1.03	1.23	0.71	1.06	1.46	1.15
K_2O/Na_2O	0.30	0.64	0.60	0.21	0.75	1.04	0.59

9 - sample 3001, spilite, Cape Three Stone;
10 - sample 2997, spilite, Bogataya River;
11 - sample 3075, spilite, Bogataya River;
12 - sample 2776, spilite, Ugolnoye, Schmidt Peninsula;
13 - sample 2779, Cape Maria, Schmidt Peninsula;
14 - sample 2987n, spilite, Cape Maria, Schmidt Peninsula.

Let us take first the peculiarities of chemical composition of the volcanogenic formations of Sakhalin (Table 1). The average CaO content varies from 9.5-10% in Middle Miocene and Late Miocene-Pliocene rocks, to 8% in Early Miocene and 6-7% in Mesozoic. The mean Na_2O content increases from 3.2-3.4% in Neogene rocks to 4.4-4.6% in Mesozoic. The average content of K_2O in Neogene volcanic rocks makes up 0.4-0.6%, if the high content of this constituent in the lavas of Schmidt Peninsula is disregarded. In Jurassic-Early Cretaceous rocks it is equal to 0.9% and increases sharply (up to 2.5%) in Late Cretaceous lavas; the increased content of K_2O is characteristic of the latter. The ignition losses, the main components of which are bound water and carbon dioxide, increase from 1% in Late Miocene-Paleocene rocks to 1.5% in Early and Middle Miocene and 2.7-2.8% in Mesozoic rocks.

These peculiarities in behavior of the average contents of individual oxides are reflected in the variation of the values of their ratios. Particularly, the ratio of Na_2O to CaO in Miocene rocks is less than half that for Mesozoic rocks.

Table 1 (continued)

	Early Miocene				
	15	16	17	18	Average
SiO_2	49.92	51.19	50.90	52.41	
TiO_2	0.37	0.34	0.59	0.50	
Al_2O_3	20.21	20.43	19.62	20.55	
Fe_2O_3	4.08	3.45	4.13	3.54	
FeO	4.73	4.10	3.42	2.50	
MnO	0.14	0.10	0.31	0.17	
MgO	5.12	3.67	1.67	2.34	
CaO	9.49	8.13	8.29	6.26	8.04(8.81)
Na_2O	2.91	3.03	3.83	4.04	3.45(2.97)
K_2O	0.33	0.53	2.17	2.70	1.43(0.42)
P_2O_5	0.39	0.39	0.59	0.60	
S	0.03	0.08	0.08	0.14	
loss on ignition	0.38	1.73	2.05	1.55	1.55
H_2O	1.94	2.60	2.24	2.44	
total	100.54	99.27	99.89	99.74	
Na_2O/CaO	0.31	0.37	0.46	0.65	0.45(0.34)
Na_2O+K_2O/CaO	0.34	0.44	0.72	1.08	0.65(0.36)
K_2O/Na_2O	0.11	0.18	0.57	0.67	0.38(0.15)

15 - sample 2859b, basalt, Cape Choyndzho;
16 - sample 2850b, basalt, Cape Choyndzho;
17 - sample 2991, trachybasalt, Vodopadnaya River, Schmidt Peninsula;
18 - sample 2991g, trachybasalt, Vodopadnaya River, Schmidt Peninsula.

The chemical composition of the pillow lavas of the Kurile Islands is given in Table 2. The table shows that with increasing age of the rocks the mean amount of CaO decreases from 9.4 to 8.3% while Na_2O increases from 2.6 to 4.2%. The average amount of K_2O is at a minimum in Early Miocene pillow lavas and at its maximum in Late Cretaceous. The average content of the ignition losses is not large on the whole (1.5-1.9%) and shows practically no variation with the age of the rock. The average value of the ratio of Na_2O to CaO increases with increasing rock age from 0.28 for Pliocene rocks to 0.37 for Early Miocene and 0.53 for Late Cretaceous rocks.

These features of the chemical composition of the volcanogenic formations of Sakhalin and the Kurile Islands are due to the peculiarities of their petrographic composition. Particularly, the process of deanorthitization of plagioclase, commonest in the oldest rocks, is reflected in the decrease in the amount of CaO and the increase in that of Na_2O. In spite of the fact that the oldest rocks are the most albitized (in Sakhalin, Jurassic-Early Cretaceous rocks), the

Table 1   (continued)

	Middle Miocene						
	19	20	21	22	23	24	Average
SiO_2	47.24	49.56	49.66	49.80	50.36	53.24	
TiO_2	0.90	0.96	0.87	0.63	0.73	0.32	
Al_2O_3	16.61	19.20	17.90	18.86	19.50	18.87	
Fe_2O_3	4.67	3.16	2.72	2.81	3.16	4.80	
FeO	6.34	4.48	5.56	5.28	5.30	1.91	
MnO	0.16	0.08	0.11	0.26	0.20	0.17	
MgO	5.46	4.54	4.65	4.84	4.03	2.43	
CaO	10.70	9.88	10.32	10.83	10.71	9.46	10.32
Na_2O	3.60	3.34	4.72	2.97	2.97	2.98	3.43
K_2O	0.51	0.46	0.49	0.46	0.53	0.22	0.45
P_2O_5	0.19	0.17	0.09	0.14	0.14	0.16	
S	0.04	0.17	0.04	-	-	0.08	
loss on ignition	3.34	0.96	1.19	1.20	1.00	2.25	1.66
H_2O	0.75	2.63	2.19	2.08	1.76	3.37	
total	100.51	99.59	100.51	100.16	100.41	100.26	
Na_2O/CaO	0.34	0.35	0.46	0.27	0.28	0.32	0.34
Na_2O+K_2O/CaO	0.38	0.38	0.50	0.32	0.38	0.34	0.38
K_2O/Na_2O	0.14	0.14	0.10	0.15	0.18	0.07	0.13

19 - sample 3048b, basalt, Cape Markevichs;
20 - sample 3037a, basalt, Moneron Island;
21 - sample 3031, basalt, Moneron Island;
22 - sample 2636a, basalt, Moneron Island;
23 - sample 2633, basalt, Moneron Island;
24 - sample 3022, andesite-basalt, Makarov River.

amount of CaO remains rather high. This circumstance, in our opinion, is connected, not with the fact that these rocks are incompletely albitized, but with the presence of secondary minerals which contain calcium (carbonates, epidote, zeolites, etc.). The possibility is not excluded that the increased amount of K_2O in the Late Cretaceous rocks of Sakhalin is due not only to the peculiarities of the parental magma, but also to the occurrence of potassium metasomatism. Finally the widespread development of minerals of the palagonite-chlorophaeite series may be the cause of relatively high ignition losses in the most recent (Pliocene) pillow lavas of the Kurile Islands.

Thus, in spite of the fact that the pillow lavas of Sakhalin and the Kurile Islands were outpoured under submarine conditions, only a small part of them consists of spilites. Such rocks are the oldest (Jurassic-Early Cretaceous and partly Late Cretaceous) pillow lavas that have subsided at some time to considerable depth and that bear traces of the superposed process of propylitization. More recent pillow lavas were mildly affected by the secondary alterations, as expressed in slight

Table 1 (continued)

	Late Miocene - Pliocene				
	25	26	27	28	Average
SiO_2	49.80	51.59	52.25	53.08	
TiO_2	0.70	0.66	0.57	2.10	
Al_2O_3	17.00	16.66	16.11	19.47	
Fe_2O_3	3.52	4.35	2.79	1.71	
FeO	8.12	4.50	8.22	5.35	
MnO	0.08	0.10	0.12	0.19	
MgO	3.91	2.87	5.46	3.96	
CaO	10.85	9.28	8.59	9.99	9.68
Na_2O	3.03	3.05	3.21	3.50	3.20
K_2O	0.63	0.53	0.58	0.73	0.62
P_2O_5	0.37	0.39	0.34	0.23	
S	-	0.30	0.17	0.03	
loss on ignition	1.00	2.06	0.49	0.33	0.97
H_2O	1.32	4.01	1.70	0.41	
total	100.33	100.35	100.70	100.09	
Na_2O/CaO	0.28	0.33	0.37	0.35	0.33
Na_2O+K_2O/CaO	0.34	0.39	0.44	0.42	0.40
K_2O/Na_2O	0.21	0.17	0.18	0.21	0.19

25 - sample 220b, basalt, Lamanons massif;
26 - sample 2964, basalt, Lamanons massif;
27 - sample 2964a, basalt, Lamanons massif;
28 - sample 2626, andesite-basalt, Criljon Peninsula.

All samples belong to SHILOV's collection.

All analyses were made in the analytical laboratory of SahKNJJ.

albitization, chloritization, etc. The most recent pillow lavas are practically unaltered volcanic rocks.

In an earlier article (SHILOV and KALISHEVICH, 1958) we discussed the origin of rocks of spilitic-keratophyric formation and concluded that for their formation, the outflow of lavas in submarine conditions, at depths of up to 100 m, is insufficient under any circumstances. ARUSTAMOV and FISHMAN (1963) by mistake included us among the supporters of the transvaporation hypothesis. The data given above on species composition of the faunas allow us to state that albitization processes do not occur at greater depths either. The results of oceanographic investigations show (YAGI, 1960) that the rocks from the greatest depths of the world's ocean, and also those from bores to depths of more than 200 m below the ocean floor (ENGEL and ENGEL, 1961) are not albitized.

Table 2. Chemical composition of the pillow lavas from the Kurile Islands

	Late Cretaceous				
	29	30	31	32	Average
SiO_2	51.23	50.40	51.46	51.81	
TiO_2	0.69	0.38	0.43	1.06	
Al_2O_3	16.18	18.50	20.69	18.19	
Fe_2O_3	6.82	3.65	3.40	5.07	
FeO	4.78	3.77	3.50	2.94	
MnO	0.26	0.14	0.16	0.18	
MgO	3.61	5.00	3.94	3.62	
CaO	9.77	6.95	7.89	8.33	8.30
Na_2O	3.40	4.85	4.18	4.45	4.22
K_2O	1.25	1.81	1.05	1.02	1.28
P_2O_5	0.22	0.25	0.23	0.26	
S	–	0.07	–	0.04	
loss on ignition	0.52	2.64	1.31	1.64	1.53 (1.86)
H_2O	1.55	1.81	1.49	1.15	
total	100.28	100.22	99.53	100.46	
Na_2O/CaO	0.35	0.70	0.53	0.53	0.53
Na_2O+K_2O/CaO	0.48	0.96	0.66	0.66	0.69
K_2O/Na_2O	0.37	0.37	0.25	0.23	0.31

29 - sample 320, basalt, Crab Bay, Shikotan Island;
30 - sample 317, basalt, Crab Bay, Shikotan Island;
31 - sample 2538, Chromovs Bay, Shikotan Island;
32 - sample 2538b, basalt, Chromovs Bay, Shikotan Island.

Samples 39-43 are from PISKUNOV's collection, sample 29, from SOLOVIOCA's collection, sample 30, from KAZAKOVA's collection and the rest are from SHILOV's collection.

All analyses were made in the analytical laboratory of SahKNJJ.

These data show that the peculiarities of mineralogical and chemical composition of spilites are not due to the presence of specific spilitic magmas, but are the consequence of modification of common volcanic rocks in the specific situation of folded geosynclinal areas (KORZHINSKY, 1962).

References

ARUSTAMOV, A.A., FISHMAN, I.L. (1963): Transvaporation or metamorphism? Trudy Laboratorii Paleovulkanologii, vip. 2.

ENGEL, C.G., ENGEL, A.E.J. (1961): Composition of basalt cored in Mohole Project (Guadalupe site). Bull. Amer. Ass. Petr. Geol. 45, N. II.

Table 2 (continued)

	Early Miocene				
	33	34	35	36	Average
SiO_2	52.29	54.36	53.04	51.94	
TiO_2	0.49	0.31	0.40	0.19	
Al_2O_3	17.69	15.03	17.28	21.47	
Fe_2O_3	6.68	4.39	3.86	3.45	
FeO	4.33	5.57	6.05	1.49	
MnO	0.07	0.05	0.19	0.11	
MgO	4.30	6.15	3.93	4.83	
CaO	9.19	6.53	8.77	9.31	8.45
Na_2O	2.37	3.07	3.37	3.37	3.04
K_2O	0.50	0.24	0.36	0.82	0.48
P_2O_5	0.22	0.29	0.43	0.24	
S	0.01	0.44	0.22	-	
loss on ignition	1.46	3.56	1.45	0.97	1.86
H_2O	0.59	0.48	0.83	1.35	
total	100.19	100.47	100.18	99.54	
Na_2O/CaO	0.26	0.47	0.38	0.36	0.37
Na_2O+K_2O/CaO	0.31	0.51	0.43	0.45	0.42
K_2O/Na_2O	0.21	0.08	0.11	0.24	0.16

33 - sample 2893, basalt, Zalivnaya River, Kunashir Island;
34 - sample 2908, andesite-basalt, Zalivnaya River, Kunashir Island;
35 - sample 2893-1, andesite-basalt, Zalivnaya River, Kunashir Island;
36 - sample 2521b, basalt, Cape Kapari, Paramushir Island.

EROKHOV, V.F., ZHIDKOVA, L.S., LITVINENKO, A.U., SIRIK, I.M., SHILOV, V.N. (1968): New data on geology of Moneron Island (North-East part of the Japan Sea). DAN SSSR, t. 182, N. 3.

Geologia SSSR, t. XXXI. Kamchatka, Kuril'skie i Komandorskie ostrova. M., Izd-vo "Nedra". (1964)

GOLOVINSKIY, V.I. (1963): Volcanogenic formations of the Kril'on Cape. Trudy VNIGRI, vip. 224.

KORZHINSKIY, D.S. (1962): The spilite problem and the transvaporation hypothesis in view of new oceanological and volcanological data. Izv. AN SSSR, ser. geol., N. 9.

PISKUNOV, B.N. (1966): Pillow lavas and hyaloclastites of the Urup Island/Kurile Islands. DAN SSSR, t. 168, N. I.

SERGEEV, K.F. (1966): Geological structure and development of the area of North Group of the Kurile Islands. M., Izd-vo "Nauka".

Table 2 (continued)

	Pliocene							
	37	38	39	40	41	42	43	Average
SiO_2	55.94	51.82	49.73	50.89	51.78	52.21	53.00	
TiO_2	0.32	0.25	0.91	0.39	0.54	0.66	0.85	
Al_2O_3	17.77	18.43	18.35	18.15	18.23	17.72	17.90	
Fe_2O_3	3.05	3.08	4.62	3.54	4.40	4.63	4.24	
FeO	5.30	6.68	3.07	4.69	4.93	4.67	3.94	
MnO	0.19	0.14	0.09	0.18	0.12	0.14	0.13	
MgO	4.08	4.58	4.59	4.71	4.74	4.48	3.90	
CaO	8.53	10.53	10.48	8.22	9.06	9.21	9.81	9.40
Na_2O	2.43	2.53	2.56	2.70	2.94	2.80	2.53	2.64
K_2O	0.30	0.60	1.14	0.96	0.90	0.81	1.05	0.82
P_2O_5	0.07	0.01	0.01	0.02	-	-	-	
S	0.50	0.03	0.06	0.31	0.10	0.20	0.07	
loss on ignition	0.92	0.93	3.26	1.91	1.44	1.64	2.10	1.74
H_2O	0.24	0.48	2.10	3.46	0.36	0.60	1.03	(1.49)
total	99.64	99.99	100.44	100.13	99.54	99.77	100.55	
Na_2O/CaO	0.29	0.24	0.24	0.33	0.33	0.30	0.26	0.28
Na_2O+K_2O/CaO	0.32	0.30	0.35	0.44	0.42	0.39	0.36	0.37
K_2O/Na_2O	0.12	0.24	0.45	0.36	0.31	0.29	0.42	0.31

37 - sample 2574, andesite, Ozernaya River, Kunashir Island;
38 - sample 2804a, basalt, Gornaya River, Jturup Island;
39 - sample 556b, basalt, Cape Tiger, Urup Island;
40 - sample 503b, basalt, Cape Ostrovnoy, Urup Island;
41 - sample 503b, basalt, Cape Ostrovnoy, Urup Island;
42 - sample 503a, basalt, Cape Ostrovnoy, Urup Island;
43 - sample 511, andesite-basalt, Cape Ostrovnoy, Urup Island.

SHILOV, V.N., KALISHEVICH, O.K. (1958): On the conditions of formation of the rocks of spilite-keratophyric formation. DAN SSSR, t. 122, N. 5.

SHUVAEV, A.S. (1969): Stratigraphy of the Upper Cretaceous deposits and history of geological development of Sakhalin in Late Cretaceous time. Autoreferat dissertacii, preds. na soiskanie uchenoi stepeni kand. geol.-miner. nauk, M.

YAGI, K. (1960): A dolerite block dredged from the bottom of the vitiar deep, Mariana trench. Proc. Jap. Acad., N. 4, 36.

The Production of Spilitic Lithologies by Burial Metamorphism of Flood Basalts from the Canadian Keweenawan, Lake Superior

R. E. Smith

Albitization of thin mafic lava flows from the Canadian Keweenawan Series of Lake Superior has resulted in spilitic lithologies. In one part of the succession, flows less than one meter in thickness are completely spilitized. Thicker flows have spilitized amygdaloidal flow tops and flow bases, but the flow centers are basaltic. The observation that the amygdaloidal bases are spilitized in addition to the flow tops appears to rule out the possibility that spilitization represents exchange reactions with overlying water during extrusion. The relation of maximum spilitization to thinnest flows is contrary to what one would expect from the autometasomatic origin. All petrographic evidence, however, is compatible with an origin based on spilitization during burial metamorphism.

Introduction

Along the northeast shore of Lake Superior, at Mamainse Point, a thick sequence of upper Precambrian basalt lavas of the Keweenawan series is exposed (Figs. 1 and 2). The thickness of Keweenawan exposed here is between 5 000 and 6 000 meters, and rests unconformably on an Archaean granite-greenstone complex. The total thickness of the Keweenawan sequence, however, is unknown as the upper part disappears beneath Lake Superior. Lava flows form approximately 75-80% of the sequence. Conglomerates largely derived from both Archaean and Keweenawan debris form the rest. Of the eruptive rocks, mafic lavas by far dominate, basalts together with a few basaltic andesites form about 95% of these. Discontinuous felsic bodies occur throughout much of the sequence, but these may increase in abundance in the upper half. Some are probably flows; others, strongly discordant, are probably volcanic plugs.

Near the bottom of the sequence occurs a series of thin mafic lavas which have albite as the chief feldspar. In addition to albite the rocks contain clinopyroxene magnetite, interstitial chlorite, and variable amounts of epidote, prehnite and hematite. Therefore, based on their present mineralogy, these mafic flows can be classified as spilites. Flows less than 1 meter thick are albitized throughout. Thicker flows, however, commonly have albitized amygdaloidal flow tops and bases, but their centers are still basaltic and contain only scattered amygdules. Alternation accompanying albitization has resulted in extremely heterogeneous flow tops and bases by the production of zones of epidotization and chloritization.

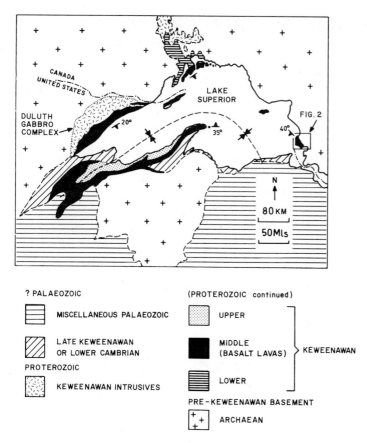

Fig. 1. General geological map of the Lake Superior region, after HALLS (1966). The inferred axis of the Lake Superior Syncline is shown dashed

Geological Setting

The Keweenawan Series of basalts represents a great accumulation of numerous flood basalt flows. The total thickness of sequence poured out in the Lake Superior Basin exceeds 10 000 meters (WHITE, 1960). Although poured out in a tectonic basin the majority of flows were probably extruded over a subhorizontal surface, flowing outwards from the source area which is believed to have been somewhere near the axis of the basin. Successive flows of very fluid magma would gradually fill any tectonic depressions (WHITE, 1960).

In the south of the basin at Keweenaw Peninsula (the large peninsula on the south shore, Fig. 1), some flows attained great thicknesses and were continuous for many miles. The Greenstone Flow for example is more than 300 meters thick over a strike distance of 43 kilometers, (WHITE, 1960). This flow, being recognized on Isle Royale (large island near U.S.-Canada border, northwest part of lake, Fig. 1) as a 300 ft+ unit, has been calculated by WHITE as representing a volume of at least 200 cubic miles.

In the area of Canadian Keweenawan being studied, several flows at least have recognizable thicknesses of about 60 meters or more. However, many thin flows are also present.

The best accessible exposures of the Canadian Keweenawan sequence are along the coastline between Mica Bay and Coppermine Point (Fig. 2), much of the coastline cutting across the strike. These exposures are very accessible because the shore relief is low and the highway is nowhere more than a few hundred meters from the coast. Along this strip the strike swings only slightly, and dips approach uniformity.

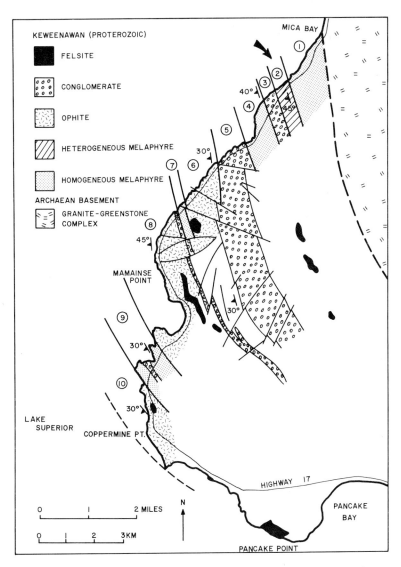

Fig. 2. Geological map of the shore line of Lake Superior at Mamainse Point, Ontario, Canada. Circled numbers refer to stratigraphic members of Fig. 3. Arrow points to member containing spilitic lavas. Base map taken from THOMSON (1954). Shore-line geology and subdivisions into members by the writer

Lava Types

The mafic lavas here can be divided into three broad types. The ophitic basalts generally show a homogeneity of appearance at any one outcrop. The main variation is simply related to thickness (cf. LANE, 1911); the ophitic mottling[1] increases with thickness of flows, i.e., it increases in the usual manner inwards from the top and bottom margins of each flow. Often accompanying this regular variation is an increase in brown coloration in the top 50 or 100 cms of each flow

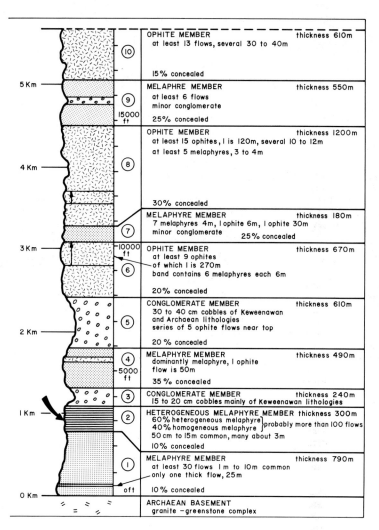

Fig. 3. Stratigraphic column compiled from shore-line and road-cut exposures of Fig. 2. Large black arrow indicates member containing spilitic lavas

---

[1] Due to subophitic to ophitic clinopyroxene growth, 2mm to +5mm diameter.

due to an increase in the abundance of hematite. This feature, seen in many recent lava sequences, is no doubt an oxidation feature of lava cooling. The second group, the homogeneous gray melaphyres, is also essentially homogeneous in appearance, except for slight vertical decreases in grain size towards the flow tops and bottoms. These flows, too, show the increase in brown coloration towards the top. The third group, the heterogeneous melaphyres, has distinct flow tops characterized by epidotized patches of yellow-green color, 50 to 100 cm across, sitting in a general gray background, which itself varies less strikingly in color (Figs. 4, 5 and 6). The heterogeneous flow top contrasts strongly with the essentially homogeneous gray layer forming the central main part of the flow. (This homogeneous gray central layer resembles the homogeneous gray melaphyres mentioned

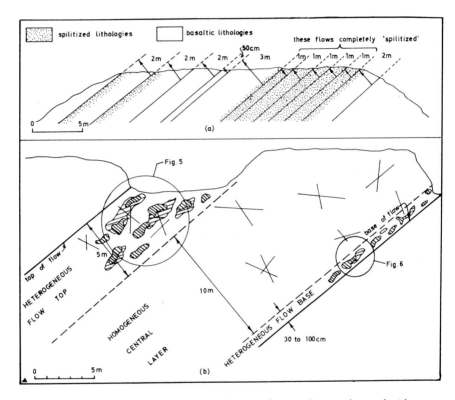

Fig. 4a. Typical series of flows seen in road cuttings through the heterogeneous melaphyres (member 2, Figs. 2 and 3). Arrows indicate individual flows, arrow head points towards the top. Stippling indicates heterogeneous flow tops and bases. As albitization accompanies heterogeneites (see albitized domain = gray domain of Figs. 5 and 6), completely stippled flows have spilite characters. The two unstippled flows in the center of this series are homogeneous melaphyres.
4b. Generalized cross section through a 15 m heterogeneous melaphyre flow from member 2 exposed at the lake shore. Heterogeneities are shown only diagrammatically here -- details of the flow top and base are shown in Figs. 5 and 6. Intersecting straight lines indicate the type of fracture pattern commonly crossing most flows. Slight marginal epidotization is common surrounding the fractures, and within the central layer a zone of albitization commonly surrounds these thin epidotized domains (Fig. 6). Flow tops are eroded along the shore line leaving the homogeneous central layer in relief as shown

above.) At the base of each flow, especially the thick flows, may
be a thin amygdaloidal zone which can also have epidotized domains
and essentially resembles the heterogeneous flow tops.

Petrography of the Heterogeneously Altered Lavas

The heterogeneously altered lavas occur in a 300 meter band, located
800 meters from the base of the Keweenawan sequence (Fig. 2). These
heterogeneous lavas aggregate 60% of the 300 meter band. Homogeneous
gray melaphyres form the rest. Altogether about 50 heterogeneous
flows are well exposed along the shore line and in road cuttings, but
probably about twice this number of flows are present in the 300 meter
band. In a 70 m part of this band good exposures enabled the recognition of 22 individual heterogeneous flows. Flows thicker than

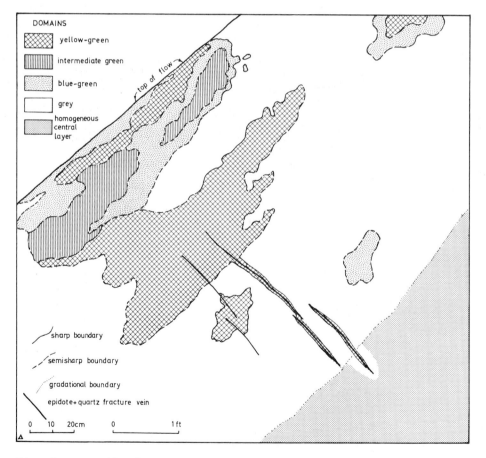

Fig. 5. Detail of a heterogeneous flow top. The yellow-green domain
owes its color to epidotization, the intermediate green is partly
epidotized; blue-green is chloritized; and in the gray domain most
plagioclase is albitized. In the homogeneous central layer relict
calcic plagioclase is common

about 2 meters show the separate feature of a heterogeneous flow top and the more uniform gray "central layer" which is the least altered part of each flow. In thinner lavas the whole flow becomes heterogeneously altered, and these have essentially "spilite" characters.

A section through a typical heterogeneous flow is shown in Fig. 4b. Such flows can be divided into two and sometimes three main layers: the heterogeneous flow top, the central homogeneous layer, and, where present, the heterogeneous flow base.

Lavas showing heterogeneous flow tops vary in thickness from about 1 meter up to 10 or 15 meters. A systematic relation between flow thickness and thickness of the various layers is shown in Fig. 7. Both the upper and (where present) the lower heterogeneous layers are more amygdaloidal than the central layer, but small scattered amygdules also may occur in the central layer, especially in thinner flows, those up to about 3 meters thick.

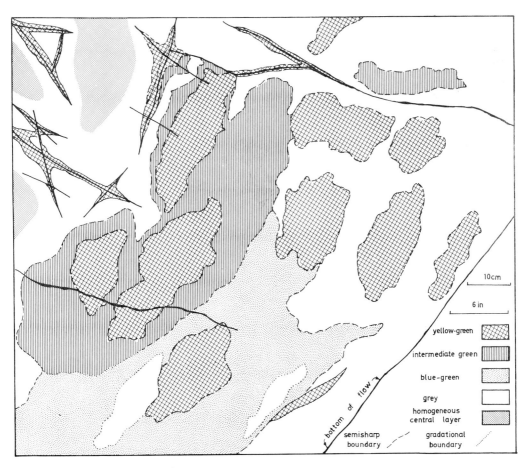

Fig. 6. Detail of a heterogeneous flow base. Same scheme as in Fig. 5: the yellow-green domain is epidotized; intermediate green is partly epidotized; blue-green is chloritized; gray is albitized. Thin epidotized domains are shown about fractures which cross the edge of the central layer at top left

Heterogeneous Flow Tops and Bases

Inspection of many flow tops led to the recognition of systematic variations in the alteration lithologies based on hand specimen characteristics - largely color - of the altered rock. The appearance of the rock types obviously reflected variations in the abundance of the alteration minerals present, particularly epidote and chlorite, and the degree of preservation of the original fabric.

In this way flow tops were divided into the following lithological domains: the albitized, chloritized, intermediate green, and the epidotized domains (the gray, blue-green, intermediate green and yellow green domains of Figs. 5 and 6). The same domain types are present in both flow tops and flow bases. However, as shown in Fig. 7, the amygdaloidal flow tops are always much thicker than the amygdaloidal bases. Amygdules are common in the flow tops and flow bases. In the tops these are usually subspherical and reach 1 or 2 cm in diameter. In the bases subspherical amygdules reach about 1 cm in diameter. Present also in the bases of some flows are pipe amygdules. These reach 5 or 10 cm in length and about 5 mm in diameter.
In specimens from the albitized domains (gray domain) very little of the primary minerals remains. The original calcic plagioclase microlites (0.05 to 0.15 mm in length) are largely replaced by albite although sporadically relict islands within the albite are visible. In some specimens scattered opaque grains (0.01 to 0.05 mm) represent the remains of original oxide phases. Pyroxene may or may not survive in the albitized domains. Where present this clinopyroxene forms small

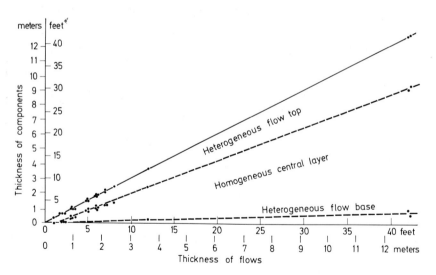

Fig. 7. Plot showing relation between total thickness of individual flows and thickness of each component layer. Due to construction method the top line (solid) must be a straight line. The upper dashed line shows that the relation between the thickness of the heterogeneous flow top and flow thickness is linear and is proportional to the thickness of each flow. Similarly the thickness of the heterogeneous flow base is proportional to the thickness of each flow but does not increase as rapidly as the thickness of the flow top. The homogeneous central layer disappears in very thin flows, those less than about 1 meter thick. These flows are albitized throughout and have the characteristics of spilites

equant subhedra (0.02 to 0.05 mm in length) and show partial alteration to chlorite, carbonate, and hematite. Scattered pseudomorphs after olivine can be recognized in some specimens but these are rare. As is the case for all rocks of the sequence, no olivine remains but is completely replaced by green layer silicated together with hematite and in some cases carbonate. Material interstitial to the plagioclase microlites now consists of chlorite heavily charged with hematite, and of sporadic patches of epidote, carbonate and sphene.

In the albitized domains the primary microlitic texture is preserved almost completely despite formation of the new minerals. Epidote is common within the albite. Sphene, another Ca-bearing mineral, commonly outlines the feldspar sites. Both these minerals probably contain part of the Ca liberated by the albitization process.

Specimens from the chloritized domains (blue-green domains) have a dull earthy blue-green color due largely to chlorite. Rocks within this category consist of plagioclase (extensively albitized), interstitial chlorite charged with variable amounts of hematite, and interstitial patches of fine quartz aggregates. Typically as one traverses from the albitized domains into the chloritized domains the interstitial chlorite becomes free of inclusions, mainly at the expense of hematite, and eventually encroaches upon sites originally held by plagioclase (Fig. 7b). Within the chloritized domain chlorite may form up to 40% of the mode. As the amount of chlorite increases, so too does the amount of quartz, forming clusters about albite microlites, and also lining the amygdules.

Specimens from epidotized domains (yellow-green domain) have a relatively simple mineralogy. They consist essentially of epidote with subordinate quartz and minor carbonate. Epidote forms a saccharoidal aggregate with a grain size of about 0.1 mm containing scattered granules of anhedral quartz of similar size. The relict microlitic pattern remains partly visible within the epidote aggregates of many specimens even though all plagioclase has been replaced. A brown dusting of very fine dark material (less than 0.001 mm) coinciding with former interstices darkens the epidote, while the epidote on former plagioclase sites is relatively clear. Amygdules in the epidotized domains generally contain quartz with subhedral epidote, but amygdules containing quartz and carbonate, or epidote alone are also common. Less commonly, chlorite may be present.

Located between the epidotized domain and the general background of albitized rock, there is occasionally a transitional zone called the intermediate green domain which shows a gradual increase in the quantity of epidote towards the epidotized domains. In specimens close in color to the gray albitized material, epidote begins to appear, lining the interstices between the albite microlites. This interstitial epidote increases in quantity at the expense of chlorite and hematite, until the interstices are filled. Epidote then encroaches upon the feldspar sites. Eventually all feldspar is replaced, and the rock is placed in the epidotized category.

Homogeneous Central Layer

The homogeneous central layer most closely resembles the inferred original mineralogy of the flows. Relict calcic plagioclase is the typical feldspar and the clinopyroxene is preserved in variable degrees. Lath-shaped microlites of plagioclase reach 0.15 or 0.20 mm in length, pyroxene about 0.03 mm. Material interstitial to the plagioclase and pyroxene consists of chlorite, sphene, minor carbonate, relict opaque oxides (up to 0.03 mm) and specks of hematite. The texture is intersertal with no evident flow orientation of microlites. Scattered amygdules are present in some specimens but are not common. These can reach 0.5 mm in diameter. Where present, the amygdules are generally filled with chlorite with a lining of small isolated granules of anhedral quartz. In addition hematite is present in some amygdules.

Although the central layer has characteristics distinct from the albitized domain, the boundary is very gradational. The main difference, of course, is the degree of albitization of the feldspar. In outcrop the most noticeable difference is the easily eroded nature of the albitized domains which leaves the central layer standing in relief (Fig. 4b).

Fracture veins bearing epidote and quartz can cross any of the domain types. In the central layer these veins are surrounded by thin domains of epidotized material extending generally 2 or 3 cm outwards. Extending about 5 cms further from the fracture is a zone in which partial albitization of the calcic plagioclase is generally observed

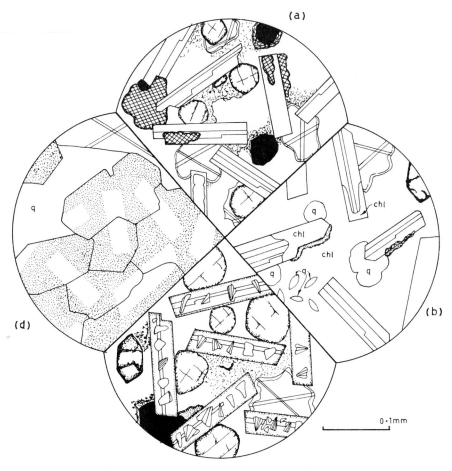

(Fig. 6), the intensity of albitization decreasing away from the fracture.

Discussion

The observation that the amygdaloidal flow bases are albitized in addition to the flow tops appears to rule out the possibility that albitization represents exchange reactions with any overlying water which might have been present at the time of extrusion. Many examples show a heterogeneously altered and albitized flow base resting directly on top of another lava flow with no intervening sediment that might have retained water.

Late magmatic reaction between a hydrated mafic lava and its volatile fluids is said to be capable of albitization of plagioclase, the process being called autometasomatism (NIGGLI, 1952; AMSTUTZ, 1954, 1958; BATTEY, 1956; DONNELLY, 1966). Such a process should be more effective in thick flows than in thin, because of the greater retention of heat, and the consequent longer cooling period would allow more extensive alteration. The relation of thickness of flow to the extent of albitization or alteration in these Canadian Keweenawan lavas (Fig. 7) is the opposite to what one would expect if autometasomatism had been the driving force behind albitization, since thin flows are completely albitized. It is difficult to envisage one-foot-thick flows retaining enough heat to cause the complete alteration described above. In addition, the environment of extrusion was likely to have been subaerial or perhaps shallow water. Abyssal environments would not fit the geological setting. High confining pressures due to superincombent water, which would favor autometasomatism by retention of volatiles, could not have been present.

◄ Fig. 8a-d) from flow top; c) from central layer.
a) Albitized domain. Microlites of clear albite - some examples containing granular epidote (cross-hatched). A partly albitized microlite is shown at the bottom of this field. Equant clinopyroxene granules (shown with relief and cleavage) are partly altered to a colorless carbonate (with oblique twinning). Magnetite granules (black) are partly altered to hematite. Remaining interstices are filled with chlorite together with subordinate epidote (cross-hatched). The lath shapes of the albite are inherited from the former plagioclase microlites which had a calcic composition
b) Chloritized domain. Microlites of albite set in a chloritized matrix. The chlorite (chl) is free from most inclusions except for quartz grains (q) which are either set against microlite sites or occur as groups of lensical grains. Chlorite has partly encroached upon microlite sites. Sphene granules (with relief) commonly cluster along microlite margins
c) Central layer. Original calcic plagioclase microlites are charged with flakes of fine white mica, stippling shows slight marginal zoning of plagioclase. Clinopyroxene granules (with relief and cleavage) are largely unaltered, but former olivines (left) are completely altered to chlorite and hematite. Chlorite fills interstices and is commonly dusted with fine hematite specks. Towards the margins of the central layer clinopyroxene in some examples breaks down to carbonate
d) Epidotized domain. Epidote is stippled to show the distribution of very fine dusty inclusions corresponding to former interstices between microlites. White lath shapes within epidote aggregate represent clear epidote on sites of former plagioclase

Another possibility should now be considered. If thick successive flows erupted and accumulated rapidly on top of each other, it is likely that enough heat would gradually build up that it could add appreciably to or perhaps even dominate the heat component supplied by depth of burial. In the Canadian sequence being studied, the succession underlying the spilitic lavas consists of thin flows, generally 1 to 10 meters thick. The thickest flow is only 25 meters. Likewise, the flows associated with the spilitic lavas are thin (1/2 to 15 m). There are certainly no thick flows in the overlying 200 meters sequence, or in the underlying 800 meters. The closest thick flow is probably the 50 m thick ophite in member 4 being approximately 700 meters above the band containing the heterogeneous flows (Fig. 3). It therefore seems unlikely that much of the temperature increase arose from retention of original magmatic heat of individual flows within the volcanic pile.

One is left with the conclusion that albitization and the related alteration must represent reactions due to burial metamorphism. Amygdule minerals such as quartz, prehnite, epidote, carbonate, and chlorite are common in the lower part of the section with laumontite being present in addition at the top. These minerals, together with minor pumpellyite, would indicate a grade of metamorphism apparently at about the beginning of the prehnite-pumpellyite facies.

An obvious problem presents itself: why did certain lavas become heterogeneously altered and albitized while associated lavas of similar thickness and otherwise of probable similar major characteristics (the homogeneous melaphyres) remain unalbitized?

The heterogeneously altered and albitized flows or portions of flows represent parts of the sequence which reached a more advanced degree of adjustment to burial metamorphism than the rest of the sequence. It is generally accepted that a more extensive reaction may be facilitated by the infiltration of a hydrous fluid phase during burial. If this is the situation here, the following factors are likely to have played a dominant role in channelling the fluid phase and hence largely controlling the degree of reaction: 1. a high original vesicularity; 2. the presence of fractures or brecciation which would act as permeable channel-ways; 3. any permeability along the interface that separates successive lavas; 4. a high glass content of the groundmass, which by hydration may become relatively permeable by diffusion under metamorphic conditions; and 5. a high volatile content frozen into a glass may render such a glass more reactive than another with a lower volatile content.

The answer does not appear to lie in differences of original vesicularity because both the heterogeneous and homogeneous melaphyres appear to be comparable in vesicularity. Nor do the heterogeneous flows appear more fractured. Further the interface separating successive lavas does not appear to be significantly different in the two cases. Provided that the answer is not simply a random coincidence, one is left with the conclusion that the different degrees of metamorphic reaction of certain flows may represent either variation in the amount of original interstitial glass in the groundmass, or possibly variations in the composition of this glass. As an attempt to assess these possible factors, a comparative study of the three lava types - the ophytes, the homogeneous melaphyres, and the heterogeneous melaphyres - is being undertaken by the writer.

Conclusion

Albitization of calcic plagioclase in thin basalt lavas from part of the Canadian Keweenawan has resulted in spilite characteristics. Albitization was accompanied by epidotization and chloritization resulting in a marked heterogeneous appearance of the spilitic lavas. Flows less than 1 meter thick were albitized throughout and can be called spilites in the sense that they are mafic lavas with albite as the chief feldspar. Thicker flows have albitized flow tops and flow bases, but their central parts retain the original calcic plagioclase and are therefore still basaltic.

Albitization by exchange reactions with overlying water during extrusion seems unlikely because flow bases as well as flow tops were albitized. Autometasomatism also seems unlikely because the thinnest flows are most completely albitized. All observations, however, are compatible with an origin based on albitization and related reactions taking place during burial metamorphism.

The heterogeneous flow tops and bases represent lithologies which show a more advanced adjustment to burial metamorphic conditions. These heterogeneous flow tops and bases must represent zones which were rendered permeable to hydrous fluids at some stage during burial. The controlling influence which rendered some flows permeable and prone to heterogeneous alteration but left others homogeneous is not known but may have been related to the character and quantity of glass formerly present in the groundmass. The grade of metamorphism was apparently at about the beginning of the prehnite-pumpellyite facies, because, although prehnite is commonly present, pumpellyite is very rare. Where preserved the character of the central parts of the thicker flows shows that the original lavas were fine-grained aphyric pyroxene basalts.

Acknowledgements

The research for this paper was initiated at the University of Manitoba while supported by a Post-doctoral Fellowship sponsored jointly by the National Research Council of Canada and the University of Manitoba. Final preparation for publication was carried out at the State University of New York at Binghamton, during tenure of an Organized Research Program in Solid Earth Geology Post-doctoral Fellowship. The writer wishes to thank W.T. JOLLY and B. DOOLAN for reading the manuscript.

References

AMSTUTZ, G.C. (1954): Geologie und Petrographie der Ergußgesteine im Verrucano des Glarner Freiberger. Vulkaninstitut Imm. Friedlaender. Publ. No. 5, Zürich.

AMSTUTZ, G.C. (1958): Spilitic rocks and mineral deposits. Bull. Miss. School of Mines and Metallurgy, Tech. ser., no. 96, 11.

BATTEY, M.H. (1956): The petrogenesis of a spilitic rock series from New Zealand. Geol. Mag. 93, 89-110.

DONNELLY, T.W. (1966): Geology of St. Thomas and St. John, U.S. Virgin Islands. Geol. Soc. Am. Memoir 98, 310. H.H. Hess, ed.

HALLS, H.C. (1966): A review of the Keweenawan geology of the Lake Superior Region. In: The earth beneath the continents. J.S. Steinhart and T. Jefferson Smith, editors. Geophysical Monograph 10.

LANE, A.C. (1911): The Keweenawan series of Michigan. Michigan Geol. & Biol Survey, pub. 6, Geol. Ser. 4, 983, 2 vols.

NIGGLI, P. (1952): The chemistry of the Keweenawan lavas. Am. Jour. Sci., Bowen Volume, pt. 2, 381-412.

THOMSON, J.E. (1954): Geology of Mamainse Point Copper Area. Ontario Dept. Mines 62nd Ann. Rept., pt. 4, 1-25.

WHITE, W.S. (1960): The Keweenawan lavas of Lake Superior, an example of flood basalts. Am. Jour. Sci., v. 258, 367-374.

A New Appraisal of Alpine Spilites

M. Vuagnat

Abstract

Alpine spilites were thought to have a primary origin for two main reasons. They occur in regions which were considered at one time to lie outside the domain of Alpine regional metamorphism, and they exhibited fresh primary textures of volcanic rocks. These two criteria have been invalidated; regional metamorphism in the Alps is much more extensive than was thought and some fresh-looking spilitic rocks have been demonstrated to develop from basaltic andesites in the laumontite and the prehnite-pumpellyite facies. Spilitic assemblages have been looked for in vain in rather shallow continental non-orogenic pillow lavas. The rare spilites dredged from the bottom of the ocean have been interpreted as resulting from the metamorphism under a pile of lava flows of normal basalts (CANN, 1969).

The idea of primary (sensu lato) spilites should not yet be discarded. One can conceive that magma intruding into soft uncompacted sediment full of water could consolidate into a rock remaining hot long enough in the presence of water to undergo conditions similar to those of light burial metamorphism.

Introduction

About 20 years ago, it seemed that, in the Alps at least, the primary origin of spilites was the most likely hypothesis. Petrographic work done by the author on alpine pillow lavas and sills of Mesozoic age in regions considered to be outside the zone of regional metamorphism led to the conclusion that assemblages such as albite-augite, albite-chlorite, albite-chlorite-hematite$^{\pm}$ carbonate were primary, either truly magmatic or autometasomatic (VUAGNAT, 1946).

I was led to this conclusion not only by the apparent freshness of "secondary" minerals, but above all by the fact that the finest textures, for instance the subvariolitic texture of the outer part of the pillows, were entirely preserved without showing any trace of mechanical deformation. The absence of secondary geometric features is part of what has been called by AMSTUTZ the "congruence or symmetry of similarity between geometric features and compositional spaces" and considered by this author as the strongest evidence in favor of a primary origin of spilites (AMSTUTZ, 1968, p.744).

Other petrographic investigations on older alpine rocks not related to the penninic geosyncline also led to the same conclusion. I have in mind here the spilites or so-called melaphyres of the Permian Verrucano of Eastern Switzerland (AMSTUTZ, 1954) and the melaphyres of the Triassic sediments transgressing on the Pelvoux Massif in France

(VUAGNAT, 1947). It is interesting to note that in these two cases the spilitic rocks generally do not show a pillowy structure and are not part of thick sedimentary series characteristic of many geosynclinal environments. Especially in the case of the Pelvoux melaphyres, the Triassic series is a thin epicontinental sequence of limestones and dolomites.

The results of all of these investigations on alpine spilites supported the conclusions of other geologists working on similar rocks of older orogens, for instance those of LEHMANN (1941) on the volcanics of the Lahn-Dill basin in Germany.

Use of the Term Spilite

My purpose in this paper is to see to what extent recent developments in Alpine geology have borne out these ideas. Before doing this, I would like to make clear the definition of the term spilite which will be used here.

Spilitic rocks should combine geometrical features (texture, fabric) typical of eruptive (mostly but not only volcanic) rocks with an assemblage of minerals some of which at least are low temperature species found in the metamorphic rocks of the zeolite facies or of the green-schist facies. The most common assemblages have been listed by VALLANCE (1960, p.31). They all have albite as the main component. I still consider, as I did 24 years ago, that we can speak of a spilitic trend, i.e., that we have a gradation from rocks not very different from ordinary basalts to highly spilitic ones like those characterized by albite and chlorite to the exclusion of lime-silicates. This latter assemblage is probably the most common and the most typical among spilitic assemblages.

Most spilites show a SiO_2 content corresponding to the chemical composition of basalts or basaltic andesites. I do not quite agree with AMSTUTZ (1968, p.745) that Na enrichment in spilites is due to a sampling bias. This factor may sometimes play a part, but, by and large, I am of the opinion that at least typical spilites, with the common albite-chlorite assemblage, have a true deficiency in Ca and excess in Na by comparison with the mean composition of basalts; I think that these differences increase with the accentuation of the spilitic character.

In opposition however with my ideas of twenty years ago and for reasons which will be made clear below, I completely agree with the opinion according to which the term spilite should be used in a purely descriptive sense and bear no genetic meaning. Thus the albite-chlorite assemblage for instance could have a magmatic, a metasomatic or a metamorphic origin; as long as the primary eruptive texture is preserved ("clean" fabric of AMSTUTZ, 1968, p.745), we should speak of spilites.

In his 1968 paper AMSTUTZ listed seven distinct theories advanced to explain the genesis of spilites. Here we shall oversimplify and consider only the two main opposite opinions. On one side we lump together all hypotheses which consider that the spilitic assemblage is contemporaneous or pene-contemporaneous with the crystallization of the rock and which include the hypothesis of primary origin (sensu lato) of spilites. This takes care of IV to VII of the AMSTUTZ list. On the other side, we group the theories which maintain that spilites are truly secondary rocks which acquired their present composition after a geologically significant lapse of time (often millions of years) subsequent to the crystallization of the original magma. This hypothesis of a secondary origin of spilites covers theories I to III of the list mentioned above. However I consider numbers I and II (surface-weathering; dry or wet diffusion) very unlikely and hence I will treat a secondary origin as more or less synonymous with metamorphic or diagenetic origin.

Extension of the Domain of Regional Metamorphism in the Alps

The occurrence of spilites in supposedly unmetamorphic zones of the Alps and the congruence between composition and texture were the two criteria which led, in the late forties and early fifties, to the conclusion that alpine spilites had a primary origin sensu lato. During the last decade, however, several observations indicated that the extension of regional metamorphism was greater than was thought before and overlapped on the "Massifs cristallins externes", on the Helvetic zone of the High Calcareous Alps in Switzerland and on the Subalpine Chains in France (MARTINI and VUAGNAT, 1969). For instance, the mineral stilpnomelane was found in the pre-Alpine nappes (SALIMI, 1965) and in the Mesozoic deposits of the High Calcareous Alps (STRECKEISEN and NIGGLI, 1958; DELALOYE, 1966). Another clue was the very young potassium-argon and rubidium-strontium ages found for some of the Paleozoic rocks of the "Massifs cristallins externes" (Aar Massif, Mont-Blanc/Aiguilles-Rouges Massif, etc.)(JÄGER and FAUL, 1959; KRUMMENACHER and EVERNDEN, 1960). These apparent ages give evidence of leakage of argon and strontium due to reheating of these rocks during the Tertiary phase of Alpine regional metamorphism.

More important was the discovery that the Taveyanne sandstones of the Helvetic Nappes and of the Subalpine Chains were metamorphosed into the zeolite facies (MARTINI and VUAGNAT, 1965; MARTINI, 1968). This formation of Upper Eocene-Lower Oligocene age is made up of graywackes containing up to 90% volcanic fragments (basaltic andesites) grading into pelitic rocks. It overlies a Mesozoic and Lower Tertiary sequence of limestones, shales and calcareous sandstones which does not appear, and was never thought to be, metamorphic.

Detailed studies of the Taveyanne sandstone between the River Arve and the River Giffre (Haute-Savoie, France) and comparison with rocks of the same formation elsewhere have shown that three main types of graywackes may be recognized: laumontite type, prehnite-pumpellyite type, "green" type. The first two types correspond to the two zones of light regional metamorphism which have been recently added above the classical epizone (COOMBS et al. 1959). They are characterized by abundant secondary lime-silicates developed usually but not only in the plagioclases which have been concomitantly albitized.

The typical mineral assemblage of the "green" graywackes is mainly albite, chlorite, and titanite as an accessory constituent. The laumontite type is more or less restricted to the western part of the domain of distribution of the Taveyanne formation whereas the pumpellyite-prehnite type predominates further east, a fact probably to be correlated with a thicker overburden of nappes. The "green" type may be found anywhere; although its mineral assemblage is also stable in the greenschist facies, it does not seem to be related to an increase in burial metamorphism.

Thus it has become evident that rocks which do not exhibit signs of mechanical deformation and which do not show the slightest crystallization schistosity, have in fact been extensively transformed in the uppermost zones of regional metamorphism sometime during the Tertiary age. These rocks are not exceptional occurrences found in a few places only, but are part of an extensive zone reaching from the Rhine Valley in eastern Switzerland to the Dauphiné in France. They are located either on the "Massifs cristallins externes" or to the north and to the west of those, giving evidence that Alpine regional metamorphism went far beyond the domain of the Penninic nappes.

All this means that the first argument in favor of a primary origin of Alpine spilites does not hold. Furthermore, the Mesozoic ophiolitic spilites (pre-Alps, "Aroserschuppenzone", Montgenèvre), like the Permo-Triassic ones, are found in zones where the influence of regional metamorphism is either proved or very probable.

For some time it seemed that there might be some way to escape from this awkward situation in studying ophiolites of the Apennines where spilitic rocks are known. However, the very recent discovery (ELTER et al. 1969) in this mountain chain of Tertiary volcanic graywackes very similar to the Taveyanne sandstones and also transformed into the zeolite facies, casts strong doubt on the "unmetamorphic" state of Apenninic ophiolites.

New Appraisal of the "Geometrical" Criterion

Now we shall turn to the second criterion of the primary origin of spilites, that of congruence between geometric features and compositional spaces or, for shortness sake, the "geometrical" criterion. This too seems no longer reliable. If we turn to the same Taveyanne graywackes, we see that many volcanic fragments, especially in the "green" type have a spilitic composition with albite and chlorite as main components. The primary volcanic fabrics of these fragments have been perfectly well preserved even in the fine-grained groundmass which exhibits hyalopilitic, pilotaxic, trachytic, etc. textures. Twenty-five years ago I thought that these fragments were bona fide primary spilites, that they already had this albite-chlorite composition at the time of their deposition and that they came from the erosion of spilitic subaerial flows (VUAGNAT, 1943, 1952). That it is not so is demonstrated by the existence of fragments with normal basic plagioclase in calcitized margins about one inch thick frequently found at the top and basis of graywacke beds, the plagioclases being thoroughly albitized elsewhere in the rock. The fact that at least a good part of the spilites of the Taveyanne sandstones were normal basaltic andesites is confirmed by the occurrence of basic plagioclase in calcite-rich concretions or pseudo-pebbles found in many of these graywackes (MARTINI, 1968). Finally it is relevant to mention here that in the Grès de Clumanc (Basses-Alpes, France) only normal basaltic andesites with calcic plagioclases are found. There is little doubt that these sandstones are the southern extension of the Taveyanne sandstones in a more external zone of the Alps where there was no tectonic overburden sufficient to induce light regional metamorphism.

That spilitic rock fragments abound in the "green" graywacke type is manifest, but neither are they lacking in other types. The typical laumontite graywackes are spotted ("grès mouchetés") with white spots of a few mm in diameter surrounded by a green groundmass. Under the microscope it can be seen that the white spots coincide with laumontite-rich spaces. The zeolite has invaded the plagioclase, most of the other minerals and even the interstices between the grains. The primary textures may still be recognized but are rather vague, more or less erased by the secondary growth of laumontite; clearly there is no congruence between primary geometric features and compositional spaces and the rock looks deeply altered. The green groundmass has a quite different aspect: there is little or no zeolite; the plagioclases are albitized but look reasonably fresh due to the absence of lime-silicate inclusions; the augite and (or) hornblende may still persist or may be transformed into chlorite. The primary geometric features are very well preserved, as well as in the "green" type. The difference in aspect between white spots and groundmass is striking, not less striking is the fact that the limit between the two can go across a single element. Thus we see in the same rock, in very close association, two different kind of transformations, one in which the main secondary mineral does not respect the boundary of the primary components, the other of a pseudomorphic type resulting in spilitic fragments.

The conclusion from these observations is that an important part of the spilitic grains of the Taveyanne graywackes acquired their peculiar composition after the deposition of the sediment and were once common basaltic andesites. It is not yet possible to exclude the possibility that some volcanic pebbles were already spilites prior to the time of their deposition; this could be the case, for instance, in the uppermost beds of the formation (grès de Taveyanne IV, VUAGNAT, 1952) where the green type is predominant. The fragments of spilitic pillow lavas found in the Val d'Illiez sandstones, graywackes slightly younger than the Taveyanne sandstones, might also be derived from spilitic submarine flows because the Val d'Illiez sandstones are systematically devoid of secondary lime-silicates.

Recent petrological work on the Taveyanne formation has thus cast doubt on the validity not only of the first criterion of primary origin of spilites but also on the second one, the geometrical criterion. Pseudomorphic transformations render this argument invalid.

Of course what is demonstrated is not the non-existence of primary spilites but only that some Alpine spilites thought to be primary by myself and by others are either certainly secondary (like most spilitic fragments of the Taveyanne graywackes) or could be secondary as a result of the greater extension of the domain of regional metamorphism in the Alps; and that the geometrical criterion can no longer be used to eliminate this possibility.

Evidence from Non-Orogenic Rocks

Almost all known continental spilites are found in orogenic zones where regional metamorphism is developed as a matter of fact. It is thus unlikely that the study of continental spilites will ever solve the problem of their origin; even with very fresh looking rocks, there will always be the possibility that they have acquired their present composition during a phase of light metamorphism.

In order to avoid this dilemma we shall have to look at other rocks. It is well known that by far the most common occurrence of spilitic mineral assemblages is found in submarine volcanics exhibiting pillow structures. Typical pillow lavas do not occur only in orogenic zones, they are fairly frequent in different environments. There are numerous pillow basalts flows in the Columbia Plateau basalts, in the palagonite formation of Iceland, in the pre-Aetnean lavas of Sicily (RITTMANN, 1958; VUAGNAT, 1958) and in the volcanics of Grand Canary Island. In all of these cases no typical spilitic assemblage has been described, and my personal observations also point to the fact that these pillows are made up of more or less common basalts with basic plagioclases and show a glassy rim. Of course it can be argued that all of these non-orogenic pillows are part of shallow subaquatic, or, in the case of Iceland, subglacial flows, whereas orogenic pillow lavas flowed on the bottom of geosynclinal seas and consolidated at a much greater depth, possibly under pressures higher than the critical pressure of water. It could be that very high water pressure is necessary to develop primary spilitic assemblages of minerals. Alternatively, one might also argue that in orogenic regions the original basaltic magma has opportunity to evolve, in geosynclinal environments, either by differentiation or by assimilation (including contamination by sea water) in such a way as to favor the formation of spilites.

If it is difficult to discard rapidly the second hypothesis, the first one does not seem very strong as more and more normal basaltic pillows are dredged, sometimes from great depths (see for instance ENGEL and ENGEL, 1964). Effusion under high water pressure is thus certainly insufficient to account for the genesis of spilitic assemblages.

Among the few oceanic spilites which have been reported, the Carlsberg Ridge find, in the Indian Ocean, is probably the best documented (CANN, 1969). In contrast to the albite greenstones described by MELSON and VAN ANDEL (1966) from the Mid-Atlantic Ridge, the spilites from the Carlsberg Ridge do not show any structural signs of dynamic metamorphism and the pillows exhibit the variolitic structure so common in alpine spilitic pillows.

In his 1969 paper CANN considers that the Carlsberg Ridge spilites are probably basaltic rocks metamorphosed under a thick pile of younger lava flows. He postulates a reasonable mechanism to explain the necessary transformations. However the arguments he uses against a primary origin are not derived from new evidence gathered from the study of these submarine rocks; they could, for instance, also be used in the case of alpine spilites. Nevertheless it is interesting to point out that among other rocks hauled from the same area metagabbro and serpentinites were found (CANN, 1969, p.2). The association with deep-seated rocks brought to the bottom of the sea by some as yet obscure faulting mechanism agrees well with the hypothesis of a metamorphic origin of these spilites but does not prove it.

It seems thus that to be certain of the primary origin (sensu lato) of oceanic spilites, we should find these rocks where it can be clearly demonstrated that they were emplaced near or on the bottom of the sea; recent flows would be ideal in this respect.

It is maybe not out of place to discuss here an important feature of Alpine spilitic pillows that might be used as an argument in favor of a primary origin of their composition. I refer to the variation in the content of some chemical elements, mostly an increase in Na, from the center to the rim of the pillow where the variolitic matrix begins. One of the first records of such a variation is from spilitic pillows from eastern Switzerland (VUAGNAT, 1946). Since then it has been found in several other occurrences of orogenic pillows but not in fresh basaltic ones. It must be made clear that this increase in sodium is not due to a more thorough albitization of the outer part, but to an augmentation of the quantity of albite crystals. This means that the increase in plagioclase content toward the exterior was a primary feature and, if the albite is not primary, then it must have replaced labradorite or andesine selectively (see also CANN, 1969, p.16).

It is interesting to mention that pillows showing good radial differentiation generally also have a typical variolitic border made up of albite-rich radiating spherulites dispersed in a chloritic matrix. Here again, shallow water pillows of basaltic composition do not show this structure although some may exhibit a slight tendency toward it. I do not follow those who think that the variolitic structure is a secondary metamorphic feature. For one thing there is a kind of continuity in the geometrical pattern from the variolitic structure to the intersertal divergent texture of the core of many pillows and nobody ever doubted that this last type of texture was primary. However, there is a possibly more cogent reason for viewing with scepticism a metamorphic origin of variolites. In some less spilitic pillows there is still fresh augite, which is also present as fine needles in the varioles. It would be highly unlikely to have augite crystallizing under the conditions of low temperature and pressure corresponding to those of the green schist or to zeolite facies. The varioles are doubtless primary structures resulting either from the crystallization in the liquid state or from devitrification of a hot basic glass that has just consolidated.

The existence of radial differentiation and of variolitic rims of spilitic pillows does not prove the primary origin of their composition but points to the fact that it is probably incorrect to think that pillows from orogenic belts are completely comparable to shallow water non-orogenic pillows.

Conclusions

Spilites are rocks in which secondary-looking mineral assemblages coexist with primary textures; they might have a primary or a secondary origin. At first both working hypotheses were equally acceptable as a starting point. Then some typical spilites, as are found for instance in fragments in the Taveyanne formation, have been demonstrated to be the result of metamorphism in the zeolite facies zone. It seems that the burden of proof has been shifted to the supporters of the primary origin hypothesis. However, it is difficult to see on which criteria of "primariness" we should agree. The geometrical criterion is invalid because of the existence of pseudomorphic transformations; it seems probable that we cannot exclude the possibility of some metamorphism in most zones of the Alps where spilites occur and the situation is probably similar in older orogens; if we turn to oceanic spilites it is difficult to prove that these rocks have not been dragged from some deeper zone to the surface by some faulting process (MELSON and VAN ANDEL, 1966). The question will still remain open for a long time.

We may look at this problem from a different standpoint. We know that spilites may derive from basaltic rocks when they are submitted to metamorphic conditions corresponding to the laumontite and to the prehnite-pumpellyite facies; that means a temperature of at least $200^{\circ}C$ corresponding, under normal conditions, to an overburden of about 5 500 m. The transformations have probably evolved in the presence of abundant water, connate water in the case of the Taveyanne sandstones, in a porous medium. In any place where these conditions are obtained, we may expect that basic eruptive rocks can be spilitized. CANN (1969) expects that in regions of high heat-flow, at midoceanic ridge crests, it is conceivable that temperatures of 200° to 250°C could be reached at depths of 1 1/2 - 2 km.

One may wonder if similar conditions cannot be realized in some cases just after consolidation. There are clues that the most typical Alpine Mesozoic spilites are found in sills or in flows of moderate thickness in contact with sediments, often with calcareous shales. Could it not be possible that in the case of the flows also, the magma did not reach the sediment/water interface, but was intruded into soft, uncompacted sediments full of water. The eruptive rock could remain hot under a more or less impervious cover long enough to enable the spilitic transformation; the necessary water could be partly of magmatic origin, partly derived from the sediments themselves. We can mention here that some sills in geosynclinal series seem to have pillowy margins (see for instance LOUBAT, 1968).

Of course I am quite conscious that we may also suppose that the proximity of sedimentary rocks during regional metamorphism was one of the factors inducing the formation of spilitic assemblages (COOMBS, verbal communication).

I am sure that both possibilities should be seriously explored. Depending on the results of this exploration, it might be that the gap between autometasomatic and truly regional metamorphic spilites is not as wide as was first thought.

In any case the discovery of oceanic spilites, if they are metamorphic, means that all Alpine spilites do not necessarily result from the late regional phase of metamorphism due to the overburden of tectonic nappes. They may also have acquired their present composition much earlier during preorogenic processes of the geosynclinal phase. We could have here an analogy with rodingites deriving from gabbro or diabase dikes in peridotite that have probably been transformed before the piling up of the nappes during the process of cold intrusion of slices of the Upper Mantle (VUAGNAT, 1967).

Acknowledgements

I wish to thank all those with whom I have discussed the spilite problem, in particular D.S. COOMBS of the University of Otago who was kind enough to read and correct the manuscript of this paper, H. LOUBAT of Lakehead University, J. MARTINI of the University of Geneva, J. MOORE of Carleton University, D.M. SHAW of McMaster University and H. WILLIAMS of the University of California at Berkeley. I am also indebted to the Swiss National Foundation for its financial help in research projects related to the spilite problem.

References

AMSTUTZ, G.C. (1954): Geologie und Petrographie der Ergußgesteine im Verrucano des Glarner Freiberges. Vulkaninstitut Immanuel Friedlaender, Publ. 5, 150, Zürich.

AMSTUTZ, G.C. (1968): Spilites and spilitic rocks. In: Basalts. The Poldervaart Treatise on rocks of basaltic composition. New York, London and Sydney: Interscience, Vol. 2, 737-753.

CANN, J.R. (1969): Spilites from the Carlsberg Ridge, Indian Ocean. J. Petr. 10, 1-19.

COOMBS, D.S., ELLIS, A.J., FYFE, W.S., TAYLOR, A.M. (1959): The zeolite facies with comments on the interpretation of hydrothermal syntheses. Geochim. Cosmochim. Acta 17, 53-107.

DELALOYE, M. (1966): Contribution à l'étude des silicates de fer sédimentaires. Le gisement de Chamoson (Valais). Mat. Géol. Suisse, Sér. géotechn. 9/13, 71.

ELTER, P., GRATZIU, C., MARTINI, J., MICHELUCCINI, M., VUAGNAT, M. (1969): Remarques sur la ressemblance pétrographique entre les grès de Petrignacola (Apennin) et les grès de Taveyanne des Alpes franco-suisses. Arch. Sci., Genève, 22, 187-212.

ENGEL, C.G., ENGEL, A.E.J. (1964): Igneous rocks of the East Pacific Rise. Science, N.Y. 146, 477-486.

JÄGER, E., FAUL, H. (1959): Age measurements on some granites and gneisses from the Alps. Bull. Geol. Soc. Amer. 70, 1553-1558.

KRUMMENACHER, D., EVERNDEN, J.F. (1960): Déterminations d'âge isotopiques faites sur quelques roches des Alpes par la méthode K-Ar. Bull. suisse Min. Pétr. 40, 267-278.

LEHMANN, E. (1941): Eruptivgesteine und Eisenerze im Mittel und Oberdevon der Lahnmulde, Wetzlar. Herausgegeben von der Bezirksgruppe Wetzlar der Fachgruppe Eisenerzbergbau.

LOUBAT, H. (1968): Etude pétrographique des ophiolites de la "zone du Versoyen" (Savoie, France et Province d'Aoste, Italie). Arch. Sc., Genève, 21, 265-455.

MARTINI, J. (1968): Etude pétrographique des Grès de Taveyanne entre Arve et Giffre (Haute-Savoie, France). Bull. suisse Min. Pétr. 48, 539-654.

MARTINI, J., VUAGNAT, M. (1965): Présence du faciès à zéolites dans la formation des "grès" de Taveyanne (Alpes franco-suisses). Bull. suisse Min. Pétr. 45, 281-293.

MARTINI, J., VUAGNAT, M. (1969): Metamorphose niedrigst temperierten Grades in den Westalpen. Fortschr. Miner. 47, 52-64.

MELSON, W.G., VAN ANDEL, Tj.H. (1966): Metamorphism in the Mid-Atlantic Ridge, 22 N. latitude. Mar. Geol. 4, 165-186.

RITTMANN, A. (1958): Il meccanismo di formazione delle lave a pillows e dei cosidetti tufi palagonitici. Boll. Ac. Gioenia Sc. nat. Catania (4), 4, 311-318.

SALIMI, F. (1965): Etude pétrographique des roches ophiolitiques des Préalpes romandes. Bull. suisse Min. Pétr. 45, 189-279.

STRECKEISEN, A., NIGGLI, E. (1958): Über einige neue Vorkommen von Stilpnomelan in den Schweizer Alpen. Bull. suisse Min. Pétr. 38, 76-82.

VALLANCE, T.G. (1960): Concerning spilites. Proc. Linnean Soc. N.S. Wales 85, pt. 1, 8-52.

VUAGNAT, M. (1943): Les grès de Taveyannaz du Val d'Illiez et leurs rapports avec les roches éruptives des Gets. Bull. suisse Min. Pétr. 23, 353-436.

VUAGNAT, M. (1946): Sur quelques diabases suisses. Contribution à l'étude du problème des spilites et des pillow lavas. Bull. suisse Min. Pétr. 26, 116-228.

VUAGNAT, M. (1947): Sur le caractère spilitique des mélaphyres de la région du Pelvoux. Cpte rendu séances Soc. Phys. Hist. nat. Genève 64, 63-65.

VUAGNAT, M. (1952): Pétrographie, répartition et origine des microbrèches du Flysch nordhelvétique. Mat. carte géol. Suisse, Nlle sér. 97, 103.

VUAGNAT, M. (1958): Les basaltes en coussins d'Aci Castello et du Val di Noto. Rend. Soc. Miner. Ital. 15, 311-322.

VUAGNAT, M. (1967): Quelques réflexions sur les ophisphérites et les rodingites. Rend. Soc. Min. Ital. 23, 471-482.

Bibliography

A) Alphabetical List

1) Literature on Spilites and Keratophyres

ADITYA, S. (1955) Studies about an occurrence of albite-epidote rock from near Shusina, Manbhum District, Bihar. Quart. Journ. Geol., Min., Met. Soc. India, XXVII, p. 111-115.

ALEKSANDROV, A.I. (1956) Shungite from rocks of the spilite-albitophyre formation in the region of Krasnouralsk (Middle Urals). Izvest. Akad. Nauk SSSR, Ser. Geol. no. 5, p. 99-103.

AMSTUTZ, G.C. (1950) Kupfererze in den spilitischen Laven des Glarner -Verrucano. Schweiz. Min. Petr. Mitt. 30, p. 182-191.

AMSTUTZ, G.C. (1953) Geochemistry of Swiss lavas. Geochim. Cosmochim. Acta 3, p. 157-168.

AMSTUTZ, G.C. (1954) Geologie und Petrographie der Ergußgesteine im Verrucano des Glarner Freiberges. Publ. No. 5, Vulkaninstitut Immanuel Friedlaender, Zürich, 150 p.

AMSTUTZ, G.C. (1958) Spilitic rocks and mineral deposits. Bull. Missouri School of Mines, Tech. Ser. No. 96, 11 p.

AMSTUTZ, G.C. (1959) Syngenese und Epigenese in Petrographie und Lagerstaettenkunde. Schweiz. Min. Petr. Mitt. 39, p. 1-84.

AMSTUTZ, G.C. (1963) Space, time, and symmetry in zoning. Symposium - Problems of Postmagmatic Ore Deposition, Prague, Vol. I, p. 33-37.

AMSTUTZ, G.C. (1965) Some comments on the genesis of ores: Problems of magmatic ore deposition. Symposium, Prague, Vol. II, p. 147-150.

AMSTUTZ, G.C. (1968) Spilites and spilitic rocks. In: The Poldervaart Treatise on Rocks of Basaltic Composition, H.H. HESS and A. POLDERVAART, eds. Interscience (Wiley), New York, p. 737-753.

AMSTUTZ, G.C. (1968) Les laves spilitiques et leurs gîtes minéraux. Geol. Rundsch. 57, p. 936-954.

AMSTUTZ, G.C., and PATWARDHAN, A.M. (1972) Zur Genese der Strukturen und des Mineralbestandes der 'karbonatithaltigen' Spilite im Verrucano (Kt. Glarus, Schweiz). Fortschr. Mineralogie 50, p. 7-8.

ANGEL, F. (1919) Die Quarzkeratophyre der Blasseneckserie. Jb. Geol. B.A. Wien, 68, p. 29-62.

ANGEL, F. (1955) Über Diabasformen aus dem Bereich des Murauer Paläozoikums. Verh. Geol. B.A. Wien, Heft 3, p. 173-180.

ANGEL, F. (1957) Über die spilitisch-diabasische Gesteinssippe in der Grauwackenzone Nordtirols und des Pinzgaues. R.v. Klebelsberg-Festschrift, Geol. Gesellschaft Wien, Bd. 48 d. Mitteilungen 1955, p. 1-15.

ASHGIREJ, G.C. (1958) Das Problem des Aufsuchens verborgener Blei-Zink-Lagerstätten in Nordossetien. Materialien der Allunionsberatung zur Ausarbeitung wiss. Grundlagen für das Aufsuchen verborgener Erzkörper, Moskau, p. 74-78.

ASKLUND, B. (1949) Apatitjärnmalmernas differentiation. Geol. Fören. Förhandl. 71, p. 127-176.

BACKLUND, H.G. (1930) Die Magmagesteine der Geosynklinale von Nowaja Semlja. Rep. of the Sci. Results of the Norwegian Exped. Now.Sem. 1921, Oslo, p. 23-61.

BAILEY, E.B., and GRABHAM, G.W. (1909) Albitisation of basic plagioclase feldspars. Geol. Mag. 6, p. 250-256.

BAILEY, E.B. (1912) The Loch Awe Syncline (Argyllshire). Quart. Jour. Geol. Soc. 69, p. 280-307.

BAILEY, E.B., and McCALLIEN (1960) Some aspects of the Steinmann trinity: mainly chemical. Q.J.G.S. (Lond.) 116, p. 365-395.

BAILEY, E.H., BLAKE, M.C., Jr., and JONES, D.L. (1970) On-land Mesozoic oceanic crust in California coast ranges. U.S. Geol. Survey Prof.Paper 700-C, p. 70-81.

BAMBA, T., and SAWA, T. (1967) Spilite and associated manganiferous hematite deposits of the Tokoro district, Hokkaido, Japan. Geol. Survey of Japan, Rep. 221, p. 1-21.

BAMBA, T., and MAEDA, K. (1969) Pillow lava and spilitization in the Tokoro district, Hokkaido, Japan. Jour. Geol. Soc. Japan, 75, p. 173-181 (in Japanese).

BARTH, V. (1964) Fazielle Entwicklung des vulkanischen Komplexes im südlichen Teil des devonischen Zuges Konice-Mladec im Drahany-Hügelland. Acta Univ. Palack. Olomucensis, Fac. Rer. Nat. 17, p. 13-56.

BARTH, V. (1966) The initial volcanism in the Devonian of Moravia. Paleovolcan. Bohemian massif,(Praha), p. 115-125.

BARTRUM, J.A. (1935) Metamorphic rocks and albite-rich igneous rocks from Jurassic conglomerates at Kawhia. Trans. Roy. Soc. N. Zealand 65, pt. 2, p. 95-107.

BARTRUM, J.A. (1936) Spilitic rocks in New Zealand. Geol. Mag. 73, p. 414-423.

BATTEY, M.H. (1950) The geology of Rangiawhia Peninsula, Doubtless Bay, North Auckland (N. Zealand). Rec. Auck. Inst. Mus. 4, No. 1, p. 35-59.

BATTEY, M.H. (1951) Notes to accompany a topographical map and a provisional geological map of Great Island, Three Kings Group (N. Zealand). Rec. Auck. Inst. Mus. 4, No. 2, p. 93-98.

BATTEY, M.H. (1954) The occurrence of babingtonite in spilite from Three Kings Islands. Rec. Auck. Inst. Mus. 4, p. 263-266.

BATTEY, M.H. (1955) Alkali metasomatism and petrology of some keratophyres. Geol. Mag. XCII, No. 2, p. 104-126.

BATTEY, M.H. (1956) The petrogenesis of a spilitic rock series from New Zealand. Geol. Mag. XCIII, No. 2, p. 89-110.

BEDER, R. (1909) Über basische Eruptivgesteine im ostschweizerischen Verrucano. Diss. Zürich, 28 p.

BEDERKE, E. (1959) Probleme des permischen Vulkanismus. Geol. Rundsch. 48, p. 10 - 18.

BELYANKIN, D.S. (1911) On the Albite Diabase of Krasnaya Polyana. Trans. Polytechn. Inst. 15.

BELYANKIN, D.S. (1912) Contribution to the Petrography of the Central Caucasus. Arkhot Mountain Pass on the Grusinian Military Road. Trans. Polytechn. Inst. 18.

BELYANKIN, D.S. (1914) Orthoclase Diabase from Gekal-Don. Trans. Polytechn. Inst. 21, no. 2.

BENSON, W.N. (1913) Spilite lavas and radiolanian rocks in New South Wales. Geol. Mag. 60, p. 17-21.

BENSON, W.N. (1913-1915) The geology and petrology of the great serpentine-belt of New South Wales. Proc. Linnean Soc. N.S. Wales 38, p. 496-497; and esp. IV: The dolerites, spilites and keratophyres of the Nundle District. Proc. Linnean Soc. N.S. Wales 40, p. 121-173.

BENSON, W.N., and HOLLOWAY, J.T. (1940) Notes on the geography and rocks of the ranges between the Pyke and Matukituki Rivers, North-west Otago. Trans. Roy. Soc. N. Zealand, 70, pt.1, p. 1-24.

BERIDZE, M.A. (   ) Phénomènes de contact liés aux diabases spilitiques de la région supérieure de Ratcka. Goobshch. Akad. Nauk gruz. SSSR, vol. 32, no. 3, p. 581-588.

BERNAUER, F. (1943) Kugelbasalte und ihre Begleitgesteine. Zeitschr. Dt. Geol. Ges., Bd. 95, p. 77.

BERTRAND, J. (1970) Etude pétrographique des ophiolites et des granites du Flysch des Gets (Haute-Savoie, France). Arch. Sci. Genève 23, p. 279-542.

BERTRAND, J., and DELALOYE, M. (1970) Dosages géochimiques sur quelques laves en coussins du domaine alpin. C.R. des Séances SPHN Genève, No. 5, p. 122-128.

BESKOW, G. (1929) Södra Storfjället im südlichen Lappland. Sver. Geol. Undersökn. Arsb. 21 (1927), Ser. C, No. 350, 334 p.

BILIBIN, J.A. (1960) Die geochemischen Typen orogener Zonen. Z. angew. Geologie, Bd. 6, H. 11, p. 545-549.

BILLINGS, M.P. (1937) Regional metamorphism in the Littleton-Moosilanke area, New Hampshire. Geol. Soc. Am. Bull. 48, p. 463-566.

BLOXAM, T.W. (1959) Pillow stracture in spilite lavas at Dawnan Point, Ballantrae. Trans. Geol. Soc. of Glasgow, vol. XXIV, pt. 1, p. 19-26.

BLOXAM, T.W., and LEWIS, A.D. (1972) Ti, Zr, and Cr in some British pillow lavas and their petrogenetic affinities. Nature 237, p. 134-136.

BLYTH, F.G.H. (1935) The basic intrusive rocks associated with the Cambrian inlier near Malvern. Quart. Jour. Geol. Soc. 91, p. 463-478.

BLYTH, F.G.H. (1945) A Geology for Engineers. London, 329 p.

BOTSCHENKOAREW, W.F., and KOBLENZ, E.L. (1961) Querschnittverringerung von Bemusterungsschlitzen. Z. angew. Geol., Bd. 7, p. 571-572.

BOULADON, J., MACHAIRAS, G., and PROUHET, J.-P. (1963) Sur la découverte de kératophyres et de tufs kératophyriques intercalés dans les lydiennes de la base du viséen à Las-Cabesses (Ariège). C.R. Acad. Sci. Paris, 257, p. 191-192.

BOULADON, J., KRYLATOV, S., PASSAQUI, B., and PROUHET, J.-P. (1965) Sur l'existence d'un volcanisme du Dévonien supérieur dans la zone manganésifère de Las-Cabesses (Ariège). C.R. Somm. Séances Soc. Géol. France, 18 janv. 1965, p. 11-12.

BRODERICK, T.M. (1935) Differentiation in lavas of the Michigan Deweenawan. Bull. G.S.A. 46, p. 503-558.

BROUWER, H.A. (1947) Geological explorations in Celebes, summary of the results. Amsterdam (North-Holland Publishing Company), p. 1-64.

BROWN, W.R. (1958) Geology and mineral resources of the Lynchburg Quardrangle, Virginia. Virginia Div. of Min. Res. Bull. 74, 111 p.

BROWN, M., and ROACH, R.A. (1972) Precambrian rocks south of Erquy and around St. Cast, Côtes-du-Nord. Nature 236, p. 77-79.

BROWN, M., and ROACH, R.A. (1972) Comments on "Precambrian rocks south of Erquy and around St. Cast, Côtes-du-Nord" - a reply. Nature 239, p. 74-75.

BRUNN, J.H. (1954) Les éruptions ophiolitiques dans le NW de la Grèce; leurs relations avec l'orogenèse. C.-R. 19 Sess. Int. Geol. Congress, Algiers, fasc. 17, p. 19-27.

BUDDINGTON, A.F. (1926) Submarine pillow lavas of Southeastern Alaska. Jour. Geol. 34, p. 824-828.

BURGUNKER, M.E. (1964) Geology and ore genesis in volcanics of Siberia. Trans. Siberian Industr. Min. Geol. Geoph. Res. Institute. State Geol. Survey U.S.S.R. 35, 228 p. (Transl. Review in Int. Geol. Review 8, p. 109-111).

BURRI, C., and NIGGLI, P. (1945) Die jungen Eruptivgesteine des mediterranen Orogens. Publ. Nr. 3, Vulkaninstitut I. Friedlaender, Zürich, Teil I, 578 p.

CADISCH, J. (1953) Geologie der Schweizeralpen. Basel, 480 p.

CAMPANA, B., and KING, D. (1963) Paleozoic tectonism, sedimentation and mineralization in West Tasmania. Jour. Geol. Soc. Australia 10, p. 1-54.

CANN, J.R. (1969) Spilites from the Carlsberg Ridge, Indian Ocean. Jour. Petrol. 10, p. 1-19.

CANN, J.R., and VINE, F.J. (1966) An area on the crest of the Carlsberg Ridge: Petrology and Magnetic Survey. Philos. Trans. Roy. Soc. London, A, 259, p. 198-217.

CAPPS, S.R. (1915) Some ellipsoidal lavas on Prince William Sound, Alaska. Jour. Geol. 23.

CARVALHOSA, A. (1961) Contribuçao para estudo dos ofiolitos do Baino Alentejo - Espilitos da regiao de Castro Verde-Messejana. Comunicaçoes dos Serviços Geologicos de Portugal, t. XLV, p. 371-390.

CARSTENS, C.W. (1923) Der unterdevonische Vulkanhorizont in dem Trondhjem-Gebiet mit besonderer Berücksichtigung der in ihm auftretenden Kiesvorkommen. Norsk Geol. Tidskrift 7, p. 185-269.

CARSTENS, C.W. (1924) Der unterordovicische Vulkanhorizont im Trondhjem-Gebiet. Norsk Geol. Tidskrift, 17.

CARSTENS, H. (1957) Investigations of titaniferous iron ore deposits. Part I. D.K.N.V.S. Skrifter, No. 3, p. 1-67.

CATER, F.W., and WELLS, F.G. (1953, 54) Geology and mineral resources of the Gasquêt quadrangle, California - Oregon. U.S. Geol. Surv. Bull. 995-C, p. i-iv, 79-133.

CELIKAN, A. (1899) Die Schalsteine des Fichtelgebirges, aus dem Harz von Nassau und aus den Vogesen. Sitz.ber. K.K. Akad. Wiss. Wien, 108, p. 785.

CHIHARA, K. (1954) Spilitic basalts in the inner zone of the north-eastern Japan. Res. Bull. Geol. Min. Inst., Tokyo Univ. Education, no. 3, p. 215-225 (in Japanese).

CHUMAKOV, A.A. (1940) On the Origin of Sodium in Spilites. Izv. Akad. Nauk SSSR, Ser. Geol., no. 2, p. 40.

CHUMAKOV, A.A. (1940) Concise Geological-Petrographical Sketch of the Western Ridge of the Southern Mugodzhar Mountains. Leningrad State Univ. Studies, Ser. Geol., Issue 9, no. 49.

CHUMAKOV, A.A. (1939) Contributions to the Petrography of the Mugodzhar Mountains. I. Southern Mugodzhar (Dzhaman-Tau and Dzhaksy-Tau). Leningrad State Univ. Studies, Ser. Geol., Issue 5, no. 26.

CHUMAKOV, A.A. (1939) II. Southern Mugodzhar (Bokhtybay Ayryuk). Leningrad State Univ. Studies, Ser. Geol., Issue 9, no. 34.

CIOFLICA, G., PATRULIUS, D., IONESCU, J., and UDUBASA, G.G. (1966) Les ophiolites triasiques allochtones des Monts Persani. Rev. Roum. Géol. 10, p. 75-97.

CIRIC, B., and KARAMATA, S. (1960) L'évolution du magmatisme dans le géosynclinal dinarique au mésozoique et au cénozoique. Extr. Bull. Soc. géol. de France, 7e sér. p. 376-380.

CLAPP, C.H. (1921) Geology of the igneous rocks of Essex County, Massachusetts. U.S. Geol. Surv. Bull. 704, p. 30-31, 58-60, 69-71.

CLEMENTS, J.M., and SMITH, H.L. (1899) The Crystal Falls iron-bearing district of Michigan. U.S. Geol. Surv. , Mon. 36.

COGULU, E. (1967) Etude pétrographique de la région de Mihaliccik (Turquie). Thèse no. 1408, Univ. Genève, 824 p.

COOMBS, D.S. (1954) The nature and alteration of some Triassic sediments from Southland, New Zealand. Trans. Roy. Soc. N. Zealand, 82, Pt. 1, p. 65-109.

COOMBS, D.S., HORODYSKI, R.J., and NAYLOR, R.S. (1970) Occurrence of prehnite-pumpellyite facies metamorphism in Northern Maine. Am. Jour. Sci. 268, p. 142-156.

CORNELIUS, H.P. (1933) Ein albitreiches Gestein in der Untertrias bei Neuberg im Mürztal (Steiermark). Verh. Geol. Bundesanstalt 7/8, p. 112-115.

CORNWALL, H.R. (1951a) Differentiation in magmas of the Keweenawan series. Jour. Geol. 59, p. 151-172.

CORNWALL, H.R. (1951b) Differentiation in lavas of the Keweenawan series and the origin of the copper deposits of Michigan. Bull. Geol. Soc. Am. 62, p. 159-202.

CORNWALL, H.R. (1951c) Ilmenite, magnetite, hematite, and copper in lavas of the Keweenawan series. Econ. Geol. 46, p. 51-67.

COX, A.H. (1913) Note on the igneous rocks of Ordovician age. Rep. Brit. Ass. Bgham 496.

COX, A.H., and JONES, O.T. (1913) On various occurrences of pillow-lavas in North and South Wales. Geol. Mag. 10, p. 516-517.

COX, A.H. (1915) The geology of the district between Abereiddy and Abercastle (Pembrokeshire). Quart. Jour. Geol. Soc. 71, p. 328.

COX, A.H. (1920) On the Lower Paleozoic rocks of the Arthog-Dolgelley District (Merionethshire). Quart. Jour. Geol. Soc. 76, p. 254-324.

COX, A.H. (1925) The geology of the Cader Idris Rand, Merioneth. Quart. Jour. Geol. Soc. 81, p. 564.

CRISTI, J.M. (1958) Reconocimiento geologico en la parte S.W. de la provincia de Atacama. Anal. Fac. Cienc. Fisic. Mat., vol. 14-15, 11, p. 123-152.

CROOK, K.A.W. (1960) Petrology of Tamworth Group, Lower and Middle Devonian, Tamworth - Nundle District, New South Wales. Jour. Sedim. Petrol. 30, No. 3, p. 353-369.

DALY, R.A. (1903) Variolite pillow-lavas from Newfoundland. Am. Geol. 32, no. 2.

DALY, R.A. (1914) Igneous rocks and their origin. McGraw-Hill, New York - London, p. 339-340.

DALY, R.A. (1933) Igneous rocks and the depth of the earth. McGraw-Hill, London.

DEINES, P. (1970) The carbon and oxygen isotopic composition of carbonates from the Oka carbonatite complex, Quebec, Canada. Geochim. Cosmochim. Acta, 34, p. 1199-1225.

DE JONG, J.D. (1941) Geological investigations in West Wetar, Lirang and Solor (Eastern Lesser Soenda Islands). Diss., Amsterdam, 136 p.

DE JONG, J.D. (1942) Hydrothermal metamorphism in the Lowo Ria Region, Central Flores. Geol. Exp. to the Lesser Sunda Islands, IV, p. 319-343.

DE KONING, G. (1957) Géologie des IDA OU ZAL (Maroc). Stratigraphie, pétrographie et téctonique de la partie SW du bloc occidental du massif ancien du Haut Atlas. Diss. Univ. Leiden, 205 p.

DELESSE, A. (1851) Sur la variolite de la Durance. An. Mines, Sér. 4, 17, p. 116.

DE ROEVER, W.P. (1940a) Geological investigations in the Southwestern Moetis Region (Netherlands Timor). Diss., Amsterdam, 244 p.

DE ROEVER, W.P. (1940b) Ueber Spilite und verwandte Gesteine von Timor. Proc. Konink. Nederl. Akad. Wetenschappen XLIII, 5, p. 630-634.

DE ROEVER, W.P. (1941) Die permischen Alkaligesteine und die Ophiolithe des timoresischen Faltengebirges. Proc. Nederl. Akad. Wetenschappen XLIV, 8, p. 993-995.

DE ROEVER, W.P. (1942) Olivine-basalts and their alkaline differentiates in the Permian of Timor. Geol.Exp. to the Lesser Sunda Islands IV, p. 209-289.

DEN TEX, E. (1950) Les roches basiques et ultrabasiques des Lacs Roberts et le Trias de Chamrousse (Massif de Belledonne). Leidse Geol. Med. XV, p. 1-204.

DEWEY, H. (1910) The Geology of the Country around Padstow and Camelford. Mem. Geol. Surv. Engl. Wales 42, Expl. of sheets 335-336.

DEWEY, H., and FLETT, J.S. (1911) On some British pillow-lavas and the rocks associated with them. Geol. Mag. 563, p. 202-209, 241-243.

DEWEY, H. (1914) On Geology of North Cornwall. Proc. Geol. Ass. 166.

DEWEY, H. (1948) British regional geology: South-West England (end ed.). Geol. Survey and Museum, London, p. 22, 23, 28.

DEY, A.K. (1942) On the albitite S-W of Shusina, India, a lens-shaped rock body. Mem. Geol. Soc. India 69, Pt. II.

DICKINSON, W.R. (1962) Metasomatic quartz keratophyre in Central Oregon. Am. Jour. Sci. 260, p. 249-266.

DIMITROFF, S. (1929) Die Diabasgesteine im Isker-Durchbruch zwischen den Eisenbahnhaltstellen Bow und Lakatnik. Ann. Univ. Sofia 1928/29. Fac. Sci., p. 175-236.

DIMITROFF, S. (1934) Geologische und petrographische Untersuchungen an den südöstlichen Abhängen der Witoscha und an den nördlichen Teilen der Plana Planina (S.W. Bulgarien), mit besonderer Berücksichtigung der Kontakthöfe der Intrusivgesteine. Ann. Univ. Sofia, Fac. Sci, t. XXX, p. 41-130.

DIMITROFF, S. (1960) Magmenentwicklung und Verteilung der Erzlagerstätten in Bulgarien. Z. angew. Geol., Bd. 6, p. 306-310.

DIMROTH, E. (1971) The evolution of the central segment of the Labrador geosyncline. Part II: The ophioliti c suite. N. Jb. Geol. Paläont. Abh. 137, p. 209-248.

DIVLJAN, M., and KARAMATA, S. (1960) Terminologija stena dijabzno-spilitsko-keratofirske asocijacije. Simp. Alpisko Incijal. Magma, Ilidza - Vares X, 1-6.

DJORDJEVIC, P., and KARAMATA, S. (1972) Observations on the Desmositic and Spilositic Rocks in the Dinarides. Contr. Mineral. Petrol. 34, p. 326-335.

DOLAR-MANTUANI, L. (1941) Keratofirske kamenine v kamniski in korski dolini. Zbor. prir. dr., 2, p. 52-56.

DOLAR-MANTUANI, L. (1950) Ein Beitrag zur Charakteristik der Porphyrgesteine von Sv. Antun bei Lepoglava (Kroatien). Tscherm. Min. Petr. Mitt. 2, p. 93-104.

DONNELLY, T.W. (1959) (Spilites and keratophyres in Venezuela.) Ph.D. thesis, Princeton.

DONNELLY, T.W. (1962) Wairakite in West Indian spilitic rocks. Am. Min. 47, p. 794-802.

DONNELLY, T.W. (1963) Autometasomatism and alteration of West Indian keratophyres. Geol. Soc. Am.., Ann. Meeting 1963, p. 48 A (abstract).

DONNELLY, T.W. (1963) Genesis of albite in early orogenic volcanic rocks. Am.Jour. Sci. 261, p. 957-972.

DONNELLY, T.W. (1966) Geology of St. Thomas and St. John, U.S. Virgin Islands. Mem. Geol. Soc. Am. 98, p. 85-176.

DUBERTRET, L. (1939) Sur la genèse et l'âge des roches vertes syriennes. C.R. Acad. Sc. 209, Paris.

DUBERTRET, L. (1954) Basaltes et roches vertes du Liban, de Syrie et du Hatay. (Ancien Sandjak d'Alexandrette, Turquie). Trans. XIX Internat. Geol. Congr. (Algers), Section XV, p. 29-36.

DUDEK, A., and FEDIUK, F. (1955) Felswand im Moldautal bei Kralup a.d.M. Univ. Carol., Geologica I, No. 2, p. 187-228.

DUPARC, L., and PEARCE, F. (1900) Sur les andésites et les basaltites albitisées du Cap Marsa. C.R. 130.

DUSCHATKO, R., and POLDERVAART, A. (1955) Spilitic intrusion near Ladron Peak, Socorro County, New Mexico. Bull. Geol. Soc. Am. 66, p. 1097-1108.

ECKERMANN, H. von (1936) The Loos-Hamra region. Geol. Förenh. F. 58,

ECKERMANN, H. von (1938) A contribution to the knowledge of late sodic differentiates of basic eruptives. Jour. Geol. 46, p. 412-437.

ECKHARDT, F.-J. (1968) Vorkommen und Petrogenese spilitisierter Diabase des Rotliegenden im Weser-Ems-Gebiet. Geol. Jb. 85, p. 227-264.

EDWARDS, A.B. (1953) The Heemskirk-Zeehan mineral field. In: Geology of Australian ore deposits. Australas. Inst. Min. Met., p. 1166-1178.

EINBERG, L.F. (1926) Igneous rock types in the neighbourhood of Tyuya-Muyuna. Trans. on study of radium, 2.

ERDMANNSDÖRFER, O. (1901) Über die systematische Stellung der Harzer Keratophyre. Cbl. Min. 53.

ERDMANNSDÖRFFER, O. (1904) Die devonischen Eruptivgesteine und Tuffe bei Harzburg und Umwandlung im Kontakt des Brockenmassivs. Jb. Preuss. Geol. Land.-anst. Bergakademie, 25.

ESKOLA, P. (1925) On the petrology of eastern Fennoscandia I. The mineral development of basic rocks in the Karelian formation. Fennia 45, 19, p. 1-93.

ESKOLA, P. (1934a) Über die Bottenmeerporphyre. Compt.Rend. Soc. Géol. Finlande, 8, p. 111-127.

ESKOLA, P. (1934b) Tausend Geschiebe aus Lettland. Ann. Ac. Sci. Fenn. 39, 5, p. 1-41.

ESKOLA, P., VUORISTO, U. and RANKAMA, K. (1935) An experimental illustration of the spilite reaction. Compt.Rend. Soc. Géol. Finlande 9, Bull. Com. Géol. Finl. 119, p. 61-68 (1938).

EUGSTER, H.P. (1951) Petrographische Untersuchungen im Gebiete des Val Russein (Aarmassiv - Ostende). Schweiz. Min. Petr. Mitt. 31, p. 1-131.

FABIAN, H.J., and MÜLLER, G. (1962) Zur Petrographie und Alterstellung präsalinarer Sedimente zwischen der mittleren Weser und der Ems. Fortschr. Geol. Rheinld. u. Westf. 3, 3, p. 1115-1140.

FABIAN, H.J., GAERTNER, H., and MÜLLER, G. (1962) Oberkarbon und Perm der Bohrung Oberlanger Tenge Z 1 im Emsland. Fortschr. Geol. Rheinld. u. Westf. 3, 3, p. 1075-1096.

FAIRBAIRN, H.W. (1934) Spilite and the average metabasalt. Am. Jour. Sci. 27, p. 92-97.

FAIRBANKS, H.W. (1896) The geology of Pt. Sal. Univ. Calif. Publ., Bull. Dept. Geol. Sci. 2, p. 1-91.

FAIRBANKS, H.W. (1897) The geology of the San Francisco Peninsula. Jour. Geol. 5, p. 63-67.

FAIRBANKS, H.W. (1904) San Luis Quadrangle. U.S. Geol. Surv. Folio 101, 13 p.

FEDIUK, F. (1953) The geological and petrographical conditions in the Jizera between Spálov and Bitouchov (District of Železný Brod). Sborník Ústredního Ústavu Geol., XX, Oddíl geol. 72 p.

FEDIUK, F. (1962) Volcanic rocks of the Železný Brod metamorphic region. Rozpravy Ústredniho Ústavu Geol., Praha, v. 29, 116 p.

FIALA, F. (1965) The chemism of the Algonkian and Eocambrian volcanites in the Zelezne Hory Mts. Geochemie v Ceskoslovansku Sbornik Pract. I. geochem. Konference v. Ostravě, p. 15-29.

FIALA, F. (1966) Some results of the recent investigation of the Algonkian volcanism in the Barrandian and the Zelezne Hory areas. Palaeovolcanites of the Bohemian Massif, p. 9-29.

FIALA, F. (1966) Potash spilites in the Algonkian of the Barrandian area. Palaeovolcanites of the Bohemian Massif, p. 82-99.

FIALA, F. (1967) Algonkian pillow lavas and variolites in the Barrandian area. Sbornik geologickych ved. Geologie, Sv. 12, p. 7-65.

FISCH, W.P. (1961) Der Verrucano auf der Nordost-Seite des Sernftales (Kt. Glarus). Mitt. Naturf. Ges. Kt. Glarus, H. XI, p. 3-90.

FITCH, F.H. (1955) The geology and mineral resources of part of the Segama Valley and Darnel Bay area, North Borneo. Brit. Borneo Geol. Surv. Mem. 4.

FITCH, F.H. (1958) The geology and mineral resources of the Sandakan area, North Borneo. Brit. Borneo Geol Surv. Mem. 9.

FLAHERTY, G.F. (1934) Spilitic rocks of southeastern New-Brunswick (Canada). Jour. Geol. $\underline{42}$, p. 875.

FLETT, J.S. (1907) The geology of the Land's End District. Mem. Geol. Surv. Engl. Wales 25; Expl. of sheets 351-358.

FLETT, J.S. (1909) The geology of the seabord of Mid-Argyll. Mem. Geol. Surv. Scotland 53; Expl. of sheet 36.

FLETT, J.S. (1909) The geology of the country around Bodmin and St. Austell. Mem. Geol. Surv. Engl. Wales 49; Expl. of sheet 345.

FLETT, J.S. (1909) The geology of the country around Plymouth and Liskeard. Mem. Geol. Surv. Engl. Wales; Expl. of sheet 348.

FLETT, J.S. (1911) The geology of Knapdale Jura and Kintyre. Mem. Geol. Surv. Great Britain 92.

FLETT, J.S., and DEWEY, H. (1912) The geology of Dartmoor. Mem. Geol. Surv. Engl. Wales 19; Expl. of sheet 338.

FLETT, J.S., and RILL, J.B. (1912) Geology of the Lizard and Meneage. Mem. Geol. Surv. Engl. Wales 117; Expl. of sheet 359.

FLETT, J.S., and DEWEY, H. (1913) The geology of the country around Newton Abbey. Mem. Geol. Surv. Engl. Wales 53; Expl. of sheet 339.

FLOYD, P.A. (1972) Geochemical characteristics of spilitic greenstones from southwest England. Nature 239, p. 75-77.

FOJT, B. (1962) Petrographische Charakteristik des Gebietes der Kies- und Buntmetallerzlagerstätten von Horní Město (Bergstadt). Acta Acad. Scient. Cechoslov. Basis Brunensis XXXIV, fasc. 9, op. 433, p. 385-443.

FOJT, B., and POHANKA, J. (1965) Anmerkung zum Vorkommen von Keratophyr-Gesteinen bei Horní Benesov im Niederen Gesenke. Acta Musei Moraviae L, Sci.nat., p. 61-66.

FRASL, G. (1957) Der heutige Stand der Zentralgneisforschung in den Ostalpen. Joanneum, Mineralog. Mitteilungsblatt 2/1957, p. 41-63.

FRITSCH, W. (1961a) Saure Eruptivgesteine aus dem Raume nordwestlich von St. Veit an der Glan in Kärnten. Geologie, Jg. 10, p. 67-80.

FRITSCH, W. (1961b) Über eine keratophyrische Pillow-Lava (Kissenlava) bei St. Veit/Glan. Karinthia II, 71. Jg., p. 51-52.

FURSE, G.D. (1954) Geology of the Pearl Lake Section of the Porcupine Gold Area. Can. Min. Met. Bull. $\underline{47}$, p. 197-201.

GARDINER, C.H., and REYNOLDS, S.H. (1914) The Ordovician and Silurian Rocks of the Lough Nafooey Area (County Galway). Quart. Jour. Geol. Soc. 70, p. 104-118.

GEIJER, P. (1910) Geology of the Kiruna District, vol. 2: Igneous rocks and iron ores of Kirunavaara, Luossavaara und Tuolluvaara. Stockholm, 278 p.

GEIJER, P. (1930) Pre-Cambrian geology of the iron bearing region Kiruna-Gälltvare-Pajala. Sver. Geol. Undersökn. Arsb. 24, No. 3, Ser. C, 366.

GEIJER, P. (1930) The iron ores of the Kiruna type; geographical distribution, geological characters and origin. Sver. Geol. Unders. 367, Ser. C, p. 1-39.

GEINITZ, E. (1878) Über Variolite aus dem Dorathale bei Turin. Tscher. Min. Petr. Mitt. 1, p. 136-153.

GEIS, H.-P. (1961) Zur Spilitbildung. Geol. Rundsch. 51, p. 375-384.

GERMOVŠEK, C. (1953) Quartz-keratophyre near Velika Pirešica. Razprave Geologija Poročila, Ljubljana, kn. 1, p. 135-168.

GILLULY, J. (1935) Keratophyres of eastern Oregon and the spilite problem. Am. Jour. Sci. 29, p. 225.

GILLULY, J. (1937) The water content of magmas. Am. Jour. Sci. 33, p. 441.

GILLULY, J. (1946) The Ajo mining district. U.S. Geol. Surv. Prof. Paper 209, p. 23-24.

GJELSVIK, T. (1958) Extremely soda rich rocks in the Karelian Zone, Finnmarksvidda, Northern Norway. A contribution to the discussion of the spilite problem. Geol. Fören. Stockholm Förhandl. 80, p. 381-406.

GJELSVIK, T. (1964) Geological Exploration in the Antarctic and Sub-Antarctic Regions by Norwegian or Norwegian-led Expeditions. Antarctic Geology SCAR Proceedings (1963), p. 199.

GOETZ, H. (1937) Die Keratophyre der Lahnmulde. Tscherm. Min. Petr. Mitt. 49, p. 168-215.

GOLDICH, S.S., and RUOTSALA, A.P. (1955) Igneous rock series of Minnesota. Abstract, Am. Geoph. Union, Program 36th Annual Meeting, p. 32.

GOLDSCHMIDT, V.M. (1916) Geologisch-petrographische Studien im Hochgebirge des südlichen Norwegens. IV. Übersicht der Eruptivgesteine im kaledonischen Gebirge zwischen Stavanger und Trondhjem. Skrifter Vidensk 1.

GOLUB, L., and VRAGOVIC, M. (1960) Sodium diabase and spilite near Gotalovec in the province of Hrvatsko Zagorje. Acta Geol. II, p. 83-94.

GREEN, J.C. (1956) Geology of the Storkollen-Blankenberg Area, Kragerø, Norway. Norsk Geol. Tidsskrift 36, p. 89-140.

GREENLY, E. (1902) The origin and the associations of the jaspers of S.-Eastern Anglesy. Quart. Jour. Geol. Soc. 58, p. 425-440.

GREENLY, E. (1919) The Geology of Anglesy. Vol. I, Mem Geol. Surv., p. 54, 71; Vol. II, plate XXVIII, p. 405-406.

GREENLY, E., and MATLEY, C.A. (1928) The Pre-Cambrian complex and associated rocks of S. Western Lleyn (Carnarvonshire). Quart. Jour. Geol. Soc. 84, p. 454.

GREGORY, J.W. (1892) The variolitic diabase of the Fichtelgebirge. Quart. Jour. Geol. Soc. 47, p. 45-62.

GRINDLEY, G.W. (1958) The geology of the Eglinton Valley, Southland. New Zealand Geol. Surv. Bull. 58, 68 p.

GROUT, F.F. (1932) Petrography and Petrology. New York, 522 p.

GROUT, F.F. (1937) Petrographic study of gold prospects of Minnesota. Econ. Geol. 32, p. 56-68.

GRUNAU, H. (1945) Das Ophiolithvorkommen vom Hauen am Jaunpass (Kt. Bern). Schweiz. Min. Petr. Mitt. 25.

GRUNAU, H. (1947) Geologie von Arosa (Graubünden) mit besonderer Berücksichtigung des Radiolaritproblems. Diss. Bern.

GÜMBEL, C.W. von (1874) Die paläolithischen Eruptivgesteine des Fichtelgebirges. München.

GUNNING, H.C. (1937) Cadillac area, Quebec. Can. Geol. Surv. Mem. 206, p. 20-23, p. 42, 43.

GUNNING, H.C., and AMBROSE, J.W. (1937) Malartic area, Quebec. Can. Geol. Surv. Mem. 222, p. 36-39.

GUSTAFSON, J.K., and MILLER, F.S. (1937) Kalgoorlie geology re-interpreted. Econ. Geol. 23, p. 285-317.

GUSTAFSON, J.K. (1945) The Procupine Porphyries. Discussion Econ.Geol. vol. XL, No. 2, p. 148-152.

HACKMAN, V. (1927) Studien über den Gesteinsaufbau der Kittila-Lappmark. Bull. Comm. Géol. Finlande, No. 79.

HAILE, N.S., and WONG, N.P.Y. (1965) The geology and mineral resources of Dent Peninsula, Sabah, Borneo. Reg. Malaysia Geol. Surv. Mem. 16.

HALL, A.L., and MOLENGRAAF, G.A.F. (1925) The Vredefort Mountain Land in the southern Transvaal and the northern Orange Free State. Amst. Verh. Kon Akad. Wet., Tweede Serie, 24 (3).

HANSELMAYER, J. von (1963) Beiträge zur Sedimentpetrographie der Grazer Umgebung. XIX. Petrographie der Schotter aus der Würmterrasse von Friesach-Gratkorn. Mitteil. Naturwiss. 93, p. 137-158.

HANSELMAYER, J. von (1964) Zur Petrographie quartärer Schotter von St. Marein und Kindbergdörfl im Mürztal. Mitteil. Naturwiss. Ver. f. Steiermark, 94, p. 60-79.

HART, R. (1970) Chemical exchange between sea water and deep ocean basalts. Earth Planet. Sci. Letters 9, p. 269-279.

HATCH, F.H. (1889) On the occurrence of soda-felsites (keratophyres) in Co. Wicklow, Ireland. Geol. Mag. 6, p. 70-83.

HATCH, F.H., WELLS, A.K., and WELLS, M.K. (1961) Petrology of igneous rocks. Th. Murby & Co., London, 515 p.

HAUSEN, H. (1969) Some contributions to the geology of La Palma. Comm. Physico-Math., vol. 35, p. 1-140.

HEKINIAN, R. (1968) Rocks from the Mid-Oceanic Ridge in the Indian Ocean. Deep Sea Research, 15, p. 195-213.

HEKINIAN, R. (1971) Petrological and geochemical spilites and associated rocks from St. John, U.S. Virgin Islands. Geol. Soc. Am. Bull. 82, p. 659-682.

HENDRIKS, E.L.M. (1939) The Start-Dodman-Lizard boundary zone in relation to the alpine structure of Cornwall. Geol. Mag. 76, p. 385-402.

HENTSCHEL, H. (1952) Zur Petrographie des Diabas-Magmatismus im Lahn-Dill-Gebiet. Ztschr. Deutsch. Geol. Ges. 104/1, p. 238-259.

HENTSCHEL, H. (1953) Zur Frage der Chlorit- und Karbonat-Bildung in spilitischen Gesteinen (Fortschr. Min. 31, p. 35-37.

HENTSCHEL, H. (1960) Zur Frage der Bildung der Eisenerze von Lahn-Dill-Typ. Freiberger Forschh., 79 C, p. 82-105.

HENTSCHEL, H. (1961) Basischer Magmatismus in der Geosynklinale. Geol. Rundsch. 50, p. 33-45.

HENTSCHEL, H. (1966) Exkursion in das magmatogene Vordevon des Taunus am 9. Sept. 1964. Fortschr. Miner. 42, p. 321-333.

HENTSCHEL, H. (1966) Exkursion in das Dillgebiet. Fortschr. Miner. 42, p. 334-353.

HERITSCH, F. (1911) Beiträge zur Geologie der Grauwackenzone des Paltentales. Abschnitt IV, die Blasseneckserie. Mitt. Naturwiss. Ver. Steiermark (Österreich), p. 93-121.

HOOKER, M. (1956) Data on rock analyses - II. Bibliography and index of rock analyses in the African periodical and serial literature. Geochim. Cosmochim. Acta, 9, p. 190-213.

HOOKER, M. (1957) Data on rock analysis - III. New Zealand periodical and serial literature. Bibliography and index of rock analyses. Geochim. Cosmochim. Acta, 11, p. 130-138.

HOOKER, M. (1959) Data of rock analyses - V. Australian periodical and serial literature. Geochim. Cosmochim. Acta, 15, p. 342-369.

HOPGOOD, A.M. (1956) The stratigraphy and structure of the Basement and Tertiary rocks in the Cape Rodney - Kawan district. Thesis, Auckland University College, Geology Library.

HOPGOOD, A.M. (1957) Spherulitic jaspilite from Whangarei Heads. Trans. Royal Soc. New Zealand 85, p. 131-134.

HOPGOOD, A.M. (1962) Radial distribution of soda in a pillow of spilitic lava from the Francescan, California. Am. Jour. Sci. 260, p. 383-396.

HUDSON, S.N. (1937) The volcanic rocks and minor intrusions of the Cross Fell inlier Cumberland and Westmorland. Quart. Jour. Geol. Soc. 93, p. 368-405.

HUGHES, C.J. (1970) The late Precambrian Avalonian orogeny in Avalon, Southeast Newfoundland. Am. Jour. Sci. 269, p. 183-190.

HUGHES, C.J., and MALPAS, J.G. (1971) Metasomatism in the late Precambrian Bull Arm Formation in southeastern Newfoundland: recognition and implications. Proc. Geol. Soc. Canada 24, p. 85-93.

HUGHES, C.J. (1973) Spilites, keratophyres, and the igneous spectrum. Geol. Mag. 109, p. 513-527.

HÜGI, T. (1941) Zur Petrographie des östlichen Aarmassivs (Bifertengletscher, Limmernboden, Vättis) und des Kristallins von Tamins. Schweiz. Min. Petr. Mitt. 21.

INOSTRANTSEV, A.A. (1874) On variolite. West. Miner. Soc., 9, p. 1-28.

INOSTRANTSEV, A.A. (1875) Results of research on greenstone varieties in the district of Povenetz, Province of Olonetz. Trans. St. Petersburg Soc. Science, 6.

INOSTRANTSEV, A.A. (1878) Observations on variolites. Trans. St. Petersburg Soc. Science, 5, no. 1.

ISHII, K., and UEDA, Y. (1953) On the quartz-keratophyres from Otobe, Shiwa Country and the Pacific Coast, Shimohei County, Iwate Prefecture. Sci Rep. Tohoku Univ. Ser. III, vol. IV, p. 141-146.

ISTRATE, G., and PREDA, I. (1970) Prezenta Rocibor Spilitice in Valea Pesterii - Meziad. (Muntii Padurea Craiului). St. Cerc. geol., geof., geogr., Seria geologie 15 (Bucuresti) p. 107-120.

JAFFÉ, F.C. (1955) Les ophiolites et les roches connexes de la région du Col des Gets (Chablais, Haute Savoie). Bull. Suisse Min. Petr. 35, p. 150.

JOHANNSEN, A. (1939/1950) A descriptive petrography of igneous rocks. Vol. I, 2nd ed., Univ. Chicago Press, 318 p.

JOPLIN, G.A. (1964) A petrology of Australian igneous rocks. Angus & Robertson, Sydney, 214 p.

JURKOVIC, I. (1954) Rhyolith (quarzporphyric) des Vranica-Gebirges und Albit-Rhyolith (Quarzkeratophyr) von Sinjakovo im mittelbosnischen Erzgebirge. Bul. Serv. Geol. Geoph. Serbie (Yugoslavia) 11, p. 224-233.

JUTEAU, T., and ROCCI, G. (1965) Contribution à l'étude pétrographique du massif volcanique dévonien de Schirmeck (Bas-Rhin). Bull. Serv. Carte géol. Als.Lorr. 18, p. 145-176.

JUTEAU, T., and ROCCI, G. (1966) Etude chimique du massif volcanique dévonien de Schirmeck (vosges Septentrionales). Sci. Terre, XI, p. 68-104.

KAADEN, G. van der, and METZ, K. (1954) Beiträge zur Geologie des Raumes zwischen Dacta-Mugla-Dalaman Cay (SW-Anatolien). Bull. Geol. Soc. Turkey, vol. V, No. 1-2, p. 81, 92, 114-119, 144, 155-170.

KARAMATA, S. (1954) Versuch einer Klassifikation der magmatischen Gesteine in Jugoslavien nach ihrem geologischen Alter. C.R. Soc. Serbe Géol. 1954, p. 59-60.

KARAMATA, S. (1957) Karatofiri iz Okoline Zvornika [Der Keratophyr aus der Umgebung von Zvornik (Bosnien)]. Geol. Glasnika, Sarajevo, p. 181-183.

KARAMATA, S. (1958) Albit-Granit von Sjenica (Serbien). N.Jb. Miner. Mh. 6, p. 137-142.

KARAMATA, S. (1960) Die Melaphyre von Vares. Simpoz. Problem. Alpisk. Inicijal. Magmat, Ref. III, p. 3-20.

KARAMATA, S. (1960) Petrologische Charakteristiken des alpinen Initial-Magmatismus in den Dinariden. Simpoz. Problem. Alpisk. Inicijal. Magmat., Ref. IX, p. 1-9.

KARAMATA, S., and PAMIC, J. (1960) Gabbros, Diabase und Spilite des Gebietes von Tribija. Simpoz. Problem. Alpisk. Inicijal. Magmat., Ref. V, p. 1-17.

KARAMATA, S. (1961) Die Typen der triassichen magmatischen Tätigkeit und deren Produkte im südöstlichen Montenegro (Crna Gora). Congr. Geol. Yougoslavia (Titograd), p. 357-361.

KASSIN, N.G. (1931) Brief Geological Outline of North-Western Kazakhstan. Trans. Geol. Research Soc. SSSR, No. 165.

KASSIN, N.G., et al. (1933) General geology. Map of Kazakhstan. Description of the papers of Mid-Chenderlin and Ulenty. Boshchekul, Sary-daryr, Kadzhanchad.

KEIJZER, K.G. (1945) Outline of the geology of the eastern part of the province of Oriente, Cuba, E of 16° W.L. with notes on the geology of the other parts of the island. Ac. thesis, Utrecht.

KERN, A. (1952) Erzbergführer. Erzberg, 67 p.

KETTNER, R. (1917) Versuch einer stratigraphischen Einteilung des böhmischen Algonkiums. Geol. Rundsch. 8, p. 179.

KIRK, H.J.C. (1962) The geology and mineral resources of the Semporna Peninsula, North Borneo. Brit. Borneo Geol. Surv. Mem. 14.

KIRK, H.J.C. (1968) The igneous rocks of Sarawak and Sabah. Geol. Surv. Malaysia Bull. 5, 210 p.

KNAUER, E. (1958) Ein Beitrag zur Petrographie des Keratophyrs vom Büchenberg bei Elbingerode im Harz. Geologie 7, p. 629-638.

KNOPF, A. (1918) Geology and ore deposits of the Jerington district, Nevada. U.S. Geol. Surv. Prof. Paper No. 114, p. 13-16.

KNOPF, A. (1921) Ore deposits of Cedar Mountain Mineral County, Nevada. Bull. Geol. Soc. Am. 725-H, p. 360-382.

KNOPF, A. (1924) Geology and ore deposits of the Rochester district, Nevada. Bull. U.S. Geol. Surv. No. 762, p. 20.

KNOPF, A. (1929) The Mother Lode system of California. U.S. Geol. Surv. Prof. Paper No. 157, 88 p.

KNOPF, A., and ANDERSON, C.A. (1930) The Engels copper deposit, California. Econ.Geol. 25, p. 14-35.

KOLDERUP, N.-H. (1929) En vestnorsk kisførende kvartskeratofyr. Bergens Mus. Arbok Naturvidenskapelig rekke No. 4, p. 1-22.

KOLENKO, B.Z. (1855) Geological sketch of Zaonezh. Mathem. Geol. Russia, 13.

KOPF, M. (1961) Dichtewerte von Gesteinen des Erzgebirges und der angrenzenden Gebiete. Z. angew. Geologie, $\underline{7}$, H. 11, p. 301-302.

KORZHINSKY, D.S. (1936) Mobility and inertia of components under metasomatism. Trans. Acad. Nauk SSSR, Ser. Geol. 1, p. 35-60.

KORZHINSKY, D.S. (1940) Factors of mineral equilibria and mineralogical facies of depth. Trans. Acad. Nauk SSSR, no. 12, Ser. Petr. no. 5.

KORZNISKIJ, O.S. (1963) Das Spilitproblem und die Transvaporisationshypothese im Lichte neuer ozeanologischer und vulkanologischer Ergebnisse. Ber. Geol. Ges. DDR, Sonderh. 1, p. 89-95.

KROL, G.L. (1960) Theories on the genesis of the Kaksa. Geol. Mijnbouw, 39e, p. 437-443.

KTENAS, K.A. (1908) Contemporaneous extrusion of keratophyric and peridotitic magmas. Habil. thesis, Athens, 34 p. (original in greek).

KTENAS, K.A. (1909) Über die eruptiven Bildungen des Parnesgebirges in Attika. Cbl. Min., p. 557-558.

KTENAS, K.A. (1909) Les formations éruptives du Parnès. Bull. Soc. Géol. France, $\underline{9}$ (3), p. 6-7.

KÜNDIG, E. (1956) Geology and ophiolite problems of East Celebes. Verh. Kon. Ned. Geol. Mynb. Genootschap, Geol. Series, 16.

KUPLETSKY, B.M. (1932) The diabase rocks of the Rusanov Valley and the Cross Bay of Novaya Zemlya. Trans. Geol. Inst. Acad. Nauk SSSR, No. 1. p. 153.

KUPLETSKY, B.M. (1932) Contributions to the study of diabase rocks on Novaya Zemlya. Trans. Petr. Inst. Acad. Nauk SSSR, No. 2.

KUZNETSOV, E.A. (1939) Geology of the greenstone zone on the eastern slope of the Central Urals. Publ. Acad. Nauk SSSR.

LAGORIO, A.E. (1897) Itinéraire par le Karadag: Guide des excursions du VII Congr. Géol. Intern., p. 31.

LAGORIO, A.E. (1897) Itinéraire d'Alchouta à Sébastopol par l'Jalta, Bachtshisaray à Mangoup-Kale: Guide des excursions du VII Congr. Géol. Intern., p. 33.

LA ROCHE, H., ISNARD, P., and GRANDCLAUDE, P. (1969) Domaines et tendances ignées communes dans les diagrammes Q-Ab-Or, An-Ab,Or and Q-AbAn-Or. (Etude documentaire). Sci.Terre, XIV, p. 371-382.

LAWSON, A.J. (1895) Sketch of the geology of the San Francisco Peninsula. U.S. Geol. Surv. Ann. Rep. 15, p. 399-476.

LAWSON, A.J. (1914) San Francisco region. U.S. Geol. Surv. Folio 193.

LEBEDINSKIJ, W.I. (1963) Some general problems of genesis of spilito-keratophyre association exemplified by volcanic rocks of the High Crimea. Magmatism, Metamorphism and Metallogenesis of the Ural Mountains. Papers of the 1st Seminar on the Ural; Section on volcanic rocks, p. 93-103.

LEBEDINSKIJ, W.I. (1964) Genesis and classification of the spilito-keratophyric formations. From: Petrographic processes and problems of petrogenesis. Int. Geol. Congress, 22nd session (Lectures of the geologists of the USSR), Nauk, Moscow, p. 28-43.

LEHMANN, E. (1941) Eruptivgesteine und Eisenerze im Mittel- und Oberdevon der Lahnmulde. Wetzlar.

LEHMANN, E. (1949a) Das Keratophyr-Weilburgit-Problem. Heidelberger Beiträge Min. Petr., Bd. 2, p. 1-166.

LEHMANN, E. (1949b) Über die Genesis der Eisenerzlagerstätten vom Lahntypus. Zeitsch. f. Erzbergbau u. Metallh., II, 8, p. 239-246.

LEHMANN, E. (1952a) Über Miktitbildung. Heidelberger Beiträge Min. Petr., 3, p. 9-35.

LEHMANN, E. (1952b) Beitrag zur Beurteilung der paläozoischen Eruptivgesteine Westdeutschlands. Zeitschr. Deutsch. Geol. Ges. 104/1, p. 219-237.

LEHMANN, E. (1952c) Diskussionsbemerkung zum Thema "Weilburgit" und "Schalstein". Zeitschr. Deutsch. Geol. Ges. 104/1, p. 255-256.

LEHMANN, E. (1956) Merkmale magmatischer Infiltration und Injektion in den Keratophyrtuffen des Sauerlandes. Zeitschr. Deutsch. Geol. Ges. 106, p. 353-360.

LEHMANN, E. (1965a) Non-Metasomatic Chlorite in Igneous Rocks. Geol. Mag. 102, p. 24-35.

LEHMANN, E. (1965b) Diabasgesteine SW-Englands und damit verbundene Probleme. Zeitschr. Deutsch. Geol. Ges. 115, p. 228-276.

LEHMANN, E. (1968) Diabasprobleme und problematische Diabase II. Chemie der Erde 27, p. 39-76.

LEVINSON-LESSING, F. Yu. (1884) Die Variolite von Jalguba im Gouvernement Olonez. Tscherm. Min. Petr. Mitt. 6, p. 281-300.

LEVINSON-LESSING, F. Yu. (1888) The diabase formation of Olonetz. Trans. St. Petersburg Sci. Soc., 19.

LEVINSON-LESSING, F. Yu. (1898) Analyses of theoretical petrography in connection with the study of igneous rocks of the Central Caucasus. Trans. St. Petersburg Sci. Soc., 26, No. 5.

LEVINSON-LESSING, F. Yu. (1899) On spherulitic rocks of Mogodzhar. Trans. Mugodzhar Exped., vol. 2.

LEVINSON-LESSING, F.Yu., and DYAKONOVA-SAVELYEVA, E.N. (1933) The volcanic group of the Karadag in the Crimea. Publ. Acad. Nauk SSSR.

LEVINSON-LESSING, F. Yu. (1935) On the peculiar type of differentiation in the variolite of Jalguba. Trans. Inst. Petr. Acad. Nauk SSSR, No. 5, p. 5-11.

LEWIS, J.V. (1914) Origin of pillow-lava. Bull. Geol. Soc. Am. 25, p. 591.

LIDIAK, E.G. (1965) Petrology of andesitic, spilitic, and keratophyric flow rock, North-Central Puerto Rico. Geol. Soc. Am. Bull. 76, p. 57-88.

LODOCHNIKOV, V.N. (1938) The principal rock-forming minerals. United Scientific and Technical Publishing Houses.

LOESCHKE, J. (1973) Zur Petrogenese paläozoischer Spilite aus den Ostalpen. N. Jb. Miner. Abh. 119, p. 20-56.

LUCHITSKY, V.I. (1939) Petrography of the Crimea. Publ. Acad. Nauk, SSSR.

LUKOVIC, S. (1952) The occurrence of quartz-keratophyre in the Tara River Canyon. Zbornik rad. geol. i. rud. fakult., Beograd, 1, p. 119-125.

LUNDBOHM, H. (1910) Sketch of the geology of the Kiruna district. Geol. Fören. Förhandl. 32.

MAJER, V. (1960) Magmatic rocks in the region of Bassit between Latakia and Kessab in N.W. Syria. Fac. Sci. Univ. Skopje, ed. sp., Livre 10, p. 1-37.

MAJER, V., and MARIC, L. (1964) Pregled Petroloske Gradje Dijabaz-Loznacke Formacije u Nekim Dijelovima Denarida i Tanrida. Sveuciliste u Zagrebu. 25 Godisnjica Rudarkog Odjela Tehnoloskog Fakulteta, 1939-1964, p. 53-73.

MAJER, V., and BARIC, L. (1972) Spilitgesteine und die in ihnen vorkommenden Aderparagenesen im zentralen Gebiet Kroatiens.(Jugoslawien). Fortschr. Miner. 50, p. 60-61.

MARINOS, G., KATSIKATSOS, G., GEORGIADIS, E., and MIRKOS, R. (1971) The system of the schists of Athens. Ann. Géol. des Pays Helléniques 23, p. 183-216 (original in Greek).

MARTINI, J., and VUAGNAT, M. (1970) Metamorphose niedrigst temperierten Grades in den Westalpen. Fortschr. Miner. 47, p. 52-64.

MATTSON, P.H. (1960) Geology of the Mayagüez Area, Puerto Rico. Bull. Geol. Soc. Am. 71, p. 319-362.

McKINSTRY, H.E. (1939) Discussion: Pillow-lavas of Borabora, Soc. Islands. Jour. Geol. 47, p. 202-204.

McMATH, J.C., GRAY, N.M., and WARD, H.J. (1953) The geology of the country about Coolgardie, Coolgardie Goldfield, W.A.: I. Regional Geology (J.C. McMATH), 119 p.; II. Selected Mining Groups (N.M. GRAY and H.J. WARD), 187 p. Geol. Surv. W. Australia Bull. 107, 365 p.

MEISTER, A.K. (1908) Contributions to the petrography of the Crimea. Trans. Geol. Com. 27, No. 10, p. 669-706.

MELSON, C.W., and VAN ANDEL, T.H. (1966) Metamorphism in the Mid-Atlantic ridge, 22° N Lat. Mar. Geol. 4, p. 165-186.

MELSON, C.W., THOMPSON, G., and VAN ANDEL, T.H. (1968) Volcanism and metamorphism in the Mid-Atlantic ridge, 22° N Lat. J. Geophys. Res. 73, p. 5925-5941.

MERILÄINEN, K. (1961) Albite diabases and albitites in Enontekiö and Kittilä, Finland. Bull. Comm. Geol. Finland 195, p. 1-75.

MERLITSCH, B.W., and SPITKOWSKAJA, S.M. (1958) The Paleogene phase of volcanism in the eastern Carpathians. Geol. Sborn. Lwowsk, Geol. Obschtsch. 4, p. 171-177 (original in Russian).

METZ, K. (1951) Die stratigraphische und tektonische Baugeschichte der steirischen Grauwackenzone. Mitt. Geol. Ges. 44, p. 1-84.

MICHEL-LEVY (1877) Mémoire sur la variolite de la Durance. Bull. Soc. Géol. France 3, sér. 5, p. 232-266.

MIDDLETON, G.V. (1960) Spilitic rocks in south-east Devonshire. Geol. Mag. 97, p. 192-207.

MILOVANOVIC, B., and ILIC, M. (1960) Geologische Problematik der Diabas-Hornstein-Formation der Dinariden. Simpoz. Problem. Alpisk. Inicijal. Magmat., Ref. I, p. 3-9.

MOLENGRAAFF, G.A.F. (1902) Geological explorations in Central Borneo (1893-94). Leyden and Amsterdam, Soc. for the promotion of the scientific exploration of the Dutch colonies.

MOORE, E.S. (1930) Notes on the origin of pillow-lavas. Trans. Roy.Soc. Canada, Ser. 4, p. 137.

MOORE, J.G. (1965) Petrology of deep-sea basalt near Hawaii. Am. Jour. Sci. 263, p. 40-52.

MOORHOUSE, W.W. (1942) Gold mineralization in minor igneous intrusions. Econ.Geol. 37, p. 318-329.

MORET, L. (1929) Notice explicative d'une carte géologique de la Savoie. Trav.Lab. Univ. Grenoble.

MORRE-BIOT, N. (1970) Pétrologie des formations volcaniques du permo-carbonifère du Nord de la France. Diss. Univ. Besançon, 148 p.

MOURANT, A.E. (1936) Les roches volcaniques du Trégorrois en relation avec celles de Jersey. Mém. Soc. Géol. Min. Bretagne, III, p. 79-88.

MOUSSU, R. (1959) Géologie et gîtes minéraux de la région de l'Ounein (Haut Atlas). Notes et Mém. Service Géol. Maroc 145, 118 p.

NAREBSKI, W. (1964) Petrochemistry of pillow lavas of the Kaczawa Mountains and some general petrogenetical problems of spilites. Prace Muzeum Ziemi No. 7, p. 69-205.

NAREBSKI, W. (1968) Über die petrogenetische Bedeutung von Spurenelement-Paragenesen der Eisengruppe in den initialen Vulkaniten einiger kaledonischer Geosynklinalen. In: Probleme der Paragenese von Mineralen, Elementen und Isotopen. Freiberg. Forsch. C 231, p. 259-265.

NICHOLLS, G.D. (1959) Autometasomatism in the Lower Spilites of the Builth Volcanic Series. Quart. Jour. Geol. Soc. London, 114, p. 137-161.

NIGGLI, E. (1944) Das westliche Tavetscher Zwischenmassiv und der angrenzende Nordrand des Gotthardmassivs. Diss. Zürich; Schweiz. Min. Petr. Mitt. 24, p. 58-301.

NIGGLI, P. (1952) The chemistry of the Keweenawan lavas. Jour. Sci., Bowen volume, p. 381-412.

NOLAND, T.B. (1930) Underground geology of the western part of the Tonopah mining district, Nevada. Bull. Univ. Nevada 24, No. 4.

OFTEDAHL, C. (1958) A theory of exhalative-sedimentary ores. Geol. Fören. Stockh. Förhand. 80, p. 1-19.

OGNIBEN, G. (1956) Fenomeni di albitizzazione negli scisti e nelle apliti di Giustino (Val Rendena). Rend. Soc. Miner. Italiana XII, p. 3-7.

OIKONOMOU, K. (1973) Study of the rocks of the ophiolite complex of the Isle of Kreta. Diss. Athens (original in Greek), 112 p.

ORLANDO, C.G. da (1961) Associaçao genética dos espilitos com os jazigos de manganês do Baixo Alentejo. Estudos, Notas e Trabalhos do S.F.M. 15, p. 5-24.

PAAKKOLA, J. (1971) The volcanic complex and associated manganiferous iron formation of the Prokonen Pahtavaara area in Finnish Lapland. Bull. Comm. Géol. de Finlande No. 247, p. 1-83.

PAMIĆ, J. (1962) The spilite-keratophyre association in the area of Prozor and Jablanica. Prirodoslov. istraz / Acta geol. III, (Zagreb), 31, p. 5-94.

PAMIĆ, J., DIMITROV, P., and ZEK, F. (1964) Geological characteristics of acid volcanics of neogene in the river Bosna valley. Geoloski glasnik 10, p. 241-250.

PAMIĆ, J. (1965) Petrologic study of Mount Ozren in the north of Bosina. Acta geol. IV (Zagreb), p. 265-314.

PAMIĆ, J., and BUZALJKO, R. (1966) Middle Triassic spilites and keratophyres in the area of Cajnice (S. Bosnia). Geoloski glasnik 11, p. 55-77.

PAMIĆ, J., and PAPES, J. (1969) Petrology and geology of Middle Triassic igneous rocks in the area of Kupresko Polje (Bosnia). Anals, Geol. Peninsule Balkan, XXXIV, p. 555-576.

PAMIĆ, J. (1969) High temperature feldspars from the Middle Triassic spilite-keratophyre association of the Dinarids. Bull. Scientifique, Sec. A, 14, p. 4.

PAMIĆ, J., SĆAVNICAR, S., and MEDJIMOREC (1973) Mineral assemblages of amphibolites associated with Alpine type ultramafics in the Dinaride ophilote zone (Yougoslavia). Jour. Petrol. 14, p. 133-157.

PAPAJIANNOPOULOS, A. (1971) The volcanic rocks of the Chronia area (Euböa). Diss. Athens (original in Greek).

PAPEZIK, V.S., and FLEMING, J.M. (1967) Basic volcanic rocks of the Wahlesback area, Newfoundland. Geol. Assoc. Canada, sp. paper 4, p. 181-192.

PAPEZIK, V.S. (1970) Petrochemistry of volcanic rocks of the Harbour Main Group, Avalon Peninsula, Newfoundland. Canad. J. Earth Sci. 7, p. 1485-1498.

PARK, C.F. (1946) The spilite and manganese problems of the Olympic Peninsula, Washington. Am. Jour. Sci. 244, p. 305-323.

PARKER, D.C. (1959) A petrologic investigation of the greenstone flow at Tamarack location, Houghton County, Michigan. Master's thesis (unpubl.),Michigan College of Mining and Technology, 94 p.

PATWARDHAN, A.M., BHANDARI, A., and SINGH, G. (1970) Mandi Traps - their field disposition and geochemistry. Publ. Centre of Adv. Study in Geology, Panjab Univ., Chandigarh (India), No. 7, p. 45-54.

PATWARDHAN, A.M. (1972) Crystallisation features in ophitic spilites from Mandi, India. Fortschr. Miner. 50, p. 79-80.

PELLIZZER, R. (1954) Le spiliti di Comeglians nella media val Degano (Carnia). Soc. Miner. Itali., Rend. an. 10, p. 405-414.

PELLIZZER, R. (1955) Primi confronti tra alcune ofioliti alpine ed appenniniche. Rend. Soc. Miner. Italiana XI, p. 3-9.

PERRIN, R., and ROUBAULT, M. (1941) Quelques observations sur le spilite de Montvernier (Savoie). Bull. Soc. Hist. Nat. Toulouse, t. LXXVI, 3e trim., p. 161-171.

PETTERLONGO, J.M. (1968) Les ophiolites et le métamorphisme à glaucophane dans le massif de l'Inzecca et la région de Vezzant (Corse). Bull. B.R.G.M., 2e Série, Sec. 4, p. 17-94.

PETTIJOHN, F.J. (1943) Archean Sedimentation. Bull. Geol. Soc. Am. 54, p. 925-972.

PETTIJOHN, F.J., and BASTRON, H. (1959) Chemical composition of argillites of the Cobalt Series (Precambrian) and the problem of soda-rich sediments. Bull. Geol. Soc. Am. 70, p. 593-600.

PHILLIPS, J.A. (1876) On so-called "greenstones" of western Cornwall. Quart. Jour. Geol. Soc. 32, p. 155-179.

PHILLIPS, J.A. (1878) On the so-called "greenstones" of Central and Eastern Cornwall. Quart. Jour. Geol. Soc. 34, p. 471-497.

PICHAMUTHU, C.S. (1938) Spilitic rocks from Chitaldrug, Mysore State. Current Sci. 7, no. 2, p. 55-57.

PICHAMUTHU, C.S. (1946) Quartz-keratophyres from Galipuje, Kadur Dt., Mysore State. Quart. Jour. Geol. Min. Met. Soc. India 18,No. 4, p. 125-129.

PICHAMUTHU, C.S. (1950) Pillow structures in the lavas of Dharwar age in the Chitaldrug district, Mysore State. Current Sci. 19, p. 110-111.

PICHAMUTHU, C.S. (1957) Pillow lavas from Mysore State, India. Bull. Mysore Geol. Assoc. 15, p. 1-26.

PIISPANEN, R. (1972) On the spilitic rocks of the Karelidic belt in western Kuusamo, Northeastern Finland. Acta Univ. Ouluensis, Ser. A, Geologica no. 2, p. 1-73.

PISKIN, Ö.(1972) Etude minéralogique et pétrographique de la région située à l'est de Çelikhan (Taurus Oriental, Turquie). Mém. Minéralogie Univ. Genève, 3, 152 p.

PODOLSKII, J.V. (1963) Linear paragenesis of the major elements in rocks of the spilite-keratophyre formation. Doklady Akad. Nauk USSR, 152/4, p. 975-978 (in Russian).

PRIDER, R.T. (1939-40) A note on the age relations of the basic porphyrites and albite porphyries of the Golden Mile, Kalgoorlie, Western Australia. Jour. Roy. Soc. W. Australia 26, p. 99-101.

PRIGOROVSKY, M.M. (1914) Concise geological outline of the Mugodzhar Mountains and the adjacent parts of the Turgai and Ural Steppe. Trans. Geol. Com. 33, No. 8, p. 889-928.

QUADE, H. (1970) Der Bildungsraum und die genetische Problematik der vulkano-sedimentären Eisenerze. Clausthaler Hefte (Thienhaus-Band), 9, p. 27-65.

RAISIN, C.A. (1893) Variolite of theLleyn and associated volcanic rocks. Quart. Jour. Geol. Soc. 49, p. 145-165.

RANSOME, F.L. (1893) The eruptive rocks of Point Bonita. Bull Univ. Calif., Dept. Geol. 1, p. 77-114.

RANSOME, F.L. (1894) Geology of Angel Island. Univ. Calif. Publ., Bull. Dept. Geol. Sci. I, p. 193-234.

REDLICH, K.A., and PREDIK, K. (1930) Zur Tektonik und Lagerstättengenesis des steirischen Erzbergs. Jb. Geol. Bundesanstalt Wien, p. 231-260.

REED, J.J. (1950) Spilites, serpentines and associated rocks of the Mossburn district, Southland (New Zealand). Trans. Roy. Soc. N.Z., Wellington, 18, pt. 1, p. 106-126.

REID, C., and DEWEY, H. (1908) On the origin of the pillow-lava near Port Isaac in Cornwall. Quart. Jour. Geol. Soc. 64, p. 264-272.

REID, J.A. (1945) The Hard Rock "Porphyry" of Little Long Lac. Econ. Geol. 40, p. 509-516.

REINHARD, M., and WENK, E. (1951) Geology of the Colony of North Borneo. Geol. Surv. Dept. Brit. Terr. Borneo, Bull. 1, p. 75-79, 81.

RENZ, C. (1940) Die Tektonik der griechischen Gebirge. Pragmat. Akad. Athens, 3, 172 p.

RICHARZ, S. (1926) Quartzkeratophyre from the Porcupine gold area, Ontario. Am. Jour. Sci. 11, p. 441-442.

RICHTER, D., and VILLWOCK, R. (1960) Die Rotliegend-Vulkanite am Westrande des Fichtelgebirges. N. Jb. Miner. Abh. 95, p. 88-105.

RICHTER, D., and VILLWOCK, R. (1961) Über einen Graphit-Einschluß im Quarzkeratophyr bei Lenau (westl. Fichtelgebirge). N. Jb. Miner. Mh. 1961, 11/12, p. 247-252.

RICHTER, D. (1968) Der geosynklinale Vulkanismus im Devon von SE-Devonshire, England. Z. Deutsch. Geol. Ges. 117, p. 767-818.

RINNE, F. (1895-96) Über Diabasgesteine in mitteldevonischen Schiefern aus der Umgebung von Goslar am Harz. N. Jb., B.B. 10, p. 363-411.

RIPPEL, G. (1954) Räumliche und zeitliche Gliederung des Keratophyrvulkanismus im Sauerland. Geol. Jb. 68, p. 401-456.

RIVALENTI, G., and SIGHINOLFI, G.P. (1971) Geochemistry and differentiation phenomena in basic dikes of the Frederikstad district, S.W. Greenland. Atu Soc. Tosc. Sc. Nat. Mem. Anno, Ser. A., LXXVII, p. 358-380.

ROCCI, G., and JUTEAU, T. (1968) Spilite-kératophyres et ophiolites; influence de la traversée d'un socle sialique sur le magmatisme initial. Geol. en Mihnbouw 47, p. 330-339.

ROGOVER, G.B. (1939) The copper pyrite deposits of the Blyava. State United Publishing House of Science and Technology.

RÖSLER, H.J. (1959) Zur Petrographie, Geochemie und Genese des oberdevonisch-unterkarbonischen Magmatismus und der an ihn gebundenen Lagerstätten in Ostthüringen. Bergakademie 9/10, p. 618-619.

RÖSLER, H.J. (1960) Zur Petrographie, Geochemie und Genese der Magmatite und Lagerstätten des Oberdevons und Unterkarbons in Ostthüringen. Freiberg. Forschh. C 92, 275 p.

RÖSLER, H.J. (1960) Zum Chemismus der oberdevonischen und unterkarbonischen Karbonatgesteine in Ostthüringen. Geologie 8, p. 867-883.

RÖSLER, H.J. (1960) Bemerkungen zur Genese von Geosynklinalmagmatiten. Intern. Geol. Congr., XXI, Pt. XIII, p. 96-107.

RÖSLER, H.J. (1961) Versuch einer genetischen Deutung der Eisenerze vom Typ Lahn-Dill im Erzrevier Schleiz-Pörnitz (Ostthüringen). Geologie, Jg. 10, 1, p. 94-96.

RÖSLER, H.J. (1962) Die Variation der Suszeptibilität und ihre Ursachen in einem mächtigen Diabaslager Thüringens. Geophys. und Geologie 3, p. 3-8.

RÖSLER, H.J. (1963) Einige Beobachtungen und Gedanken zur Frage des Wassergehaltes basischer Magmen und Gesteine. Ber. Geol. Ges. DDR, Sonderh. 1, p. 97-101.

RUTLEY, F. (1891) The study of rocks. An elementary text-book of petrology. 5th Ed., London, 321 p.

RUTTEN, M.G. (1936) Geology of the northern part of the province Santa Clara, Cuba. Geogr. Geol. Med. XI, Utrecht (thesis), 59 p.

SAKSELA, M. (1960) Beiträge zur Kenntnis der sogenannten chloritischen Kupferformationen im fennoskandischen Grundgebirge. N.Jb. Miner. Abh. 94, p. 319-351.

SARGENT, H.C. (1917) On a spilitic facies of Lower Carboniferous lava-flows in Derbyshire. Quart. Jour. Geol. Soc. London 73, p. 11-23.

SAVU, H. (1968) Considérations concernant les relations stratigraphiques et la pétrologie des ophiolites mésozoiques de Roumanie. Ann. Comité d'Etat, Géologie, XXXVI, p. 143-175.

SAVU, H. (1967) Die mesozoischen Ophiolite der Rumänischen Karpaten. Acta Geol. Acad. Sci. Hungar., vol. 11 (1-3), p. 59-70.

SCHERMERHORN, L.J.G. (1970a) The deposition of volcanics and pyritite in the Iberian Pyrite Belt. Mineral. Deposita, 5, p. 273-279.

SCHERMERHORN, L.J.G. (1970b) Mafic geosynclinal volcanism in the Lower Carboniferous of South Portugal. Geol. Mynbouwkd., 49, p. 439-449.

SCHERMERHORN, L.J.G. (1973) What is keratophyre? Lithos, 6, p. 1-11.

SCHIDLOWSKI, M., STAHL, W., and AMSTUTZ, G.C. (1970) Oxygen and Carbon isotope abundances in carbonates of spilitic rocks from Glarus, Switzerland. Naturwissenschaften 57, p. 542-543.

SCHIDLOWSKI, M., and STAHL, W. (1971) Kohlenstoff- und Sauerstoff-Isotopenuntersuchungen an der Karbonatfraktion alpiner Spilite und Serpentinite sowie von Weilburgiten des Lahn-Dill-Gebiets. N.Jb. Miner. Abh. 115, p. 252-278.

SCHILOW, W.N., and KALISCHEWITSCH, O.K. (1958) On the petrogenesis of the spilite-keratophyre-formation. Doklady Akad. Nauk SSSR, 122, p. 902-904 (original in Russian).

SCHLOSSMACHER, K. (1920) Keratophyre aus dem rechtsrheinischen Vordertaunus. Jb. Preuss. Geol. Landesanst., Berlin, p. 308-348.

SCHOELL, M., and STAHL, W. (1972) Sauerstoff- und Kohlenstoff-Isotopenuntersuchungen an Karbonaten hydrothermaler Lagerstätten. Mitt. Labor f. stabile Isotope Nr. STI 2/72, Bundesanst. f. Bodenforschung, Hannover, 22 p.

SCHÜLLER, A., and DETTE, K. (1955) Magmatische Gesteine aus dem Thüringer Wald; die Stellung der Keratophyre, I. Geol., Berlin, Jg. 4, H. 5, p. 463-478.

SCOTT, B. (1951a) A note on the occurrence of intergrowth between diopsidic augite and albite, and of hydrogrossular from King Island, Tasmania. Geol. Mag. 88, p. 429-431.

SCOTT, B. (1951b) The petrology of the volcanic rocks of south-east King Island, Tasmania. Pap. and Proc. Royal Soc. Tasmania, 1950, p. 113-136.

SCOTT, B. (1954) The metamorphism of the Cambrian basic volcanic rocks of Tasmania and its relationship to the geosynclinal environment. Roy. Soc. Tasmania, Pap. and Proc. 88, p. 129-150.

SHAGAM, R. (1960) Geology of the Central Aragua, Venezuela. Bull. Geol. Soc. Am. 71, p. 260-269, 278.

SHARFMAN, V.S. (1968) On the average chemical composition of spilites. Doklady Akad. Nauk SSSR 180, p. 202-203.

SHCHERBAKOV, D.I. (1914) Contributions to the geology of the Crimea. I. The Limen eruption of igneous rocks. Trans. St. Petersb. Polytechn. Inst., 21, No. 2.

SHCHERBAKOV, D.I. (1915) Greenstone Varieties of Alupka. Collection of scientific studies dedicated to Prof. F. Yu. Levinson-Lessing in honour of his 30 years of activity in scientific research.

SHCHERBAKOV, D.I. (1924) On useful excavations in South Karelia. Trans. North Scient.-profess. Exped., No. 24.

SHIMAZU, M. (1971) On authigenic minerals in the spilitic basalts in the inner belts of Northeastern Japan. Mineral. Soc. Japan Spec. Pap. 1, p. 134-139.

SIDERIS, K. (1966) The Greek Spilites of the Ophiolite Series. Diss. Athens (original in Greek), 58 p.

SIEGEL, A. (1951) Diabase exotique dans le crêtacé des Karpates. Ann. Soc. Géol. Pologne 20, Fasc. 1-2, p. 159-168.

SKRIPCHENKO, N.S. (1965) Peculiarities of the crystallization and the autometamorphism of the spilites of the volcanic rocks of the Lower Carboniferous of the Ural Mountains. Trans. Min. Soc. USSR, part 94, Nauk, p. 288-297.

SKOUNAKIS, S. (1972) The igneous rocks of the area of Athens and their metallogenesis. Diss. Athens (original in Greek), 90 p.

SLAVIK, F. (1908) Spilitische Ergußgesteine im Präkambrium zwischen Kladno und Klattan. Archiv naturw. Landesforschung Böhmen, Bd. XIV, No. 2 (Prag).

SLAVIK, F. (1915) Über Spilite im Pribramer Algonkinen. Festschr. z. 70. Geburtstag von Vrba (Prag).

SLAVIK, F. (1928) Les "pillow-lavas" algonkiennes de la Bohème. Int. Geol. Congr., 14th sess., Spain, 1926; Madrid, p. 1389-1395.

SMITH, R.E. (1967a) Curved albite microlites in spilitic lavas from a low-grade metamorphosed sequence (abstr.). Geol. Soc. Amer. Spec. Paper 115, p. 208.

SMITH, R.E. (1967b) Segregation vesicles in basaltic lava. Am. Jour. Sci. 265, p. 696-713.

SMITH, R.E. (1968) Redistribution of major elements in the alteration of some basic lavas during burial metamorphism. Jour. Petrol. 9, p. 191-219.

SMITHERINGALE, W.G. (1972) Low-potash Lush's Bight Tholeiites: Ancient oceanic crust in Newfoundland? Canad. Jour. Earth Sci. 9, p. 574-588.

SOUSTOV, N.I. (1935) Geological-petrographic sketch of the greenstone mass to the south of the Khibin Mountain mass. Material on petr. and geochem. of the Kola peninsula, pt. 6.

SOUSTOV, N.I. (1940) The Proterozoic spilito-diabasic formation of the Imandr-Varzug on the Kola peninsula. Trans. Inst. Geol. Sci., Akad. Nauk SSSR, Petr. Series 26, No. 9.

SPADEA, P. (1968) Pillow-lavas nei terreni alloctoni dell'Appennino Lucano. Atti della Acad. Gioenia Sci. Nat. in Catania, Serie VI, vol. XX, p. 105-142.

SPRY, A. (1953) The thermal metamorphism of portions of the Woolomin group in the Armidale District, N.S.W. Part I: The Puddledock Area. Jour. and Proc. Roy. Soc. N.S. Wales, LXXXVII, p. 129-136.

STAMP, L.D., and WOOLDRIDGE, S.W. (1923) The igneous and associated rocks of Llanwrtyd (Brecon). Quart. Jour. Geol. Soc. 79, p. 16-46.

STARK, M. (1937) Zur Gauverwandtschaft der Spilitgesteine Innerböhmens. Vest.Kral. Ges. Spol Nauk Tr. 11, Praha, Ser. 1-4.

STARK, J.T. (1938) Vesicular dikes and subaerial pillow-lavas of Borabora, Soc. Islands. Jour. Geol. 46, No. 3, pt. 1, p. 225-238.

STARK, J.T. (1939) Discussion: Pillow-lavas. Jour. Geol. 47, p. 205-209.

STEINMANN, G. (1927) Die ophiolitischen Zonen in den mediterranen Kettengebirgen. C.R. Intern. Geol. Congr., 14th sess., Madrid, 2, p. 637-667.

STEPHENS, E.A. (1956) The geology and mineral resources of the Kota Belud and Kudat area, North Borneo. Brit. Borneo Geol. Surv. Mem. 5.

STEPHENSON, P.J. (1964) Some geological observations on Heard Island. Antarctic Geology, SCAR Proc. (1963), p. 14-26.

STEWART, D. (1963) Petrography of some dredgings collected by operation Deep Freeze IV. Proc. Am. Phil. Soc. 107, p. 431-442.

STRAUSS, G.K. (1965) Zur Geologie der SW-iberischen Kiesprovinz und ihrer Lagerstätten, mit besonderer Berücksichtigung der Pyritgrube Lonsal / Portugal. Diss. Univ. München, 152 p.

STRAUSS, G.K. (1970) Sobre la geología de la provincia piritifera del suroeste de la Peninsula Ibérica y de sus yacimientos, en especial sobre la mina de pirita de Lonsal (Protugal).

SUDOVIKOV, N.G. (1931) Contribution to the petrography of Central Karelia. Trans. Geol. Prospect. Admin., No. 51, p. 781-796.

SUNDIUS, N. (1912) Pillow-lava from the Kiruna district. Geol. Fören. Förhandl. 34, p. 317-332.

SUNDIUS, N. (1915) Geologie des Kirunagebietes. Vitensk. prakt. Undersök. i Lappland, Uppsala.

SUNDIUS, N. (1916) Zur Frage der Albitisierung im Kirunagebiet. Geol. Fören. Förhandl. 38, p. 444-462.

SUNDIUS, N. (1930) On the spilitic rocks. Geol. Mag. 67, p. 1-17.

SUTTON, J., and WATSON, J. (1951) Varying trends in the metamorphism of dolerites. Geol. Mag. 88, p. 25-35.

SUZUKI, J., and MINATO, M. (1958) Gotlandian and Devonian volcanic activities in the Kitakami Mountains. Proc. Jap. Acad. 34, p. 284-286.

SVITALSKY, N.I. (1925) Albite Diabase on the Cross Bay of Novaya Zemlya. Trans. Geol. and Min. Museum, Akad. Nauk SSSR.

SZADECZKY-KARDOSS, E. (1963) Wasser und Magma. Ber. Geol. Ges. DDR, Sonderh. 1, p. 49-65.

SZENTPÉTERY, S. von (1933) Physiographie und Genesis der Diabasarten des Bükker Ortásberges. Acta Chem., Min. et Phys. (Szeged), t. III, fasc. 1-2, p. 66-97.

TANE, J.-L. (1962) Géologie - Sur l'origine de certains niveaux versicolores du trias de la zone alpine externe. C.R. Séances Acad. Sci. 254, p. 3391-3392.

TANE, J.-L. (1967) Contribution à l'étude du phénomène de spilitisation. Trav. Lab. Géol. Grenoble, 43, p. 187-192.

TAYLOR, H.P., Jr., FRECHEN, J., and DEGENS, E.T. (1967) Oxygen and carbon isotope studies of carbonatites from Laacher See District, West Germany, and Alnö District, Sweden. Geochim. Cosmochim. Acta 31, p. 407-430.

TEALL, J.J. (1895) On greenstones associated with radiolarian cherts. Trans. Roy. Geol. Survey Corn., 11, p. 560.

TEALL, J.J. (1899) The Silurian rocks of Great Britain. Mem. Geol. Survey Great Britain 1 (Scotland), p. 85.

TEALL, J.J. (1907) The geology of the country around Mevagissey. Mem. Geol. Survey Engl. Wales 56, Expl. of sheet 353.

TERMIER, P. (1898) Sur l'élimination de la chaux par métasomatose dans les roches éruptives basiques de la région du Pelvoux. Bull. Soc. Géol. France 26, p. 165.

THIADENS, A.A. (1937) Geology of the southern part of the province Santa Clara, Cuba. Geogr. Geol. Med. 12, Holland, p. 1-69.

THOMAS, H.H. (1911) The Skomer volcanic series (Pembrokeshire). Quart. Jour. Geol. Soc. 67, p. 175-212.

THOMAS, H.H., and JONES, O.T. (1912) On the Pre-Cambrian and Cambrian rocks of Brawdy Haycastle and Brimaston (Pembrokeshire). Quart. Jour. Geol. Soc. 68, p. 385.

THOMAS, H. (1958) Geología de la Cordillera de la Costa entre el valle de la Ligna y la cuesta de Barriga. Inst. de Invest. geol. Bull. 2, p. 1-86.

TIMOFEYEV, V.M. (1909) On the variolite on the Island of Suisari. Trans. St. Petersburg Sci. Soc.

TIMOFEYEV, V.M. (1916a) On the residue of the surface of the lava-flow in the ancient volcanic region of Suisar. Trans. St. Petersb. Sci. Soc. 38, No. 5.

TIMOFEYEV, V.M. (1916b) On the finding of pillow-lava in the district of Olonetz. Geol. vestn., 8, No. 3, p. 128-132.

TIMOFEYEV, V.M. (1935) Petrography of Karelia. Publ. Akad. Nauk SSSR.

TISCHENDORF, G. (1959) Zur Genesis einiger Selenidvorkommen, insbesondere von Tilkerode im Harz. Freiberger Forschungsh., C 69, 168 p.

TOMITA, T. (1934) Variations in optical properties according to chemical composition in the pyroxenes of the clinoenstatite-clinohyperstene-diopside-hedenbergite system. Jour. Shang. Sci. Ing. 1, 2, p. 41-58.

TOMKEIEFF, S.I., and MARSHALL, C.E. (1940) The Killough-Ardglass dyke swarm. Quart. Jour. Geol. Soc. 96, p. 321-338.

TSCHERWJAKOWSKI, G.F. (1960) Sucharbeiten auf "verborgene" Kieskörper im Ural. Z. angew. Geol., 6, H. 10, p. 478-480.

TURNER, H.W. (1895) Geology of Mount Diablo. Bull. Geol. Soc. Am. 2, p. 384-402.

TURNER, F.J., and VERHOOGEN, J. (1960) Igneous and metamorphic petrology. 2nd Ed. McGraw-Hill, New York, p. 258-272, 316.

TUZCU, N. (1972) Etude minéralogique et étrographique de la région de Baskisla dans le Taurus occidental (Karaman, Vilâyet de Konya, Turquie). Mém. Minéralogie, Univ. Genève, 1, 106 p.

TYRRELL, G.W. (1955) Distribution of igneous rocks in space and time. Bull. Geol. Soc. Am. 66, p. 405-426.

VALLANCE, T.G. (1960) Concerning Spilites. Proc. Linn. Soc. N.S. Wales, Vol. IXXXV, Part 1, p. 8-52.

VALLANCE, T.G. (1965) On the chemistry of pillow lavas and the origin of spilites. Mineralog. Mag. 34, p. 471-481.

VALLANCE, T.G. (1968) Recognition of specific magmatic character in some Palaeozoic mafic lavas in New South Wales. Spec. Publ. Geol. Soc. Austr. 2, p. 163-167.

VALLANCE, T.G. (1969) Spilites again: Some consequences of the degradation of basalts. Proc. Linn. Soc. N.S. Wales, 94, p. 8-51.

VALLET, J.-M. (1950) Etude géologique et pétrographique de la partie inférieure du Val d'Hérens et du Val d'Hérémence (Valais). Schweiz. Min. Petr. Mitt. 30, p. 322-476.

VAN OVEREEM, A.J.A. (1948) A section through the Dal formation (S.W. Sweden). Leiden, Holland, 131 p.

VAN WEST, F.P. (1941) Geological investigations in the Miomaffo region (Netherlands Timor). Thesis, Univ. Amsterdam, 131 p.

VINCENT, E.A., and CROCKET, J.H. (1960) Studies in the geochemistry of gold. Geochim. Cosmochim Acta, 18, p. 130-142, and 143-148.

VOGT, C. (1886) Diabasporphyrite aus der Umgegend der Stadt Petrosawodsk im Gouvernement Olonetz. Tschermaks Min. Petr. Mitt. 8, p. 107-112.

VOGT, T. (1927) Sulitelmafältets geologi o petrografi. Norsk Geol. Unders. No. 121.

VOISEY, A.H. (1939a) The Upper Palaeozoic rocks between Mount George and Wingham, N.S. Wales. Proc. Linn. Soc. N.W. Wales, LXIV, parts 3-4, p. 242-254.

VOISEY, A.H. (1939b) The geology of the lower Manning District of New South Wales. Proc. Linn. Soc. N.S. Wales LXIV, parts 3-4, p. 394-407.

VUAGNAT, M. (1946) Sur quelques diabases suisses. Contribution à l'étude du problème des spilites et des pillow lavas. Min. Petr. Mitt. 26, p. 116-228.

VUAGNAT, M. (1947a) Sur la variolite de Spiss près de Viège (Valais). C.R. Soc. Phys. et d'Hist.Nat., Genève, 64, No. 2, p. 45-47.

VUAGNAT, M. (1947b) Sur le caractère spilitique des mélaphyres de la région du Pelvoux. C.R. Soc. Phys. et d'Hist.Nat., Genève, 64, p. 63-65.

VUAGNAT, M. (1948) Remarques sur trois diabases en coussins de l'Oberhalbstein. Schweiz. Min.Petr. Mitt. 28, p. 263-273.

VUAGNAT, M. (1949a) Variolites et spilites. Comparaison entre quelques pillow lavas britanniques et alpines. Archives Sci., Fasc. 2, Genève, p. 223-236.

VUAGNAT, M. (1949b) Sur les pillow lavas dalradiennes de la péninsule de Tayvallid (Argyllshire). Schweiz.Min.Petr.Mitt. 29, p. 524-536.

VUAGNAT, M. (1952) Pétrographie, répartition et origine des microbrèches du Flysch nordhélvétiques. Beitr. Geol. Karte Schweiz, N.S. 97, 103 p.

VUAGNAT, M. (1959) Les laves en coussins de l'Orthys. Grèce. Archives Sci., Genève, 12, fasc. 1, p. 118-122.

WANG, Y. (1951) A preliminary study on the so-called dolerite of Cheinshan Hengchun, Taiwan. Formosan Sci. 5, p. 39-51.

WEBER, M. (1910) Über Diabase und Keratophyre aus dem Fichtelgebirge. Cbl. Min. 37.

WEDEPOHL, K.H. (1969) Handbook of geochemistry. Vol. 1, Composition and abundance of common igneous rocks. Springer, Heidelberg - Berlin - New York, 442 p.

WEISE, G. (1966) Spilit und Pillowbrekzien des vogtländischen Oberdevons. Geologie, Jg. 15, p. 661-680.

WELLS, A.K. (1922, 1923) The nomenclature of the spilitic suite. Part I: The keratophyric rocks. Geol. Mag. LIX, p. 346-354. Part II: The problem of the spilites. Geol. Mag. LX, p. 62-74.

WELLS, A.K. (1925) The geology of the Rhobell Fawr district, Merioneth. Quart. Jour. Geol. Soc. 81, p. 463.

WENK, E. (1949) Die Assoziation von Radiolarienhornsteinen mit ophiolitischen Erstarrungsgesteinen als petrographisches Problem. Experientia 5, p. 226-232.

WHITE, D. (1954) The stratigraphy and structure of the Mesabi Range, Minnesota. Min.. Geol. Survey Bull. 38, p. 65.

WHITE, W.S. (1960) The Keweenawan lavas of Lake Superior, an example of flood basalts. Am. Jour. Sci. 258, p. 367-374.

WHITLEY, N. (1849) On the remains of ancient volcanoes on the north coast of Cornwall in the Parish of St. Minver. 30th Ann. Rep. Roy. Inst. Corst., 60.

WIDMER, H. (1948) Zur Geologie der Tödigruppe. Diss., Zürich, 97 p.

WILLIAMS, D. (1930) The geology of the country between Nant Peris and Nant Francon (Snowdonra). Quart. Jour. Geol. Soc. 86, p. 208.

WILLIAMS, G.H. (1890) The greenstone schist areas of the Menominee and Marguette region of Michigan. Bull. U.S. Geol. Surv. no. 62, p. 167.

WILLIAMS, H. (1927) The geology of Snowdon (N. Wales) Quart. Jour. Geol. Soc. 83, p. 406.

WILLIAMS, H., TURNER, F.J., and GILBERT, C.M. (1954) Petrography. An introduction to the study of rocks in thin sections. San Francisco, p. 58-60.

WILSON, R.A.M. (1961) The geology and mineral resources of the Banggi Island and Sugut River area, North Borneo. Brit. Borneo Surv. Mem. 15.

WILSON, W.F., and ALLEN, E.P. (1968) Spilitic amygdaloidal basalt flow rocks and associated pillow structure in Orange County, North Carolina. In: Southeastern Geology, Duke Univ. Durham, N. Carolina, 9, p. 133-141.

WILLSON, M.E. (1914) Kewagama Lake map area, Quebec. Mem. Geol. Survey Canada, 39.

WINCHELL, A.N. (1933) Elements of optical mineralogy. Part 2.

WINCHELL, A.N. (1935) Further studies in the pyroxene group. Am. Min. 20, p. 567.

WOLDRICH, J. (1913) Geologische und montanistische Studien in den Karpaten nördlich von Dobschau. Archiv f. Lagerstättenforsch., H. 11, Kgl. Preuss. Geol. Landesanstalt, 108 p.

YEGOROVA, E.N. (1926) Greenstone varieties on the Onego-Belomorsk watershed. Trans. Olonetz Science Exped. State Hydrolog. Inst., part 3, no. 1.

YELISEYEV, N.A. (1925) On diabases in the region of Tulomozero. Trans. Leningrad Nat. Science Soc., 5, no. 1.

YELISEYEV, N.A. (1928) On spilites of Segozero. Western Miner. Soc., 57, no. 1, p. 105-121.

YELISEYEV, N.A. (1938) Petrography of the ore-bearing Altai and Kalba. Petrografiya SSSR, no. 6.

YEREMINA, E.V., and LEVINSON-LESSING, F. Yu. (1905) Contribution to the petrography of the Mugodzhar Mountains. Trans. St. Petersburg Sci. Soc., 33, no. 5.

YEREMINA, E.V. (1912) The group of the Bokhtybay Mountains. Trans. St. Petersburg Sci. Soc., 35, no. 5.

YODER, H.S., Jr. (1967) Spilites and Serpentinites. Yearb. Carnegie Instn. Washington, 65, p. 269-279.

ZAVARITSKY, A.N. (1935) On one important petrochemical regularity. Leningrad Mining Inst., 9, no. 2.

ZAVARITSKY, A.N. (1936a) The pyrite deposit of the Blyava in the South Urals and the pyrite beds of the Ural mountains in general. Trans. Geol. Inst. Akad. Nauk SSSR, 5, p. 29-64.

ZAVARITSKY, A.N. (1936b) On the study of the chemistry of rock varieties with the help of a diagram. Collection dedicated to the 50 years of scientific and pedagogical activity of the Academ. V.J. Vernadsky. Vol. 2, p. 1041-1058.

ZAVARITSKY, A.N. (1941) Check of chemical analyses of igneous rock varieties and definition of their chemical types. Publ. Akad. Nauk SSSR.

ZAVARITSKY, V.A. (1946 / 1960) The spilite-keratophyre formation in the region of the Blyava deposit in the Ural Mountains. Trudy Inst. Geol. Nauk, 71, Petrogr. Ser. 24, p. 1-83. [Transl. from Russian: Int. Geol. Rev. 2, No. 7, p. 551-581, No. 8, p. 645-687; 1960]

ZAYTSEV, A.M. (1908, 1910) On the petrography of the Crimea. Annals of Geology and Mineralogy of Russia, 1908, 10, no. 5-6; 1910, 12, no. 3-4, 7-8.

ZBINDEN, P. (1949) Geologisch-petrographische Untersuchungen im Bereich südlicher Gneise des Aarmassivs. Schweiz. Min. Petr. Mitt. 29, p. 221-356.

ZERVAS, S. (1972) On the ophiolitic intrusion in the Argolis area. Contribution to the knowledge of the geological situation and the petrology of ophiolites. Ann. Géol. des Pays Hélléniques, 24, p. 1-108 (original in Greek).

Supplement:

AMSTUTZ, G.C. (1948) Lavaströme im Glarner Freiberg. Leben und Umwelt, p. 90-94.

AMSTUTZ, G.C. (1957a) The spilite problem. Anal.Intern.Geol.Congr. Mexico 1956 (abstract).

AMSTUTZ, G.C. (1957b) Spilitisation - the missing link in rock and ore genesis. Naturwissenschaften, Jg. 44, H. 18, p. 490.

AMSTUTZ, G.C. (1957c) The genesis of spilitic rocks and mineral deposits (abstract). Bull. Geol. Soc. Am. 68, Spec. issue, p. 1695-1696.

AMSTUTZ, G.C. (1958b) The genesis of the Lake Superior copper deposits (abstract). Instit. on Lake Superior Geology, Program, p. 25.

BRÖGGER, W.C. (1898) Eruptivgesteine des Kristianiagebietes. III.

BRONGNIART, A. (1827) Classification et caractères minéralogiques des roches homogènes et hétérogènes. Levrault, Paris, 144 p.

HAILE, N.S. (1957) The geology and mineral resources of the Lupar and Saribas Valleys, West Sarawak. Geol. Survey Dept., Brit. Territories in Borneo. Memoir 7, 123 p.

HYNES, A.J., NISBET, E.G., SMITH, A.G., WELLAND, J.P., and REX, D.C. (1972) Spreading and emplacement ages of some ophiolites in the Othris region (eastern central Greece). Z. Deutsch. Geol. Ges. 123, p. 455-468.

JOVANOVIC, Z.Z. (1973) Late Cretaceous Volcanogenic Flysh and Diabase-chert Formations in the Yugoslav Inner Dinarides. Sedimentary Geology, 9, p. 117-147.

LEHMANN, E. (1972) On the Source of the Iron in the Lahn Ore Deposits. Mineral. Deposita 7, p. 247-270.

NAKAZAWA, K., and KAPOOR, H.M. (1973) Spilitic Pillow Lava in Panjal Trap of Kashmir, India. Mem. Fac. Sci., Kyoto Univ., Ser. Geol. & Min. XXXIX, p. 83-98.

2) Related Literature (obviously only a selection)

AGRELL, S.O. (1939) The adinoles of Dinas Head, Cornwall. Min. Mag. 25

ALBEE, A.L. (1962) Relationship between the mineral association, chemical composition and physcial proterties of chlorite series. Am. Miner. 47, p. 851-870.

ANDERSSON, L.H., and LINDQUIST, B. (1956) Some experiments on the interaction between feldspars and salt solutions. Geol. Foren. Forh. 78, p. 459-461.

ANGEL, F. (1954a) Waldsteinit: ein Na-metasomatischer, eisenerzdurchstäubter diabasisch-tonschieferiger Metatuffit aus der Steiermark. Tschermaks Min.Petr. Mitt. 4, p. 440-453.

ANGEL, F. (1954b) Die Entstehung des "Österreichischen Traß" (Gossendorfit) und seine Stellung im Gleichenberger Vulkanismus. Joanneum, Miner. Mitt.bl., p. 9-11.

ARBENZ, K. (1947) Geologie des Hornfluhgebietes (Berner Oberland). Beitr. Geol. Karte Schweiz, N.F. 89.

ARNI, P. (1942) Materialien zur Altersfrage der Ophiolithe Anatoliens. Maden Tetkîk ve Arama, Ankara, Turkey, sene 1, sayi 3/28, p. 472-480 (Turkish), p. 481-488 (German).

BARAGAR, R.A. (1960) Petrology of basaltic rocks in part of the Labrador Trough. Bull. Geol. Soc. Am. 71, p. 1589-1644.

BENSON, W.N. (1922) An outline of the geology of New Zealand. Jour. Geol. 30, no.1, p. 1-17.

BENSON, W.N. (1941) Cainozoic petrographic provinces in New Zealand and their residual magmas. Am. Jour. Sci. 239, p. 537-552.

BENSON, W.N. (1943) The basic igneous rocks of eastern Otago and their tectonic environment. Part IV, Sect. A. Trans. Roy.Soc. New Zealand, 72, pt. 2.

BENSON, W.N. (1944) The basic igneous rocks of eastern Otago and their tectonic environment. Part IV, Sect. B. Trans.Roy. Soc. New Zeland, 71, pt. 1, p. 71-123.

BENSON, W.N. (1945) The basic igneous rocks of eastern Otago and their tectonic environment. Part IV, Sect. C. Trans. Roy. Soc. New Zeland, 75, pt. 2, p. 288-318.

BENSON, W.N. (1946) The basic igneous rocks of eastern Otago and their tectonic environment. Part V. Trans. Roy. Soc. New Zealand, 76, pt. 1, p. 1-36.

BERTRAND, J. (1969) Sur la présence de hyaloclastites dans la région des Gets (Haute-Savoie). C.R. des Séances, SPHN Genève, NS. 3, p. 112-121.

BIRCH, F., and LeCOMTE, P. (1960) Temperature-pressure plane for albite composition. Am. Jour. Sci. 258, p. 209-217.

BLOXAM, T.W. (1960) Jadeite-rocks and claucophane-schists from Angel Island, San Francisco Bay, California. Am. Jour. Sci. 258, p. 555-573.

BOTTKE, H. (1965) Die exhalativ-sedimentären devonischen Roteisensteinlagerstätten des Ostsauerlandes. Beih. Geol. Jahrb., H. 63, 147 p.

BOWEN, N.L. (1915) The crystallization of haplobasaltic, haplodioritic and related magmas. Am. Jour. Sci., Ser. 4, 40, p. 161-185.

BOWEN, N.L., and SCHAIRER, J.F. (1938) Crystallization equilibrium in nepheline-albite-silica mixture with fayalite. Jour. Geol. 46, No. 3, pt. 2.

BOWEN, N.L. (1945) Phase equilibria bearing on the origin and differentiation of alkaline rocks. Am. Jour. Sci. 243-A, p. 75-89.

CAPEDRI, S. (1970) New evidence on secondary twining in albite plagioclases. Contr. Mineral. Petrol. 25, p. 289-296.

CHAYES, F. (1950) On a distinction between late magmatic and post magmatic replacement reactions. Am. Jour. Sci. 248, p. 52-66.

COOMBS, D.S., ELLIS, A.F., FYFE, W.S., and TAYLOR, A.M. (1960) Lower grade mineral facies in New Zealand. 21st Intern. Geol. Congr. (Copenhagen) Reports, Part 13, p. 339-351.

DAHM, K. (1967) Pillowlaven in der DDR - Zeichnerische Darstellung einiger typischer Beispiele. Ber. deutsch. Ges. Geol. Wiss. A 3, p. 257-265.

DAVIES, R.G. (1959) The Cader Idris granophyre and its associated rocks. Quart. Jour. Geol. Soc. London, 115, p. 189-216.

DAVIS, E.F. (1918) Radiolarian cherts of the Franciscan group. Cal. Univ. Dept. Geol. Sci. Bull. 11, p. 235-432.

DE VORE, G.W. (1956) Surface chemistry as a chemical control on mineral association. Jour. Geol., 64, p. 31-55.

DICKSON, F.W., and TUNELL, G. (1954) The Saturation Curves of Cinnabar and Metacinnabar in the System $HgS-Na_2S-H_2O$ at 25 °C. Science 119, p. 467-468.

DIETRICH, V. (1969) Die Oberhalbsteiner Talbildung im Tertiär. Ein Vergleich zwischen den Ophiolithen und deren Detritus in der ostschweizerischen Molasse. Eclog. geol. Helvet. 62, p. 637-641.

DOUGHERTY, E.Y. (1939) Some geological features of Kolar, Porcupine and Kirkland Lake. Econ. Geol. 34, p. 622-653.

EITEL, W. (1954) The physical chemistry of the silicates. Univ. Chicago Press, 1592 p.

ERNST, W.G. (1961) Stability relations of Glaucophane. Am. Jour. Sci. 259, p. 735-765.

ESKOLA, P. (1938) Prehnite amygdaloid from the bottom of the Baltic. C.R. Soc. Géol. Finlande 8, p. 1-13.

FAWCETT, H.J., and YODER, H.S. (1966) Phase relationships of chlorites in the system $MgO-Al_2O_3-SiO_2-H_2O$. Am. Min. 57, p. 332-360.

FEARNSIDES, W.G., and TEMPLEMAN, A. (1932) A boring through Edale shales to carboniferous limestone and pillow lavas at Hope Cement Works, near Castleton, Derbyshire. Proc. Yorks. Geol. Soc., 22, p. 100-121.

FEDIUK, F. (1962) Vulkanity. Zeleznobrodskeho krystalinika. Rozpravy, Svazek 29, 116 p.

FENNER, C.N. (1933) Pneumatolytic processes in the formation of minerals and ores. In: Ore Deposits of Western States, Am. Inst. Min.Met. Eng., p. 58-106.

FLASCHEN, S.S., and OSBORN, E.F. (1957) Studies of the system iron oxide - silica - water at low oxygen partial pressures. Econ. Geol. 52, p. 923-943.

FULLER, R.F. (1932) Concerning basaltic glass. Am. Mineral. 17, no. 3, p. 104-107.

GALE, G.H., and ROBERTS, D. (1972) Palaeogeographical implications of greenstone petrochemistry in the southern Norwegian caledonides. Nature, 238, p. 60-61.

GANSSER, A. (1959) Ausseralpine Ophiolitprobleme. Eclog. geol. Helv. 52, p. 659-680.

GEES, R.A. (1956) Ein Beitrag zum Ophiolit-Problem, behandelt an einigen Beispielen aus dem Gebiet von Klosters Davos (Graubünden). Schweiz. Min. Petr. Mitt. 36, p. 454-488.

GEIJER, P. (1916) Notes on albitization and the magnetite syenite porphyries. Geol. Fören. Stockholm Förh. 28, p. 243-264.

GEIJER, P. (1930a) Gällivare Malmfält. Geologisk Beskrivning. Sver. Geol. Undersök., Ser. C, No. 22, 115 p.

GEIJER, P. (1930b) The iron ores of the Kiruna type. Sver. Geol. Undersök., Ser. C, No. 367, Aarsbok 24, no. 4, p. 1-39.

GEIJER, P. (1967) Internal features of the apatite-bearing magnetite ores. Sver. Geol. Undersök., Ser. C, Nr. 624, p. 1-32.

GILAROVA, M.A. (1959) Pillow lavas of the Suissari region of southern Karelia and the problem of genesis of pillow lavas. Uczonyje Zapiski L.G.U. nr. 268, Leningrad, p. 3-69.

GJELSVIK, T. (1960) The differentiation and metamorphism of the West Norwegian dolerites. D.K.N.V.S. Skrifter, No. 3, p. 3-11.

GLASSER, F.P., WARSHAW, I., and ROY, R. (1960) Liquid immiscibility in silicate systems. Physics and Chemistry of Glasses, v. 1, p. 39-45.

GOLDING, H.G. (1961) Leucoxene terminology and genesis (Discussion). Econ. Geol. 56, p. 1139-1149.

GORANSON, R.W. (1936) Silicate-water systems: The solubility of water in albite-melt. Trans. Am. Geophys. Union, 17th Ann. Meeting Rep. and Papers, Volcanology, p. 257-259.

GREENWOOD, H.J. (1961) The system $NaAlSi_2O_6-H_2O-Argon$; Total pressure and water pressure in metamorphism. Jour. Geophys. Res., Vol. 66, p. 3923-3946.

GUPPY, E.M., and SABINE, P.A. (1956) Chemical analyses of igneous rocks, metamorphic rocks and minerals. 1931-1954. Mem. Geol. Surv. Great Britain.

HARKER, A. (1917) Discussion of SARGENT's paper (1917). Quart. Jour. Geol. Soc. 73, p. 23-25.

HERGET, G. (1966) Archaische Sedimente und Eruptiva im Barberton-Bergland (Transvaal, Südafrika). N. Jb. Miner. Abh. 105, p. 161-182.

HESS, H.H. (1955) Serpentines, orogeny, and epeirogeny. In: The crust of the earth. Geol. Soc. Am. Special Paper 62, p. 391-408.

IRWIN, W.P. (1972) Terranes of the western Paleozoic and Triassic belt in the southern Klamath Mountains, California. U.S. Geol. Surv. Prof. Paper 800-C, p. 103-111.

JONES, O.T., and PUGH, W.J. (1948) The form and distribution of dolerite masses in the Builth-Llandrindod Inlier, Radnorshire. Quart. Jour. Geol. Soc. 104, p. 71-98.

JOVANOVIC, R. (1960) Diabase-Hornstein- Formation in Bosnien und Hercegovina. Simpoz. Problem. Alpisk. Inicijal. Magmat., Ref. II, p. 1-17.

KAADEN, G. van der (1951) Optical studies on natural plagioclase feldspars with high- and low-temperature optics. Diss. Univ. Utrecht, 105 p.

KARL, F. (1954) Über Hoch- und Tieftemperaturoptik von Plagioklasen und deren petrographische und geologische Auswertung am Beispiel einiger alpiner Ergußgesteine. Tschermaks Min. Petr. Mitt., Ser. 3, 4, p. 320-328.

KHITAROV, N.I. (1958 / 1960) On the reaction of oligoclase with water under conditions of high temperature and pressure. Akad. Nauk SSSR, Inst, Khimi Silikatov, 1958, p. 208-213. [Transl. from Russian: Int. Geol. Rev. 2, No. 4, p. 322-326; 1960]

KORZHINSKY, D.S. (1963) Das Verhalten des Wassers bei magmatischen und postmagmatischen Prozessen. Ber. Geol. Ges., Sonderh. 1, p. 67-79.

KUSHIRO, I. (1964) Petrology of the Atumi dolerite, Japan. Jour. Fac. Sci., Univ. Tokyo, Sect. II, Vol. XV, pt. 2, p. 135-202.

KUSHIRO, I. (1968) Composition of magmas formed by partial melting of the earth's upper mantle. Jour. Geol. Res. 73, p. 619-630.

KUSHIRO, I. (1969) The system forsterite-diopside-silica with and without water at high pressure. Am. Jour. Sci. 267, p. 269-294.

LEHMANN, E. (1952) The significance of hydrothermal stage in the formation of igneous rocks. Geol. Mag. 89, p. 61-69.

LEHMANN, E. (1965a) Der "Kugeldiabas" bei Giebringhausen (Diemel). N. Jb. Miner. Abh. 103, p. 99-115.

LEHMANN, E. (1965b) Zur 'Anchimetamorphose' von Eruptivgesteinen, insbesondere Diabasen. N. Jb. Miner. Mh., Jg.1965, H. 6, p. 184-190.

MACKENZIE, WM. S. (1957) The crystalline modifications of $NaAlSi_3O_8$. Am. Jour. Sci. 255, p. 481-516.

MATHEWS, W.H. (1964) Settling of olivine in pillows of an icelandic basalt. Am. Jour. Sci. 262, p. 1036-1040.

McBIRNEY, A.R. (1963) Factors governing the nature of submarine volcanism. Bull. Volcanol. 26, p. 455-469.

MIROSLAV, T. (1947) Albite-dolerite from Nakop Brook in Požega Mountain. Geol. Vjesn. Geol.-rdarsk. instit. Zagreb, p. 182-189.

MITCHELL, R.C. (1955) The ages of the serpentinized peridotites of the West Indies I. Proc. Koninkl. Nederl. Akad., Ser. B, 58, p. 194-212.

MITCHELL, A.H., and READING, H.G. (1971) Evolution of Island Arcs. Jour. Geol. 79, p. 253-284.

MIYASHIRO, A. (1966) Some aspects of peridotite and serpentinite in orogenic belts. Jap. Jour. Geol. Geogr. XXXVII, no. 1, p. 45-61.

MIYASHIRO, A., and SHIDO, F. (1970) Progressive metamorphism in zeolite assemblages. Lithos 3, p. 251-260.

MOREY, G.W. (1924) Relation of crystallization to the water content and vapour pressure of water in a cooling magma. Jour. Geol. 32, p. 291-295.

MOREY, G.W., and CHEN, W.T. (1956) Pressure - temperature curves in some systems containing water and salt. Jour. Am. Chem. Soc. 78, p. 4249-4252.

MOREY, G.W. (1957) The system water - nepheline - albite: a theoretical discussion. Am. Jour. Sci. 255, p. 461-480.

MUIR, J.D., and TILLEY, C.E. (1966) Basalts from the northern part of the Mid-Atlantic Ridge. Jour. Petrol. 7, p. 193-201.

NASHAR, B., and BASDEN, R. (1965) Solubility of basalt under atmospheric conditions of temperature and pressure. Min. Mag. 35, p. 408-411.

NICHOLS, G.D. (1934) Basalts from the deep ocean floor. Min. Mag. 34 (TILLEY Volume), p. 373-388.

NIGGLI, P. (1941) Das Problem der Koexistenz der Feldspäte in den Eruptivgesteinen. Schweiz. Min. Petr. Mitt. 21, p. 183-193.

OFTEDAHL, C, (1968) Greenstone volcanoes in the Central Norwegian caledonides. Geol. Rundsch. 57, p. 920-930.

ORVILLE, P.M. (1963) Alkali ion exchange between vapour and feldspar phases. Am. Jour. Sci. 261, p. 201-237.

OSBORN, E.F. (1959) Role of oxygen pressure in the crystallization of basaltic magma. Am. Jour. Sci. 257, p. 609-647.

PACKHAM, G.H., and CROOK, K.A.W. (1960) The principle of diagenetic facies and some of its implications. Jour. Geol. 68, p. 392-407.

PALIVCOVÁ, M. (1959) Ke genesi některých typu vyvreluve středočeském plutonu. Čas. min. geol. 4, p. 163-166.

PAMIĆ, J., and MAKSIMOVIĆ, V. (1968) Middle Triassic diabases in Skythian sediments in the neighbourhood of Konjic (Hercegovina). Geol. Glasnik 12, p. 131-137.

PAWLOW, D.I., and RJABTSCHIKOW, I.D. (1968) Dolerite, die in Salzschichten erstarrt sind. Trans. Akad. Nauk SSSR, ser. geol., no. 2, p. 52-63.

PELLIZZER, R. (1955) Ricerche sulle ofioliti della zona tra la Futa e la Raticosa. Bull. Serv. Geol. Italia, 77, p. 605-681.

PELLIZZER, R. (1957) Tranzformazioni sperimentali alle condizioni pneumatolitiche e idrotermali di rocce ofiolitiche appenniniche. R.C. Soc. Miner. Italiana, XIII, p. 3-31.

PICHAMUTHU, C.S. (1960) Albite-chlorite schists from Lingadhalli, Mysore State, India. Bull. Mysore Geol. Assoc. 19, p. 3-17.

POLDERVAART, A. (1953) Metamorphism of basaltic rocks - A review. Bull. Geol. Soc. Am. 64, p. 259-274.

RAASE, P., and KERN, H. (1969) Über die Synthese von Albiten bei Temperaturen von 250 bis 700 °C. Contr. Mineral. Petrol. 21, p. 225-237.

RAGLAND, P.C., ROGERS, J.J.W., JUSTUS, P.S. (1968) Origin and differentiation of Triassic dolerite magmas, North Carolina, U.S.A. Contr. Mineral. Petrol. 20, p. 57-80.

RIDGE, J.D. (1951) Water in primordial and derivate magma and its relation to the oreforming fluid. Am. Jour. Sci. 249, p. 512-532.

RINGWOOD, A.E. (1959) Genesis of the basalt trachyte association. Beitr. Min. Petr. 6, p. 346-351.

ROBSON, J. (1953-54) The Cornish 'greenstones'. Trans. Roy. Geol. Soc. Cornwall, XVIII, p. 475-492.

ROEVER, W.P., De (1955) Genesis of jadeite by low grade metamorphism. Am. Jour. Sci. 253, p. 293-298.

ROUTHIER, P. (1963) Les gisements métallifères. Masson, Paris, p. 643, 651, 654.

RUCHIN, L.B. (1958) Grundzüge der Lithologie. Akademie-Verlag Berlin, p. 519-527 [transl. from Russian].

SAND, L.B., ROY, R., and OSBORN, E.F. (1957) Stability relations of some minerals in the $Na_2O-Al_2O_3-SiO_2-H_2O$ system. Econ. Geol. 52, p. 169-179.

SATTERY, J. (1941) Pillow lavas from the Dryden-Wabigoon Area, Kenora District, Ontario. Univ. Toronto Studies. Geol. Series 46, Contributions to Capadian Mineralogy, p. 119-136.

SCHAIRER, J.F., and BOWEN, N.L. (1947) Melting relations in the systems $Na_2O-Al_2O_3-SiO_2$ and $K_2O-Al_2O_3-SiO_2$. Am. Jour. Sci. 245, p. 193-204.

SCHNEIDERHÖHN, H. (1952) Genetische Lagerstättengliederung auf geotektonischer Grundlage. N. Jb. Miner. Mh. 2 & 3, p. 47-89.

SCHÜLLER, A. (1960) Das Jadeitproblem vom petrogenetischen und mineralfaziellen Standpunkt. N. Jb. Miner. Abh. 94, p. 1295-1308.

SCHÜRMANN, H.M.E. (1956a) Beiträge zur Glaukophanfrage (3). N. Jb. Miner. Abh. 89, p. 41-85.

SCHÜRMANN, H.M.E. (1956b) The geology of the glaucophane rocks in Turkey and Japan: a summary. Geol. Mijnbouw 4, ser. 18e, p. 119-122.

SCHÜRMANN, H.M.E. (1958) The geology of the glaucophane rocks in Taiwan, India, Iran, Iraq and New Caledonia. Geol. Mijnbouw 5, ser. 20e, p. 133-145.

SHAND, S.J. (1944) The terminology of late magmatic and post magmatic processes. Jour. Geol. 52, p. 342-350.

SHAW, H.T. (1964) Theoretical solubility of H_2O in silicate melts: Quasicrystalline models. Jour. Geol. 72, p. 601-617.

STEEN, D.M. (1972) Etude géologique et pétrographique du complexe ophiolitique de la Haute-Ubaye (Basses-Alpes, France). Mem. Mineralogie, Univ. Genève 2, 235 p.

STEVENSON, J.S. (1945) Geology of twin 'J' mine. Trans. Can. Inst. Min. Met. XLVII, p. 294-308.

SUZUKI, J., and SUZUKI, Y. (1959) Petrological study of the Kamuikotan metamorphic complex in Hokkaido, Japan. Jour. Fac. Sci. Univ. Hokkaido, Ser. IV, Vol. X, no. 2, p. 349-446.

SZÁDECZKY-KARDOSS, E., and ERDÉLYI, J. (1957) Über die Zeolithbildung der Basalte der Balatongegend. Bull. Hungar. Geol. Soc. 87, 3, p. 302-308.

TAJDER, M. (1956) Albitski riolit od Blackog u Požeškoj Gori. Geol. Vjesnik, p. 191-196.

TALIAFERRO, N.L., and HUDSON, F.S. (1943) Genesis of the manganese deposits of the coast ranges of California. In: Manganese in California, State Div. of Mines Bull. no. 125, San Francisco, p. 217-276.

TALIAFERRO, N.L. (1943) Manganese deposits of the Sierra Nevada, their genesis and metamorphism. In: Manganese in California, State Div. of Mines Bull. no. 125, San Francisco, p. 277-332.

TANEDA, S. (1966) The petrogenetic significance of the vapour pressure in magmas. Mem. Fac. Sci. Kyushu Univ., Ser. D, XVII, p. 311-330.

TOBI, A.C. (1962) Characteristic patterns of plagioclase twinning. Norsk Geol. Tids. 42, p. 264-271.

TUTTLE, O.F., and BOWEN, N.L. (1950) High temperature albite and contiguous feldspars. Jour. Geol. 58, p. 572-583.

VANCE, J.A. (1961) Polysynthetic twinning in plagioclase. Am. Miner. 46, p. 1097-1119.

VOZÁR, J. (1967) Zur petrographischen Charakteristik der Melaphyre der Kleinen Karpaten. Geol. práce, Zprávy 41, Bratislava, p. 153-165.

WAGER, L.R. (1960) The relationship between the fractionation stage of basalt and the temperature of the beginning of its crystallisation. Geochim. Cosmochim. Acta 20, p. 158-160.

WAGNER, P. (1939) Differentiationserscheinungen in Diabasen des ostthüringischen Hauptsattels. Min. Petr. Mitt. (Leipzig) 50, p. 107-180.

WALKER, G.P.L. (1959) Some observations on the Antrim basalts and associated dolerite intrusions. Proc. Geol. Assoc., 70, p. 179-205.

WALKER, G.P.L. (1960) The amygdale minerals in the Tertiary lavas of Ireland. III - Regional distribution. Min. Mag. 32, p. 503-527.

WASSERBURG, G.J. (1957) The effects of H_2O in silicate systems. Jour. Geol. 65, p. 15-23.

WINKLER, H.G., and PLATEN, H. von (1958) Experimentelle Gesteinsmetamorphose - II. Geochim. Cosmochim. Acta 15, p. 91.

WRIGHT, G.M. (1955) Geological notes on central district of Keewatin Northwest Territories. Geol. Surv. Canada, Paper 55-17, p. 1-17.

WYLLIE, P.J., and TUTTLE, O.F. (1959a) Melting of calcite in the presence of water. Am. Miner. 44, p. 453-459.

WYLLIE, P.J., and TUTTLE, O.F. (1959b) Effect of carbondioxide on the melting of granite and feldspars. Am. Jour. Sci. 257, p. 648-655.

WYLLIE, P.J., and TUTTLE, O.F. (1960) Experimental investigation of silicate systems containing two volatile components. Part I, Geometrical considerations. Am. Jour. Sci. 258, p. 498-517; Part II, Am. Jour. Sci. 259, p. 128-143 (1961); Part III, Am. Jour. Sci. 262, p. 930-939 (1964).

YODER, H.S., Jr. (1950) The jadeite problem. Am. Jour. Sci. 248, p. 225-248, p. 312-334.

Sections B, C, and D refer only to the literature listed under A 1).

The geographical and geological classifications contain only those references which could be assigned to definite occurrences, or periods resp. Consequently, they are not complete, but will serve as a guide to pertinent literature on occurrences in specific countries and regions, or during certain geological periods.

(References found in the supplement are marked "sup.").

B) Geographical Classification of Occurrences

Africa

General:
MELSON and VAN ANDEL, 1966 (Atlantic Ocean); MELSON et al., 1968 (Atlantic Ocean).

Egypt:
v. ECKERMANN, 1936.

Morocco:
DE KONING, 1957; MOUSSU, 1959.

South Africa:
HALL and MOLENGRAAF, 1925.

America

- Central America:

 Cuba:
 KEIJZER, 1945; RUTTEN, 1936; THIADENS, 1937.

 West Indies:
 DONNELLY, 1962, 1963; LIDIAK, 1965; MATTSON, 1960.

- North America

 Canada:
 DALY, 1903; DIMROTH, 1971; FLAHERTY, 1934; FURSE, 1954; GUNNING, 1937; GUNNING and AMBROSE, 1937; GUSTAFSON, 1945; HUGHES, 1970; HUGHES and MALPASS, 1971; MOORE, 1930; PAPEZIK and FLEMING, 1967; PAPEZIK, 1970; PETTIJOHN, 1943; PETTIJOHN and BASTRON, 1959; RICHARZ, 1926; SMITHERINGALE, 1972; WILLSON, 1914.

 U S A:
 AMSTUTZ, 1958, 1958b (sup.), 1959, 1965, 1968; BAILEY et al., 1970; BILLINGS, 1937; BRODERICK, 1935; BROWN, 1958; BUDDINGTON, 1926; CAPPS, 1915; CATER and WELLS, 1953, 54; CLAPP, 1921; CLEMENTS and SMITH, 1899; COOMBS et al., 1970; CORNWALL, 1951a, 1951b, 1951c; DICKINSON, 1962; DONNELLY, 1966 (Virgin Islands); FAIRBANKS, 1856, 1897, 1904; GEIJER, 1930; GILLULY, 1935, 1946; GOLDICH and RUOTSALA, 1955; GROUT, 1937; HEKINIAN, 1971 (Virgin Islands); HOPGOOD, 1962; KNOPF, 1918, 1921, 1924, 1929; KNOPF and ANDERSON, 1930; LAWSON, 1895, 1914; NIGGLI, P., 1952; NOLAND, 1930;

U S A (continued)

PARK, 1946; PARKER, 1959; RANSOME, 1893, 1894; WHITE, D., 1954; WHITE, W.S., 1960; WILLIAMS, G.H. 1890; WILSON and ALLEN, 1968.

- South America

General:

MELSON and VAN ANDEL, 1966 (Atlantic Ocean); MELSON et al., 1968 (Atlantic Ocean).

Chile:

CRISTI, 1958.

Venezuela:

DONNELLY, 1959; SHAGAM, 1960.

Arctic and Antarctic

Antarctic:

GJELSVIK, 1964; STEPHENSON, 1964; STEWART, 1963.

Asia

China:

WANG, 1951.

East Indies:

BROUWER, 1947; DE JONG, 1941, 1942; DE ROEVER, 1940a, 1940b, 1941, 1942; FITCH, 1955, 1958; HAILE, 1957 (sup.); HAILE and WONG, 1965; KIRK, 1962, 1968; KÜNDIG, 1956; MOLENGRAAF, 1902; REINHARD and WENK, 1951; STEPHENS, 1956; VAN WEST, 1942; WILSON, R.A.M., 1961.

India:

ADITYA, 1955; CANN and VINE, 1966 (Indian Ocean); DEY, 1942; HEKINIAN, 1968 (Indian Ocean); NAKAZAWA and KAPOOR, 1973 (sup.); PATWARDHAN et al., 1970; PATWARDHAN, 1972; PICHAMUTHU, 1938, 1946, 1950, 1957.

Japan:

BAMBA and SAWA, 1967; BAMBA and MAEDA, 1969; CHIHARA, 1954; ISHII and UEDA, 1953; SHIMAZU, 1971; SUZUKI and MINATO, 1958.

Syria:

DUBERTRET, 1939, 1954; MAJER, 1960.

Turkey:

COGULU, 1967; v.d. KAADEN and METZ, 1954; PISKIN, 1972; TUZCU, 1972.

U S S R (Asian Part):

ASHGIREJ, 1958; BACKLUND, 1930; BERIDZE; BURGUNKER, 1964; CHUMAKOV, 1940, 1939; GOLUB and VRAGOVIC, 1960; KASSIN, 1931; KASSIN et al., 1933; KOLENKO, 1855; KUPLETSKY, 1932, 1939; LAGORIO, 1897; LEVINSON-LESSING,

U S S R (continued)
1899; PODOLSKII, 1963; PRIGOROVSKY, 1914; ROGOVER, 1939; SCHILOW and KALISCHEWITSCH, 1958; SVITALSKY, 1925; YELISEYEV, 1938; YEREMINA and LEVINSON-LESSING, 1905; YEREMINA, 1912.

Australia and Oceania

Australia:
CAMPANA and KING, 1963; CROOK, 1960; EDWARDS, 1953; GUSTAFSON and MILLER, 1937; HOOKER, 1969; JOPLIN, 1964; McMATH et al., 1953; PRIDER, 1939-40; SCOTT, 1951a, 1951b, 1954; SPRY, 1953; VALLANCE, 1968; VOISEY, 1939a, 1939b.

New Zealand:
BARTRUM, 1935, 1936; BATTEY, 1950, 1951, 1954, 1955, 1956; BENSON, 1913, 1913/1915; BENSON and HOLLOWAY, 1940; COOMBS, 1954; GRINDLEY, 1958; HOOKER, 1957; HOPGOOD, 1956, 1957; REED, 1950; VALLANCE, 1960.

Oceania:
McKINSTRY, 1939; MOORE, 1965 (Hawai); STARK, 1938.

Europe

General:
BURRI and NIGGLI, 1945; STEINMANN, 1927.

Austria:
ANGEL, 1919, 1955, 1957; CORNELIUS, 1933; FRASL, 1957; FRITSCH, 1961a, 1961b; HANSELMAYER, 1963, 1964; HERITSCH, 1911; KERN, 1952; LOESCHKE, 1973; METZ, 1951; REDLICH and PREDIK, 1930.

Bulgaria:
DIMITROFF, 1929, 1934, 1960.

Czechoslovakia:
BARTH, 1964, 1966; DUDEK and FEDIUK, 1955; FEDIUK, 1953, 1962; FIALA, 1965, 1966, 1967; FOJT, 1962; FOJT and POHANKA, 1965; KETTNER, 1917; SLAVIK, 1908, 1915, 1928; STARK, 1937; WOLDRICH, 1913.

Finland:
BESKOW, 1929; ESKOLA, 1925, 1934a, 1934b; ESKOLA et al., 1935; HACKMAN, 1927; MERILÄINEN, 1961; PAAKOLA, 1971; PIISPANEN, 1972; SAKSELA, 1960.

France:
BERTRAND, 1970; BERTRAND and DELALOYE, 1970; BOULADON et al., 1963; BOULADON et al., 1965; BROWN and ROACH, 1972; CELIKAN, 1899; DELESSE, 1851; DEN TEX, 1950; JAFFÉ, 1955; JUTEAU and ROCCI, 1965, 1966; LA ROCHE et al., 1969; MARTINI and VUAGNAT, 1970; MICHEL-LÉVY, 1877; MORET, 1929; MORRE-BIOT, 1970; MOURANT, 1936; PERRIN and ROUBAULT, 1941; PETTERLONGO, 1968; ROCCI and JUTEAU, 1968; TANE, 1962, 1967; TERMIER, 1898; VUAGNAT, 1947b, 1949a.

Germany:

BEDERKE, 1959; CELIKAN, 1899; ECKHARDT, 1968; ERDMANNSDÖRFER, 1901, 1904; FABIAN and MÜLLER, 1962; FABIAN et al., 1962; GOETZ, 1937; GREGORY, 1892; v. GÜMBEL, 1874; HENTSCHEL, 1952, 1953, 1960, 1961, 1966; KOPF, 1961; LEHMANN, 1941, 1949a, 1949b, 1952a, 1952b, 1952c, 1956, 1965a, 1968, 1972 (sup.); QUADE, 1970; RICHTER and VILLWOCK, 1960, 1961; RINNE, 1895-96; RIPPEL, 1954; RÖSLER, 1959, 1960, 1961, 1962, 1963; SCHIDLOWSKI and STAHL, 1971; SCHLOSSMACHER, 1920; SCHÜLLER and DETTE, 1955; TISCHENDORF, 1959; WEBER, 1910; WEISE, 1966.

Great Britain:

BAILEY and GRABHAM, 1909; BAILEY, 1912; BAILEY and McCALLIEN, 1960; BLOXAM, 1969; BLOXAM and LEWIS, 1972; BLYTH, 1935; COX, 1913, 1915, 1920, 1925; COX and JONES, 1913; DEWEY, 1910, 1914, 1948; DEWEY and FLETT, 1911; FLETT, 1907, 1909, 1911; FLETT and DEWEY, 1912, 1913; FLETT and RILL, 1912; FLOYD, 1972; GREENLY, 1902, 1919; GREENLY and MATLEY, 1928; HENDRIKS, 1939; HUDSON, 1937; LEHMANN, 1965b, 1968; MIDDLETON, 1960; MOURANT, 1936; NICHOLLS, 1959; PHILLIPS, 1876, 1878; RAISIN, 1893; REID and DEWEY, 1908; RICHTER, 1968; SARGENT, 1917; STAMP and WOOLDRIDGE, 1923; SUTTON and WATSON, 1951; TEALL, 1895, 1899, 1907; THOMAS, 1911; THOMAS and JONES, 1912; TOMKEIEFF and MARSHALL, 1940; VUAGNAT, 1949a, 1949b; WELLS, 1922/23, 1925; WHITLEY, 1849; WILLIAMS, D., 1930; WILLIAMS, H., 1927.

Greece:

BRUNN, 1954; HYNES et al., 1972 (sup.); KTENAS, 1908, 1909; MARINOS et al., 1971; OIKONOMOU, 1973; PAPJIANNOPOULOS, 1971; RENZ, 1940; SIDERIS, 1966; SKOUNAKIS, 1972; VUAGNAT, 1959; ZERVAS, 1972.

Greenland:

RIVALENTI and SIGHINOLFI, 1971.

Hungary:

v. SZENTPETERY, 1933.

Ireland:

GARDINER and REYNOLDS, 1914; HATCH, 1889.

Italy:

GEINITZ, 1878; OGNIBEN, 1956; PELLIZZER, 1954, 1955; SPADEA, 1968.

Norway:

BRÖGGER, 1898 (sup.); CARSTENS, 1923, 1924, 1957; GEIS, 1961; GJELSVIK, 1958; GOLDSCHMIDT, 1916; GREEN, 1956; KOLDERUP, 1929; OFTEDAHL, 1958; VOGT, 1927.

Poland:

NAREBSKI, 1964, 1968.

Portugal:

CARVALHOSA, 1961; ORLANDO, 1961; SCHERMERHORN, 1970b; STRAUSS, 1965, 1970.

Romania:

CIOFLICA et al., 1966; ISTRATE and PREDA, 1970; MERLITSCH and SPITKOWSKAJA, 1958; SAVU, 1967, 1968; SIEGEL, 1951.

Spain:
HAUSEN, 1969; SCHERMERHORN, 1970a; STRAUSS, 1965, 1970.

Sweden:
ASKLUND, 1949; v. ECKERMANN, 1938; GEIJER, 1910, 1930, 1930; LUNDBOHM, 1910; SUNDIUS, 1912, 1915, 1916, 1930; VAN OVEREEM, 1948.

Switzerland:
AMSTUTZ, 1948 (sup.), 1950, 1953, 1954, 1959, 1965, 1968; AMSTUTZ and PATWARDHAN, 1972; BEDER, 1909; CADISCH, 1953; EUGSTER, 1951; FISCH, 1961; GRUNAU, 1945, 1947; HÜGI, 1941; NIGGLI, E., 1944; SCHIDLOWSKI et al., 1970; SCHIDLOWSKI and STAHL, 1971; VALLET, 1950; VUAGNAT, 1946, 1947a, 1948, 1949a, 1952; WIDMER, 1948; ZBINDEN, 1949.

U S S R (European Part):
ALEKSANDROV, 1956; BELYANKIN, 1911, 1912, 1914; GEIJER, 1930; INOSTRANTSEV, 1874, 1875, 1878; LAGORIO, 1897; LEBEDINSKIJ, 1963, 1964; LEVINSON-LESSING, 1884, 1888, 1898, 1935; LEVINSON-LESSING and DYAKONOVA-SAVELYEVA, 1933; LUCHITSKY, 1939; MEISTER, 1908; PODOLSKII, 1963; SCHILOW and KALISCHEWITSCH, 1958; SHCHERBAKOV, 1914, 1915, 1924; SKRIPCHENKO, 1965; SOUSTOV, 1935, 1940; SUDOVIKOV, 1913; TIMOFEYEV, 1909, 1916a, 1916b, 1935; TSCHERWJAKOWSKI, 1960; VOGT, 1886; YEGEROVA, 1926; YELISEYEV, 1925, 1928; ZAVARITSKY, 1936a, 1946/1960; ZAYTSEV, 1908, 1910.

Yugoslavia:
CIRIC and KARAMATA, 1960; DIVLJAN and KARAMATA, 1960; DJORDJEVIC and KARAMATA, 1972; DOLAR-MANTUANI, 1941, 1950; GERMOVSEK, 1953; JOVANOVIC, 1973 (sup.); JURKOVIC, 1954; KARAMATA, 1954, 1957, 1958, 1960, 1961; KARAMATA and PAMIC, 1960; LUKOVIC, 1952; MAJER and MARIC, 1964; MAJER and BARIC, 1972; MILOVANOVIC and ILIC, 1960; PAMIC, 1962, 1965, 1969; PAMIC et al., 1964; PAMIC and BUZALJKO, 1966; PAMIC and PAPES, 1969.

C) Classification According to the (proposed) Geological Age (as far as reported)

Precambrian

ADITYA, 1955; AMSTUTZ, 1957a, b, c, (sup.), 1958 (sup.), 1959, 1963, 1968; ASKLUND, 1949;

BESKOW, 1929; BILLINGS, 1937; BLOXAM and LEWIS, 1973; BRODERICK, 1935; BROWN and ROACH, 1972;

CLEMENTS and SMITH, 1899; CORNWALL, 1951 a, b, c;

DALY, 1903; DEY, 1942; DIMROTH, 1971; DUDEK and FEDIUK, 1955;

ESKOLA, 1925, 1934a, b;

FIALA, 1965, 1966a, b, 1967; FLAHERTY, 1934; FURSE, 1954;

GEIJER, 1910, 1930a, b; GJELSVIK, 1958; GOLDICH and RUOTSALA, 1955; GREEN, 1956; GREENLY, 1902, 1919; GREENLY and MATLEY, 1928; GUNNING, 1937; GUNNING and AMBROSE, 1937; GUSTAFSON and MILLER, 1937; GUSTAFSON, 1945;

HUGHES, 1970; HUGHES and MALPAS, 1971;

KETTNER, 1917; KNOPF and ANDERSON, 1930 (-Palaeozoic); KOLDERUP, 1929;

LUNDBOHM, 1910;

McMATH et al., 1953; MERILÄINEN, 1961;

PAKKOLA, 1971; PARKER, 1959; PATWARDHAN et al., 1970; PATWARDHAN, 1972; PICHAMUTHU, 1938, 1946, 1950, 1957; PIISPANEN, 1972; PRIDER, 1939-40;

REID, 1945; RICHARZ, 1926;

SAKSELA, 1960; SLAVIK, 1908, 1915, 1928; SMITHERINGALE, 1972; SUNDIUS, 1912, 1915, 1916, 1930;

THOMAS and JONES, 1912;

VAN OVEREEM, 1948;

WHITE, 1954, 1960; WILLIAMS, 1890.

Palaeozoic

Undifferentiated:
ANGEL, 191, 1955, 1957; BROWN, 1958; CARVALHOSA, 1961; COGULU, 1967; COX, 1920; DEWEY, 1910; DEWEY and FLETT, 1911; DEWEY, 1914, 1948; GRINDLEY, 1958; van der KAADEN and METZ, 1954; KARAMATA, 1954, 1958; KROL, 1960; KUZNETSOV, 1939; LEBEDINSKIJ, 1963, 1964; LOESCHKE, 1973; SPRY, 1953; VOISEY, 1939a; WILSON and ALLEN, 1968.

Cambrian:
ALEKSANDROV, 1956; BLYTH, 1935; KAMPANA and KING, 1963; DE KONING, 1957; EDWARDS, 1953; MOURANT, 1936; MOUSSU, 1959; NAREBSKI, 1964, 1968; SCOTT, 1951 a, b, 1954.

Ordovician:
BLOXAM, 1959; CARSTENS, 1957; COOMBS et al., 1970 (-Devon.); COX, 1913; FLETT, 1907, 1909a, b, c, 1911; FLETT and DEWEY, 1912, 1913; FLETT and RILL, 1912; GARDINER and REYNOLDS, 1914; HATCH, 1889; HUDSON, 1937; METZ, 1951; NICHOLLS, 1959; PAPEZIK and FLEMING, 1967; PAPEZIK, 1970; REID and DEWEY, 1908; SMITH, 1968; TEALL, 1895, 1907; THOMAS, 1911; VALLANCE, 1965, 1968.

Silurian:

FEDIUK, 1953, 1962; GARDINER and REYNOLDS, 1914; KERN, 1952 (-Devon.);
PRIGOROVSKY, 1914 (-Devon.); TEALL, 1899; TSCHERWJAKOWSKI, 1960 (-Devon.);
VOGT, 1927 (-Devon.); ZAVARITSKY, 1936a, b, 1946/60.

Devonian:

BARTH, 1964, 1966; BENSON, 1913, 1915; BLOXAM and LEWIS, 1972; BOULADON
et al., 1963; BOULADON et al., 1965;* BURGUNKER, 1964; CARSTENS, 1923, 1924;
CLAPP, 1921; CROOK, 1960; DIMITROFF, 1960; ERDMANNSDÖRFFER, 1901, 1904;
FLOYD, 1972 (-Carbon.); FOJT, 1962; FOJT and POHANKA, 1965; FRITSCH, 1961a,
1961b; GOETZ, 1937; HENDRIKS, 1939; HENTSCHEL, 1952, 1953, 1960, 1961, 1966;
JUTEAU and ROCCI, 1965, 1966; KNAUER, 1958; LEHMANN, 1941, 1949a, b, 1952a,
1952b, 1952c, 1956, 1965a, b, 1968, 1972 (sup.); MIDDLETON, 1960; ORLANDO,
1961 (-Carbon.); RICHTER, 1968; RINNE, 1895-96; RIPPEL, 1954; ROCCI and
JUTEAU, 1968; RÖSLER, 1959, 1960a, b, 1961, 1962 (-Carbon.); SCHIDLOWSKI
and STAHL, 1971; SCHLOSSMACHER, 1920; SCHÜLLER and DETTE, 1955 (-Carbon.);
STRAUSS, 1965, 1970; SUZUKI and MINATO, 1958; TISCHENDORF, 1959; VOISEY,
1939b; WEISE, 1966. * BRÖGGER, 1898;

Carboniferous:

CLAPP, 1921; ECKHARDT, 1968; PELLIZZER, 1954, 1955; SARGENT, 1917;
SCHERMERHORN, 1970 a, b; SKRIPCHENKO, 1965.

Permo-Carboniferous:

AMSTUTZ, 1948 (sup.), 1950, 1953, 1954, 1957 a, b, c (sup.), 1958, 1959, 1968 a, b;
AMSTUTZ and PATWARDHAN, 1972; BEDER, 1909; HÜGI, 1941; KARAMATA, 1957;
MORRE-BIOT, 1970; NAKAZAWA and KAPOOR, 1973 (sup.); E. NIGGLI, 1944;
P. NIGGLI, 1952; SCHIDLOWSKI et al., 1970; SCHIDLOWSKI and STAHL, 1971;
VUAGNAT, 1946; WIDMER, 1948; ZBINDEN, 1949.

Permian:

BEDERKE, 1959; DE JONG, 1941; DE ROEVER, 1940a, 1941, 1942; EUGSTER, 1951;
FABIAN and MÜLLER, 1962 (Permo-Triassic); FABIAN et al., 1962 (Permo-Triassic);
FISCH, 1961; GILLULY, 1935; KTENAS, 1908, 1909 (-Triassic); (?) REED, 1950.

Mesozoic

Undifferentiated:

BAILEY et al., 1970? BAMBA and SAWA, 1967; BAMBA and MAEDA, 1969; BARTRUM,
1936; BATTEY, 1950, 1951, 1954, 1956; BRUNN, 1954; BURRI and P. NIGGLI, 1945;
CHIHARA, 1954; DIVLJAN and KARAMATA, 1960; DUSCHATKO and POLDERVAART,
1955 (or Tertiary); FRASL, 1957; GOLUB and VRAGOVIC, 1960; GRUNAU, 1945, 1947;
HYNES et al., 1972 (sup.) (Lower Cretaceous or earlier); MAJER, 1960; MARTINI and
VUAGNAT, 1970; OIKONOMOU, 1973; SAVU, 1967, 1968; SKOUNAKIS, 1972;
VUAGNAT, 1946.

Triassic:

CADISCH, 1953; CIOFLICA et al., 1966; COOMBS, 1954; CORNELIUS, 1933; DEN
TEX, 1950; DOLAR-MANTUANI, 1950; DUBERTRET, 1939, 1954; GERMOVŠEK, 1953;
GJELSVIK, 1964; KARAMATA, 1961; KNOPF, 1921 (-Tertiary); MORET, 1929;
PAMIC and BUZALJKO, 1966; PAMIC and PAPES, 1969; PAMIC, 1969; SIDERIS, 1966
(-Cretaceous); TANE, 1962, 1967; TERMIER, 1898; VUAGNAT, 1947b; ZERVAS, 1972
(-Tertiary).

Jurassic:

ASHGIREJ, 1958; BARTRUM, 1936; CATER and WELLS, 1953-54; CIRIC and KARAMATA, 1960; DJORDJEVIC and KARAMATA, 1972; FAIRBANKS, 1896, 1897, 1904; HOPGOOD, 1962; LAWSON, 1895, 1914; PAPAJIANNOPOULOS, 1971 (-Cretaceous); RENZ, 1940; SPADEA, 1968 (-Cretaceous).

Cretaceous:

BROUWER, 1947; CRISTI, 1958; DONNELLY, 1962 (Pre-Upper Cretaceous); FITCH, 1955, 1958 (-Tertiary); GILLULY, 1946; HAILE, 1957 (sup.); HAILE and WONG, 1965; JOVANOVIC, 1973 (sup.); KIRK, 1962, 1968 (-Tertiary); LIDIAK, 1965 (see also Tertiary); MARINOS et al., 1971; PARK, 1946 (? Tertiary); PISKIN, 1972; REINHARD and WENK, 1951 (-Tertiary); SIEGEL, 1951; THOMAS, 1958; TUZKU, 1972; VAN WEST, 1941.

Tertiary

BERTRAND, 1970; BERTRAND and DELALOYE, 1970; BROUWER, 1947; CATER and WELLS, 1953-54; DE JONG, 1942; DELESSE, 1851; HAILE, 1957 (sup.); HAILE and WONG, 1965; HOPGOOD, 1956; LIDIAK, 1965; MERLITSCH and SPITKOWSKAJA, 1958; MICHEL-LEVY, 1877; SHIMAZU, 1971; TOMKEIEFF and MARSHALL, 1940; VUAGNAT, 1952.

Recent

CANN, 1969; CANNE and VINE, 1966; HEKINIAN, 1968; (?) MELSON and VAN ANDEL, 1966; (?) MELSON et al., 1968; STEPHENSON, 1964.

D) Classification According to the Years of Publication

1827 BRONGNIART, A. (sup.)
1849 WHITLEY, N.
1851 DELESSE, A.
1855 KOLENKO, B.Z.
1874 GÜMBEL, C.W. von; INOSTRANTSEV, A.A.
1875 INOSTRANTSEV, A.A.
1876 PHILLIPS, J.A.
1877 MICHEL-LEVY
1878 GEINITZ, E.; INOSTRANTSEV, A.A.; PHILLIPS, J.A.
1884 LEVINSON-LESSING, F.Yu.
1886 VOGT, C.
1888 LEVINSON-LESSING, F.Yu.
1889 HATCH, F.H.
1890 WILLIAMS, G.H.
1891 RUTLEY, F.
1892 GREGORY, J.W.
1893 RAISIN, C.A.; RANSOME, F.L.
1894 RANSOME, F.L.
1895 LAWSON, A.J.; RINNE, F.; TEALL, J.J.; TURNER, H.W.
1896 FAIRBANKS, H.W.
1897 FAIRBANKS, H.W.; LAGORIO, A.E.
1898 BRÖGGER, W.C. (sup.); LEVINSON-LESSING, F.Yu.; TERMIER, P.
1899 CELIKAN, A.; CLEMENTS, J.M., and SMITH, H.L.; LEVINSON-LESSING, F.Yu.; TEALL, J.J.
1900 DUPARC, L, and PEARCE, F.

1901 ERDMANNSDÖRFFER, O.
1902 GREENLY, E.; MOLENGRAFF, G.A.F.
1903 DALY, R.A.
1904 ERDMANNSDÖRFFER, O.; FAIRBANKS, H.W.
1905 YEREMINA, E.V., and LEVINSON-LESSING, F.Yu.
1907 FLETT, J.S.; TEALL, J.J.
1908 KTENAS, K.A.; MEISTER, A.K.; REID, C., and DEWEY, H.; SLAVIK, F.; ZAYTSEV, A.M.
1909 BAILEY, E.B., and GRABHAM, G.W.; BEDER, R.; FLETT, J.S.; KTENAS, K.A.; TIMOFEYEV, V.M.
1910 DEWEY, H.; GEIJER, P.; LUNDBOHM, H.; WEBER, M.; ZAYTSEV, A.M.

1911 BELYANKIN, D.S.; DEWEY, H., and FLETT, J.S.; HERITSCH, F.; THOMAS, H.H.

1912 BAILEY, E.B.; BELYANKIN, D.S.; FLETT, J.S., and RILL, J.B.; SUNDIUS, N.; THOMAS, H.H., and JONES, O.T.; YEREMINA, E.V.

1913 BENSON, W.N.; COX, A.H., and JONES, O.T.; WOLDRICH, J.

1914 BELYANKIN, D.S.; DALY, R.A.; DEWEY, H.; FLETT, J.S., and DEWEY, H.; GARDINER, C.H., and REYNOLDS, S.H.; LWASON, A.J.; LEWIS, J.V.; PRIGOROVSKY, M.M.; SHCHERBAKOV, D.I.; WILLSON, M.E.

1915 BENSON, W.N.; CAPPS, S.R.; COX, A.H.; SHCHERBAKOV, D.I.; SLAVIK, F.; SUNDIUS, N.

1916 GOLDSCHMIDT, V.M.; SUNDIUS, N.; TIMOFEYEV, V.M.

1917 KETTNER, R.; SARGENT, H.C.

1918 KNOPF, A.

1919 ANGEL, F.; GREENLY, E.

1920 COX, A.H.; SCHLOSSMACHER, K.

1921 CLAPP, C.H.; KNOPF, A.

1922 WELLS, A.K.

1923 CARSTENS, C.W.; STAMP, L.D., and WOOLDRIDGE, S.W.

1924 CARSTENS, C.W.; KNOPF, A.; SHCHERBAKOV, D.I.

1925 COX, A.H.; ESKOLA, P.; HALL, A.L., and MOLENGRAAF, G.A.F.; SVITALSKY, N.I.; WELLS, A.K.; YELISEYEV, N.A.

1926 BUDDINGTON, A.F.; EINBERG, L.F.; RICHARZ, S.; YEGOROVA, E.N.

1927 HACKMANN, V.; STEINMANN, G.; VOGT, T.; WILLIAMS, H.

1928 GREENLY, E., and MATLEY, C.S.; SLAVIK, F.; YELISEYEV, N.A.

1929 BESKOW, G.; DIMITROFF, S.; KNOPF, A.; KOLDERUP, N.-H.; MORET, L.

1930 BACKLUND, H.G.; GEIJER, P.; KNOPF, A., and ANDERSON, C.A.; MOORE, E.S.; NOLAND, T.B.; REDLICH, K.A., and PREDIK, K.; SUNDIUS, N.

1931 KASSIN, N.G.; SUDOVIKOV, N.G.

1932 GROUT, F.F.; KUPLETSKY, B.M.

1933 CORNELIUS, H.P.; DALY, R.A.; KASSIN, N.G. et al.; LEVINSON-LESSING, F.Yu., and DYAKONOVA-SAVELYEVA, E.N.; SZENTPÉTERY, S. von; WINCHELL, A.N.

1934 DIMITROFF, S.; ESKOLA, P.; FAIRBAIRN, H.W.; FLAHERTY, G.F.; TOMITA, T.

1935 BARTRUM, J.A.; BLYTH, F.G.H.; BRODERICK, T.M.; ESKOLA, P. et al.; GILLULY, J.; LEVINSON-LESSING, F.Yu.; SOUSTOV, N.I.; TIMOFEYEV, V.M.; WINCHELL, A.N.; ZAVARITSKY, A.N.

1936 BARTRUM, J.A.; ECKERMANN, H. von; KORZHINSKY, D.S.; MOURANT, A.E.; RUTTEN, M.G.; ZAVARITSKY, A.N.

1937 BILLINGS, M.P.; GILLULY, J.; GOETZ, H.; GROUT, F.F.; GUNNING, H.C.; GUNNING, H.C., and AMBROSE, J.W.; GUSTAFSON, J.K., and MILLER, F.S.; HUDSON, S.N.; STARK, M.; THIADENS, A.A.

1938 ECKERMANN, H. von; LODOCHNIKOV, V.N.; PICHAMUTHU, C.S.; STARK, J.T.; YELISEYEV, N.A.

1939 CHUMAKOV, A.A., DUBERTRET, L.; HENDRIKS, E.L.M.; JOHANNSEN, A.; KUZNETSOV, E.A.; LUCHITSKY, V.T.; McKINSTRY, H.E.; PRIDER, R.T.; ROGOVER, G.B.; STARK, J.T.; VOISEY, A.H.

1940 BENSON, W.N., and HOLLOWAY, J.T.; CHUMAKOV, A.A.; DE ROEVER, W.P.; KORZHINSKY, D.S.; RENZ, C.; SOUSTOV, N.I.; TOMKEIEFF, S.I., and MARSHALL, C.E.

1941 DE JONG, J.D.; DE ROEVER, W.P.; DOLAR-MANTUANI, L.; HÜGI, T.; LEHMANN, E.; PERRIN, R., and ROUBAULT, M.; ZAVARITSKY, A.N.

1942 DE JONG, J.D.; DE ROEVER, W.P.; DEY, A.K.; MOORHOUSE, W.W.; VAN WEST, F.P.

1943 BERNAUER, F.; PETTIJOHN, F.J.

1944 NIGGLI, E.

1945 BLYTH, F.G.H.; BURRI, C., and NIGGLI, P.; GRUNAU, H.; GUSTAFSON, J.K.; KEIJZER, K.G.; REID, J.A.

1946 GILLULY, J.; PARK, C.F.; PICHAMUTHU, C.S.; VUAGNAT, M.; ZAVARITSKY, A.N.

1947 BROUWER, H.A.; GRUNAU, H.; VUAGNAT, M.

1948 AMSTUTZ, G.C. (sup.); DEWEY, H.; VAN OVEREEM, A.J.A.; VUAGNAT, M.; WIDMER, H.

1949 ASKLUND, B.; LEHMANN, E.; VUAGNAT, M.; WENK, E.; ZBINDEN, P.

1950 AMSTUTZ, G.C.; BATTEY, M.H.; DEN TEX, E.; DOLAR-MANTUANI, L.; PICHAMUTHU, C.S.; REED, J.J.; VALLET, J.-M.

1951 BATTEY, M.H.; CORNWALL, H.R.; EUGSTER, H.P.; METZ, K.; REINHARD, M., and WENK, E.; SCOTT, B.; SIEGEL, A.; SUTTON, J., and WATSON, J.; WANG, Y.

1952 HENTSCHEL, H.; KERN, A.; LEHMANN, E.; LUKOVIC, S.; NIGGLI, P.; VUAGNAT, M.

1953 AMSTUTZ, G.C.; CADISCH, J.; CATER, F.W., and WELLS, F.G.; EDWARDS, A.B.; FEDIUK, F.; GERMOVSEK, C.; HENTSCHEL, H.; ISHII, K., and UEDA, Y.; McMATH, J.C. et al.; SPRY, A.

1954 AMSTUTZ, G.C.; BATTEY, M.H.; BRUNN, J.H.; CHIHARA, K.; COOMBS, D.S.; DUBERTRET, L.; FURSE, G.D.; JURKOVIC, I.; KAADEN, G. van der, and METZ, K.; KARAMATA, S.; PELLIZZER, R.; RIPPEL, G.; SCOTT, B.; WHITE, D.; WILLIAMS, H., et al.;

1955 ADITYA, S.; ANGEL, F.; BATTEY, M.H.; DUDEK, A., and FEDIUK, F.; DUSCHATKO, R., and POLDERVAART, A.; FITCH, F.H.; GOLDICH, S.S., and RUOTSALA, A.P.; JAFFÉ, F.C.; PELLIZZER, R.; SCHÜLLER, A., and DETTE, K.; TYRRELL, G.W.

1956 ALEKSANDROV, A.I.; BATTEY, M.H.; GREEN, J.C.; HOOKER, M.; HOPGOOD, A.M.; KÜNDIG, E.; LEHMANN, E.; OGNIBEN, G.; STEPHENS, E.A.

1957 AMSTUTZ, G.C. (sup.); ANGEL, F.; CARSTENS, H.; DE KONING, G.; FRASL, G.; HAILE, N.S. (sup.); HOOKER, M.; HOPGOOD, A.M.; KARAMATA, S.; PICHAMUTHU, C.S.

1958 AMSTUTZ, G.C.; AMSTUTZ, G.C. (sup.); ASHGIREJ, G.C.; BROWN, W.R.; CRISTI, J.M.; FITCH, F.H.; GJELSVIK, T.; GRINDLEY, G.W.; KARAMATA, S.; KNAUER, E.; MERLITSCH, B.W., and SPITKOWSKAJA, S.M.; OFTEDAHL, C.; SCHILOW, W.N., and KALISCHEWITSCH, O.K.; SUZUKI, J., and MINATO, M.; THOMAS, H.

1959 AMSTUTZ, G.C.; BEDERKE, E.; DONNELLY, T.W.; HOOKER, M.; MOUSSU, R.; NICHOLLS, G.D.; PARKER, D.C.; PETTIJOHN, F.J., and BASTRON, H.; RÖSLER, H.J.; TISCHENDORF, G.; VUAGNAT, M.

1960 BAILEY, E.B., and McCALLIEN; BILIBIN, J.A.; CIRIC, B., and KARAMATA, S.; CROOK, K.A.W.; DIMITROFF, S.; DIVLJAN, M., and KARAMATA, S.; GOLUB, L., and VRAGOVIC, M.; HENTSCHEL, H.; KARAMATA, S.; KARAMATA, S., and PAMIC, J.; KROL, G.L.; MAJER, V.; MATTSON, P.H.; MIDDLETON, G.V.; MILOVANOVIC, B., and ILIC, M.; RICHTER, D., and VILLWOCK, R.; RÖSLER, H.J.; SAKSELA, M.; SHAGAM, R.; TSCHERW-JAKOWSKI, G.F.; TURNER, F.J., and VERHOOGEN, J.; VALLANCE, T.G.; VINCENT, E.A., and CROCKET, J.H.; WHITE, W.S.

1961 BOTSCHENKOAREW, W.F., and KOBLENZ, E.L.; CARVALHOSA, A.; FISCH, W.P.; FRITSCH, W.; GEIS, H.-P.; HATCH, F.H. et al.; HENTSCHEL, H.; KARAMATA, S.; KOPF, M.; MERILÄINEN, K.; ORLANDO, C.G.; RICHTER, D., and VILLWOCK, R.; RÖSLER, H.J.; WILSON, R.A.M.

1962 DICKINSON, W.R.; DONNELLY, T.W.; FABIAN, H.J., and MÜLLER, G.; FABIAN, H.J., et al.; FEDIUK, F.; FOJT, B.; HOPGOOD, A.M.; KIRK, H.J.C.; PAMIC, J.; RÖSLER, H.J.; TANE, J.-L.

1963 AMSTUTZ, G.C.; BOULADON, J., et al.; CAMPANA, B., and KING, D.; DONNELLY, T.W.; HANSELMAYER, J. von; KORZNISKIJ, O.S.; LEBEDINSKIJ, W.I.; PODOLSKII, J.V.; RÖSLER, H.J.; STEWART, D.; SZADECZKY-KARDOSS, E.

1964 BARTH, V.; BURGUNKER, M.E.; GJELSVIK, T.; HANSELMAYER, J. von; JOPLIN, G.A.; MAJER, V., and MARIC, L.; NAREBSKI, W.; PAMIC, J., et al.; STEPHENSON, P.J.

1965 BOULADON, J., et al.; FIALA, F.; FOJT, B., and POHANKA, J.; HAILE, N.S., and WONG, N.P.Y.; JUTEAU, T., and ROCCI, G.; LEHMANN, E.; LIDIAK, E.G.; MOORE, J.G.; PAMIC, J.; SKRIPCHENKO, N.S.; STRAUSS, G.K.; VALLANCE, T.G.

1966 BARTH, V.; CANN, J.R., and VINE, F.J.; CIOFLICA, G., et al.; DONNELLY, T.W.; FIALA, F.; HENTSCHEL, H.; JUTEAU, T., and ROCCI, G.; MELSON, C.W., and VAN ANDEL, T.H.; PAMIC, J., and BUZALJKO, R.; SIDERIS, K.; WEISE, G.

1967 BAMBA, T., and SAWA, T.; COGULU, E.; FIALA, F.; PAPEZIK, V.S., and FLEMING, J.M.; SAVU, H.; SMITH, R.E.; TANE, J.-L.; TAYLOR, H.P., Jr., et al.; YODER, H.S., Jr.

1968 AMSTUTZ, G.C.; ECKHARDT, F.-J.; HEKINIAN, R.; KIRK, H.J.C.; LEHMANN, E.; MELSON, C.W., et al.; NAREBSKI, W.; PETTERLONGO, J.M.; RICHTER, D.; ROCCI, G., and JUTEAU, T.; SAVU, H.; SHARFMAN, V.S.; SMITH, R.E.; SPADEA, P.; VALLANCE, T.G.; WILSON, W.F., and ALLEN, E.P.

1969 BAMBA, T., and MAEDA, K.; BLOXAM, T.W.; CANN, J.R.; HAUSEN, H.; LA ROCHE, H., et al.; PAMIC, J., and PAPES, J.; PAMIC, J.; VALLANCE, T.G.; WEDEPOHL, K.H.

1970 BAILEY, E.H., et al.; BERTRAND, J.; BERTRAND, J., and DELALOYE, M.; COOMBS, D.S., et al.; DEINES, P.; HART, R.; HUGHES, C.J.; ISTRATE, G., and PREDA, I.; MARTINI, J., and VUAGNAT, M.; MORRE-BIOT, N.; PAPEZIK, V.S.; PARWARDHAN, A.M., et al.; QUADE, H.; SCHERMERHORN, L.J.G.; SCHIDLOWSKI, M., et al.; STRAUSS, G.K.

1971 DIMROTH, E.; HEKINIAN, R.; HUGHES, C.J., and MALPAS, J.G.; MARINOS, G., et al.; PAAKKOLA, J.; PAPJIANNOPOULOS, A.; RIVALENTI, G., and SIGHINOLFI, G.P.; SCHIDLOWSKI, M., and STAHL, W.; SHIMAZU, M.

1972 AMSTUTZ, G.C., and PATWARDHAN, A.M.; BLOXAM, T.W., and LEWIS, A.D.; BROWN, M., and ROACH, R.A.; DJORDJEVIC, P., and KARAMATA, S.; FLOYD, P.A.; HYNES, A.J., et al. (sup.); LEHMANN, E. (sup.); MAJER,

1972 V., and BARIC, L.; PATWARDHAN, A.M.; PIISPANEN, R.; PISKIN, Ö.;
      SCHOELL, M., and STAHL, W.; SKOUNAKIS, S.; SMITHERINGALE, W.G.;
      TUZCU, N.; ZERVAS, S.

1973 HUGHES, C.J.; JOVANOVIC, Z.Z. (sup.); LOESCHKE, J.; NAKAZAWA, K.,
      and KAPOOR, H.M. (sup.); OIKONOMOU, K.; PAMIC, J., et al.;
      SCHERMERHORN, L.J.G.

Subject Index

Accessory minerals, 88, 107, 201
Adinolization, 203
Al$_2$O$_3$, 42, 101, 141, 143f., 147, 151, 153f., 237f., 355
   variance, 55
Albite, 9, 11, 16-18, 20, 25, 30, 44, 71, 73-79, 104, 107, 119-121, 154, 165, 176f., 180, 185-187, 192, 195, 197-199, 201-203, 212, 219, 221-224, 233f., 240, 243, 245f., 248, 261f., 267f., 270, 273, 276, 281-283, 323, 334-339, 341, 343f., 349, 351, 354, 366, 370, 373f., 377, 379f., 392f., 403, 419, 423
  Ala, 338
  antiperthitic, 261
  assemblages (see also metamorphic facies)
    albite-amphibole-pyroxene, 199f.,
    albite-augite, 177, 180
    albite-augite-chlorite, 177
    albite-augite-chlorite-epidote, 338
    albite-biotite, 198
    albite-carbonate, 75, 78
    albite-chlorite, 9, 11, 17f., 74
    albite-chlorite-biotite, 199f.
    albite-chlorite-calcite, 336, 373, 382
    albite-chlorite-clinopyroxene, 66, 365
    albite-chlorite-epidote, 195
    albite-chlorite-pyroxene, 180
    albite-chlorite-titanite, 377
    albite-chlorite-titanite-hematite ± calcite, 335
    albite-chlorite-titanite ± pyroxene, epidote and calcite, 335
    albite-hornblende, 201f.
  authigenic, 334, 344
  autometasomatic, 129
  Baveno, 338
  checker-board structure, 118f.
  clouded, 234
  contemporaneity albite-pyroxene, 186
  high-temperature, 129, 156
  Karlsbad, 338
  low-temperature, 59, 64, 129, 231, 233, 377
  microlites, 177, 338, 367
  phenocrysts, 171, 338

Albite (continued)
  primary, 162, 187, 344
  secondary, 66, 118, 180
  spilitic, 66f.
  veinlets, 71, 338, 343, 393
Albitite, 78, 103, 110, 130, 162, 166, 180, 186, 192, 198f., 203
Albitization, 9, 20, 30, 55, 83, 88, 98, 101f., 118f., 124f., 130, 153, 155, 171, 238, 248, 344, 359, 367, 391f., 397-399, 407-410, 412-414, 419, 421f.
Albitophyres, 13, 19
Alkali, 11, 15, 17f., 36, 101, 104, 109, 121, 153, 180f., 236, 239, 244, 298, 390, 394f.
  ratio, 118
Alpinotype (volcanism), 353
Alteration (see also under individual alteration, e.g. sericitization), 17f., 33, 83, 91, 98, 102f., 118, 176f., 180, 196, 201, 261, 334, 338, 344, 388f., 392, 398, 409f., 414f.
  autometamorphic, 359
  deuteric, 19, 94, 109, 180, 342, 345, 373
  diagenetic, 342
  postmagmatic, 344
  processes, 366
  of the sedimentary rocks, 333
  secondary, 37, 392, 398
  wallrock alteration, 98, 101f.
Amphibole, 29, 44, 107, 165, 192, 196, 199f., 203, 276, 335, 339
  actinolite, 153, 264, 274, 276, 282, 339, 343, 373, 377
  assemblages
    amphibole-albite, 199f.
    amphibole-chlorite, 199f.
    amphibole-chlorite-albite-epidote, 196f.
    amphibole-epidote-albite, 196f.
    amphibole-labradorite, 196f.
  hornblende, 88, 195, 198, 201f., 212f., 216, 218-224, 254, 276, 338f., 343, 421
  assemblages
    hornblende-albite, 201
    hornblende-chlorite, 198
    hornblende-epidote-albite, 195, 198
    hornblende-labradorite, 195
  nephrite, 108, 110
  uralite, 153, 195, 221, 274, 276, 339, 344

Amphibolite (see hornblendite), 11f., 20, 86, 129, 161, 200
Amygdales or Amygdules, 10, 16f., 33, 71, 116, 121, 177, 261f., 268, 270, 276, 278, 281f., 285f., 288, 336-338, 341, 344, 369, 380, 391-394, 403, 408-410, 412, 414
Amygdaloids, 9, 11, 13, 15, 33, 71, 74, 77-79, 115f., 118f., 121, 164f., 175, 195, 197, 333, 391, 393
Andesite, 9, 14, 19f., 37, 51, 85, 161, 165, 181, 281, 391, 400f., 420f.
  spilitic, 261, 264, 266
Andesito-basalts or Andesite-basalts, 14f., 151, 389, 393, 395, 397f., 400f., 403
Anorthosite, 196-198
Apatite, 31, 199, 201, 264, 274, 276, 335-337, 343
Assimilation, 325, 422
Augitite, 180
Autohydrothermal (see autometamorphism)
Autolysis (see autometamorphism)
Autometamorphism, 19, 118, 180f., 368, 374, 381, 413, 415, 423
Autometasomatism (see autometamorphism)

Babingtonite, 369
Basalt, 11, 15-18, 20, 43f., 48f., 51f., 87, 92, 101, 109f., 161, 165, 180f., 185f., 192, 195, 203, 229, 281, 291, 301, 345, 350, 388-391, 393-399, 401, 403f., 422
  albite-, 232, 234f., 239, 242, 245ff., 365
  albitized, 229, 232-235, 237-240, 242
  alkali, 18, 56, 92, 180f., 241, 248, 323
  alkali olivine, 92, 327
  alteration of, 154
  aphyric pyroxene, 415
  average compositions, 60
  clinopyroxene, 92
  conversion of solid basalt to spilite, 67
  dolerite, 332
  flow, 404
  high-alumina, 92, 101, 147
  hydrolysis of, 67
  Hy-Ol-normative, 61
  labradorite, 232, 234
  non spilitic, 368
  normal, 186, 192, 203
  olivine, 15, 389
  olivine-basalt formation, 141
  ophitic, 406

Basalt (continued)
  plateau, 181
  sodic, 365
  solubility of water in basalt, 376
  spilitic, 36
  submarine, 241, 248
  suite basalt-spilite-keratophyre, 9
  tachylitic, 229
  tholeiite, 15, 92f., 109f., 181, 186, 223, 227-232, 234ff., 238, 247, 366f., 369
  tholeiitic, weak metamorphism of, 365
  transformation basalt-spilite, 154
  types, basaltoid, 11, 19
  unspilitized, 375
Basanite, 51
Bowlingite, 75
Breccias, brecciation (see also spilite and pillow), 10, 71f., 78f., 195, 257, 264, 276, 281, 333-339, 344f., 414

Calc-alkali series, 93, 149
Calcite (see carbonate), 16f., 25, 27f., 33, 48-52, 72, 75, 77f., 115f., 118, 154f., 165, 171, 176f., 180, 187, 192, 201, 232-235, 244, 261f., 264, 268, 270, 282f., 283, 285f., 288, 323, 333, 336-339, 341, 343f., 349, 351, 354, 360, 373, 391ff.
  assemblage
  calcite-pumpellyite, 333
Calcitization (see also carbonatization), 334, 344
CaO, 19, 134, 141, 143, 147, 151, 153f., 235, 237-239, 243f., 285
Carbonates (see calcite, dolomite), 17, 30f., 36, 44, 47, 71, 74, 77-79, 116f., 177, 199, 201-203, 208, 334, 336, 342, 390ff., 394, 398, 411, 413f.
Carbonatite, 78
Carbonatization, 11, 23, 25, 48f., 51f., 115f., 153, 366, 392
Ca-Tschermak molecule, 64
Chalcedony, 261, 264, 276, 281, 392
Chemical analyses, 10-15, 30, 34ff., 90-93, 96, 100-103, 106ff., 167, 181-185, 195ff., 199f., 360, 394-401
Chemistry of rocks, 166
  of the spilite-keratophyre suite, 284-321
Chert, 83f., 86f., 94, 97, 109f., 331, 333f.
Chilled margin, 18, 98, 101, 104
Chlorite, 9, 17, 20, 25-30, 33, 36, 44, 47-51, 53, 65, 71-75, 77ff., 88, 94, 104, 107, 116, 118, 121, 124f., 153, 165, 176f., 180, 185ff., 195f., 198-201, 212f., 219-223, 233ff., 243, 245, 248, 261f.,

Chlorite (continued)
  264, 268, 270, 278, 281ff.,
  285f., 288, 323, 332, 336-341,
  342ff., 349, 356, 360, 366f.,
  369, 373f., 377, 380, 389-392,
  394, 403, 411, 414, 419
  assemblages
    chlorite-calcite, 379
    chlorite-calcite-muscovite, 337
    chlorite-diopside-anorthite, 376
    chlorite-diopside-forsterite, 376
    chlorite-leucoxene, 392
    chlorite-pumpellyite, 345
  chemical composition, 342
    average, 141
    of basaltic, 343
    of chlorites in spilites and diabases, 340
    of gabbroic, 343
  clinochlore, 29
  diabantite, 27, 33
  pennine, 262, 264, 276, 281
Chloritization, 16, 23, 25, 30, 52, 101,
  153f., 172, 359, 388, 390, 393,
  399, 403, 408-411, 415
Chlorophaeite, 393, 398
Cr, 141, 143, 149, 151
Chromite, 102f., 198
Clay minerals, 17, 177, 337, 342, 380, 392
CO_2, 238
Congruency (of minerals, e.g.), 75f.,
  79, 197, 417, 419
Contamination, 19f., 149, 326, 391
Correlation analyses, 151-153
Cu, 97, 198, 201f.

Dacite, 9, 14, 51, 165
Deanorthitization, 153, 397
Deuteric (see alteration)
Devitrification, 27, 392
Diabase (see also dolerite), 9f., 14ff.,
  17-20, 30, 32f., 35ff., 46,
  51f., 83-87, 92-98, 100-104,
  107, 113, 125, 130, 161, 192,
  195-198, 200-203, 210, 212,
  220f., 223, 258f., 261, 264,
  266, 270, 276, 279, 281, 282,
  287, 291, 305, 314, 319, 324,
  332, 338-344, 387, 392, 394
  albite, 129f., 149, 162, 165f., 192,
    199-202, 307
  alkali, 35, 83, 94f., 97, 104, 107f.,
    109
  aphanitic, 10, 14f., 18f., 259
  doleritic, 14, 16, 18-20
  essexitic, 15
  greenschist-albite diabase, 147
  greenstone, 196-198
  norms, 101
  quartz, 46, 51f., 130, 219, 276, 279,
    282, 290, 305, 326

Diabase-chert formation, 161
Diagenesis, 79, 185, 332, 342, 344
Diagrams
  $(Al/3 - K)/(Al/3 - Na)$, 40, 42-47, 53ff.
  $Al_2O_3/(Na_2O + K_2O)$, 92
  $Al_2O_3 - CaO - (Fe, MgO)$, 378
  $(CaO/CaO + Na_2O + K_2O)/SiO_2$, 185
  $Fe_2O_3 - CaO - (Mg, Fe)O$, 379f.
  feldspar normative, 94, 96, 107, 180f., 243
  FMA, 93
  isotopes, C/O, 78
  Jung (calc-alkalinity), 295, 297f.
  Kuno $[Al_2O_3/(Na_2O + K_2O)]/SiO_2$, 92, 101
  major elements = $f(SiO_2)$, 237
  Na-K-Ca, 298f.
  Na_2O/K_2O, 15
  $(Na_2O - K_2O)/SiO_2$, 13, 15, 92, 181, 242
  $(Na_2O/Na_2O + K_2O)/SiO_2$, 185
  Na-Di-Ol-Hy-Qz, 60f.
  Niggli, variation, 12, 293-296
  Nockolds, 184, 239, 242
  $[Si/3 - (K + Na + 2Ca/3)]/[K - (Na + Ca)]$, 42, 48-51, 53ff.
  triangle, γ-Mg triangle, 167-170
    K-π triangle, 167-170
    QLM triangle, 167-170
  variation of major elements, 90
  Zavaritsky, 10, 14
Differentiation, 10-15, 18ff., 40, 103,
  108, 186, 198, 201, 226, 244ff.,
  392, 422
  differentiation series, 201, 203
  filter flow, 226
  fractional crystallization, 129, 141,
    151, 153, 220, 225
  gravitative, 74, 124, 129
  hydrothermal, differentiation fractionation, 79
  igneous, 42
  "magmatic differentiation", 72, 74, 326
  primary, 18, 20
  radial within individual pillows, 370
Diffusion, alkali-volatile, 129
Dike, 16, 19, 94, 97f., 103, 108ff., 192,
  195, 198, 201ff., 213, 216, 220f.,
  224, 226, 257, 343
Diorite, 162, 166
Dipyre, 264
Dolerite (see also diabase), 15, 36, 51f.,
  92, 130, 212, 220f., 223, 257,
  261, 264, 267, 274, 276, 279,
  287, 291, 307, 324, 365
  albitic, 61, 288, 374, 376, 381, 382
  altered, 212
  alkali dolerite intrusion, 380
  grained, 319
  intrusive, 46
  modes, 92
  quartz, 368

Dolerite (continued)
   sill, 16
   tholeiitic, 365
Dolomite (see carbonate), 194

Effusion, 257, 261, 266
   geosynclinal, 155
   keratophyric submarine, 260
   porphyric, 259
   spilitic submarine, 260
Enrichment
   in Ca and Si, 245, 246
   in Fe, 334
   in Na, 40, 129
   in Si, 56, 153
Epidote, 13, 30, 33, 44, 47f., 51-53, 64, 67, 103f., 106f., 125, 153, 165, 176f., 180, 197, 195ff., 199, 201, 212, 234, 264, 274, 276, 282, 335f., 338f., 341-344, 349, 359, 369, 374f., 377, 379, 392f., 398, 403, 407, 411, 414
   assemblages
      epidote-albite, 197f.
      epidote-chlorite, 345
   clinozoisite, 165, 171, 341, 375
   pistacite, 216, 339, 341
   pseudozoisite, 341
   zoisite, 44, 341
Epidotization, 13, 103, 152, 403, 407, 408-410, 415
Evolution (magmatic or metasomatic) 55, 129, 147, 153
Extrusion, 244, 262, 335, 413

Fe and oxides, 75, 149, 151, 180, 202, 236, 239, 244, 261f., 281f., 373
   enrichment, 124
   ferric, 143
   ferrous, 143
   oxidation of, 342
   trivalent, 141
Fe/Mg ratio, 238
Fe_2O_3/FeO ratio, 151, 238, 243
Feldspar, 116, 118, 120, 124f., 186, 245, 279, 339, 342, 352, 415
   alkali-feldspar (see albite), 123, 243
      adularia, 379
      microperthite, 118-123
      orthoclase, 42, 44, 48f., 115, 118, 240, 243, 283
         albitized, 273f.
         Na orthoclase, 115
         perthitic, 268
      potash feldspar, 166, 177, 266, high temperature form, 166, 171
   character, 121
   early crystallized, 119
   fibrous, 88, 336

Feldspar (continued)
   filled, 154
   framework, 154
   initial, 118
   normative, 118, 248
   plagioclase, 16ff., 42, 48ff., 53, 64, 74f., 88, 94f., 98, 100, 104, 107, 115f., 186, 191f., 196, 203, 216, 231, 261, 264, 270, 278, 338, 388-394, 397, 411, 419
      antiperthitic, 268
      basic, 16, 389, 392
      calcic, 154, 180, 217, 366, 369, 408, 410, 415, 420
      composition of, 213, 217
      high temperature form, 165, 171, 377
      microlites, 389, 411, 413
      phenocrysts, 16ff., 388f.
      zoned, 218

      albite-oligoclase, 16f., 165, 276, 392
      oligoclase, 11, 16f., 165, 217, 274, 392
      andesine, 217, 248, 276, 288, 423
         keratophyre, 165, 171
         saussuritized, 276
         spilite, 165, 171
      labradorite, 217, 231, 234, 244, 246, 276, 351, 354, 355, 389ff., 393f., 423
      labrador-bytownite, 44, 390
      bytownite, 165, 217, 389f., 393
      anorthite, 44, 118, 154, 193, 203, 240, 243f., 291, 367, 376
Feldspathisation, sodipotassic, 51
Feldspatoïd, 48, 380
Flow, 98, 164, 175, 181, 210, 403, 408ff., 413f., 422, 424
Flowage, 72, 336
Frequency distribution of elements, 132-134, 141f., 144ff., 150

Gabbro, 33, 86, 96f., 161, 162, 166, 194, 196, 198f., 208, 212, 217, 222f., 320, 344, 353, 355
   albite, 199f.,
   altered, 212
   altered hornblende-, 342
   eucritic, 217
   hornblende, 343
   layered, 225
   pyroxene, 196f.
   unaltered, 210
Garnet, 103, 366, 369
Gas and vapor phases, 37, 333
Geosyncline, 17f., 36, 83f., 86, 94, 98, 109, 156, 163f., 193, 243, 248, 326, 331, 344, 351, 360, 387, 400, 422
   eugeosyncline, 109, 164, 167
   miogeosynclinal, 164, 167
   geosynclinal volcanism, 9, 17, 130

Glass, 17-20, 23, 87f., 90f., 98, 100, 104, 125, 177, 235, 266, 344, 369f., 377, 397, 389-393, 415, 422f.
Globules, 27, 28
Greenschist (see metamorphic facies) 20, 104, 129, 147, 154, 192, 195
Greenstone, 11, 20, 130, 192, 195, 196-198, 422
Greywacke, 98, 194, 382, 419
Groundmass (see matrix)

Hawaiite, 156
Hematite, 73, 85-87, 94f., 110, 212, 222, 226, 276, 33f., 336, 337f., 341, 343f., 349, 360, 377, 403, 407, 411-413
  alteration to, 411
  -epidote, 379
  -epidote-calcite, 379
  skeletal, 337
  -spilite, 76f.
Hercynotype (volcanism), 130, 149, 292, 353,
Hornblendite, 199f.
Hornfels, 86, 96f.
Hyaloclastite, 15, 18, 210, 234f., 381
Hydrothermal
  solutions (see differentiation) 74, 94 95, 109f.
  stage, 74

Iddingsite, 75, 388, 390
Ilmenite, 88, 94f., 104, 107, 222, 225f., 341, 343, 392
  leucoxenized, 341
Ilmeno-magnetite, 221, 222, 224, 225
Injecta, 23, 28, 113, 115-121, 123ff.
  oversaturation of injectum, 121
Isotopes
  C, 31, 78
  O, 31, 78

Karjalite, 192
Keratophyre, 9f., 19f., 46f., 51f., 88, 90-93, 95, 102ff., 107, 109f., 129f., 150, 156, 161, 165f., 171, 257, 259, 260, 279, 281, 289, 295, 315, 326, 370f., 374, 399
  alcaline, 302
  aphanitic, 261, 268, 273f., 290, 303
  classification, 281
  microlitic, 282
  porphyric, 267, 290
  porphyric-quartzitic, 274, 304
  quartz-, 164-166, 308f.
  siliceous, 302
  sodic, 129

Keratophyre (continued)
  saussuritized, 276
  trachytic, 52, 289
Keratospilite, 130, 156
K_2O, 42, 141, 147, 238f., 242, 244

Lamprophyre, 14, 355
  transition towards spilite, 354
Lapilli, 23, 25f.
Lava, 16, 83-88, 90-96, 98, 102ff., 106f., 109f., 130, 164, 193, 195, 333-338, 340, 342, 344f., 403, 406f., 415
  autoclastic, 333-335, 337, 340, 342, 344
  basaltic, 383, 403, 415
  basic, 331
  breccia, 207, 212, 332f.
  extruded, 342
  flows, 98, 403
  heterogeneously altered, 408
  hydrated mafic, 413
  selvedge, 365
  subaqueous, 98, 109, 374
Leucoxene, 16, 20, 119, 199, 224f., 262, 264, 282, 392
Limonite, 75, 334

Magma, 121 236
  basaltic, 16, 20, 27, 33, 108, 186, 324, 326, 327, 345
  basic, 36
  chamber, 20, 129
  diabasic, 36
  femic, 33
  hydromagmas, 18, 23, 33, 36, 74, 79, 121, 124
  keratophyric, 326
  olivine-basaltic, 16, 130, 156, 203
  peridotitic, 33
  spilitic, 23, 60, 98, 186, 203, 324, 326, 345, 387, 400
  tholeiitic, 92, 156, 203, 244, 248, 365
  ultrafemic, 33
  weilburgitic, 29, 33
Magmatism
  geosynclinal, 254, 256
  initial, 279
  initial hercynotype, 52f.
  preorogenic, 322
Magmatic processes, 163
  late, 59, 67, 130, 171, 180
  post, 59, 67, 171, 180, 187
Magnetite, 202, 222, 225f., 274, 343, 360, 367, 379, 390, 392, 413
Major elements
  in basalts, 139
  in spilites, 135
Mantle, 30, 108, 144, 208, 280, 359

Matrix, 13, 26f., 75, 218, 234f., 264, 270, 281ff., 390, 392ff., 421
  aphanitic, 177
  breccia, 71f., 78f., 334
  chlorite, 16, 72, 177, 261, 332
  -feldspar, 118f.
  palagonitic, 27
Melaphyre, 161, 359, 361, 407
  heterogeneous, 408, 414, 417
  homogeneous, 414
Metabasalt, 10f., 17, 20, 181, 187, 373
  spilitic, 379
Metabasite, 129, 141, 147, 153, 192, 195
Metamorphic facies, 20, 193, 203, 414
  albite-epidote-amphibolite, 193, 195,
  amphibolite, 382
  blueschist, 382
  glaucophane schist, 378
  greenschist, 59, 193, 195, 203, 345, 373, 382, 419, 423
  laumontite, 419, 423
  lawsonite-albite-chlorite, 378
  prehnite-pumpellyite, 379, 419, 423
  pumpellyite-actinolite schist, 373, 378f., 381f.
    mineral assemblages in the Al_2O_3-poor portion, 380
  zeolite, 345, 367, 369, 371, 373, 378, 381, 419f., 423
Metamorphism, 101, 103, 192, 331, 338, 345, 366, 387
  burial, 23, 37, 345, 367, 369, 373f., 383, 414f., 419
  contact, 103, 163
  dynamo-, 337
  hydrothermal, 374, 383
  post-consolidational, 383
  regional, 83, 345, 374, 383, 423
  thermal, 334
Metasomatism, 14, 19, 30, 33, 83, 108, 192, 203, 398
Metaspilite, 20
MgO, 75, 147, 151, 180, 236-239, 244
Micas, 29, 193, 354
  biotite, 30, 196, 199, 201, 270, 354, 370, 380
  muscovite, 27, 177, 274, 337
  paragonite, 201
  sericite (damourite), 44, 94, 165, 276, 334, 342f., 354
  stilpnomelane, 419
  white mica, 335f., 413
Mineral deposits (see ore deposits)
Mineralization (see ore genesis), 86, 94
Mn, 149, 151
  Mn/Fe ratio, 148-151
Mugearite, 156, 185

Na_2O, 42, 134, 141, 144, 149, 151, 153f., 186, 238f., 244, 248, 352, 355

Na_2O (continued)
  -enrichment, 147, 149, 154f., 186
  Na_2O/CaO, 394
  Na_2O/K_2O, 394
  $Na_2O + K_2O$, 239, 242
  $Na_2O + K_2O/CaO$, 394
Nepheline, 48
Ni, 141, 143, 149, 151, 198
Nomenclature, 23f., 155, 335
Noncongruency, 180, 187
Norms, 91, 101

Oceanic ridge, 208, 380, 381
Olivine, 15f., 23, 29, 44, 48-50, 52, 71f., 177, 196, 199, 201, 203, 225, 270, 283, 323, 325-327, 336, 341, 349f., 354-356, 366, 377, 388ff., 392f., 413
  forsterite, 367, 376, 388
  pseudomorphs after, 350, 411
  serpentinized, 392
Opaques (see deposit and ore), 71f., 74f., 176f., 234, 262, 264, 334, 341
Ophiolite, 149, 162, 207f., 210, 324, 331, 332, 353, 420
Ophites, 51, 276, 291, 307, 414,
Ore (see deposit and opaques), 44, 47-50, 53, 202, 391
  Cu-Ni, 198
  degradation of carbonate Fe-Mn, 42
  iron, 202
  manganese, 202, 332, 344f.,
  Ti-iron, 198
  V-bearing Ti-iron, 198
Ore deposits
  cupriferous pyrite, 85f., 97f.,
  exhalative, 98, 202
  iron, 95, 109, 149, 198, 202
  iron-manganese, 13, 85ff., 202
  manganese, 192, 202
  manganiferous hematite, 87, 94, 110
  sedimentary-exhalative, 98, 202
Ore genesis (see mineralization)
Ore minerals, 180, 199, 389, 391, 394
Orthokeratophyre, 47, 51, 274, 290, 302, 304
Oxygen pressure, 226

Palagonite, 176, 235, 422
  formation, 370
  tuff, 234
Palagonitization, 365, 367, 369ff.
Paleo- (see under rock name, andesite e.g.)
Paragenesis, 349
  spilitic, 283
Pectolite, 108, 110
Pentlandite, 202
Peridotite, 103f., 108, 194, 196, 212, 270, 291, 314, 324, 353, 355
  amphibole, 23
  of the Upper Mantle, 208

Permeability, between successive lava layers, 414f.
Phase
    deuteric, 74
    hydrous, 74, 345, 369, 374, 414
    primary, 64, 366
    secondary, 64
Phase equilibria, relevant to the spilite problem, 375
Phonolite, 51
Picotite,
    chromium, 222
Picrite, 15, 51
Pillow, 4f., 10, 15, 17, 47, 50, 53f., 83, 100, 113-115, 118f., 121, 124f., 232-234, 240, 243, 245, 281, 317, 332, 336, 387-391, 422f.
    accumulation, 115, 121
    border, 43, 47, 53, 106, 125, 390f., 393
    breccia, 381
    core, 43, 47, 53ff., 106, 118, 268, 312, 423
    crust, 262
    glassy, 365
    interpillow, 10, 19, 47, 53f.
    selvedge, 17f., 371,
    spilitic, 18, 59f., 212, 216
    structure, 97f., 101, 114, 210, 235, 332, 392,
    variolitic zoned, 340
    vesicular 232
Pillow lava, 10, 15f., 18f., 47, 55, 83-88, 90-96, 98, 102ff., 106f., 109f., 164, 213, 235, 255, 257-259, 261f., 266, 268, 280, 282f., 285, 288, 294f., 311, 325, 332-334, 336, 341f., 344, 370, 381, 387-395, 398f., 422
    norms, 91
    selvedge, 369
    spilitic or spilitized, 86-90, 94, 96, 103, 107, 268
    submarine, 368
    unzoned, 341, 344
    zoned, 335, 337, 341, 344
Poenite (see potassic spilites)
Porphyrite, 9f., 13-16, 19, 30, 51, 161, 217, 220, 291, 336, 359, 361
    spilitic, 13, 16, 282
Postmagmatic events (see also phase and autometamorphism), 15, 17, 36f., 98, 185, 343
Prehnite, 30, 33, 37, 65, 116, 125, 165, 171, 233f., 369, 373f., 377, 379f., 392, 403, 414
Propylite, propylitization, 187, 398
Pumice flow deposits, 85, 86

Pumpellyite, 30, 37, 74, 335, 338f., 341, 344f., 360, 369, 373, 374-377, 379-381, 414
    -prehnite, 360, 378, 381-383, 414f.
Pyrite, 85f., 97f., 102, 201f., 234, 335, 360, 393
Pyroclastic rocks, 23, 25, 51, 83f., 86, 106f., 109f., 130, 161, 164, 229, 260, 279, 291, 332, 390
Pyroxene, 16, 44, 61, 64, 125, 153, 165, 177, 180, 186, 199, 200f., 203, 213, 231, 233ff., 243, 248, 262, 270, 279, 287f., 336, 339, 342, 344, 355, 369, 373f., 376, 381, 392, 410, 412
    augite, 16f., 34, 48f., 51, 53, 64, 71-73, 94f., 98, 103f., 107, 109, 165, 176f., 180, 185ff., 234f., 243, 246, 262, 264, 270, 274, 276, 282, 323, 336, 338f., 343f., 369, 388-394, 421, 423
        assemblage: augite-chlorite, 376
        diopsidic, 34, 94f., 103f., 107, 109
        titaniferous, 264, 339, 392
    bastite, 276
    clinopyroxene, 32, 62, 64-66, 88, 92, 94, 98, 100, 104, 107, 212f., 216, 220f., 365-367, 370, 377, 388-390, 403, 406, 410, 413
    diopside, 154, 196, 198, 202f., 376
    endiopside, 217
    enstatite, 23, 29, 367
    hypersthene, 48f., 196
    orthopyroxene, 221, 393
Pyroxenite, 196f., 199f.
Pyrrhotite, 102, 201f.

Quartz (see also SiO_2), 13, 25, 42, 44, 48-50, 52, 75, 86, 94, 98, 100, 103, 109, 115, 118f., 176f., 180, 196, 199, 201f., 212, 219, 224, 232-234, 261, 264, 266-268, 273f., 278f., 335-337, 343, 349, 354, 374f., 377, 392, 411-414
Quartz-diorite, 166
Quartz-keratophyre, 10, 13, 19, 161, 164-167, 289
Quartz-porphyrite, 164, 219
Quartz-porphyry, 9, 14
Quartz, trachyte, 51

Rhyolite, 9, 14, 19, 51, 85, 165, 229, 281, 370
    metasomatism of original rhyolite, 371
Rodingite, rodingitization, 103, 108, 110, 367
Rutile, 337, 341

Sampling error (Na content), 40, 47, 59, 418

Saussuritization, 154, 335f., 343, 366, 392
Schalstein, 23, 30f., 37, 52, 78f., 266, 270, 276, 279f., 292, 310, 316, 327
Sericitization, 95, 98, 101, 115, 119, 392
Serpentine, 84, 86, 96, 102ff., 107-110, 196, 199, 201, 248, 291, 335, 337
  antigorite, 270
  saponite, 221
Serpentinite, 196, 200, 331f., 382, 422
Serpentinization, 108
Silicification, 48f., 51f., 115, 130, 294
Sill, 16, 18, 20, 213, 257, 259, 370, 382, 424
SiO_2 (see also quartz), 44, 47, 49, 118, 121, 141, 143, 147, 149, 151, 153f., 181, 185, 236-238, 242-244, 285, 352, 355f.
Solid-state transformation, 367
Sphene (see titanite)
Spheralite, 201, 261f., 274, 336f., 423
Spilite (not exhaustive - see also magma)
  albitized, 147
  amygdaloidal, 274, 279, 281, 285, 307, 310, 317
  amygdaloidal brecciated, 287
  associations, different mineralogical 59, 67, 78, 180, 186, 336f., 340, 342, 376, 417
  association, spilite-keratophyre, 54, 66, 102, 108f., 161, 165, 167, 171, 254f., 279f., 293, 323, 399
  autometasomatic (autometamorphic) origin (especially Part 3), 251-362
  basalt, relationship with, 43
  breccia, 60, 264, 276
  classification, 281
  clinopyroxene fraction, 65
  continental, 421
  conversion of basalt to, 67
  definition, 418
  effusive, 300f.
  evolution, 129
  feeder dikes, 222
  genesis (see also primary and secondary), 66, 203, 417f.
  hercynotype, 254
  heterogeneity, 60, 186
  heterogenetism, 127
  intersertal, 166
  intersertal microlitic, 261, 268, 276, 279, 287, 313, 318
  intrusive, 368
  lava, 90, 339, 405f., 414
  magmatic, 155f.
  massive, 51f., 59f., 258
  metamorphic, 155

Spilite (continued)
  metasomatic, 23, 33, 36
  micrograined, 282
  microlitic, 261, 274, 281, 306, 317
  mineral facies of, 377
  mineralogy, 216, 369, 375, 377
  Na-content, 366, 422
  oceanic, 186, 194, 422
  ophitic, 165f.
  parentage, definition of, 60
  pillow, 186, 421
  polygenetism, 127, 155
  polyorogenetic history, 147
  potassic, 10-13, 19, 125, 164, 166, 180, 183, 284, 287, 290, 365
  primary origin (especially Part 2: 69-250)
    magma, 92
    problem, 79, 367
  protospilite, 366
  pyroxene bearing, 61, 66, 336f., 340f.
  reaction, 151, 153
  residual liquids, 207, 220, 222
  residual melt, 226
  secondary origin (especially Part 4: 363-426)
  semi-spilite, 32-34
  shallow water conditions, 18, 248
  sill, 212
  stability field of minerals, 373f.
  stratiform bodies, 198, 201
  texture of, 367
  "true", 210
  typical, 16
  unmetamorphosed, 374
  unoxidized, 340
  variolitic potash, 10
  vesicular, 51f.
  weathering, 342
Spilitization, 15f., 55f., 83, 86, 90, 94f., 102, 243, 248, 280, 322, 344, 349, 350f., 353, 355, 357, 359, 381
  reaction, 154
  tendency, 46f., 49, 51
Stilpnomelane, 12f., 19, 27, 33, 419

Talc, 196, 201
T-ratio, 143-145
Telescoping, 352, 355
Teschenite, 15
Texture (see also amygdales, globules, spherulites, varioles and vesicles) 72, 79, 322, 332, 334, 337, 368
  aphyric, 332, 336, 344, 391f.
  dendritic, 177
  diktytaxitic, 336
  fluidic, 72f., 267
  granular, 177
  holohyaline, 393
  hyalopilitic, 165, 388ff., 392ff.
  hypidiomorphic, 201

Texture (continued)
  hypohyaline, 393
  intergranular, 233
  intersertal, 15f., 18-20, 72, 165, 177, 210, 233, 264, 270, 274, 276, 336f., 389, 391-394, 412, 423
  micrograined, 264
  microlitic, 261f., 267, 270, 273f., 393, 411
  microporphyritic, 388, 390
  oligophyric, 391f.
  ophitic, 16, 20, 32, 36, 72, 83, 86, 91, 93f., 98, 100f., 104, 107, 177, 180, 195, 201, 210, 221, 226, 233, 270, 406
  pegmatitic, 210
  perlitic, 391
  pilotaxitic, 165, 201, 336, 391ff.
  poikilitic, 270
  poikilo-ophitic, 177
  porphyritic, 20, 116, 165, 201, 210, 261f., 267, 322, 336, 389-391, 393
  sideranitic, 392
  subophitic, 64f., 75, 87f., 90f., 98, 177, 267, 276, 336f.
  tortoise-shell, 88
  trachytic, 267
  variolitic, 87f., 90f., 98, 100, 104, 106, 210, 336-338, 376, 388
  vitrophyric, 388, 393
Tholeiite, 56, 92, 101, 109, 156, 194-198, 201, 237ff., 242-248, 345
  dry, 248
  normative olivine, 66
  quartz, 65
  series, 49, 56, 93
  tendency, 292
  wet, 248
TiO_2, 75, 141, 143f., 149, 238, 339
Ti/Fe ratio, 143, 147-150
Titanite, 72, 176, 196, 199, 221, 262, 264, 274, 282, 336-338, 341, 343f., 369, 373, 376f., 411, 413, 419
Titanomagnetite, 198, 270, 392
Tourmaline, 201
Trace element, 143
Trachybasalt, 143, 395
Trachyte, 156, 181
Transition
  albite-orthoclase, 48
  anorthite-albite, 41f., 48f., 51, 53
  basic rocks-acid rocks, 41f.
Transvaporization, 16, 18, 20, 399
Traps, 175
Trondhjemite, 193
Tuff, 10f., 15f., 23, 25f., 28, 30f., 52, 86, 97, 100, 102f., 113-115, 124, 129f., 166, 193, 257-259, 261, 264, 266, 272,

Tuff (continued)
    276, 292, 321, 333, 338f., 344, 359, 369
  diabasic, 16, 83, 84-86, 97, 102f.
  keratophyric, 113ff., 123, 259, 261, 266, 276, 309
  spilitic, 261, 266
  weilburgitic, 23

Ultrabasic (ultramafic, ultrafemic) rocks 14f., 29, 33, 51, 53, 103, 108, 161, 191, 198, 201-203, 208, 270, 280, 291, 331
Uranium, 192

Varioles, variolites (see also variolitic texture), 10, 88, 199f., 422f.
  albitic, 327
Veins, veinlets, 18, 71f., 78, 88, 109, 177, 187, 333, 336-338, 341, 344, 369, 392, 394, 412
  calcite, 337
  calcite-chlorite, 337
Vermiculite, 278
Vesicles and vesiculation, 18, 66, 72, 210, 213, 226, 261f., 323, 376f., 414
Volatiles, 19, 74, 121, 124f., 186, 203, 237, 325, 339, 413
  sodic, 124
Volcanic
  activity, quiet, 335
  associations, 40
  bomb, 266
  chimney (vent), 185, 354
  episode, 334
  pile, 414
  plug, 403
  rocks, 164, 166f., 198, 210
Volcanism
  contemporary, 66
  final, 147, 359
  initial preorogenic, 43, 50, 128, 156
  submarine, 9, 15, 18-20, 94, 185f., 344, 359, 387f., 398f., 422
  two-stage, 129
Volcano-sedimentary rocks, 162, 164, 210, 292, 332f., 373, 380, 389

Water, 238f., 349, 351f., 356, 413
  depth of water, 243
  interaction
    with basement, 352
    with magma, 20, 27, 29, 351f.
  juvenile, 108
  pressure of water, 238, 244, 422
  subaerial or shallow, 413
Wehrlite, 270, 291, 314
Weilburgite, 23, 27, 30, 33-36, 78, 116-118, 121-124, 166, 180, 199, 284

Weilburgite (continued)
  common, 33
  hydromagma, 23
  massive, 25
  normal, 32
  pillowy, 120
  potash, 122
  pyroxene-bearing, 33f., 36
  semi-spilitic, 36
  soda, 122

X-ray data on
  albite, 338
  chlorite, 339f.
  epidote, 341
  hematite, 341
  ilmenite, 341
  opaque or semiopaque minerals, 341
  pumpellyite, 341
  pyroxene, 339

Zeolite, 177, 180, 212f., 232-234, 381, 391, 394, 398
  apophyllite, 234,
  calcic, 244
  laumontite, 233f., 374f., 377f., 414, 421
  mesotype, 276
  wairakite, 378
Zeolitization, 367
Zircon, 141

Minerals, Rocks and Inorganic Materials

Monograph Series of Theoretical and Experimental Studies

Edited by W. von Engelhardt, Tübingen; T. Hahn, Aachen; R. Roy, University Park, Pa.; P. J. Wyllie, Chicago, Ill.

Vol. 1: W. G. Ernst
Amphiboles
Crystal Chemistry, Phase Relations and Occurrence
With 59 figures. X, 125 pages. 1968
Cloth DM 30,—; US $12.30

Vol. 2: E. Hansen
Strain Facies
With 78 figures, 21 plates. X, 208 pages. 1971
Cloth DM 58,—; US $23.80
Distribution rights for U.K., Commonwealth, and the Traditional British Market (excluding Canada): Allen & Unwin, Ltd., London

Vol. 3: B. R. Doe
Lead Isotopes
With 24 figures. IX, 137 pages. 1970
Cloth DM 36,—; US $14.80

Vol. 4: O. Braitsch
Salt Deposits — Their Origin and Composition
Translated from the German edition by P. J. Burek and A. E. M. Nairn in consultation with A. G. Herrmann and R. Evans
With 47 figures. XIV, 297 pages. 1971
Cloth DM 72,—; US $29.60

Vol. 5: G. Faure, J. L. Powell
Strontium Isotope Geology
With 51 figures. IX, 188 pages. 1972
Cloth DM 48,—; US $19.70

Vol. 6: F. Lippmann
Sedimentary Carbonate Minerals
With 54 figures. VI, 228 pages. 1973
Cloth DM 58,—; US $23.80

Vol. 7: A. Rittmann
Stable Mineral Assemblages of Igneous Rocks
A Method of Calculation
With contributions by V. Gottini, W. Hewers, H. Pichler, R. Stengelin
With 85 figures, XIV, 262 pages
1973. Cloth DM 76,—; US $31.20

Vol. 8: S. K. Saxena
Thermodynamics of Rock-Forming Crystalline Solutions
With 67 figures. XII, 188 pages. 1973
Cloth DM 48,—; US $19.70

Vol. 9: J. Hoefs
Stable Isotope Geochemistry
With 37 figures. IX, 140 pages, 1973
Cloth DM 39,—; US $16.00

Prices are subject to change without notice

Springer-Verlag
Berlin Heidelberg New York
München Johannesburg London
New Delhi Paris Rio de Janeiro Sydney
Tokyo Wien

A critical presentation, by approximately 70 specialists, of important facts about the distribution of the chemical elements and their isotopes in the earth and the cosmos

Executive Editor:
K. H. Wedepohl

Editorial Board:
C. W. Correns
D. M. Shaw
K. K. Turekian
J. Zemann

Handbook of Geochemistry

The work consists of two volumes, published in installments.
A subscription price is applicable on orders for the complete handbook, valid until the last installment is published.
Each installment is available separately at list price.
One installment probably remains to be published.

The Handbook of Geochemistry offers a critical selection of important facts about the distribution of the chemical elements and their isotopes in the earth and the cosmos. Approximately 70 specialist authors have made this selection from the flood of information which resulted from improved investigative methods. The data are set out in the main part of the Handbook (Vol. II) in tables and diagrams as an integral part of extensive discussions on abundance, distribution, and behavior of the elements. As this book clearly shows, geochemistry and cosmochemistry are intimately linked with a number of other disciplines.

With the exception of the introductory part (Vol. I), the work is arranged according to the atomic numbers of the elements, each chapter being organized in the same way. Thus, the reader will find such frequently needed information as the crystal chemical properties of an element or its occurrence in meteorites or metamorphic rocks under the same section in each of the 65 chapters. The loose-leaf system enables the contributions to be published in random order, regardless of their position in the book, and revisions to be made as desired.

Vol. II/3

Third Installment
With 142 figures
IV, 845 pages. 1972
Loose-leaf binder
DM 258,—; US $105.80
For subscribers to the complete Handbook
DM 206,40; US $84.70

The fourth installment (in preparation) will complete the handbook

Vol. I

With 60 figures. XV, 442 pages. 1969

Vol. II/1

With 172 figures. X, 586 pages. 1969
Loose-leaf binder. Vols. I and II/1 comprise the first installment and are not sold separately.
Boxed DM 224,—; US $91.90
For subscribers to the complete Handbook
DM 179,20; US $73.50

Vol. II/2

Second Installment. With 105 figures
IV, 667 pages. 1970. Loose-leaf binder DM 212,—; US $87.00
For subscribers to the complete Handbook
DM 169,60; US $69.60

Prices are subject to change without notice

Springer-Verlag Berlin · Heidelberg · New York
München Johannesburg London New Delhi Paris Rio de Janeiro
Sydney Tokyo Wien